Textile Science and Clothing Technology

Series Editor

Subramanian Senthilkannan Muthu, SgT Group & API, Hong Kong, Kowloon, Hong Kong

This series aims to broadly cover all the aspects related to textiles science and technology and clothing science and technology. Below are the areas fall under the aims and scope of this series, but not limited to: Production and properties of various natural and synthetic fibres; Production and properties of different yarns, fabrics and apparels; Manufacturing aspects of textiles and clothing; Modelling and Simulation aspects related to textiles and clothing; Production and properties of Nonwovens; Evaluation/testing of various properties of textiles and clothing products; Supply chain management of textiles and clothing; Aspects related to Clothing Science such as comfort; Functional aspects and evaluation of textiles; Textile biomaterials and bioengineering; Nano, micro, smart, sport and intelligent textiles; Various aspects of industrial and technical applications of textiles and clothing; Apparel manufacturing and engineering; New developments and applications pertaining to textiles and clothing materials and their manufacturing methods; Textile design aspects; Sustainable fashion and textiles; Green Textiles and Eco-Fashion; Sustainability aspects of textiles and clothing; Environmental assessments of textiles and clothing supply chain; Green Composites; Sustainable Luxury and Sustainable Consumption; Waste Management in Textiles; Sustainability Standards and Green labels; Social and Economic Sustainability of Textiles and Clothing.

K. Murugesh Babu · Abera Kechi Kabish · Getnet Belay Tesema · Bizuayehu Kerisew Semahagn
Editors

Cotton Sector Development in Ethiopia

Challenges and Opportunities

Springer

Editors
K. Murugesh Babu
Ethiopian Institute of Textile and Fashion Technology
Bahir Dar University
Bahir Dar, Ethiopia

Abera Kechi Kabish
Ethiopian Institute of Textile and Fashion Technology
Bahir Dar University
Bahir Dar, Ethiopia

Getnet Belay Tesema
Ethiopian Institute of Textile and Fashion Technology
Bahir Dar University
Bahir Dar, Ethiopia

Bizuayehu Kerisew Semahagn
Department of Biology, College of Science
Bahir Dar University
Bahir Dar, Ethiopia

ISSN 2197-9863 ISSN 2197-9871 (electronic)
Textile Science and Clothing Technology
ISBN 978-981-99-9148-8 ISBN 978-981-99-9149-5 (eBook)
https://doi.org/10.1007/978-981-99-9149-5

This Springer imprint is published by the registered company Springer Nature Singapore Pte Ltd.
The registered company address is: 152 Beach Road, #21-01/04 Gateway East, Singapore 189721, Singapore

Paper in this product is recyclable.

Foreword

This book presents the experiences and knowledge now available pertaining to the cultivation of cotton, the value chain, cross-cutting issues, opportunities, and challenges influencing the sector's development in Ethiopia. The compilation of this book strives for methodical and collective approach that it would give a useful and encouraging platform for ongoing study and development into overall cotton sector challenges in Ethiopia. With 30% of all fibres produced globally for textile manufacture still being used, cotton is the most commonly used natural fibre. Despite the fact that numerous synthetic fibres are produced and utilized to make textiles today, the need for cotton fibre is still rising due to its vital features, particularly when it comes to fabric used for clothing.

Furthermore, the byproduct of cotton seeds is a key component in the creation of edible oil, an essential component of both human and animal diets. A byproduct of the extraction of edible oils is cakes which is used for animal feed. Africa is experiencing an increase in demand for animal feed, notably cottonseed cakes. As a result, cotton has gained importance as a cash crop on a global scale for both economic and social reasons. Accordingly, up to a billion people rely heavily on cotton for their livelihood and income, of which 250 million work in the cotton-processing industry and 100 million are cotton farmers. The remaining people are involved in businesses related to cotton, such as those that produce textiles, machinery, and agricultural supplies. Over 2.5% of the world's arable land is used to grow cotton, which is actively traded across over 150 countries worldwide, with each country either importing or exporting cotton. The international trade in cotton and cotton yarn exceeded a total value of 26.3 billion USD. In rural areas of developing countries in Africa, Asia, and Latin America, cotton production contributes to increased life expectancy and greater food security.

Ethiopia, which is thought to be one of the regions where the cotton fibre originated, has a long history of cotton farming. Currently, cotton is cultivated on 90,000 hectares of land in Ethiopia, 70% of which is used for commercial agriculture and 30% of which is used for smallholder agriculture utilizing both irrigation and rainfall. Due to the country's favourable agro-ecological zones and widely accessible land resources, which have an estimated area of over 2.6 million hectares suited for cotton cultivation, the potential for cotton production in the nation is enormous. In light of its enormous potential, Ethiopia is therefore among the world's potential cotton-producing nations. Despite its poor performance in cotton production and marketing, Ethiopia still has a unique opportunity in the cotton sub-sector because it can serve as a foundation for the country to transition to high value-added technological transformation thanks to its reliance on strong backward and enhanced forward links with various sectors and its provision of employment opportunities for a significant number of rural poor people. The Ethiopian government also wants to make a focussed effort and take action to encourage the development and potential of this sub-sector in order to make cotton one of the nation's main commercial crops.

Given the country's current socioeconomic condition, increased traditional weaving, the erection of spinning mills and integrated industrial parks, and the rising demand for cotton on the global market, cotton production must increase both in quality and quantity. Due to increased traditional weaving, the establishment of spinning mills and integrated industrial parks, as well as the rising demand for cotton on the global market, the country's socioeconomic condition currently calls for cotton to be produced in greater quantities and of higher quality. The federal government is doing its utmost to alter plans to take into account the elements of the assistance package for cotton growers in order to help the cotton sector moving ahead. In order to effectively connect the value actors along the value chain, the policy framework must be set up. Furthermore, cotton's research and development (R&D) activity must be improved, and favourable conditions must be created to draw significant foreign direct investment (FDI) to the cotton production sector as well as the value chain industries.

I value the efforts put out by the Ethiopian Institute of Textile and Fashion Technology at Bahir Dar University to prepare a textbook on *Cotton Sector Development in Ethiopia: Challenges and Opportunities* that has 3 sections and 19 chapters. The required and significant information on the prospects for Ethiopian cotton production and the challenges that the industry faces is contained in each section and chapter. The book, which is in the first edition, was created with the help of all the stakeholders in the cotton industry in the nation who are involved in cotton value chains. As a result, the book will serve as a vehicle and framework for bringing together important players in the cotton sector's value chain in order to improve Ethiopia's cotton output while promoting R&D activities. In general, this book will be a useful resource for scientists working on cotton research; students; professionals in the cotton spinning, weaving, processing, and fashion sectors; academics; policymakers; and consumers. I hope that the effort will be maintained with the next series of editions.

H. E. Mr. Melaku Alebel
Minister
Ministry of Industry
The Federal Democratic Republic of Ethiopia
Addis Ababa, Ethiopia

Preface

Cotton is one of the oldest fibre crops in Ethiopia and cotton clothing is a part of the Ethiopian tradition. The cotton production regions in Ethiopia include: the Awash Valley in the Northeast; Arba Minch/Sille, Woito, Omerate in the South; Gambella and Beles in the West; Metema and Setit Humera in the North and Northwest; and Gode in the East. Additionally, there are large potential regions for cotton production in the Western, Northern, Southern, and Eastern parts of the country. In most of these regions, smallholder farmers operate rain-fed and irrigated cotton farms, but mostly in agro-ecologies ranging from 500 to 1500 metres above sea level (masl). Traditional cotton cultivation is also widely practiced in the mid-altitude regions ranging from 100 to 1800 masl, typically under a biannual or perennial system of production.

Ethiopia is well suited for cultivating this renewable raw material due to favourable natural conditions (climate, fertile soils, and water availability) and a long-standing tradition in cotton and textile fabric production. Despite these favourable climatic and soil conditions, the potential of the local cotton industry is still largely untapped. Currently, the country cultivates only 3% of the total 2.6 million ha that is potentially suitable for cotton production.

The predominant species of cotton grown in Ethiopia is Gossypium hirsutum L., commonly known as upland cotton. Several varieties of upland cotton have been obtained or introduced, tested, and registered in the National Variety Register. Even after 30 years since their release, DP-90 and Acala SJ-2 are still commercially cultivated in many areas of Ethiopia. Currently, one variety (DP-90) is grown on over 90% of the cotton farms. The staple length of DP-90 fibre ranges from 26 to 28 mm with a Ginning Out Turn (GOT) of about 38%. There are plans to cultivate extra-long staple (ELS) cotton in the future. ELS cotton is defined as cotton with a staple length greater than 34 mm. There is a premium demand for ELS cotton, which is used in higher value products.

The editors of this book acknowledge that individuals employed in the Ethiopian textile companies should have knowledge about the working conditions in agriculture. Lecturers at textile institutes should not only concentrate on techniques, information and a functional unit, but also acknowledge the fundamental understanding of the textile sector. Furthermore, they should strive to tackle the challenges and opportunities of cotton sector development in Ethiopia. In addition to demanding a fair price for their cotton, farmers should also show a keen interest in comprehending the difficulties faced by spinning factories.

To tackle the challenges and opportunities of cotton sector development in Ethiopia, the authors of this book sought out authors for each chapter and finally found the solution in the structure presented in the book.

Part I of the book focusses on cotton agriculture, covering topics such as cotton agronomy and production as a scientific matter, cotton genetics and breeding, cotton protection, cotton biotechnology, irrigated and mechanized cotton production, pre-harvest and harvest management, and post-harvest handling and storage.

Part II provides significant details on the demand and supply of cotton fibre in Ethiopia, the presence of ginning industries in the country, the testing and characterization of cotton quality, trends and grading in the industry, the requirements for spinning cotton, the technology used in fabric manufacturing, the chemical processing of cotton textiles, and the role of cotton fabrics in Ethiopian fashion design. The main focus of this section is on cotton quality and testing.

Part III also discusses sustainable cotton production and processing. This section covers not only environmental aspects but also touches on food and nutrition, social and economic factors. The topics addressed in this section include technology for cotton seed production; the use of cotton as a feed and food crop; research, extension, and promotion of cotton; challenges in cotton production, quality, and future prospects; the cotton value chain (CVC) and economics; adapting sustainable cotton production for the better livelihood of farmers; and sustainable landscapes.

The intended reader of this book encompasses a wide range of qualifications including the following: (1) undergraduate and postgraduate students in textile and fashion institutions who are studying textile science, textile engineering, apparel engineering, and product development; (2) textile researchers and engineers working in interdisciplinary projects of traditional and modern textiles; (3) textile technologists working in supervisory positions or as a worker in textile processing factories; (4) non-textile engineers who are involved in different projects containing cotton fibrous material; (5) product developers of textile and fashion products; (6) textile and fashion researchers who are seeking new research ideas; 7) college professors who are involved in science, technology, engineering, and mathematics (STEM) programmes and interdisciplinary research activities.

The editors would like to extend their sincere appreciation to all the authors who have devoted significant amounts of time and effort in contributing to this book. Their work and expertise make this book very special. Each chapter is an authoritative treatise on its specific topic and can be read independently. Vivid diagrams and illustrations in each chapter, as well as throughout the book, enhance the accessibility of the theory and technologies described therein. All the chapters of this book were reviewed by at least one reviewer, revised by the authors according to reviewers' comments and then edited by the editors.

It is our sincere hope that this new edition will make a valuable contribution to the textile and fashion industry both in understanding key fundamental aspects and in implementing reliable opportunities for the production and processing of this renewable natural raw material “King of fibres” …Cotton.

Bahir Dar, Ethiopia
August 2023

K. Murugesh Babu
Abera Kechi Kabish
Getnet Belay Tesema
Bizuayehu Kerisew Semahagn

Acknowledgements

We would like to convey our gratitude to the numerous organizations and individuals who provided us with guidance, support, and aid in completing this challenging scientific publication. A special appreciation goes out to the Ethiopian Institute of Textile and Fashion Technology (EiTEX) for initiating the development of this book.

The following institutions played a significant role in identifying authors for the chapters in this book, for which we are extremely thankful:

(1) Department of Textile Engineering, Ethiopian Institute of Textile and Fashion Technology, Bahir Dar University, P.O. Box 1037, Bahir Dar, Ethiopia.
(2) Department of Textile Chemical Process Engineering, Ethiopian Institute of Textile and Fashion Technology, Bahir Dar University, P.O. Box 1037, Bahir Dar, Ethiopia.
(3) Department of Garment Engineering, Ethiopian Institute of Textile and Fashion Technology, Bahir Dar University, P.O. Box 1037, Bahir Dar, Ethiopia.
(4) Department of Fashion Design, Ethiopian Institute of Textile and Fashion Technology, Bahir Dar University, P.O. Box 1037, Bahir Dar, Ethiopia.
(5) Department of Plant Science, College of Agriculture and Environmental Sciences, Bahir Dar University, P.O. Box 79, Bahir Dar, Ethiopia.
(6) Department of Biology, College of Science, Bahir Dar University, P.O. Box 79, Bahir Dar, Ethiopia.
(7) Department of Plant Science, College of Agriculture and Environmental Sciences, Bahir Dar University, P.O. Box 79, Bahir Dar, Ethiopia.
(8) Department of Horticulture, Bahir Dar University, P.O. Box 79, Bahir Dar, Ethiopia.
(9) Ethiopian Institute of Agricultural Research (EIAR), Melkasa Agricultural Research Center, P.O. Box 2003, Addis Ababa, Ethiopia.
(10) Cotton Extension Researcher, Textile and Garment Industry Development and Research Center, Addis Ababa.
(11) Cotton quality research and inspection directorate, Ethiopian Textile Industry Development Institute, Addis Ababa, Ethiopia.

(12) Cotton plant research, Ethiopian Textile Industry Development Institute, Addis Ababa, Ethiopia.
(13) Cotton research projects, Solidaridad, Addis Ababa, Ethiopia.

Bahir Dar, Ethiopia
August 2023

K. Murugesh Babu
Abera Kechi Kabish
Getnet Belay Tesema
Bizuayehu Kerisew Semahagn

Contents

Cotton Agriculture

Cotton Agronomy and Production

Yayeh Bitew and Alemu Abate

Abstract In Ethiopia, there is immense potential for the cultivation of cotton, considering its suitable agro-ecological zones and the presence of water. However, there is a significant disparity between the potential and the actual production when it comes to Ethiopia's share of global production. Limited availability of research and extension services, insufficient supply of inputs, inability to produce high-quality products, and inadequate infrastructure and finance are some of the factors contributing to this gap. The following comprehensive agronomic management recommendations are proposed to achieve at least the optimal yield of the typical output: on average, it takes 20–50 heat units (degree-days) from cotton planting to 60% boll opening. Cotton thrives in deep arable soils with good drainage, organic matter, and high moisture-retention capacity. Among the different cotton varieties, G. hirsutum is the suitable and predominant species, accounting for approximately 90% of global production alone. Seed selection is a prerequisite for cotton cultivation. After selecting the seeds, purity should be above 95% and germination should be above 85%. Seed rate and plant spacing vary depending on the cotton species. Cotton is planted at a depth of 1.5−2 cm in humid conditions and 3−4 cm in dry conditions. At the field level, specific crop management practices need to be followed from sowing to emergence, including management during the seedling stage, flower and bud formation stage, flowering stage, and boll opening stage.

Keywords Production level · Cultivated species · Climatic and soil requirement

Y. Bitew · A. Abate (✉)
Department of Plant Sciences, College of Agriculture and Environmental Sciences, Bahir Dar University, P.O. Box 79, Bahir Dar, Ethiopia
e-mail: abatalem@yahoo.com

Y. Bitew
e-mail: yayehbitew@gmail.com

K. Murugesh Babu et al. (eds.), *Cotton Sector Development in Ethiopia*, Textile Science and Clothing Technology, https://doi.org/10.1007/978-981-99-9149-5_1

1 Introduction

Cotton is a significant cash crop in numerous developed and developing nations. It has a notable impact on reducing poverty, as it is grown on small family farms in areas with limited opportunities for growing other crops and low per capita income [10]. In 2007, cotton was cultivated in 90 countries. The cost of cotton cultivation in Africa is lower compared to other countries. However, African countries only contribute 12% to the global market [33]. In this continent, cotton is typically grown by small-scale farmers as the primary cash crop on rain-fed land, with minimal use of purchased inputs like chemicals and fertilizer [3]. According to [30], cotton production in sub-Saharan Africa increased by a factor of 8.5 from 200,000 tonnes per year to over 1,700,000 in 2004/05, while global production volume only tripled. However, over the past decade, yields have stagnated at approximately half of the world's production due to a lack of irrigation and inconsistency in input provision and advice across the region. Additionally, the author demonstrated that the area of land dedicated to cotton is expanding while cotton productivity in Africa remains only half of the world's production (Ibid). Ethiopia is also among the sub-Saharan African countries that cultivate and export cotton. It has a long history of cotton cultivation, with an estimated area of over 2.6 million hectares suitable for cultivation. The primary markets for Ethiopian cotton are Africa, Asia, and Europe, with Asia alone accounting for 67% of total exports [7]. Currently, the price of Ethiopian cotton is determined by the textile industry development institute as Grade "A" ($1.47/kg), Grade "B" ($1.43/kg), and Grade "C" ($1.40/kg). However, Ethiopia only contributes 5% to the total production in Africa (Ethiopian Investment Agency [7, 8]). Based on these reports, despite having ample land for abundant cotton production, Ethiopia has performed poorly in terms of cotton production and marketing.

1.1 Production Levels and Major Production Areas in Ethiopia

Cotton is one of the few agricultural products whose production and consumption are almost entirely global in scope. Cotton is growing in more than 70 countries, including Ethiopia, and many developed and developing countries rely on lint imports for their spinning and textile industries. Cotton production and consumption have increased significantly over the last four decades, rising from 9.8 million tonnes in 1960/61 to 18.5 million tonnes in 1998/99 and 21.1 million tonnes in 2001/02 [4].

The major cotton producers are concentrated in the developed world, with the United States far and away the largest producer, followed by China, India, Pakistan, Uzbekistan, and West African countries. Only 30% of total global production is exported annually, as most producers are becoming major consumers of their own production and even import cotton due to their expanding spinning and textile industries. This has resulted in a significant shift in trade flows away from the main exporting regions and towards the leading producers and importers of cotton, such as those in Asia [12].

Ethiopia, too, has a huge potential for cotton production due to its suitable agro-ecological zones and water availability. According to the Ministry of Agriculture, the suitable cotton production area is estimated to be 2,575,810 ha, which is comparable to Pakistan, the fourth largest producer. Despite this enormous potential, Ethiopia produces only 77,000–84,000 MT of raw cotton per year from a total area of 42,371 acres [1]. The disparity between existing potential and actual practice is more apparent when we consider Ethiopia's share of international cotton production and marketing, with an average share of only 0.13% of total cultivated land and 0.1% of cotton produced from 1998 to 2000 [14]. In terms of international trade in lint cotton, Ethiopia has a 0.1% export share and a 0.06% revenue share for the same year. There could be several reasons for the country's poor performance in cotton production and marketing. Among the few are the limited availability of research and extension services, as well as an insufficient supply of inputs and a lack of capacity to supply quality products, as well as the presence of inadequate infrastructure and finance [1].

Despite its poor performance, the cotton sub-sector remains a unique opportunity for Ethiopia in terms of serving as a bedrock upon which the country can shift to high value-added technological transformation due to its strong backward and forward linkages with various sectors and its provision of employment opportunities for a large number of rural poor. Against this backdrop, the Ethiopian government wishes to make a concerted effort and take action to stimulate the growth and potential of this sub-sector in order to make cotton one of the country's major commercial crops.

Ethiopia is ideal for cotton cultivation due to favourable natural conditions (climate, fertile soils and water availability) and a millennium-long tradition of producing cotton and textile fabrics (Fig. 1). Only 3% of the total 2.6 million ha potentially suitable for cotton production are currently under cultivation in the country. Cotton is typically grown in rotation with sesame, sunflower, sorghum and mung beans. The national seed cotton production in the 2018/2019 campaign was estimated to be 95,750 t. Lint and then yarn are made from seed cotton, as are cotton oilseeds, from which cotton oil and cake are made, and planting seeds (Fig. 2). The main variety of seed cotton cultivated in Ethiopia (DP90), like in West Africa, produces medium lint. Aside from that, cotton produced in the country is generally of low quality (heterogeneous, irregular and contaminated with foreign material) [9].

1.2 Agronomic Requirements and Practices

1.2.1 Abiotic Environment

Cotton, despite its origins in the tropics and subtropics, has come to be cultivated primarily in subtropical and warm-temperate zones, which account for more than half of global production. The species' photoperiod had to change for this geographical

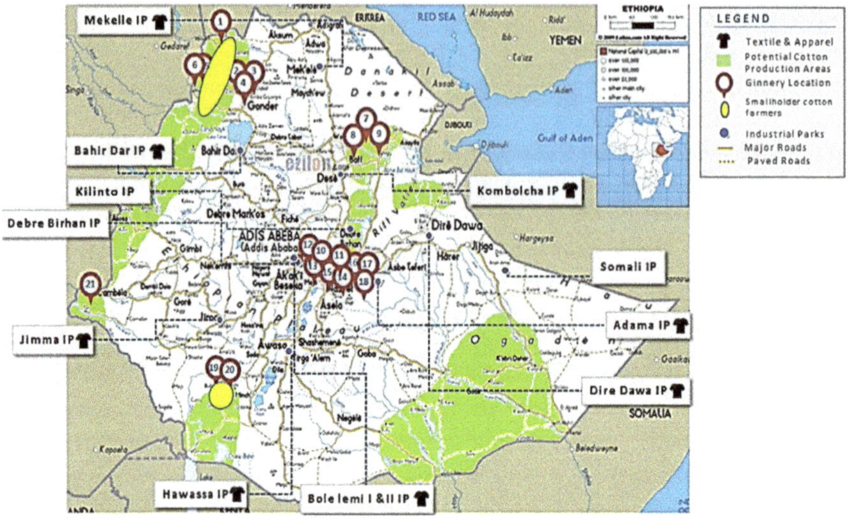

Fig. 1 Distribution of Ethiopia's Industrial Parks (IPs), potential cotton-growing sites and ginnery locations (*Source* TIDI, Overview of the Ethiopian Cotton Sector 2017)

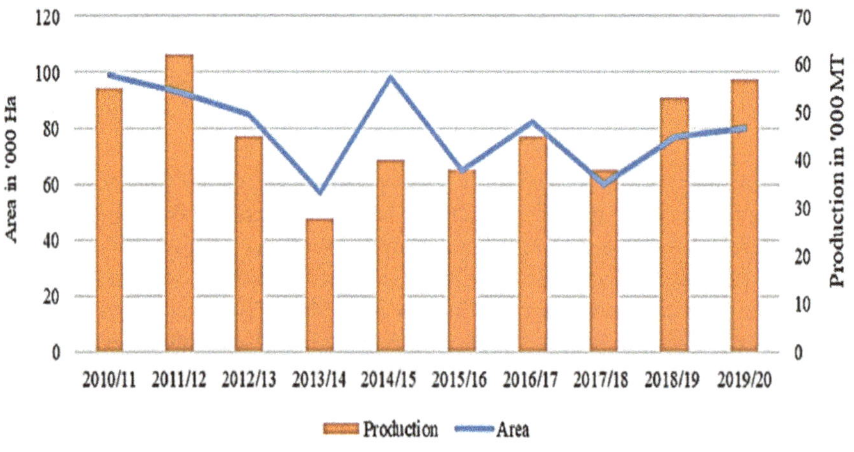

Fig. 2 Ethiopian's cotton production trends: 2010–2020

shift to be possible as a crop—the naturally short-day plant became a day-neutral plant that could be cultivated as an annual crop in the longer summers [28].

1.2.2 Climate Requirements

The main climatic factor determining the geographical range in which cotton can be grown is temperature. In general, the plant is extremely temperature sensitive

(Reddy et al. 2006). Seeds do not germinate, nor do seedlings begin to grow, until the temperature rises to 15 °C; above 38 °C, they are delayed. *Gossypium barbadense* seedling development is generally not sensitive to temperatures ranging from 15 °C to 40 °C in the first 2 weeks, but 3 weeks after emergence, the young plants are more sensitive than *G. hirsutum* (having fewer fruiting branches at 35°/27 °C than at 30°/22 °C, and none at 40°/32 °C). Nonetheless, *G. barbadense* cultivars with heat tolerance comparable to *G. hirsutum* exist [31]. The optimum daytime temperature range for *G. hirsutum* is 30–35 °C, with a loss of fruit above 35 °C and a 50% yield reduction at 25 °C [22]. For *G. hirsutum*, 180–200 frost-free days are required after planting, with an average of 150 days of suitable temperatures (i.e. 1200 heat units above 15.5 °C accumulated); for G. barbadense, 200–250 days are required [33]. Although the values vary by variety, the required minimum is 2050 heat units (degree-days or day-degrees) from cotton planting to 60% boll opening [20]. Cotton prefers full sun because it is a short-day plant. Cotton's light saturation point is 70,000–80,000 lax.

1.2.3 Soil and Water

Cotton plants are grown in a wide range of soil types, but the crop grows best in deep arable soils with good drainage, organic matter and a high moisture-retention capacity. Cotton is a salt-tolerant plant, with *G. barbadense* being more salt tolerant than *G. hirsutum.* Nonetheless, salinity stress has a negative impact on germination and emergence [2], the most common stress effect is general stunting of plant growth. Cotton crops grown on dry land (non-irrigated) require at least 500 mm of rainfall during the growing season [6].

Cotton is also grown as an irrigated crop, and sprinklers with fixed or mobile outlets with total coverage are still widely used. Irrigation enables cultivation in poor-quality soils by providing the necessary moisture and nutrients in a controlled manner. Irrigation is done primarily at ground level, flooding the furrows, which necessitates adequate field levelling. The use of drip irrigation has increased, allowing for water savings and the use of less-than-ideal soil (due to a sloping surface, a lack of fertility or an excessively high salt content). *G. barbadense* has similar water requirements to *G. hirsutum* in general; however, the longer growing season of *G. barbadense* may necessitate additional irrigation to mature its later-set bolls [26]. The application of water at the right time optimizes the plant's vegetative growth, flowering and boll production. Flower and boll formation in *G. barbadense* is aided by short periods of direct sunlight, high minimum humidity and a slow evaporation rate [25].

1.3 Cultivated Species

Cotton has four cultivated species: *Gossypium arboreum, G. herbaceum, G. hirsutum* and *G. barbadense.* The first two species are diploid (2n = 26) and indigenous to the old world. They are also known as Asiatic cottons due to their origins in Asia. The last two species are tetraploid (2n = 52) and are also known as New World Cottons. *G. hirsutum* is also known as American cotton or upland cotton, and *G. barbadense* as Egyptian cotton, Sea Island cotton, Peruvian cotton, Tanguish cotton, or quality cotton. *G. hirsutum* is the dominant species, accounting for roughly 90% of total global production. Perhaps India is the only country in the world where all four cultivated species are grown commercially (OECD 2021).

1.4 Cultivation

1.4.1 Sowing

Before beginning anything, seed selection is required for cotton production. Smaller, imperfect, impurity, disease and insect-damaged ones should be removed. After seed selection, the purity and germination should be greater than 95 and 85%, respectively. Seed treatment should be performed if necessary. Among the techniques are: sun dry—to some extent, this kills bacteria and fungi and increases water absorption ability. The acid treatment is used to delint all of the linter, make the seed shine, improve germination capacity, remove dormancy, control seed-borne disease, and save labour for replanting.

Cotton planting necessitates careful soil preparation in order to achieve sufficient moisture for favourable germination and rapid root development. Pre-prepared ridges are recommended for adequate water drainage and temperature control. Temperature determines the best sowing date. Temperature is the most important factor influencing the development and yield of cotton plants (Robertson et al. 2007). Sowing can begin when the minimum soil temperature at a depth of 10 cm exceeds 14 °C for at least 3 days in a row. Lint yield is reduced if *G. hirsutum* is planted too early (due to cold temperatures) or too late (due to a shortened growing season). Because G. barbadense prefers a longer growing season (>200 days) for increased yield, it is more sensitive to planting delays [26]. Table 1 shows how seed rate and plant spacing differ between cotton species. Cotton planted at a sowing depth of it ranges from 1.5 to 2 cm in humid conditions and from 3 to 4 cm in dry conditions. Cotton is planted manually as well as in planters.

1.4.2 Field Management

From sowing to emergence: Make the surface soil friable and loose to encourage emergence, strengthen the drainage system and avoid flooding and logging.

Management at seedling stage: Maintaining full and uniform stands encourages plant development. An early thinning out and a suitable time of final singling are required. After a uniform emergency, it is preferable to begin first thinning out at the 1–2 true leave stage and second thinning out at the 3–4 true leave stage to achieve

Table 1 Seed rate for various cotton species

Species	Growing conditions	Seed rate (kg/ha)	Spacing (cm)
G. hirsutum	Irrigated	20–22 10–15	75 × 15 75 × 30 75 × 45
	Rain fed	18–20	60 × 30
G. *arboreum*	Irrigated	10–12	60 × 30
G. *herbaceum*	Rain fed	12–15 12–15	45 × 30 60 × 30
G. barbadense	Irrigated	8–10 12–15	90 × 30 75 × 30
Hybrids	Irrigated	2–3	45 × 60 90 × 60 45 × 30
		2–3.5	120 × 40 120 × 60
		3–3.5	67.5 × 67.5
	Rain fed	3–3.5 2.5–3	150 × 60 120 × 60
Bt hybrids	Irrigated	1.5	90 × 60 120 × 40 120 × 60

a single plant. Weeding and inter-row tillage are carried out. Inter-row tillage is required once after rainfall or irrigation.

a. The first inter-row tillage and weeding is done in conjunction with thinning out. This is done at a depth of about 4–5 cm.
b. A second round of inter-row tillage and weeding is performed in conjunction with the final singling at a depth of 6–7 cm.
c. A third round of inter-row tillage and weeding is performed before squaring, with a depth of up to 7–8 cm. Fertilizer application at around 3–4 true leaves stage connected with the second time of inter-row tillage for each plant by applying 0.5–1 g of urea is critical.

Management at flower and bud forming stage: This is to control luxuriant growth and to encourage measures to achieve strong growth through various mechanisms.

Inter-Tillage and Weeding:

- To have a normal cotton plant field, deep inter-row tillage with a depth of 10 cm is required, as well as weeding.
- To obtain vigorous cotton plants, the depth of inter-row tillage must be increased.
- Shallow inter-row tillage results in weak cotton plants in cotton plant fields.
- Hilling up of the plant:

This is done from the final singling to the beginning of the flower stage. It aids in drainage, fertilizer application and removal of leaf branches. It is possible to control the luxuriant growth by.

(i) Deep inter-tillage to cut some of the root in order to reduce the plant's absorbing ability.
(ii) Pinching the top to reduce dominance and control plant growth.
(iii) Spraying an artificial plant regulator (growth regulator), such as cycagon.

Management at Flowering Stage

This is to encourage early boll formation and maturation, as well as to reduce shedding of flower buds and young bolls by.

- Applying a large amount of fertilizer when the first bolls bloom.
- Spraying the plants with 0.2% borate and 0.2% KH_2PO_4.
- Topping—the removal of the plant's terminal point (apex). The timing of topping varies depending on the species. *Hirsutum spp.*, for example, is when the plant has 15–16 fruiting branches, whereas *Barbadense spp.* is when the plant reaches a height of 1–1.2 m, or 80–90 days after emergence.
- Chemical control: apply 60–70 ppm DPC after topping.
- Clearing away any dead leaves and branches.

Management at Boll Opening Stage

This is done to promote early maturation and to prevent premature decay by

- Pruning of old leaves and empty branches to remake aerial condition.
- Spraying 0.5–1% urea on the leaves to prevent premature decay.
- Avoiding severe drought. If there is a severe drought, provide adequate and timely irrigation to the field. However, excessive vegetative growth, which delays maturity and extends the harvesting period of the plant, should be avoided.

Irrigation is required when rainfall is less than 450 mm and erratic in nature. There should be an adequate amount of moisture for good yield and growth. Furrow irrigation is widely used and effective at distributing water throughout a field.

1.4.3 Fertilizers

Phosphorus (P) and potassium (K) are applied at the base of the plants based on soil content. Nitrogen (N) is applied to the plant's base and top at a rate of up to 200 or 250 units of N, depending on environmental and crop conditions. *Gossypium barbadense* requires slightly more nitrogen, phosphorus and potassium per unit of lint produced. It is, however, more sensitive to a slight excess of N, which can stimulate increased vegetative growth and delay maturity [16]. Furthermore, deep application of compost as basal manure accounts for 50–60% of the total amount of fertilizers.

1.4.4 Crop Rotation

Cotton crop rotation is typically accomplished by alternating with other traditional crops grown in the area. Contrary to best agricultural practices, cotton is sometimes planted in the same field again, for 2 years or longer. The crop damage caused by diseases, particularly *Verticillium wilt*, limits the number of repetitions [16].

1.5 Biotic Environment

1.5.1 Vesicular Arbuscular Mycorrhizae (VAM)

Cotton crop growth in most soils is dependent on the interaction with mycorrhizal fungi [15]. Fungi (such as *Glomus mosseae*) grow intercellularly in the root cortex. They form vesicular arbuscules with the plasma membrane in cortical cells, which serve as sites of mineral exchange from the fungus to the plant and carbohydrate exchange from the plant to the fungus. The main advantage for cotton plants is improved phosphate uptake [15].

1.5.2 Insect Pests and Nematodes

Pest and disease control is expensive [19], and repeated applications of insecticides and fungicides may be used. Cotton is a favourite food for a variety of insect pests. Insects that are natural enemies of pests are encouraged as part of integrated pest management systems. Cultivation of varieties with genetically engineered insect resistance has been a significant advancement in crop management against some major pests. Arthropod pests can have an impact on boll production or fibre quality. Aphids (*Aphis gossypii, Aphis craccivora, Myzus persicae*) and the silverleaf whitefly *Bemisia tabaci* are the most common pests that affect fibre quality, producing sticky cotton with dark stains if not controlled late in the season. The pink bollworm *Platyedra gossypiella*, various Hemiptera such as Lygus bugs and various mites such as the two-spotted spider mite, *Tetranychus urticae,* all reduce fibre yield and quality. Cotton bollworms (*Helicoverpa armigera, H. punctigera*) and the spiny bollworm are major pests affecting boll production. *Earias insulana* primarily reduces fibre production. In some areas, the cotton boll weevil *Anthonomus grandis* is a very aggressive pest. Other significant pests include the leafhopper, *Empoasca lybica* (the cotton jassid) [18].

Nematodes that cause damage in some regions or areas include particularly the root-knot nematodes—*Meloidogyne incognita* (as well as M. *acronea), reniform* nematode—*Rotylenchulus reniformis*, lance nematodes—Hoplolaimus columbus (and several other spp.) and sting nematode—*Belonolaimus longicaudatus* [24], as well as associated ring nematodes—*Criconemella spp.*, spiral nematodes—*Helicotylenchus spp.*, needle nematode—*Longidorus africanus*, stunt nematodes—*Merlinius spp.* and *Tylenchorhynchus spp.*, stubby-root nematodes—Paratrichodourus *spp.*, pin nematode—*Paratylenchus hamatus*, lesion nematodes—*Pratylenchus spp.*, spiral nematodes—*Scutellonema spp.* and American dagger nematodes—the *Xiphinema americanum* group [24].

In Africa, there are approximately 150 insect species that attack cotton. The following are the most serious:

I. Leaf suckers (*Aphis gossypii*).
II. Jassids (*Empoasca spp.*).
III. American bollworms (*Heliothis armigera*).
IV. Red bollworms (*Diparopsis and castanea*).
V. Spiny bollworms (*Pectinophora and gossypiella*).
VI. Stainers (*Dysdercus spp.*).

Control measures include crop rotation and field sanitation. Supplementary measures, such as the application of appropriate pesticides, are usually required.

1.6 Diseases

The most common cotton disease is Verticillium wilt, which is caused by Verticillium dahliae [11]. This fungal disease is widely distributed in areas where G. hirsutum is grown, conventionally bred resistant varieties are available [20]. Other diseases, such as damping off, are caused by a complex of pathogens that have a significant impact on the crop. The primary pathogens are Rhizoctonia solani, Pythium ultimum, Thielaviopsis basicola and Fusarium spp. Many other fungi have been linked to cotton diseases, either as primary agents or secondary invaders: *Alternaria spp., Ascochyta gossypii, Aspergillus flavus, Brasilomyces malachrae, Cladosporium herbarum, Fusarium spp.* (e.g. *F. oxysporum f. sp. vasinfectum*), *Glomerella gossypii (anamorph Colletotrichum gossypii), Lasiodiplodia theobromae (synonym Diplodia gossypina), Leveillula taurica (anamorph Oidiopsis haplophylli [synonyms O. gossypii, O. sicula]), Macrophomina phaseolina, Mycosphaerella spp., Nematospora spp., Phakopsora gossypii, Phymatotrichopsis omnivora, Phytophthora spp., Puccinia cacabata* and *P. schedonnardi, Pythium spp.* and *Sclerotium rolfsii.*

Boll rot caused by these diseases causes significant losses in production. Damage is more severe in crops grown in high humidity and low light intensity environments, and it worsens if the bolls have mechanical lesions. The main problem that these fungi cause is fibre contamination, especially if open bolls are exposed to rain or high humidity for an extended period of time. These agents, in addition to causing undesirable discoloration of the fibre, may cause enzyme degradation in some basic components, as is common in cellulose. Cotton diseases caused by bacteria, such as Xanthomonas campestris pv. malvacearum, and viruses, such as abutilon mosaic geminivirus, cotton leaf crumple geminivirus, cotton leaf curl geminivirus, cotton yellow mosaic geminivirus and cotton anthocyanosis virus, are also common. The etiology of cotton bunchy top, cotton leaf mottle and cotton leaf roll diseases is unknown [18].

Disease is much less important in Ethiopian cotton-growing areas than insects. Cotton's main diseases are.

a. Bacterial blight—*Xanthomonas malvacearum.* This can result in the death of leaves and branches, the shedding of young bolls and the premature opening of bolls. The primary control methods are seed dressing and clean cultivars.
b. Fusarium wilt—Fusarium oxysporum causes Fusarium wilt. It has the potential to kill or stunt the plant. The most effective method of control is to select highly resistant cultivars.
c. Leaf curl—Leaf curl is caused by a virus. It is spread by the white fly (*Bemisia tabaci*). To control this disease, resistance cultivars and clean cultivars are the best options.

1.7 Weeds

Weed control in cotton fields is critical, and it is done mechanically by passing through the crop rows, as well as chemically [18]. Many different herbicides are used in cotton cultivation, with their application occurring during pre-sowing and/or pre-emergence of seedlings, or, less frequently, post-emergence. Various herbicides can be applied during land inclusion and pre-sowing, pre- and post-sowing, immediately post-sowing, pre-sowing and pre-emergence, pre-emergence, pre- and post-emergence, immediately post-sowing and post-emergence, and post-emergence [16]. Crop rotations and farm hygiene are examples of integrated weed management measures that include crop rotations and farm hygiene to prevent weed seed spread [5, 23]. The cultivation of herbicide-tolerant varieties developed through genetic engineering has also significantly improved crop weed management. The most common and troublesome weeds vary greatly depending on region and management practices. Table 2 lists genera that frequently have species of significant concern in areas.

1.8 Harvest and Processing (Ginning, Crushing)

To facilitate harvest and subsequent ginning (freeing of fibres from seed to obtain lint), the plant is chemically defoliated. This improves both the cleanliness and the quality of the fibres. Spindle picker machines in two or four rows are used for mechanized harvest. The cotton is then ginned in saw gins to produce bales classified according to grade and length of fibre. The separated cottonseed is further processed by first separating the hulls from the kernels. The kernels are crushed, and the oil extracted and processed for use in human food or other products. The hulls are used for livestock feed or industrial products, while the remaining kernel (high in protein) is converted into cottonseed meal for livestock. In the case of *G. hirsutum*, the fuzzy

Table 2 Genera of weeds in cotton

Dicotyledons	Monocotyledons
Abutilon	*Alopecurus*
Achyranthes	*Cenchrus*
Alternanthera	*Commelina*
Amaranthus	*Cynodon*
Boerhavia	*Cyperus*
Capsella	*Dactyloctenium*
Celosia	*Digitaria*
Chamaesyce (Euphorbia)	*Echinochloa*
Chenopodium	*Eleusine*
Convolvulus	*Leptochloa*
Croton	*Lolium*
Datura	*Panicum*
Desmodium	*Paspalum*
Diplotaxis	*Poa*
Fumaria	*Rottboellia*
Geranium	*Setaria*
Heliotropium	*Sorghum*

seed (i.e. seed with linters) is delinted, which means that the linters are removed mechanically or chemically. These residual short fibres are used for a variety of applications, including cellulose bases for food and other consumer products. Picking and ginning techniques for *G. barbadense* cotton differ from those used for *G. hirsutum* cotton in order to maintain its superior fibre quality. Because *G. barbadense* does not produce linters, its seed is available in two forms: unprocessed "seed cotton" and processed black seed [16, 18].

2 Summary

Cotton is one of the scarce agricultural products whose manufacturing and usage are more or less worldwide in scope. Cotton is cultivated, including in Ethiopia, in over 70 nations, where many developed and developing nations rely on the import of lint for their spinning/fabric industries. In Ethiopia, as well, there is immense potential for the cultivation of cotton, due to its appropriate agro-ecological zones and the presence of water. Despite its underperformance, the cotton sub-sector still presents a distinctive chance for Ethiopia in terms of serving as a foundation upon which the country can transition to high-value-added technological transformation following its strong backward and forward connections with various sectors, and its provision of employment opportunities for a large number of the rural impoverished.

The nation currently cultivates just 3% of the overall 2.6 million hectares that have the potential for cotton production. Cotton is typically rotated with sesame, sunflower, sorghum and mung beans. Although the values vary among different types, around 2050 heat units (degree-days or day-degrees) are the minimum requirement from cotton planting to 60% boll opening. Cotton prefers full sunlight as it belongs to short-day plants. The light saturation point of cotton is 70,000–80,000 lax. Cotton plants are grown in a wide range of soils, but the crop thrives best in deep arable soils with good drainage, organic matter and a high moisture-holding capacity. There are four cultivated types of cotton, namely, Gossypium arboreum, G. herbaceum, G. hirsutum and G. barbadense. Temperature is the main factor that affects the development and yield of cotton plants. Sowing can begin when the minimum soil temperature at a depth of 10 cm is above 14 ℃ for at least 3 consecutive days. Seed rate and plant spacing vary depending on the type of cotton, as shown in Table 1. For cotton planted in humid conditions, the sowing depth ranges from 1.5 to 2 cm, while under dry conditions, it is from 3 to 4 cm. Cotton can be planted manually or by using planters. Phosphorus (P) and potassium (K) are applied based on soil content, at the base of the plants. Nitrogen (N) is distributed between the base and top of the plant, with an application rate of up to 200 or 250 units of N, depending on the environmental and crop conditions. Deep application of compost as basal manure makes up for 50–60% of the total amount of fertilizers. Pest and disease control is a significant expense, and repeated applications of insecticides and fungicides may be necessary.

The cultivation of varieties with genetically modified resistance to certain insects has been a significant advancement in managing the crop against major pests. Approximately 150 insect species attack cotton in Africa. The most notable disease affecting cotton is Verticillium wilt, caused by Verticillium dahlia. In Ethiopia's cotton-growing regions, diseases are less significant compared to insects. To make harvesting and ginning (the process of separating fibres from seeds to obtain lint) easier, the plant is defoliated using chemical treatment. This enhances cleanliness and fibre quality.

References

1. Agridev Consult (2004) Cotton-textile-apparel value chain reporter Ethiopia. Regional Agricultural Trade Expansion Support Program. Nairobi. Kenya
2. Ashraf M (2002) Salt tolerance of cotton: Some new advances. Crit Rev Plant Sci 21:1–30
3. Bosena F, Bekabil G, Berhanu, Dirk H (2011) Factors affecting cotton supply at the farm level in metema district of Ethiopia
4. CBI (2006) I lie outwear market in the EU. Marketing Survey
5. Charles G (2002) Managing weeds in cotton. Pp. B3.1-B3.22 in WEED pak—A guide for integrated management of weeds in cotton. Australian Cotton Cooperative Research Centre, Narrabri, New South Wales
6. Cothren JT (1999) Physiology of the cotton plant in W.C. Smith and J.T. Cothren, eds., Cotton: Origin, History, Technology and Production. John Wiley & Sons, New York, pp 207–268

7. EIA (Ethiopian Investment Agency) (2012) Investment opportunity profile for cotton production and ginning in Ethiopia. Updated Report on Cotton Ginning. Addis Ababa, Ethiopia. https://doi.org/10.1094/PDIS-11-11-0999-PDN
8. Ethiopian Science and Technology Commission (ESTC) (2006) R and D and innovation in the textile sub sector: Training Course on R and Capacity Building in the Industry Sector. June 02, 2006, Bahir Dar, Ethiopia. 77
9. FAS (Foreign Agricultural Service/USDA) (2001) Global market analysis. Available at https://public.govdelivery.com/accounts/USDAFAS/subscriber/new
10. Goreux L (2004) Cotton after cancun: Draft discussion paper to inform debate. March [Online]. Retrieved from http://www.acp-eu-trade.org/pdf
11. Kirkpatrick TL, Rothrock CS (2001) Compendium of Cotton Diseases, 2nd ed. APS Press, St. Paul, Minnesota, USA, p 100
12. Marianne L (2004) Quality standard setting in the global cotton chain and cotton sector reform in SSA. Institute for International Studies, Copenhagen
13. Ministry of Agriculture and Rural Development (MORAD) (2004) Comprehensive cotton development and marketing master plan
14. Ministry of Agriculture and Rural Development (MoARD) (2004) Crop variety register, issue no 7. Addis Ababa, Ethiopia
15. Nehl D, Allen S (2004) Symptoms of Diseases and Disorders of Cotton in Australia. Australian Cotton Cooperative Research Centre, Narrabri, Australia, p 38
16. OECD (The Organisation for Economic Co-operation and Development), (2021) OECD Economic Outlook, Interim Report March 2021. Available in December at. https://doi.org/10.1787/16097408
17. OECD (2010) "Section 1—Cotton (Gossypium spp.)", in safety assessment of transgenic organisms, vol 4. OECD Consensus Documents, OECD Publishing, Paris. https://doi.org/10.1787/9789264096158-5-en
18. Oerke E-C (2006) Crop losses to pests. J Agric Sci (Camb) 144:31–43
19. OGTR (2008) The biology of Gossypium hirsutum L. and Gossypium barbadense L. (Cotton), Ver. 2. Office of the Gene Technology Regulator, Australian Government Department of Health and Ageing, Canberra. 91—PART 1:76
20. Reddy KR, Hodges HF (2006) Exploring the limitations for cotton growth and yield. J New Seeds 8(2):1–22
21. Reddy KR, Reddy VR, Hodges HF (1992) Temperature effects on early season cotton growth and development. Agron J 84:229–237
22. Roberts G, Charles G (2002) Integrated weed management (IWM) guidelines for Australian cotton production. Pp. B2.1-B2.10 in WEEDpak—A guide for integrated management of weeds in cotton. Australian Cotton Cooperative Research Centre, Narrabri, New South Wales
23. Robinson AE (1999) Cotton nematodes in W.C, pp 595–615
24. Sawan ZM, Hanna LI, McCuistion WL (2005) Response of flower and boll development to climatic factors before and after anthesis in Egyptian cotton. Climate Res 29:167–179
25. Silvertooth JC (2001) Agronomic guidelines for pima cotton production in Arizona. University of Arizona College of Agriculture and Life Sciences, Cooperative Extension AZ1242:4
26. Silvertooth JC, Norton ER, Unruh BL, Navarro JA, Clark LJ, Carpenter EW (1995) Nitrogen management experiments for upland and Pima cotton in Cotton, A College of Agriculture Report. Univ Ariz Coll Agric Ser 99:311–326
27. Smith CW, Cantrell RG, Moser HS, Oakley SR (1999) History of cultivar development in the United States. In: Smith CW, Cothren JT (eds) Cotton: origin, history, technology and production. John Wiley & Sons, New York, pp 99–171
28. Sneyd A (2006) Cotton and poverty in Sub-Saharan Africa: A review of the factors (PhD Qualification Paper). Department of Political Science McMaster University, Hamilton
29. Srivastava A, Lu Z, Zeiger E (1995) Modification of guard cell properties in advanced lines of Pima cotton bred for higher yields and heat resistance. Plant Sci 108:125–131
30. Srivastava D (1982) Studies in the pollen biology of certain cultivated Malvaceae. Adv Pollen-Spore Res 9:165

31. Unruh BL, Silvertooth JC (1997) Planting and irrigation termination timing effects on the yield of upland and Pima cotton. J Prod Agric 10:74–79
32. Unruh BL, Silvertooth JC (1996) Comparisons between an upland and a Pima cotton cultivar: II. Nutr Uptake PartitIng Agron J 88:589–595
33. USDA (2010) The progress of West African cotton production and trade presented by H.E Tièna COULIBALY. Ambassador of Mali to the USA, unpublished

Cotton Genetics and Breeding

Anbesaw Gate, Alemu Abate, and Bizuayehu Kerisew Semahagn

Abstract Cotton (*Gossypium spp.*) is a soft, fluffy staple fibre that develops in a boll, or protective casing, around the seeds of cotton plants of the genus *Gossypium* in the mallow family *Malvaceae.* Arboreums are thought to be indigenous to India. *G. herbaceum* may have been introduced into western India from Arabia, Persia and Baluchistan, as well as race acerifolium. Cotton is the most widely used fibre in the production of clothing and other textiles. Around 25 million tonnes of cotton are currently produced worldwide. China and India produce the most cotton, followed by Brazil and the United States of America. Cotton is Ethiopia's most important cash crop, and it is vital to the country's agricultural, industrial and economic development. It provides basic raw materials (cotton fibre) to the textile industry for use in the domestic or export markets. Currently, 96% of cotton varieties grown in the country are upland, with DP 90 covering 80% of the acreage, Stam 59A (15%), Acala SJ2 (0.5%), and land races and local varieties accounting for the remaining 4%. Cotton cultivation has long been practiced in Ethiopia. Cotton is grown primarily in the Awash Valley, Gambela, Humera and Metema. Cotton has grown in many parts of the country. There are large potential areas in each region: Tigray has 269,130 ha; Amhara has 678,710 ha; South nations, nationalities and people region (SNNPR) has 600,900 ha; Oromia has 407,420 ha; Gambella has 316,450 ha; Benshangul has 303,170 ha; Afar has 200,000 ha and Somali has 225,000 ha. Germplasm enhancement, a lack of cotton seed production, and abiotic and biotic stresses are major production and cotton breeding constraints. Breeders are currently experimenting with these factors. Cotton breeding in Ethiopia has achieved varietal development. Among those released varieties, the major ones include disease resistance, high yield, pest resistance, high fibre quality, comfort to mechanical harvesting by increasing

A. Gate · A. Abate (✉)
Department of Plant Sciences, College of Agriculture, Oda Bultum University, Chiro, Ethiopia
e-mail: abatalem@yahoo.com

A. Abate
Department of Plant Sciences, College of Agriculture and Environmental Sciences, Bahir Dar University, P.O. Box 79 Bahir Dar, Ethiopia

B. K. Semahagn
Department of Biology, College of Science, Bahir Dar University, P.O. Box 79, Bahir Dar, Ethiopia

K. Murugesh Babu et al. (eds.), *Cotton Sector Development in Ethiopia*, Textile Science and Clothing Technology, https://doi.org/10.1007/978-981-99-9149-5_2

cotton height, early maturity and adaptability to harsh environments. In light of the challenges, we recommended the following feature directions for cotton breeders in Ethiopia. Work on extra-long cotton varieties, varieties suitable for mechanical harvesting, varieties for high fibre quality, varieties for biotic and abiotic resistance and upgrading breeding activities to modern biotechnological tools.

Keywords Cotton · Breeding · Ethiopian achievement · Challenges · Future direction

1 Introduction

Cotton (*Gossypium spp.*) is a soft, fluffy staple fibre that develops in a boll, or protective casing, around the seeds of cotton plants of the genus Gossypium in the mallow family Malvaceae [14]. First, it was originated in Indus River Valley, which is now in Pakistan. However, this does not mean that the plant is not grown elsewhere; it is practically grown everywhere in the world's tropical and subtropical regions, including the United States, China, India, Uzbekistan, Pakistan, Brazil and Turkey. Texas produces the most cotton in the United States, and the South Plains region in the country's north is the largest continuous cotton-growing region [26].

Cotton production is thought to have originated in Ethiopia, where the country's agricultural history is inextricably linked to that of the crop. Since 350 A.D., the Amharic word 'tet', which means cotton, has been used in writing, including during the reign of King Aizana of Axum. This suggests that Axum was long known for its raw materials, which included iron, bronze and cotton textiles [5]. Cotton has been produced and used in Ethiopia since ancient times. According to [35], one cotton species, *Gossypium herbaceum*, which is occasionally seen in Ethiopian fields and gardens, may be indigenous to this country [5].

Cotton is grown throughout Ethiopia at elevations above 1000 m and below 1400 m due to a lack of adequate rainfall in the majority of the lowlands and the need for irrigation [22]. Tendaho, in the lower Awash Valley, was one of Ethiopia's largest cotton plantations. Cotton grown with rain also grew in Humera, Bilate and Arba Minch. Ethiopia has the potential to become a significant global cotton producer due to its favourable growing conditions, abundance of raw materials and availability of land, but as of 2011, the country's cotton production lagged significantly behind that of its coffee and grain industries [36]. Ethiopia has a huge potential for cotton production due to its favourable agro-ecological zones and water availability. According to the Ministry of Agriculture, the area suitable for cotton cultivation is 2,575,810 ha, which is equivalent to the fourth-largest producer, Pakistan. Despite this enormous potential, Ethiopia currently produces only 77,000–84,000 metric tonnes of raw cotton annually from a total area of 42,371 acres [32].

Cotton, Ethiopia's most important cash crop, is critical to the country's agricultural and industrial growth. Cotton provides essential raw materials (cotton fibre) to the textile industry for both the domestic and export markets. Cotton seed, in addition to fibre, is a valuable source of high-quality protein and oil for both humans and animals [5]. It is a significant cash crop for smallholder farmers at the household level throughout the country, providing money for rural households' food security, particularly for peasant subsistence farmers in Ethiopia [42]. Cotton also supports the livelihoods of hundreds of thousands of people who work in agriculture, processing, trading and marketing [25]. Cotton contributes significantly to the overall growth of the national economy because it is a key export and import replacement crop that generates significant foreign exchange earnings when grown for export while also serving as a key import substitute crop [10].

2 Centre of Origin and Distribution of Cotton

Arboreums are thought to be indigenous to India. *G. herbaceum* may have been introduced into western India from Arabia, Persia and Baluchistan, as well as race acerifolium [35]. The barbadense appears to have begun as a wild species in South America before becoming a cultivated variety. The ratooned Tanguis cottons, which account for 90% of the Peruvian harvest, are the result of the native perennial barbadense. According to cytological studies, all four species are monophyletic (evolved from a single interspecific hybrid combination). G. hirsutum must have moved from the Diversity Center near the Guatemalan and Mexican borders.

Cotton was probably domesticated in Ethiopia, as it is one of the centres for the domestication and variation of many cultivated plants. This chapter assesses the available information and speculates on Ethiopia's potential role in the emergence and development of cotton cultivation [35]. Cotton is the most important vegetable fibre cultivated in approximately 60 countries worldwide, including some African countries such as Ethiopia. Russia, the United States, China, India, Brazil, Pakistan, Turkey, Egypt, Mexico and Sudan account for approximately 85% of total cotton production [1].

3 Taxonomy and Botany of Cotton

The term 'cotton' refers to four species in the genus *Gossypium* (family Malvaceae) (Table 1), namely, *G. hirsutum L.*, *G. barbadense L.*, *G. arboreum L.*, and *G. herbaceum L.*, that were domesticated independently as sources of textile fibre. The *Gossypium* genus contains about 50 species worldwide [19].

Cotton plants have a taproot system that grows quickly and can reach a depth of 20–25 cm before the seedling has even emerged above ground [45]. Lateral roots begin to form when cotyledons emerge and expand, they initially extend horizontally, then downwards. The taproot is rapidly spreading. The final depth of the root system is typically between 180 and 200 cm but can vary depending on soil moisture, aeration, temperature and variety [20]. The cotton plant has an upright primary stem and

Table 1 Taxonomy and botany of cotton

Kingdom	Plantae
Division	Magnoliophyta
Class	Magnoliopsida
Order	Malvales
Family	Malvaceae
Genus	*Gossypium*

numerous lateral branches. The stem has an apical bud and a growing point at the tip. Furthermore, unlike most types, the leaves are usually hairy. Hairy leaves are easier to harvest mechanically and more resistant to jassids, but they also harbour more white flies, which appear to find more shelter in the leaf hairs [7].

4 Genetics of Cotton

The taxonomic and evolutionary history of the genus Gossypium is extensive [19]. Gossypium consists of 52 species, 46 of which are diploids (2n = 2x = 26) and 5 of which are allotetraploids (2n = 4x = 52) [49]. Only four of these have been domesticated. There are two old world diploids (*G. arboreum L.* and *G. herbaceum L.*) and two new world allopolyploids (*G. hirsutum L.* and *G. barbadense L.*) in total. Four domesticated species contribute to the global production of natural fibre [37].

5 Cotton Genetic Diversity

Genetic diversity denotes the existence of various biological form variations or the degree of morphological and physiological characteristics of organisms within populations (commonly referred to as traits), which are required for biological individuals to survive and respond positively to rapid environmental change. Due to the susceptibility of genetically homogeneous varieties to various biotic and abiotic stresses, the lack of genetic diversity or its narrowness in different types of crops poses a potential threat to plant productivity. As a result, the wide genetic variety of crops has the potential to protect them from new diseases, pests and unanticipated changes in the global environment [2]. Thus, the genus Gossypium, which occupies a wide range of ecological and geographical niches, has a wide range of morpho-biological and genetic variety that has been conserved in cotton breeding centres (in situ centres of origin) around the world, as well as collections of cotton germplasm and ex situ centres of origin [28]. These resources can be successfully used in cotton breeding programmes to introduce economically beneficial features from wild species to cultivated genotypes in order to develop competitive promising cultivars.

6 Mode of Pollination and Reproduction

The cotton plant continues to develop vegetatively and reproductively for the remainder of the growing season due to its irregular growth pattern. The development of floral buds in the plant's apical portion, which gives rise to flowers and later bolls, signals the start of reproductive growth. Cotton generally reproduces sexually, but it can also reproduce asexually through root cutting. Cotton does not reproduce vegetatively in nature, although rooting has occurred under experimental conditions. Cuttings of G. barbadense (also known as G. vitifolium) can be grown in a laboratory environment, but significant rooting occurs only when the cuttings have numerous internodes and the parent plants are between 6 and 10 weeks old. Tannic acid or naphthaleneacetic acid (NAA) is used in this process [18]. Every cotton progeny receives two copies (or alleles) of each gene, one from each parent. Their characteristics are influenced by the combination of alleles they receive. Gene variations influence the length, quantity and quality of the fibre, as well as its colour and fire resistance.

Cotton is mostly self-pollinating, but it has a 5–50% chance of cross pollination through natural mechanisms, and it produces large, sticky, heavy pollen that is difficult for the wind to transport [31]. The flowers are large, noticeable and appealing to insects [23], so it is a chance out-crosser when insects are present in a cotton flower. Insect pollinators such as honeybees are thought to be the most likely insects responsible for any cross pollination in cotton. Honeybees visit cotton flowers to collect nectar during which time the honeybee transfers pollen grains from the anther of one flower to the stigma of another flower, resulting in the crops not crossing.

7 Cotton Production Trends in Ethiopia

Ethiopia has been producing cotton for a very long time. Prior to the revolution, large-scale commercial cotton plantations were established in the Awash Valley and the Humera areas [48]. Tendaho, in the lower Awash Valley, was one of Ethiopia's largest cotton plantations. Rain-fed cotton was also grown by Humera, Bilate and Arba Minch. According to 1867 reports, a 362-km (225-mile) route from Ethiopia's cotton plantations to the Red Sea was planned to facilitate trade with Egypt and Turkey [35]. Since the revolution, state farms with irrigation have grown the majority of commercial cotton, primarily in the Awash Valley region. Production was 43,500 tonnes in 1974–75 and 74,900 tonnes in 1984–85. Similarly, cultivation increased from 22,600 hectares in 1974–1975 to 33,900 hectares in 1984–1985 [8].

Ethiopia has the potential to become a significant global cotton producer due to its favourable growing conditions, abundance of raw materials and availability of land, but as of 2011, the country's cotton production lagged significantly behind that of its coffee and grain industries [36]. However, there are significant barriers to the industry's growth in Ethiopia because the country lacks administrative agencies to oversee and accredit agricultural methods and to process cotton in factories on

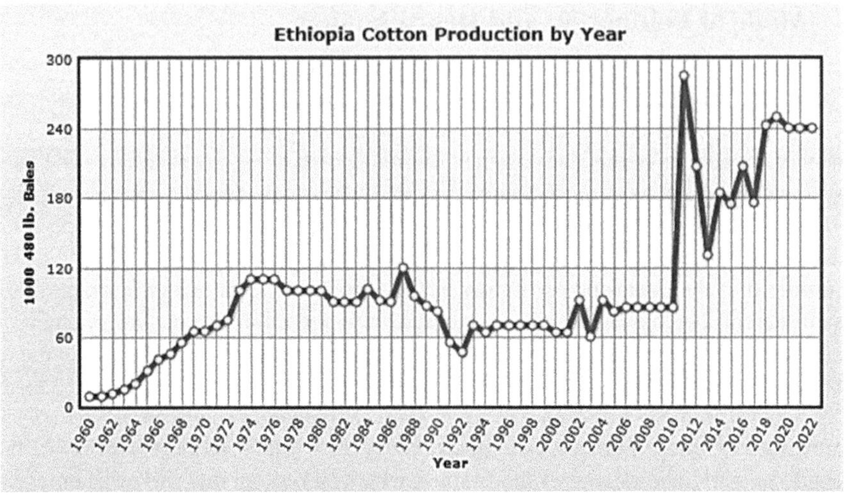

Fig. 1 Lint cotton production trends in Ethiopia (1960–2022) (*Source* http://www.indexmundi.com/agriculture/)

a large-scale commercial basis. However, the Ethiopian government prioritizes the growth of the textile industry in their economic plan, and in 2006 they launched a significant privatization initiative to attract both international and private businesses to the sector. Despite its lack of administration, Ethiopia's cotton sector is self-sufficient, producing approximately 50,000 tonnes of cotton per year for the country's textile industry as of 2002.

Cotton farming has been practiced in Ethiopia for many years. Cotton cultivation has spread throughout the country, with the majority of cotton grown in the Awash Valley, Gambella, Humera and Metema [43]. Each region has a large potential area: 269130 ha in Tigray, 678 710 ha in Amhara; 600 900 ha in the South Nations, Nationalities, and People region (SNNPR); 407420 ha in Oromia; 316 450 ha in Gambella; 303 170 ha in Benshangul; 200,000 ha in Afar and 225,000 ha in Somalia. Sixty-five percent (65%) are located in 38 cotton-growing regions with high potential, while the remaining 35% are spread across 75 districts with medium potential. Cotton is grown on 22% of state-owned farms, 45% of private farms and 33% of smallholder farms [6]. Ethiopia recently cultivated 3% of the total area suitable for cotton cultivation [48] (Fig. 1).

8 Importance of Cotton

Cotton (*Gossypium L.*), one of the most important plant species for humans, has been a valuable natural resource since the dawn of many civilizations. Cotton has a wide range of applications, including pigments, animal feed derivatives and various oil extraction methods [40]. Cotton is mostly used for medicinal purposes. Cottonseed oil has a 2:1 polyunsaturated to saturated fatty acid ratio and is typically composed of 65–70% unsaturated fatty acids, including 18–24% monounsaturated (oleic), 42–52% polyunsaturated (linoleic) and 26–35% saturated (palm tic and stearic) [3]. Cotton's most significant health benefits include treating respiratory disease, skin conditions, wounds, assisting nursing mothers, curing scorpion and rat bites, curing eye and joint pains, reducing swelling in the legs, and treating alternative medical conditions such as cancer, HIV and other diseases [40].

Cottonseed oil is primarily produced from *Gossypium hirsutum* and *Gossypium herbaceum*, which are also grown for cotton fibre and animal feed [12]. Gossypol is one of the most efficient components in both conventional pharmaceutical processes and alternative modern drug formulations. It is a hazardous polyphenolic bisesquiterpene with antiviral and antifertility properties. Cotton's potential as a natural resource in the pharmaceutical sectors is supported by the findings.

It is the most important natural fibre, widely used in clothing and home décor, and contributes significantly to the export earnings of some of the world's poorest countries [46]. There are numerous applications for it, ranging from shoe strings to blue denim. Cotton plant parts are all useful. Lint, which is used to make cotton fabric, is the most important component. Linters, or the short fuzz on the seed, are a source of cellulose that is used to make plastics, explosives and other products. Furthermore, linter is converted into batting for use as cushioning in beds, furniture and automobiles, as well as being incorporated into high-quality paper products. Cottonseed can be crushed to extract oil, meal and hulls. Cottonseed oil is most commonly used in the production of shortening, cooking oil and salad dressing. The remaining meal and hulls are used as fertilizer and animal, poultry and fish feed, either separately or in combination. Cotton plant stalks and leaves are thrown onto the ground to improve the soil. Cottonseed is also used as a high-protein concentrate in baked goods and other foods [38].

9 Cotton Production Constraints in Ethiopia

Cotton (Gossypium *hirsutum L.*) is currently grown in over one hundred countries, meeting one-third of the world's demand for natural fibre [33]. The high prevalence of pests and diseases, weed pressure and the development of herbicide resistance in weeds, salt and soil degradation, and climatic anomalies such as drought, floods, and heat waves limit cotton production globally, regardless of where it is grown. Crop production potential and constraints can differ between countries.

The primary production barriers in Ethiopia are a lack of technical expertise in biotechnologies and breeding techniques; an insufficient supply of better variety seeds; an inability to obtain loans to invest in large tracts of land; and insect, disease, and pest infestations. Lack of suggested products and procedures [11]. Insect and mite infestations are a major barrier to Ethiopian cotton production. Bollworms, aphids, whiteflies, jassids, leaf worms, red spider mites, trips, smaller armyworms, flea beetles, and stainers are some of the most significant insect pests of cotton among the 68 pest species [44]. Another issue is a lack of employment opportunities; many workers reportedly choose to work on government-owned sugar plantations, where they can earn more money and receive benefits such as housing, transportation, and health care [44].

Cotton exports were restricted from 2010 to 2012, and farmers planted fewer cotton plants instead of more profitable sesame or other cash crops due to a scarcity of quality inputs such as seed and fertilizer [9]. Pink bollworm, mealy bug, flea beetle, and bacterial blight have all caused significant cotton loss. These losses have frequently occurred as a result of repeated pesticide applications combined with unfavourable weather conditions.

10 Cotton Breeding History in Ethiopia

Egyptian cotton, Gossypium barbadense L, was historically used in the 1901–2010 Italian and 1928–28 German studies on cotton in Ethiopia's Upper Awash [47]. The project was cancelled after 7 years of research yielded no results. The Italians restarted the third phase of cotton research, and after a few years of work, they demonstrated that Ethiopia could produce significant amounts of cotton. Cotton breeding was restarted for the fourth time in 1964, with the assistance of the Food and Agricultural Organization, at the then-Melka Werer Agricultural esearch Station (now WARC) as a research department under the Ethiopian Government's Ministry of Agriculture (FAO). Later, from 1966 to 1988, the Institute of Agricultural Research (IAR) conducted cotton research under the Field Crops Department. Because it was deemed important, the crop was elevated to commodity status in research projects in 1989 [16].

Werer Agricultural Research Center currently serves as the organizational hub for cotton breeding research that is multidisciplinary, team-led, and commodity-based. This is still true for irrigated cotton research. According to an EIAR decision in 2016, the national rain-fed cotton research coordination was recently transferred from WARC to Assosa Agricultural Research Center (AARC) in western Ethiopia. Cotton, despite being a very important commodity, has lacked its own research institutions for the past 50 years. WARC instead collaborated with MOA and then-State Farms to test and demonstrate technology at various locations across the country, selecting cotton varieties and types for both irrigated and rain-fed crops [47].

Table 2 Released and commercial cotton varieties in Ethiopia [24]

Variety name	Altitude (m.a.s.l) (metres above sea level)	Time of mature (days)	Ecology	Seed cotton production (qt/ha)	
				Research field	Farmers field
DP 90	300–1400	140–150	Irrigation	–	15–25
Stam-59 A	300–1400	140–170	Irrigation	–	29–32
Raba	1000	150–170	Irrigation	15–30	15–30
Raba B 50	1000	140–170	Rain feed	15–25	15–25
A-333–57	1000	140–170	Rain feed	27–35	12–25
Bulk-202	1000	140–170	Rain feed	20–25	20–25
Albar 637	300–1400	150–180	Rain feed	15–25	15–25
Acala 1517/70	300–1000	180	Irrigation	15–25	15–25
Tate (cuokra)	750	130	Irrigation	48	25–38
Enat	750	10–140	Irrigation	48	28–40
Cucurova	750	129–140	Irrigation	50	28–40

11 Cultivated Cotton Varieties in Ethiopia

Cotton is one of Ethiopia's most important cash crops, and it is widely grown in the lowlands thanks to agricultural irrigation plans. On small farms, rain-fed agriculture is also used to cultivate it. Werer Agricultural Research Center, which is responsible for selecting and developing genotypes appropriate for the country's various agro-ecologies, has released 22 varieties and 7 hybrids for irrigated areas and 5 varieties for rain-fed areas since 1964, though only a small number of them have found commercial success [5]. Currently, 96% of cotton types planted in the country are upland, with DP 90 covering 80% of the area, followed by Stam 59A (15%) and Acala SJ2 (0.5%), with the remaining 4% covered by land races and local variations [41]. Gossypium hirsutum accounts for 98% of the nation's yearly commercial cotton production. This species is of superior quality due to its broad adaptability and high yield potential. Table 2 lists Ethiopian cotton varieties.

12 Progress of Cotton Breeding in Ethiopia

Cotton breeding in Ethiopia is currently focussed on providing farmers with more cotton varieties that outperform obsolete varieties. These released varieties are high yielders, have tolerance to salt environments, are disease resistant, and work on providing early maturity varieties. Furthermore, cotton breeders are currently focussing on quality performance of cotton such as fibre strength, fibre length and colour. Breeders work on developing biotic and abiotic resistance varieties, as well as organic cotton production [30].

Introduction and adoption of genetically modified (GM) cotton is another advancement in cotton breeding [21]. Although the use of insecticides has decreased as a result of the adoption of GM cotton, and broad-spectrum weed control has improved due to the flexibility of herbicide-based weed management, the sustainability of GM cotton may be jeopardized by resistance evolution in insect and weed biotypes. Cotton breeding in Ethiopia began with the introduction of cotton varieties from outside the country. Breeding methods have now progressed to conventional breeding and genetic engineering. WARC has used the following breeding methods or procedures: introduction, adaptation (*BT* cotton), hybridization (hybrids) and pedigree selection.

13 Achievement of Cotton Breeding in Ethiopia

The achievement of cotton breeding in Ethiopia is varietal development. Some of the benefits or purposes of this newly released variety include disease resistance, high yield, pest resistance, high fibre quality, comfort with mechanical harvesting by increasing cotton height, early maturity and adaptability to harsh environments.

13.1 Varietal Development

Werer Agricultural Research Center has been responsible for using traditional breeding techniques to create genotypes suitable for various agro-ecologies in order to meet the needs of farmers and the textile industries nationwide since its inception. To date, 26 varieties and 7 hybrids have been released for irrigated areas, and 5 varieties have been released for rain-fed areas, based on their merits in terms of seed cotton output and fibre quality characteristics. As shown in Table 3, the Cotton Research Department has released a total of 40 convectional varieties. These varieties have been made available based on methods for ranking genotype mean performances at specific locations and overall genotype mean performances, such as introduction, adaptation and line-hybridization for the creation of recombinants, which is frequently followed by pedigree selection.

Because cotton is primarily self-pollinated but can occasionally experience increased cross-pollination of up to 30% [34], the most common breeding technique used by WARC has been hybridization of lines for recombinants, followed frequently by pedigree selection to find superior genotypes. The cotton research team is currently expanding their cultivar collection, testing them for adaptation and evaluating them with a focus on fibre quality indicators [47]. Furthermore, because cotton is more susceptible to pests, the cotton research team's primary research goal focusses on pest management techniques and improved agronomic practices. During the 2017 cropping season, Ethiopia began a *Bt* cotton adaptation trial in a confined field, and depending on the results, two hybrid *Bt* cotton varieties are recommended

Table 3 Cotton varieties that are released by Werer Agricultural research Center (WARC) for production since 1966

No	Variety name	Released year	Seed cotton yield kg/ha	Fibre length mm	Fibre strength lb/sq inch	Recommended for
1	A-333–57	1960s	2930	–	–	Rain-fed
2	Acala 1517/70	1975	3890	–	–	Irrigated
3	Albar 637	1960s	2060	–	–	Rain-fed
4	Acala 1517C	Before 1970	–	–	–	Irrigated
5	Acala 1517D	Before 1968	–	–	–	Irrigated
6	AMS1(70)	1974	2590	–	–	irrigated
7	Werer 1–84	1984	2860	–	–	Irrigated
8	La Okra Leaf 2	1986	2730	–	–	Irrigated
9	Acala 4.42	1974	2350	–	–	Irrigated
10	Reba B-50	1960s	1800	–	–	Rain-fed
11	Acala SJ2	1986	3250	28.6	79.3	Irrigated
12	Arba	1987	3000	30.2	79.7	Rain-fed
13	Bulk 202	1989	3340	28.1	78.3	Rain-fed
14	Deltapine90	1989	3860	27.7	77.3	Irrigated
15	Cucurova 1518	1994	4170	26.9	74.6	Irrigated
16	Cu-Okra	1994	3760	26.1	75.7	Irrigated
17	Carolina queen	1994	4180	27.2	77.6	Irrigated
18	Malkasadi	2019	4700	29.1	30.9	–
19	Sille-91	1997	3860	27.9	72.7	Irrigated
20	Sille-13	2020	4800	27.8	28.8	–
21	Stam59A	2007	3340	29.8	32.5	Irrigated
22	Ionia	2008	2890	30.0	31.4	Irrigated
23	YD-206	2011	4200	34.4	36.5	Irrigated
24	YD-223	2011	4130	33.8	36.6	Irrigated
25	YD_211	2011	4220	34.2	36.6	Irrigated
26	YD-670	2013	4000	32.0	34.8	Irrigated
27	YD-195	2013	3370	31.7	35.2	Irrigated
28	VBCHB 1203	2013	2470	30.7	32.2	Irrigated
29	VBCH 1527	2013	2430	29.9	34.0	Irrigated
30	STG-14	2014	3880	30.0	31.7	Irrigated
31	Candia	2014	4060	29.0	30.20	Irrigated
32	Claudia	2014	3840	30.9	32.4	Irrigated
33	Gloria	2014	4260	29.4	31.96	Irrigated

(continued)

Table 3 (continued)

No	Variety name	Released year	Seed cotton yield kg/ha	Fibre length mm	Fibre strength lb/sq inch	Recommended for
34	WARC-CC1	2015	4070	28.8	25.9	Irrigated
35	WARC-AC2	2015	4300	27.7	29.5	Irrigated
36	WARC-GU3	2015	4620	26.1	29.5	Irrigated
37	Were-12	2019	4400	28.4	26.0	–
38	Were-13	2020	4900	28.7	27.5	–
39	Were-ls1	2019	2500	33.9	38.7	–
40	WARC-LS2	2019	2500	32.9	38.4	–
41	JKCH 1947 (*Bt* cotton)	2018	3056.2	27.78	27.75	Irrigated
42	JKCH 1050 (*Bt* cotton)	2018	3049.6	28.44	28.59	Irrigated

Source (MOARD, 2009, 2014, 2018 and WARC Progress Report, 2014, 2015, 2018 Unpublished) and [8]

for commercial production in 2018. Cotton production is hampered by insect pests, which cause significant yield losses and quality degradation in addition to convectional varieties [15]. Some of these varieties were not used by producers for the following reasons: some of the varieties had been replaced by other varieties and were no longer in use, some of the varieties are hybrids suggested through introduction and adaptation; producers did not import seed and produce due to high seed prices; and some of the varieties were not popularized or promoted to producers. As a result, only a few cultivars are now being produced in the country [4].

14 Breeding Impact of Cotton in Ethiopia

A variety of biotic, abiotic, social and economic variables have a significant impact on cotton production and productivity in Ethiopia [15]. Insect pests are the primary biotic constraints on production, resulting in severe yield and quality losses. More than 60 insect species and 2 mite species are wreaking havoc on cotton production in the country. The main pests are the pink bollworm (Pectinophora gossypiella) and the African bollworm (ABW) (Helicoverpa armigera) [8]. Bollworms alone can result in yield losses of up to 60% [15]. Cotton growers must now use pesticides at least five times during the crop's life cycle to manage ABW. When chemical pesticides proved ineffective, growers turned to the most dangerous (toxic) pesticides, which had negative effects on people's health. Due to the lack of a gene for native insect resistance in the cotton gene pool, efforts have been focussed on using the *Bt* gene to genetically modify crops to increase their tolerance to insects. Transgenic cotton

(*Bt* cotton) has proven to be an effective technology for reducing the number of boll worms in cotton production [15].

Cotton production in Ethiopia was low in cropping years 2015/16 to 2017/18 due to an increase in pest (disease and insect) incidence on the country's potential production area. According to the TIDI and FAS Addis Ababa Forecast report, cotton production in MY 2019/20 is expected to be 262,000 bales or 57,000 metric tonnes, an increase of 8% over the previous year's production estimate. The forecast is primarily based on the anticipated increase in cotton harvested area from 77,000 to 80,000 hectares, as well as the recent cultivation of GMO cotton. The current attractive domestic market prices, approval of *Bt* cotton seed varieties for commercial cultivation and thriving textile industrial parks will encourage existing and new commercial farms to boost cotton production and increase the area of cultivation by controlling the effect pests, the adoption of hybrid varieties and the recent cultivation of GMO (*Bt*) cotton in the country. The two *Bt* cotton varieties JKCH-1050 and JKCH-1947, imported from India, are being cultivated for the first time in Gambela in 2019. However, they are currently being produced in limited areas in the Gambela Region by commercial farmers, with JKCH-1050 GMO cotton varieties producing 30.5 quintal per hectare and JKCH-1947 *Bt* cotton varieties producing 30.6 quintal per hectare [29]. Figure 2 depicts the increase in harvested area (Ha) and cotton production (metric tonne) over the last 2 years when compared to the previous year.

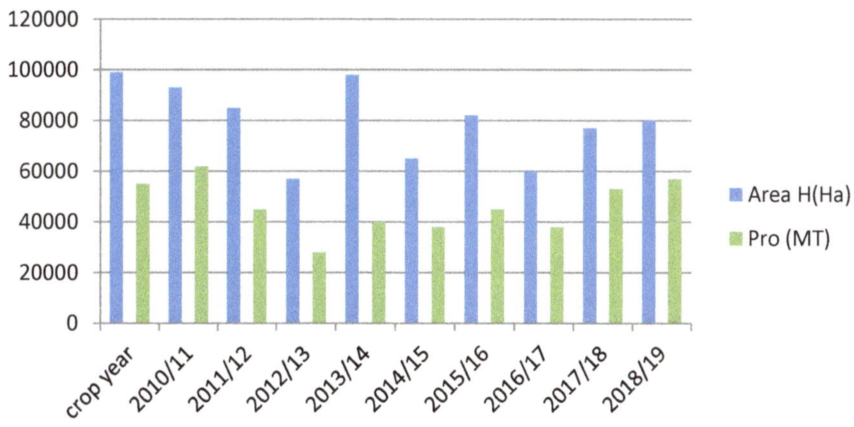

Fig. 2 Increase in harvested area (Ha) and cotton production (metric tonne) over the last 2 years (*Source* TIDI and FAS Addis Ababa Forecast)

15 Challenges for Cotton Research in Ethiopia

Cotton production in Ethiopia is challenged by a variety of biotic and abiotic factors from year to year in terms of yield, productivity and area coverage. So breeders try to solve problems that limit production by conducting research on agronomic aspects, disease aspects, insect aspects and other abiotic factors. However, breeders and experts are being challenged to conduct cotton research in Ethiopia. Among these constraints, germplasm enhancement, limited capacity of human labour facility and finance, lack of cotton seed production, supply, and delivery system, weak cotton research-extension, and biotic and abiotic stress factors are major factors that limit research to do well in solving production problems.

15.1 Germplasm Enhancement

Cotton genetic resources (germplasm) play an important role in the cotton improvement programme [39]. Cotton genetic resources are collected and conserved by plant breeders because they provide a source of genetic variability for improving yield and quality traits in hybrid cotton, varieties suitable for rain feed and irrigation, varieties resistant to both biotic and abiotic stresses, and dwarf varieties suitable for mechanical harvesting. However, some of the major constraints for plant breeders conducting cotton research in Ethiopia include: a scarcity of germplasm (limited germplasm resources), narrow genetic bases for different traits of interest at researchers hands, lack of varieties that combine high yielding ability with good quality characters, lack of hybrid cotton varieties suitable to different production systems and agro-ecologies, a scarcity of long and extra-long cotton germplasm (Gossypium barbadense/Egyptian cotton), shortage of varieties for rain-fed cotton-growing areas, lack of varieties resistance/tolerant to biotic (diseases and pests) and abiotic (salinity, drought and temperature) stresses, a scarcity of cotton varieties suitable for mechanical harvesting and lack of plant biotechnology research to assist conventional cotton breeding [24].

15.2 Limited Capacity (Human, Facility and Financial) at the Research Centre

Cotton is regarded as a cash commodity and a government interest crop, so funders have no interest in funding financial and technological research in the field. As a result, the minimal public funding for research is fully supported. Furthermore, the crop is farmed and studied in a difficult environment, and the research system has a significant impact on the effectiveness of skilled personnel.

15.3 *Lack of Cotton Seed Production, Supply and Delivery System*

There have always been constraints with regard to cotton seed for planting. Ethiopia's limited quantitative and qualitative seed cotton production is due to the poor quality and insufficient supply of planting seeds (cotton seed) offered to seed cotton growers [43]. There are no seeds available in sufficient quantity or quality. Except for the Werer Agricultural Research Center (WARC) and the informal seed channel provided by some private commercial farms, private seed companies, private traders, ginneries and smallholder farmers, Ethiopia has no cotton seed production, processing or delivery mechanism.

15.4 *Weak Cotton Research-Extension (RE) Linkage System*

Cotton is largely ignored because the extension system is primarily focussed on promoting food crops. The majority of small-scale cotton growers rely on unimproved varieties, limited inputs and poor management techniques, resulting in low output and poor-quality lint. When compared to other important crops, the level of extension service for cotton development has been dishearteningly low.

15.5 *Biotic and Abiotic Stress Factors*

Cotton insect pests pose a serious production challenge because they cause significant productivity losses and quality degradation. Bollworms can reduce yield by 36–60%, while sucking pests and cotton aphids can reduce seed yield by 22–14% on farms that grow cotton under irrigation [24]. As a result of climate change, new pests such as the mealy bug are emerging, which is becoming a problem. Another major source of concern is the dynamic transformation of some minor pests, such as white flies, thrips and spider mites, into major pests in the cotton environment. The research system must pay close attention to the unchecked spread of these pests because they pose a serious threat to cotton production. Climate change and the growing salt problem, on the other hand, will be abiotic constraints on cotton grown in drought-prone areas.

16 Future Direction of Cotton Breeding

16.1 Breeding for Long Staple Cotton Varieties

At the moment, all of Ethiopia's cotton is medium-length, which meets domestic industrial demand. The country's limited supply of Gossypium barbadense germplasm is the main impediment to the development of extra-long staples. Nonetheless, breeders are working to develop long cotton varieties from existing germplasm [24].

16.2 Developing Varieties Suitable for Mechanical Harvesting

Ethiopia has made significant progress in both area expansion and cotton production since the country's transition to large-scale commercial farming in the 1970s. As a result, cotton output quickly shifted from manual cultivation and production to semi-mechanized farming. Nonetheless, cotton harvesting is still done by hand and has not changed. Nowadays, fully matured cotton crops are frequently left exposed to the sun, dust, weathering and a variety of other environmental conditions for extended periods of time. This trend is likely to continue for some time. Because of the severe manpower scarcity that occurs during harvesting season, some private investors are currently requesting enhanced and/or potentially promising varieties that are adapted to mechanical harvesting. As a result, breeders planned to develop mechanical harvesting-ready varieties in a short period of time [24].

16.3 Breeding of Abiotic-Tolerant Varieties

Water scarcity and salinity are major issues not only in Ethiopia, but worldwide. Research on salinity, drought resistant/tolerant, and low water requirement types is needed for effective irrigation water use and addressing the growing salt problem [47].

16.4 Hybrid Cotton

Using hybrid vigour also increases quality. Werer Agricultural Research Center (WARC) conducts cotton hybrid creation research on a regular basis [17].

16.5 Transgenic Cotton

Transgenic cotton has been shown to be environmentally friendly in addition to being effective at controlling insect pests, particularly bollworms. The use of *Bt* cotton reduces the need for pesticides, resulting in lower production costs and higher farmer profits. In order to develop a domestic transgenic cotton plant for commercial production in the country, extensive research is required. The Holeta Agricultural Research Center's (HARC) Biotechnology Department is requesting permission to study cotton that has been genetically modified. According to [24], the *Bt* genes will be added to regionally available commercial cultivars in collaboration with this Department.

16.6 Breeding for Coloured Cotton

Coloured cotton is an important quality trait for cotton, with high economic values and is required by various textile industries for fibre quality. Cotton in brown, dark brown, light brown and light green colours will be developed by breeders [24].

17 Summary

Ethiopia has enormous cotton production potential in both rain-fed and irrigated systems. However, due to a variety of production constraints, the amount of cotton produced in the country is inconsistent and insufficient to meet demand in a sustainable manner. Cotton production is hampered by a lack of improved seed varieties, technical inputs such as labour, a lack of extension services and limited irrigation practices. Cotton research in Ethiopia is still done in the traditional way. As a result, it requires some modernization, particularly in the breeding system, such as the use of marker-assisted selection technology, an improved breeding system, a breeding cycle, breeding for long staple cotton, breeding varieties suitable for mechanical harvesting, breeding for drought-tolerant varieties, transgenic cotton, genomic studies, resistance breeding, hybrid technologies, mutation breeding, breeding for coloured cotton and so on. Werer Agricultural Research Center has taken responsibility for improving cotton crop. Plant breeders at Werer Agricultural Research Center create genotypes that are suitable for national agro-ecologies, and they have released around 36 genotypes that are suitable for irrigation and rain feed areas through the introduction of *Bt* cotton from India, adaptation, hybridization and pedigree selection. Among these, hybridization is the most common method of breeding at WARC.

References

1. Abdel-Salam M, Negm M, Ardabb CS (2009) The Egyptian Cotton; current constraints and future opportunities. Textile Industries Holding Co., Modern Press-Alexandria-Egypt
2. Abdurakhmonov IY, Kushanov FN, Djaniqulov F, Buriev ZT, Pepper AE, Fayzieva N, Mavlonov GT, Saha S, Jenkins JN, Abdukarimov A (2007) The role of induced mutation in conversion of photoperiod dependence in cotton. J Hered 98:258–266
3. Abu E-H, Mahmoud AAT (2017) Effect of refining process on the quality characteristics of soybean and cotton seed oils. Int J Curr Microbiol App Sci 6:207–222
4. Abuhay A, Mengie W, Tesfaye T, Gebino G, Ayele M, Haile A, Yillie D (2021) Opportunities for new biorefinery products from Ethiopian ginning industry by-products: Current status and prospects. J Bioresour Bioprod 6:195–214
5. Addis T, Kachi A, Wang J (2021) A review of current state and future directions of cotton production in Ethiopia. Cogent Food & Agric 7:1880533
6. Admassie A, Seid N, May JF, Megquier S, Moreland S (2015) The demographic dividend: an opportunity for Ethiopia's transformation. Population Reference Bureau and Ethiopian Economics Association, Washington, DC
7. Alawad MAH (2022) Impact of grazing dairy cows on bt-cotton crop residues on some milk and blood characteristics and rumen metabolism in Gezira. University of Gezira, Sudan
8. Balcha M, Bekele Y, Gurmessa D, Dejen T, Damtew S, Kebede M, Tegegn F, Seid N, Getahun S, Alemu Z (2022) Cotton production guideline in Ethiopia. Wcrer Agricultural Rcscarch Center, Ethiopian Institute of Agricultural Research
9. Bayrau A, Bekele F, Assefa B, Mihiretu M (2014) An institutional assessment of the cotton and sugarcane commodities in Ethiopia: the climate change perspective. Addis Ababa, Ethiopia
10. Bedane Gudeta SD, Egziabher AG, Gurmessa D (2019) Evaluation of yield and fibre qualities performance of long-staple cotton (Gossypium barbadense L.) Genotypes tested on multi locations of potential cotton-growing areas of ethiopia under irrigation conditions. Results of Crop Improvement and Management Research 2018
11. Belay G, Mussema R (2020) Production and marketing constraints of cotton under smallholders: the case of Arbaminch Zuria District, Southern Ethiopia
12. Bellaloui N, Stetina SR, Turley RB (2015) Cottonseed protein, oil, and mineral status in near isogenic Gossypium hirsutum cotton lines expressing fuzzy/linted and fuzzless/linted seed phenotypes under field conditions. Front Plant Sci 6:137
13. Burhan HA (2022) performance evaluation of top ten cotton producing countries by a critic based mairca method
14. Chaudhry MR, Guitchounts A (2003) Cotton facts, International Cotton Advisory Committee Washington, DC, USA
15. Daba T, Daksa J, Shonga E, Mekuria M, Tulu L, Muliye G, Gudeta B (2019) Bt-cotton production guideline
16. Debela S (1987) Research and Development in agriculture
17. Desalegn Z, Ratanadilok N, Kaveeta R, Pongtongkam P, Kuantham A (2004) Evaluation of F1 and F2 generations for yield and yield components and fibre quality parameters on cotton (Gossypium hirsutum L.) under Werer, Ethiopia condition. Agric Nat Resour 38:176–182
18. Fadl M, El-Ghandour M (1976) Effect of some growth regulators on the rooting of cotton cuttings. Egypt J Botany
19. Fryxell PA (1992) A revised taxonomic interpretation of Gossypium L. (Malvaceae). Rheedea 2:108–165
20. Gagiano R (2022) Effect of soil forms on soil moisture and dryland cotton production. North West University (South Africa).
21. Gebretsadik K, Kiflu A (2018) Challenges and opportunities of genetically modified crops production; future perspectives in ethiopia, review. Open Agric J., 12
22. Geremew M (2013) Statistical analysis of e ect of genotype x environment interaction and stability of cotton yield at some selected locations in Ethiopia

23. Green JM, Jones MD (1953) Isolation of cotton for seed increase 1. Agron J 45:366–368
24. Gudeta B, Egziabher AG (2019) Cotton production potential areas, production trends, research status, gaps and future directions of cotton improvement in Ethiopia. Greener J Agric Sci 9:163–170
25. Kamski B (2016) The Kuraz Sugar Development Project (KSDP) in Ethiopia: between 'sweet visions' and mounting challenges. J East Afr Stud 10:568–580
26. Khadi B (2008) Biotech cotton: issues for consideration. ICAC (available in https://www.icac.org/tis/regional_networks/documents/asian
27. Khan MA, Wahid A, Ahmad M, Tahir MT, Ahmed M, Ahmad S, Hasanuzzaman M (2020) World cotton production and consumption: An overview. Cotton Prod Uses: Agron, Crop Prot, Postharvest Technol, 1–7
28. Kushanov FN, Turaev OS, Ernazarova DK, Gapparov BM, Oripova BB, Kudratova MK, Rafieva FU, Khalikov KK, Erjigitov DS, Khidirov MT (2021) Genetic diversity, QTLmapping, and marker-assisted selection technology in cotton (Gossypium spp.). Front Plant Sci, 2971
29. Lançon J, Woldu B (2020) The Ethiopian cottonseed system-Description, analysis and recommendations
30. Maiti R, Kumari CA, Huda AKS, Mandal D, Begum S (2020) Harvest and postharvest technology and factors affecting fibre quality. Advances in Cotton Science. Apple Academic Press
31. Mcgregor S (1976) Insect pollination of cultivated crop plants, Chapter 9: crop plants and exotic plants: cotton. US Dep. of Agriculture. Agricultural Research Service, Washington DC. (http://www.beeculture.com/content/pollination_handbook/cotton.html)
32. Merima A, Ayele G (2008) Agri-chain analysis of cotton sub-sector in Ethiopia. Mitchell CC (2008) Development of cotton from the old world to alabama: chronological highlights in alabama cotton production
33. Mollaee, M, Mobli A, Mutti NK, Manalil S, Chauhan BS (2019) Challenges and opportunities in cotton production. Cotton Prod:371–390
34. Muhammad W, Ahmad M, Ahmad I (2020) Pollination behavior of cotton crop and its management. Cotton Production and Uses: Agronomy, Crop Protection, and Postharvest Technologies, pp 163–175
35. Nicholson GE (1960) The production, history, uses and relationships of cotton (Gossypium spp.) in Ethiopia. Econ Bot 14:3–36
36. Partzsch L, Kemper L (2019) Cotton certification in Ethiopia: Can an increasing demand for certified textiles create a 'fashion revolution'? Geoforum 99:111–119
37. Percival AE, Kohel RJ (1990) Distribution, collection, and evaluation of Gossypium. Adv Agron 44:225–256
38. Rogers GM, Poore MH, Paschal JC (2002) Feeding cotton products to cattle. Vet Clin: Food Anim Pract 18:267–294
39. Shaheen T, Tabbasam N, Iqbal MA, Ashraf M, Zafar Y, Paterson AH (2012) Cotton genetic resources. A review. Agron Sustain Dev 32:419–432
40. Shahrajabian MH, Sun W, Cheng Q (2020) Considering white gold, cotton, for its fibre, seed oil, traditional and modern health benefits
41. Sheikh RA, Lanjewar UA (2010) Decision support system for cotton bales blending using genetic algorithm. Int J Comput Appl 8:28–34
42. Tedros T (2014) Contribution of small holders' irrigation to households income and food security: a case study of gum-selasa and shilena irrigation schemes, Hintalowejerat, South Estern Zone of Tigray, Ethiopia
43. Tena E, Dejen T, Tegegn F, Bekele Y (2020) Review on agronomy research in increasing the yield and resource use efficiency of cotton crop in Ethiopia
44. Terefe G, Shonga E (2006) Cotton protection handbook
45. Ugwu K, Ogbuagu N, Egwuagu O (2015) Optimization and performance evaluation of cottonseed delinting machine. Asian J. Sci. Technol. 6:1762–1765
46. Voora V, Larrea C, Bermudez S (2020) Global market report: cotton, JSTOR

47. Workie A (2019) Cotton research in Ethiopia: Achievements, challenges, opportunities and prospects
48. Zeleke M, Adem M, Aynalem M, Mossie H (2019) Cotton production and marketing trend in Ethiopia: A review. Cogent Food & Agric. 5:1691812
49. Zou C, Chen W, Liang X, Shang H, Liu W, Shi C, Xiao G, Gou C, Ye W, Xu X (2014) Genome sequence of the cultivated cotton Gossypium arboretum. Nat Genet, 46:567572Li

Cotton Protection

Zigyalew Gashaw Belachew and Abaynew Jemal Jenber

Abstract Ethiopia has enormous untapped cotton production potential due to the availability of land with suitable climate conditions, trainable manpower, thriving textile industries, and long cotton culture with attractive government incentives indicating that there is huge potential for further cotton production development plan. However, Ethiopia now accounts for only 5% of total cotton production in Africa, having recently cultivated only 4% of the total suitable due to a variety of limiting factors. Because current national cotton production is much lower than its potential, but textile factories are increasing in number, the country is forced to import from other producing countries. Furthermore, the most common cotton production limiting biotic factors and management options in Ethiopia have not been well documented. As a result, this chapter provides a detailed explanation of the most important biotic limiting factors (diseases, insect pests, and weeds) as well as possible management options for future interventions for the development of Ethiopia's textile industry as well as cotton production itself.

Keywords Cotton protection · Insects · Diseases · Weeds · Ethiopian cotton

1 Introduction

Cotton (Gossypium spp.) is an ancient fiber and one of the oldest recorded plants grown as a profit crop in the world [44]. It is the king of natural fibers, and one of the few agricultural products whose production and consumption are more or less global in scope [12, 54]. Cotton is an annual cultivated fiver crop that is produced commercially in over 80 countries worldwide, primarily in tropical and subtropical regions [59].

Z. G. Belachew (✉)
Department of Horticulture, Bahir Dar University, Bahir Dar, Ethiopia
e-mail: zigyalewgashaw@gmail.com

A. J. Jenber
Department of Plant Sciences, Bahir Dar University, Bahir Dar, Ethiopia

K. Murugesh Babu et al. (eds.), *Cotton Sector Development in Ethiopia*, Textile Science and Clothing Technology, https://doi.org/10.1007/978-981-99-9149-5_3

Cotton is the most important currency-earning crop, with the longest value chain from raw cotton production to final wear. This value chain employs more than half of total industrial labor and accounts for more than 60% of total exports [2]. Cotton is highly resistant to high temperatures and drought-like conditions due to its vertical taproot [1]. It can also grow on a variety of soil types, from sandy to clay, and is one of the most water-efficient crops, producing one of the highest amounts of dry matter per liter of water [39]. Cotton crop production is currently an appealing trade for both foreign and domestic investors, and it has the potential to help Ethiopia provide job opportunities as well as a source of foreign exchange earnings [28, 67]. It is used in the production of textiles and garments, edible oil, soap, and livestock feeds, as well as providing income for millions of people [7, 15, 35].

Despite its importance to the global economy, cotton production faces significant sustainability challenges. Cotton pests are among the most complex in terms of sustainability challenges. However, there is a scarcity of data on cotton biotic factors in Ethiopia. As a result, this chapter will describe the major cotton pests and possible management options in Ethiopia.

2 Production Status of Cotton (*Gossypium Spp.*) in Ethiopia

Cotton contributes to poverty reduction in many countries, particularly developing countries, and is regarded as an important engine of economic growth. Currently, it is produced in over 100 countries to meet the domestic needs of the fiber and textile industries as well as as an international trading commodity [12]. Ethiopia is one of the African countries that produces and exports cotton, and it has a long tradition of cotton cultivation, with an estimated three million hectares suitable for cultivation [10].

Ethiopia has enormous untapped cotton production potential in various agro-ecological zones, but only about 4% of potential cotton production areas are currently under cultivation [10]. Cotton is primarily grown in many areas of Ethiopia, including the Awash Valley, Arbaminch-Sile, Abaya, Woito, Omorate, and North Bale in the south; Gambella in the West Beles in the north; and Metema and Humera in the northwest [72].

Ethiopia has an estimated three million hectares of land, with 1.9 million hectares (66%) in high potential cotton producing areas and the remaining one million hectares (34%) in medium potential areas. In addition, of the total land under cotton cultivation, 33% is cultivated by smallholders, 45% by private farms, and 22% by state-owned farms [67]. Furthermore, Dp 90 cotton varieties cover 90% of the total acreage under cultivation, with the remainder planted with local varieties and Stam 59A [4]. However, Ethiopia accounts for only 5% of total cotton production in Africa, as it cultivates only 4% of the total suitable land. As a result, current national cotton production is significantly below potential [29]. However, the number of Ethiopian textile factories is increasing, while production and productivity are not being met with an adequate supply of cotton fiber, resulting in imports from other producing countries [62]. From 2000 to 2018, Ethiopia produced an average of 33,842.11 metric tons of cotton [73]. Meanwhile, cotton production and productivity are severely hampered by a variety of biotic, abiotic, social, and economic constraints (Fig. 1).

At full production capacity, the total cotton consumption of Ethiopian factories is expected to be around 111,081 tons of lint cotton [36]. Every 5 years, there is a rising trend (Fig. 1). Ethiopia currently produces 230,000 tons of seed cotton per year and faces a yearly lint deficit of 70,000 tons to meet domestic demand in the developing textile sector. As a result, there is a large gap between the country's industrial production and demand (Fig. 2).

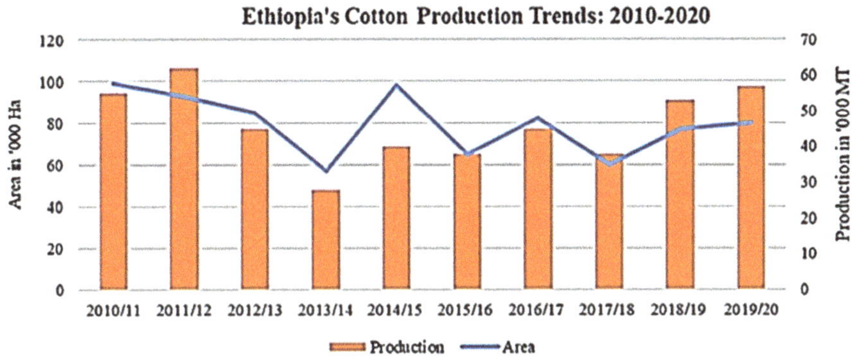

Fig. 1 Ethiopia's cotton production Trends (Ethiopia Cotton Production Annual Report 2019)

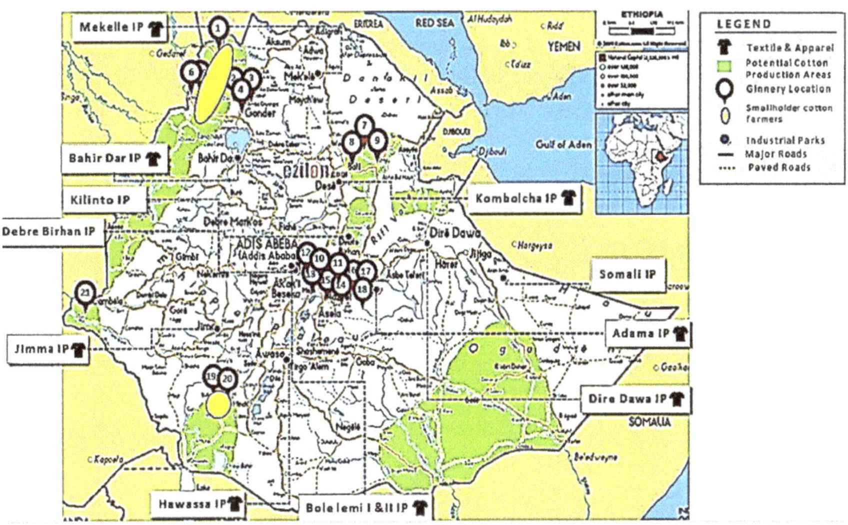

Fig. 2 Distribution of Ethiopia's Industrial Parks (IPs), Potential cotton growing sites (Ethiopia cotton production annual report 2019)

3 Common Cotton Pests in Ethiopia

Cotton production in Ethiopia is currently being impacted by both abiotic and biotic stresses. Drought, salinity, and heat stress are major abiotic stresses affecting cotton production, while insect pests, diseases, and weeds are major biotic factors [47].

Cotton is thought to be a museum for various insect pests due to the variety of insect herbivore species that feed on it. In Ethiopia, 70 different insect and mite pests are known to attack cotton [11]. The major pests among these are African bollworm (*Helicoverpa armigera* Hübner), Spiny bollworm (*Earias insulana* Boisd.), Pink bollworm (*Pectinophora gossypiella* Saunders), Sudan bollworm (*Diparopsis watersi* Roths), Mealybug (*Phenacoccus solenopsis*), Aphid (*Aphis gossypii* Glover), Jassid (*Empoasca lybica*), Whitefly (*Bemisia tabaci* Genn.), Stainers (*Dysdercus* spp.), Flea beetle (*Podagrica puncticollis* Weise), Leaf worm (*Spodoptera littoralis* Boisd.), Thrips (*Thrips tabaci* L.), and red mites (*Tetranychus* spp.), [5, 67, 72]. Insect pest infestation causes significant losses in cotton production, which can reach up to 70% in the absence of pest control measures [57].

Cotton insects are classified as sucking insects (Jassids, thrips, aphids, mealybug) and fruit, foliar, and stem feeders (African, spiny, pink, Sudanus, and spotted and Spodoptera caterpillars).

*3.1 Cotton Bollworm (Hubner) (*Helicoverpa Armigera*) (Lepidoptera: Noctuidae)*

Cotton bollworm (Fig. 3) is a type of Lepidoptera in the Noctuidae family. Cotton bollworm, corn earworm, African bollworm, or scarce bordered straw are all names for a polyphagous chewing insect pest of cotton. The larvae feed on a diverse range of plants, including many important cultivated crops. It is a major impediment to cotton production and productivity in Ethiopia [7]. In Ethiopia, the Bollworm Complex is a major threat. It occurs during the fruit formation and peak phase of cotton, which lasts 12–14 weeks after planting. In severe cases, American bollworms can cause crop losses ranging from 40 to 100% [22]. Because of the high resistance of American bollworms to insecticides, scientists developed transgenic cotton hybrids popularly known as *Bt* Cotton.

Cotton larvae cause damage to leaves, tender shoots, apical tips, flower buds, and pods. It results in significant yield loss, both in terms of quality and quantity. Larvae chew holes in boll bases and may hollow out locks. Larvae in their first instar feed on soft leaves, creating small holes. They can penetrate fruit through a small hole bored near the stalk when they reach the second instar. Most fruits are damaged during development by caterpillar mining.

When the insect population is below the threshold level, a preliminary action that should be taken is handpicking of living bollworm larvae from *Bt*-cotton fields. It is critical to deep plow residual pupae in *Bt*-cotton fields immediately after harvest

Fig. 3 Cotton bollworm

during peak bollworm infestation time. Furthermore, *Bt*-cotton hybrids have excellent control of the American bollworm, which means that insecticides may not be required to control this bollworm.

*3.2 Flea Beetles (***Podagrica Puncticollis Weise***)*

Cotton flea beetle (Fig. 4) is the most common species found on cotton in Ethiopia and neighboring countries, and it has proven to be a major pest on occasion causing extensive losses. It is one of the most damaging insect pests of cotton, particularly in the northwestern part of Ethiopia [5]. The most important factors influencing the prevalence of the flea beetle are climatic conditions and the timing of cotton sowing [3]. The recent increase in rainfall during the main cotton growing season resulted in a severe attack of cotton flea beetle on cotton seedlings due to a lack of weeds for the flea beetle when they emerged from aestivation.

The female lays her small yellow eggs in the soil near the stem base of the host plants. The larvae hatch after 7–11 days and feed for 11–28 days on the rootlets

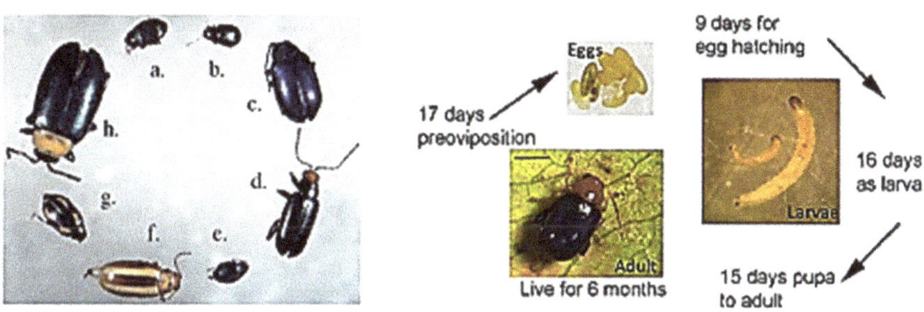

Fig. 4 Cotton flea beetles

of volunteer crops and weeds, then move to newly planted crops as they emerge; pupation occurs in the soil. Adults emerge from pupae after 10–17 days. During the course of a season, this pest produces several generations. Adults remain on the host plant after rainfall for as long as they can find suitable food. The adults always favor young cotton plants. When the cotton plants are harvested and dry, the beetles migrate into soil cracks or beneath plant debris, where they spend the dry season.

Flea beetles can be found on a variety of host plants. Flea beetles are present in cotton fields at all stages of development, and their attack during the seedling stage is more damaging than later infestation. Early-sown cotton is more vulnerable to flea beetle attacks than later-sown cotton. Cotton seedlings are especially sensitive when there is a lack of moisture due to insufficient rainfall or irrigation during the first week of sowing [6].

Damage Symptoms

The appearance of the majority of adult flea beetles that harm the cotton plant is distinct. They feed by chewing a small hole in a leaf, moving a short distance, and then chewing a new hole, and so on. The result is a number of "gunshot holes" in the leaf. Cotton plants eventually develop a grayish appearance as a result of this damage (Fig. 5). Flea beetles feed on the cotyledons and leaves of growing plants, destroying the upper layers of leaf tissue and severely limiting photosynthesis and assimilation, resulting in stunted growth. Cotton growers in the area were forced to substitute sesame and sorghum as a result of flea beetle pressure. In addition, in the Metema district, northwestern Ethiopia, untreated cotton yielded 75% less than cotton grown from treated seeds and sprayed with insecticide 5 days after seedling emergence [5].

Management Options

The best way to manage this insect pest is to provide good nutrition management and to create favorable growing conditions. It aids in the prevention of plant stresses and the strengthening of cotton plants against flea beetle attack. Weeding in and around cotton fields aids in the elimination of flea beetle habitats and breeding grounds, thereby reducing crop damage. Maintaining plant diversity on the farm will help to boost natural enemy populations. Spray botanicals such as neem, rotenone,

Fig. 5 Damage symptoms of flea beetles

pyrethrin, sabadilla, garlic, onion, and mint extracts have been recommended for flea beetle control, either alone or in combination. Furthermore, insecticidal soap has been shown to provide partial control of flea beetles. Sprays containing rotenone and insecticidal soap, on the other hand, are thought to be very effective. Although diatomaceous earth and rock powders have been shown to reduce flea beetle populations, applications must be reintroduced on a regular basis after rainfalls. Seed treatment with Cruiser combined with foliar spray with Carbaryl or Endosulfan 5 days after seedling emergence and using only Cruiser-treated seeds is also an effective flea beetle management option [3]. Furthermore, proper sowing date selection can help to reduce flea beetle infestation and damage [3].

3.3 Thrips (Thrips TabaciL.)

These insects (Fig. 6) have piercing-sucking and rasping mouthparts and feed on almost all parts of the cotton plant and its stages, with the most severe damage occurring on seedlings from plant emergence to five true leaves [19, 55, 69].

Damage Symptoms

Thrips are found on the undersides of leaves, damaging them by piercing the epidermis of the tissues and sucking the sap oozing from wounds [60]. As a result, leaves become slivery due to the formation of white patches or streaks, which eventually cause scarring and distortion [52]. Excessive feeding injury can result in severely stunted plants, which often results in yield loss or a delay in crop maturity.

Management

Thrips control is typically provided by seed treatments until the two- to three-true-leaf stage. Systemic insecticides, in addition to treated seed, can be considered in areas with a history of repeated, heavy thrips injury. Foliar sprays are frequently applied too late to prevent damage, and studies have shown that applying foliar sprays after significant thrips damage has occurred does not increase yields. Carbamate, organophosphate, organochlorine, and pyrethroid are the most commonly used insecticide classes in Ethiopia, where they are frequently and indiscriminately applied

Fig. 6 Thrips

[49]. To manage cotton thrips, it is also recommended to use sequential applications of insecticides such as Closer 240% SC, Dimethoate 40% EC, Rectro 20% SC, and Imidacloprid.

3.4 Mealy bug

Mealybugs (Fig. 7) resemble cotton and cause significant economic damage to cotton by sucking sap from all plant parts. Honeydew secreted by the mealybug promotes the growth of black sooty mold, which reduces photosynthetic activity. Infested plants are associated with ant movement, which aids in the spread of mealybugs from one location to another. Cotton mealybug has a diverse host range ranging from herbaceous to woody plants. The mealybug is a soft-bodied insect that reproduces mostly parthenogenetically. The female lays eggs in ovisacs that contain 150–600 eggs. Under ideal conditions, hatching occurs in 3–9 days into nymphs (crawlers) that live for 22–25 days before maturing into adults in 25–30 days. They can produce hundreds of nymphs in a single generation and lay up to 6000 eggs per generation [50].

Cotton mealybug is a polyphagous sucking pest that undergoes incomplete metamorphism. It is an exotic pest with a wide host range, a waxy protective coating on the dorsal side that counteracts potential mortality factors, a high reproductive rate, and the ability to overwinter (Egg and adult female stage), all of which contribute to the insect becoming a serious pest of many commercially important crops. It attacks host plants by sucking phloem tissue cell sap and secreting honeydew, which forms sooty molds on the surface of the leaves, halting the natural process of photosynthesis and eventually killing plant tissues [20].

Damage Symptoms

Mealybug-infected cotton plants have a white fluffy mass on the underside of the leaves near the growing tips, along the veins, and on the stem, as well as distorted or bushy shoots and crinkled, twisted, or bunchy leaves. Boll opening may also be negatively impacted, resulting in significant yield losses.

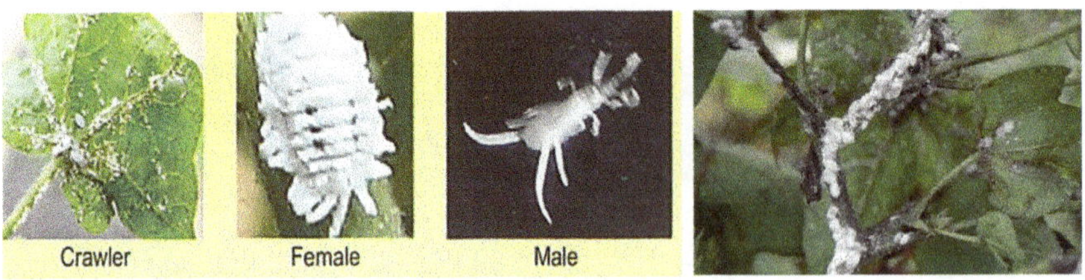

Fig. 7 Mealybug

Sucking feeding causes stunted growth and drying of the plant. The bolls are not completely opened, and the fiber quality suffers as a result. Infestation on young twigs causes internode shortening, resulting in the appearance of rosette.

Management Options

Plant parts that have been infested should be burned or buried. Another method is to remove weeds (particularly Parthenium) from within and around plantings that may harbor mealybug infestations. Also, discouraging farmers from planting okra near cotton fields can help manage this bug.

3.5 Cotton Aphids

Cotton aphids (Fig. 8) are tiny, multicolored, soft-bodied insects. Female aphids give birth to live young, so a new generation can be formed every 5 days during the summer. As the season progresses, a number of generations may overlap in the field. Fortunately, a vast array of beneficial insect parasitoids (small wasps), predators (e.g., ladybird beetles, lacewing larvae), and diseases keep aphid populations under control. A cotton aphid can have up to 51 generations per year [56].

Aphids feed on cotton terminal buds and foliage with their piercing-sucking mouthparts. Direct feeding injury occurs when aphid colonies form on stems, leaves, and heads from seedling to head filling. The extent of damage is determined primarily by the percentage of infested tillers, the number of aphids per tiller, and the duration of the infestation.

Damage Symptoms

Yellowing occurs to some extent on infested leaves, whereas severe aphid infestations cause young leaves to curl and become misshapen, stunt plant growth, and reduce the number and quality of flowers and fruits. Aside from feeding damage, aphids leave honeydew, which fungi (e.g., black mold) use as a growing medium. As honeydew

Fig. 8 Cotton aphids

Fig. 9 Cotton aphids damage symptoms

falls onto the lint, moldy growth can stain it, lowering the product's quality and value (Fig. 9).

Threshold level: Consider direct control of Aphids if their number averages 5–10 per fifth stem node leaf per plant and row during the period between open bolls and pre-harvest. To scout for aphids, inspect the undersides of 5 plants at each of 10 different locations in a field. Aphid injury is classified as low if there are less than 10 aphids per leaf, medium if there are 11–25 aphids per leaf, and high if there are 26 or more aphids per leaf.

Management Options

Management options should be chosen with the population size in mind. Crop rotation with maize or onion is required, but not with melon, cucumber, pumpkin, or tomato, which are also hosts for this aphid. Use tolerant cotton varieties (Deltapine-90, Acala SJ-2), do not apply too much nitrogen (less than 80–120 kg/ha) as it makes cotton attractive to aphids, sow early: mid-April–mid-May for middle Awash; mid-May–mid-July for lower Awash; early June–early July for rainfed areas to escape aphids and grow long-flowering plants or crops, such as mustard or broad beans at field edge to enhance beneficial ladybird beetles and parasitoids.

Manage ants by wrapping a band of sticky tape around the trunk of the crop and placing a sugar-soaked sponge at the base of the trunk or where ants are visible. Aphids are protected from natural enemies by ants. In modern ecological technologies, botanical pesticides are used as an alternative to synthetic chemical pesticides for pest control [71].

Mix 100 g ground garlic, 10 g powder soap, 12 L water, and 2 tablespoons oil; after 1 day, add 20 L of water and spray. Spray fermented animal urine, such as cow urine, diluted with water (1:10 or 1:15). Spray a Neem extract solution no more than twice (2.5–3 L/ha or 50–60 ml/20 L water; or 20–50 g neem seed cake or seed powder/liter water plus a few tablespoons of soaps). Examine the labels of marketable products. Add 30–40 hot peppers to 10 L warm (not boiling) water. Allow the liquid to steep for at least 24 h. Remove the pepper rests and fill the container with 10 L of cold water and 20 drops of dishwashing soap. Spray no more than twice in 1 week. Insecticides based on diafenthiuron, such as Polo 500SC, and carbosulfan, such as Marshal 25%EC and Marshal 25%ULV, should not be sprayed in full sunlight.

3.6 Whiteflies *(*Bemisia Tabaci*)*

Bemisia tabaci is a polyphagous insect that can harm cotton, brinjal, ladyfinger, some other vegetables, and ornamental plants. It can harm plants in two ways: first, by sucking the sap, and second, by excreting honeydew, which fosters the growth of sooty mold. The honeydew secreted onto the leaves attracts ants, which disrupts the activities of natural enemies as well as the photosynthesis activity of cotton plants. Whiteflies may also carry and transmit viral diseases such as the leaf curl Multan virus [51].

Whiteflies (Fig. 10) typically lay eggs beneath leaves. Tiny crawlers hatch and settle on leaves before molting into motionless nymphs. Adults are approximately 1 mm long, with the male being slightly smaller than the female. The whitefly's body and both pairs of wings are covered in a powdery, waxy secretion that ranges from white to slightly yellowish in color. Females can live up to 60 days. Males have much shorter lives, ranging from 9 to 17 days. Each female can oviposit over 300 eggs during her lifetime, which are often arranged in an arc around the female as she rotates on her stylet. Depending on the climatic conditions, 11–15 generations can occur in a year [25].

Damage Symptoms

Cotton whiteflies harm plants in a variety of ways, including sucking the cell sap from the leaves as nymphs and adults. Sooty mold grows on the sticky material secreted by the insect, interfering with plant photosynthesis. As a vector, it can also spread more than 350 virus species in plants [46], (Rodrguez et al. 2019). Leaf yellowing, adhesive upper surface leaf adhesion, and possibly black sooty mold from sticky honeydew produced by larval stages are the main symptoms. They are most active all year at temperatures ranging from 10 °C to 35 °C, but at lower temperatures, they simply slow their rate of development.

Fig. 10 Whiteflies

Management Options

Cultural practices such as regulating irrigation and fertilizer application can be altered to make crop fields unfriendly to insect pests. Drip irrigation reduces whitefly density and viruses transmitted by them in several crops associated with furrow/sprinkler irrigation. Sulfur-containing fertilizers have varying effects on whitefly populations in different crops. A susceptible host can avoid the pest's peak population by adjusting the crop sowing season. Maintaining crop areas free of susceptible species for 60 days during the wet summer season reduces the incidence of whiteflies and associated viruses. Neem oil 2.5 lit/ha mixed with 0.1% detergent powder can be used to control jassids, whiteflies, and aphids.

3.7 Leaf Miner

Leaf miners (Fig. 11) of the genus Liriomyza, which belongs to the family Agromyzidae and order Diptera, are a very important insect pest in most vegetables, ornamentals, and their associated weeds all over the world.

Damage Symptoms

The harm caused by Liriomyza spp. is classified into two types: direct and indirect. Larval feeding is the leading cause of serious injury. The mining activity of larvae can reduce the plant's photosynthetic capacity. A severe infestation will result in dehydration and premature leaf fall [64]. Feeding lesions caused by adult females can completely destroy seedlings and young plants. Adult females also cause harm by digging feeding holes. Fungi and bacteria can enter through the feeding punctures.

Fig. 11 Leaf miner

*3.8 Jassid (*Amrasca Biguttula Biguttula Ishida*) (Cicadellidae:Hemiptera)*

The fully developed cotton jassid (Fig. 12) is a long and slender insect measuring approximately 2.6 mm (0.1 in) in length. It is yellowish-green in color, with a visible black spot on one or both sides of the head and another near the forewing tip. The membranous wings are transparent and iridescent, and the head is pale green. The insect moves diagonally across leaf surfaces and immediately jumps and flies away when disturbed [65].

Cotton jassid is a cotton-sucking insect pest. It damages cotton plants all year and lays its eggs in soft plant tissue. Cotton jassids are more active during the hot summer months and can also live on other host plants such as potato, chili, brinjal, okra, and ornamental plants.

Females prefer to lay eggs in the leaf tissue, usually in the spongy parenchymatous layer between the vascular bundles and the epidermis. Freshly laid eggs are translucent, slightly oval-shaped, and yellow in color. They are elongated in shape, with a broad end and a tapered end. The incubation period of eggs ranged from 6 to 7 days. The nymphs of A. biguttula biguttula emerge during the morning hours and progress through five nymphal instars. The newly emerged nymphs are translucent and yellowish in color. The total nymphal period of A. biguttula biguttula ranges between 5 and 10 days [42].

Jassid is a sucking pest found on cotton throughout the growing season in all zones. During feeding, the pest injects toxins into the leaves, causing abnormal changes.

Adults are about 3 mm long and greenish yellow in the summer, turning reddish in the winter. The vertex of the forewings' hind portion has two black spots. ***Nymph*** are wedge-shaped and greenish yellow. ***Eggs*** are deposited in the parenchymatous tissue of the leaves. The nymph and adult stages have lifespans of 7–21 days and 35–50 days, respectively. A year has approximately 7–8 generations.

Damage symptoms: Infestation occurs 15–20 days after sowing. Yellowing, reddening along the leaf margin, and complete drying of the leaves. The nymphs and adults suck the sap from the leaves, causing phytotoxic indications known as

Fig. 12 Cotton Jassid

Fig. 13 Jassid damage symptoms

hopper burn, which results in complete drying of the plants and has become one of the limiting factors in cotton economic productivity (Fig. 13).

Cotton jassids are most likely found on the undersides of leaves. Plant growth is reduced at high infestation levels, the leaves curl downward, and the leaves develop an overall pale color with red margins. Turn yellowish then brownish, beginning at the margins and progressing to the midrib. Before completely drying and shedding, the leaves begin to curl. Severe incidents cause "hopper burn" injury and leaf death, eventually leading to stunting of young plants. In the latter stages, it results in lower yields and poor fiber quality.

Management

Beneficial predators of jassids include Chrysoperla carnea (lacewing), Scymnus sp., and Anagrus atomus. Botanical management techniques such as neem oil and mineral oil can be used. Dimethoate (1.5 mL/L water), Phosphamidon (2.0 mL/L water), and Lambda-cyhalothrin 5% EC (0.75 mL/L water) are recommended insecticides.

4 Common Diseases of Cotton

*4.1 Anthracnose (***Colletotrichum Gossypii***)*

Anthracnose disease (Fig. 14) is a fungal disease that can affect any part of the cotton plant in Ethiopia. Anthracnose can account for up to 80% of plant and fruit losses in nurseries and more than 50% of plant and fruit losses in the field, respectively [32, 48]. It reproduces primarily through asexual reproduction. *Glomerella gossypii* is the result of sexual reproduction, while *Colletotrichum gossypii* is the result of asexual reproduction. The pathogen is primarily transmitted via seed, with secondary transmission via air and soil-borne conidia. The fungus spreads quickly in the field during wet weather, causing seedling blight, stem lesions, and boll rot. The disease spreads much faster in wet weather than in dry weather.

Fig. 14 Anthracnose disease

Anthracnose symptoms vary depending on the host, but in general, they include irregularly shaped spots or blotches on cotton leaves, as well as dead areas that often follow the veins of the leaves. The color of affected tissue can vary, but it is usually tan or dark brown. Leaves that are severely anthracnosed will frequently curl and fall off. At the seedling stage, however, the circular red-colored spots are visible, and the seedlings may even die. Later stages of infection result in sunken, circular, reddish-to-brown spots on bolls and leaves [16, 43]. Infected bolls grow, burst, and desiccate prematurely. The lint becomes yellow or brown in color and solidifies into a brittle mass.

Management Options of Anthracnose Disease on Cotton

Alternative preventive measures include quarantine regulation, using disease-free seeds, resistant varieties, using compost to give the plant enough strength to resist the disease, avoiding excess irrigation or waterlogging, rogue weeds and possible alternative hosts, removing infected cotton plant debris from soil surface after harvest, crop rotation 2–3 years interval with non-susceptible crops, and keeping ripening fruits from touching the soil. Furthermore, seed treatment with captan, carboxin, or thiram (usually 2 g/kg of seeds) is used to treat or reduce the incidence of the disease [16]. The severity of the symptom can be reduced by foliar spraying the crop with mancozeb and copper oxychloride (2.5 ml/1L of water) during the boll formation stage.

4.2 Alternaria Leaf Spot *(*Alternaria Macrospora*)*

Alternaria leaf spot (Fig. 15), a major cotton foliar fungal disease caused by Alternaria macrospora, is one of the most common cotton foliar fungal diseases. This disease is found in almost every cotton-growing country on the planet. This disease is more dangerous to hybrids. The infection's severity causes severe defoliation of cotton, as well as a significant reduction in yield and crude fiber quality. The primary source of inoculum is from undecomposed crop residues and infected seeds. The pathogen spreads via airborne spores within leaf spots, and the infection process is aided by wet (humid) weather and temperatures around 27 °C [13].

Fig. 15 Alternaria leaf spot

Damage Symptoms

Brown irregular or round spots and blights on leaves, and more spots merge to form larger spots. In severe infections, the leaves will become weak and fall off. Small, circular brown lesions on cotyledons and seedling leaves that expand and develop a concentric pattern; necrotic areas coalesce and often have a purple margin; centers of lesions may dry out and drop from the plant creating a "shot-hole" appearance on the leaves and eventually caused cankers on the stem and infection spreads to the bolls and finally falls off. Plants stressed by drought, nutrient deficiency, or other pests are more susceptible to the disease.

Management

Plow crop residue into the soil to reduce inoculum levels; provide plants with adequate irrigation and nutrients, particularly potassium; and apply appropriate foliar fungicides to susceptible cultivars [63].

4.3 Bacterial Blight

In cotton, bacterial blight indicators are first seen as small, water-soaked injuries on the upper and lower leaf surfaces. The injuries appear angular due to the limitations imposed by veins within the cotton leaf. Bacterial blight (Fig. 16), is caused by the bacteria *Xanthomonas citri* pv *malvacearum.*

Symptoms

Bacterial blight leaf symptoms may be found on leaves and, in all likelihood, bolls later in the cotton development period. Water-soaked spots that turn black as they expand and may follow veins are the most common symptoms. Similarly, bolls appear to be water-soaked, sunken injuries. Then there's the possibility of defoliation. The disease is divided into four stages. Initial symptoms include round to elongate lesions on cotyledons that are initially deep green and water-soaked before drying to brown. Lesions on the hypocotyl are black, elongated cankers that girdle and

Fig. 16 Bacterial blight

kill the seedlings. The foliar phase of bacterial blight (Angular leaf spot and vein blight phase) can appear at any time during the season as water-soaked lesions on angular-outlined leaves. The network of fine leaf veins tends to prevent infected spots from spreading. The coloration of the older spots darkens to brown as they dry. The infection may also occur in veins causing "Vein Blight". The severely affected leaves eventually fall due to the premature formation of the abscission zone at the base of the petiole. Later on, the infection spreads to the stem, resulting in elongated grayish-to-black lesions (Black arm phase). On the bolls, the infection appears as small round water-soaked spots that eventually turn brown (Boll rot phase). Secondary organisms then enter the bolls, causing extensive boll rotting. While the leaf spot complex is commonly associated with nutrient deficiency or a different stressor. Instead, a foliar fungicide may be required to reduce defoliation and protect yield in the target spot. The primary method for distinguishing the two leaf spots is where they are located in the field and canopy. Target spot injuries typically begin in the lower canopy in the rankest/growthiest cotton areas, affecting only high-yielding cotton. Leaf spot complex injuries will be found all over the canopy and may be found in low-potash areas of the field.

Management Options of Bacterial Blight

Planting resistant varieties, removing and destroying infected plant debris from cotton fields, shifting the crop cycle by using a non-host crop, early thinning, and early earthling up with potash are all examples of measures that can be taken. Whereas Delint the cotton seeds with concentrated sulphuric acid at 100 ml/kg, carboxin or oxycarboxin at 2 g/kg, or soak the seeds overnight in 1000 ppm Streptomycin sulfate. Spray at 1.25 kg/ha with a Streptomycin sulfate + Tetracycline mixture of 100 g and Copper oxychloride [38].

4.4 Wilts

Verticillium Wilt

Verticillium wilt (Fig. 17), caused by the well-known fungal pathogen Verticillium dahliae, is one of the main limiting factors for cotton production in cotton-growing regions [76]. It grows well in a simple medium containing only sugars and amino acids, which are normally found in root exudates. It grows best at temperatures ranging from 22 to 27 °C and can be found in soil debris as inactive microsclerotia up to 40 cm below the soil surface [17]. It has a wider host range than Fusarium wilt. The fungus can survive as a microsclerotic infested plant debris and soils for up to 14 years. The density of microsclerotia is generally greatest in the top 10–30 cm of soil, and their germination can be stimulated by root exudates of both host and non-host plants [24]. The primary mode of transmission is via microsclerotia or conidia in the soil. Secondary spread occurs through the contact of diseased roots with healthy roots, as well as the dissemination of infected plant parts, irrigation water, and other farm equipment. Verticillium wilt does not usually kill the tree. Instead, it causes the tree to deteriorate gradually over time, eventually leading to death. It is most common in cooler climates, but it can also occur in warmer climates if the conditions are right.

Damage Symptoms

The symptoms are visible when the crop is in the boll development stage. Early-stage infected cotton plants can become stunted. The first signs are bronzing of leaf veins, which is followed by interveinal chlorosis and yellowing of leaves, stunting,

Fig. 17 Verticillium wilt

defoliation, and a streak of light to dark on the stem [37]. Finally, the leaves begin to desiccate, resulting in a burnt appearance. At this stage, the drying of the leaf margins and areas between veins produces a "Tiger stripe" or "Tiger claw" appearance of discoloration [31, 34]. The affected leaves fall off, rendering the branches useless. Splitting open infected stems and roots reveals a pinkish discoloration of the woody tissue, which may taper off into longitudinal lines in the upper parts and branches. Brown spots appear at the tips of the petioles on the infected leaf. Smaller bolls with undeveloped lint may be produced by the affected plants. However, unlike cotton root rot or Fusarium wilt, it does not cause decay of the root surface and cortex. The xylem of the taproot and stem of Verticillium wilt has a distinct vascular discoloration.

Management of Wilts

Plant tolerant cultivars at a higher seeding rate (3–4 seeds per foot), plant early maturing cultivars, rotate with non-host crops such as grass family crops to reduce disease occurrence, solarize the soil, reduce irrigation during flowering in late July and August to avoid lowering soil temperature, and avoid excessive nitrogen use [33]. Furthermore, tolerant varieties such as Gossypium barbadense, also known as sea-island cotton, have a high level of resistance to Verticillium wilt. However, it performs poorly in terms of fiber yield and quality. As a result, it should be used in breeding programs as an indicator of Verticillium wilt tolerance. Carboxin or carbendazim, at a rate of 2 g/kg, should be applied to the cotton seeds.

Fusarium Wilt

Fusarium wilt (Fig. 18), unlike Verticillium wilt, affects a narrow range of plant types, including trees. Fusarium wilt causes significant economic losses in cotton around the world [21, 75, 77]. The fungus enters the cotton plant via the roots and then spreads to the stem and branches, causing the tree to wilt and die. This disease is most common in warm climates, but it can occur in cooler climates as well if the conditions are favorable.

Symptoms

Fusarium wilt symptoms include general wilt, which is most noticeable on warm days, as well as yellowing and necrosis of lower leaf margins, chlorosis, defoliation, stunting, and eventually plant death [30]. The vascular system of diseased cotton

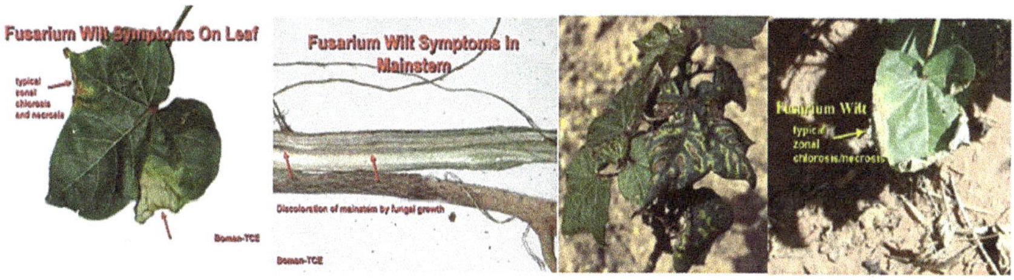

Fig. 18 Fusarium wilt

plants is discolored brown in the affected plant tissue. In addition, when we cut the lower stem, we look for red-brown streaking in the vascular tissues.

Disease Management

Fusarium wilt management is difficult, but most successful when pathogen-free planting materials and resistant cultivars are used. After inoculum has been introduced into the field, management strategies such as soil solarization and fumigation are used to keep inoculum levels under control [18]. Fusarium wilt management in cotton is difficult due to *F. oxysporum f. sp. vasinfectum's* ability to survive in soil for extended periods of time. Fusarium wilt can be controlled by fumigating the soil with methyl bromide, but this is an expensive and inefficient method for a crop like cotton that is planted over a large area [14, 27, 45].

4.5 Virus

Cotton leaf roll dwarf virus (CLRDV), vectored by the cotton aphid and capable of causing cotton blue disease [8], is the newest viral pathogen in cotton. CLRDV reduces stomatal conductance and photosynthetic activity in individual leaves, resulting in significant productivity failures in individual cotton plants (Fig. 19).

Symptoms

The disease is distinguished by downward and upward curling of the leaves, condensing of veins, and enation on the underside of the leaves. During a severe infection, all of the leaves curl, roll, deform, and grow slowly. Boll bearing capacity declines [66, 9, 53]. Severe disease infection may result in lower boll set and seed-cotton yield per plant as well as per acre [8].

Fig. 19 Virus

Management

Changes in management practices are not currently recommended. However, controlling the planting date to avoid peak vector populations, eliminating volunteer perennial cotton and alternate hosts, including malvaceous hosts like wild okra, and using the fungus *Paecilomyces farinosus,* which parasitizes B.tabaci, all help to reduce the vector population. Additionally, foliar application of neem leaf extract and 1% neem oil reduced virus transmission by 80%.

Proper nutrient management practices, as well as the incorporation of a large amount of organic matter into the soil, are critical for retaining organic carbon and increasing microbial activity. Whitefly management is recommended to prevent disease spread.

4.6 Rust

Rust (Fig. 20) is one of the most common and serious problems affecting cotton in Ethiopia's poor soils throughout the cotton belt. Cotton infected with rust does not typically produce normal growth, which is smaller in size and lacks a healthy green color. Around the middle of the season, the leaves begin to show a yellowish-mottled appearance, with the parts near the veins remaining green the longest, and the tissues between the veins, and thus farthest from the food supply, turning yellow. They eventually turn a reddish-brown color, curl up, and drop off, leaving the stalk bare. The problem is typically identified in definite spots or irregular areas in affected cotton fields, and unless corrective measures are implemented, it may reoccur in the same locations year after year.

Rust is a reaction of the cotton plant to soil environments that are unfavorable to normal growth and development. It is not vulnerable to disease-causing organisms or insect damage. The most common causes of rust are a lack of humus in the soil, a lack of potassium, and a lack of drainage. Many fields with naturally light or poor soils are planted to cotton year after year, with little or no attention paid to maintaining the supply of vegetable matter required for vigorous growth and the formation of humus

Fig. 20 Rust

in the soil. As a result, the longer the practice continues, the more rust there will be. Such soils are usually low in potash, and rust will develop if it is not supplemented with potash-rich fertilizers. Rust frequently forms in poorly drained soils during the growing season as a result of heavy and continuous rains. The occurrence of rust disease is aided by high humidity, leaf wetness, and moderate to warm temperatures. The first signs of rust in cotton will appear on older leaves. They are mostly visible as minute, bright yellow to orange lesions on the upper leaf surface.

Management of Rust

Planting early and using early maturing cultivars, using wider row spacing to accelerate canopy drying, and weeding out alternative hosts such as gramma grasses are all preventive measures. Cotton fields with general and severe rusting usually require crop rotation with other crops to replenish the humus supply depleted by continuous cultivation. Plowing under green-manure crops like rye, cowpeas, and velvet beans, or applying a liberal amount of stable manure, will help to improve soil conditions and prevent rust. Rust damage can be reduced by using potash-containing furthers [26].

4.7 Weeds

Cotton does not provide good canopy cover in its early stages and cannot compete well with weeds. Even though hand weeding is time-consuming and difficult, it is a very effective and practical method in the early stages of growth. Because of the presence of diverse plant species in agro-ecology, insects have a better chance of selecting favorable hosts for feeding and oviposition, allowing cotton insect pests to maintain a higher population [68] recorded 16 weed species at Middle Awash 26, Metema 45, and Humera. Broad-leafed weeds (Datura stramonium and Parthenium hystrephorou), grass weeds (Sudan grass and *Cyperus rotundus*), and sedges are the three types of cotton weeds. Due to a crop's lack of canopy cover in the early stages, hand-weeding cotton within 35 days of crop emergence is appropriate. Weed competition is most severe 30 days after crop emergence. Delaying weeding for 60 and 75 days after crop emergence can result in yield losses of 35–88% and 56–94%, respectively [10]. Weeding twice after crop emergency, at 20 and 34 days, results in higher cotton yield and yield components. Cotton growers should align weeding time to these schedules in order to increase crop yield [70]. Stomp 455 CS, a chemical, was registered in Ethiopia for the control of grass and broad-leaf weed species in cotton [10].

Summary

Ethiopia has enormous potential for cotton production in both rainfed and irrigated systems. However, the amount of cotton produced in the country is inconsistent and insufficient to meet demand in a sustainable manner. As a result, it is necessary to improve Ethiopian cotton's competitiveness in the global market by mitigating the impact of potential biotic factors (disease, insect, and weed). Diseases such as bacterial blight, antracnose, rust, Alternaria leaf spot, cotton leaf roll dwarf virus, and wilts (Verticillium wilt and fusarium wilt) are the most important biotic factors. Cotton boll worm, Aphid, flea beetel, whitefly, thrip, mealybug, and jassid are also common insect pests that harm Ethiopian cotton production and productivity. In order to overcome these biotic factors, integrated pest management options such as cultural, botanical, biological, and chemical should be used. Even though the availability of land with suitable climate conditions, trainable manpower, thriving textile industries, long cotton culture, and attractive government incentives are promising for further development of cotton production in Ethiopia, due to gradual changing weather conditions and global warming caused incidence of newly introduced diseases and insects. As a result, prior emphasis should be placed on those biotic factors and their management options in a timely manner.

References

1. Abbas S (2020) Climate change and cotton production: an empirical investigation of Pakistan. Environ Sci Pollut Res 27(23):29580–29588
2. Abbas S, Waheed A (2017) Trade competitiveness of Pakistan: evidence from the revealed comparative advantage approach. Compet Rev Int Bus J 27(5):462–475
3. Abebe EA et al (2021) Susceptibility of several cotton varieties to the cotton flea beetle, Podagrica puncticollis Weise (Coleoptera: Chrysomelidae), in a hot dry tropical environment of Ethiopia. Entomologia hellenica 30(1):1–19
4. Addis T et al (2022) Analysis of the cotton varieties under production in Ethiopia. J Nat Fibres 19(4):1522–1530
5. Agegnehu E (2015) Cotton Flea Beetle, Podagrica puncticollis Weise (Coleoptera: Halticidae) in Metema District, Northern Ethiopia: Yield Loss Assessment and Management. Open Access Lib J 2(01)
6. Agegnehu E (2021) Studies on some ecological aspects and management of cotton flea beetle, Podagrica puncticollis Weise (Coleoptera: Chrysomelidae) on cotton (Gossypium hirsutum L.) in metema district,Northwestern Ethiopia. Haramaya university
7. Alemu Z, Azerefegne F, Terefe G (2021) Sequential Application of Various Insecticides for the Management of Cotton Bollworm (Hubner) Helicoverpa armigera (Lepidoptera: Noctuidae) in Cotton Production. East Afr J Sci 15(1):41–50
8. Avelar S et al (2020) Characterization of the complete genome and P0 protein for a previously unreported genotype of cotton leafroll dwarf virus, an introduced polerovirus in the United States. Plant Dis 104(3):780–786
9. Bag S, Roberts PM, Kemerait RC (2021) Cotton leafroll dwarf disease: an emerging virus disease on cotton in the US. Crops Soils 54(2):18–22
10. Balcha M et al (2022) Cotton Production Guideline in Ethiopia. Wcrer Agricultural Rcscarch Center, Ethiopian Institute of Agricultural Research
11. Bayeh M, Meisso H (2013) Assessment of cotton mealybug situation in rain fed and irrigated cotton of Ethiopia. Technical report submitted to the Food Agriculture Organization
12. Belay G, Mussema R (2020) Production and marketing constraints of cotton under small-holders: the case of Arbaminch Zuria District, Southern Ethiopia
13. Bharathi N, Nirmala RC, Ramalingam J (2017) Exploring the competence of fungicides for targeting endopolygalacturonase responsible for Alternaria leaf spot a foliar disease in cotton-an in silico approach. Trends Biosci 10(45):9275–9279

14. Bletsos FA (2005) Use of grafting and calcium cyanamide as alternatives to methyl bromide soil fumigation and their effects on growth, yield, quality and fusarium wilt control in melon. J Phytopathol 153(3):155–161
15. Brandenburg M et al (2022) Sustainability in Ethiopian textile and apparel supply chains. Africa and sustainable global value chains. Springer, pp 195–215
16. Chinara N, Mohapatra KB (2020) Present scenario of diseases in cotton and their management. In: Diseases of field crops: diagnosis and management. Apple Academic Press, pp 299–323
17. Chohan S et al (2020) Cotton diseases and their management. Cotton production and uses: agronomy, crop protection, and postharvest technologies, pp 239–270
18. Cianchetta AN, Davis RM (2015) Fusarium wilt of cotton: management strategies. Crop Prot 73:40–44
19. Cook D et al (2011) Biology, crop injury, and management of thrips (Thysanoptera: Thripidae) infesting cotton seedlings in the United States. J Integr Pest Manage 2(2):B1–B9
20. Culik MP, Gullan PJ (2016) A new pest of tomato and other records of mealybugs (Hemiptera: Pseudococcidae) from Espirito Santo, Brazil
21. Cun M, Guiliang J, Chuanlin Z (2002) The advances in cotton breeding resistance to Fusarium and Verticillium wilts in China during past fifty years. Zhongguo Nongye Kexue (China)
22. Dilnur T et al (2019) Association analysis of salt tolerance in Asiatic cotton (Gossypium arboretum) with SNP markers. Int J Mol Sci 20(9):2168
23. Eshetu A, Mulatu B, Damte T (2019) Host range of cotton flea beetle (Podagrica puncticollis) in a Hot dry tropical environment of Ethiopia. Uganda J Agricult Sci 19(1):21–31
24. Fradin EF, Thomma BPHJ (2006) Physiology and molecular aspects of Verticillium wilt diseases caused by V. dahliae and V. albo-atrum. Mol Plant Pathol 7(2):71–86
25. Gangwar RK, Gangwar C (2018) Lifecycle, distribution, nature of damage and economic importance of whitefly, Bemisia tabaci (Gennadius). Acta Sci Agricult 2(4):36–39
26. Garrett DE, Garrett DE (1996) Utilization of potash in agriculture. Potash: Deposits, Processing, Properties and Uses, pp 493–555
27. Ge A-H et al (2021) 'Microbial assembly and association network in watermelon rhizosphere after soil fumigation for Fusarium wilt control', Agriculture. Ecosyst Environ 312:107336
28. Gebregergis Z (2018) Incidence of a new pest, the cotton mealybugphenacoccus solenopsistinsley, on sesame in North Ethiopia. Int J Zool 2018:1–7
29. Gudeta B, Egziabher AG (2019) Cotton production potential areas, production trends, research status, gaps and future directions of cotton improvement in Ethiopia. Greener J Agricult Sci 9(2):163–170
30. Halpern HC et al (2018) First report of Fusarium wilt of cotton caused by Fusarium oxysporum f. sp. vasinfectum race 4 in Texas, USA. Plant Dis 102(2):446
31. Hanson LE (2000) Reduction of Verticillium wilt symptoms in cotton following seed treatment with Trichoderma virens
32. Higuera JJ et al (2019) The strawberry FaWRKY1 transcription factor negatively regulates resistance to Colletotrichum acutatum in fruit upon infection. Front Plant Sci 10:480
33. Hu J (2021) Verticillium Wilt of Cotton
34. Hu X-P et al (2015) Nondefoliating and defoliating strains from cotton correlate with races 1 and 2 of Verticillium dahliae. Plant Dis 99(12):1713–1720
35. Jjumba N et al (2016) The agri-business market scan for Uganda. Netherlands Ministry of Agriculture, Netherlands
36. Kabissa J, Elobu P, Muriithi A (2022) Cotton growing in East Africa. In: Pest management in cotton: a global perspective. CABI Wallingford UK, pp 156–184
37. Karademir E et al (2012) Effect of Verticillium dahliae Kleb. on cotton yield and fibre technological properties
38. Kemerait B et al (2017) Identification and management of bacterial blight of cotton. Cotton Incorp Bull Cotton Incorporated, Cary, NC
39. Khan MA et al (2020) World cotton production and consumption: An overview. Cotton production and uses: agronomy, crop protection, and postharvest technologies, pp 1–7

40. Khan SM et al (2019) Integrated pest and disease management for better agronomic crop production. Agronomic crops, vol 2. Management Practices, pp 385–428
41. Kumar A et al (2020) Legumes an alternative land use options for sustaining soil health. Agricult Food e-newslett 6
42. Kumar RK, Bhat NS, (2012) Biology of leafhopper,[Amrasca biguttula biguttula (Ishida)] on sunflower. Karnataka J Agricult Sci 25(4):543–544
43. Kumar S et al (2021) A comparative analysis of machine learning algorithms for detection of organic and nonorganic cotton diseases. Math Probl Eng 2021:1–18
44. Lee JA, Fang DD (2015) Cotton as a world crop: origin, history, and current status. Cotton 57:1–23
45. Li R et al (2016) Novel soil fumigation method for suppressing cucumber Fusarium wilt disease associated with soil microflora alterations. Appl Soil Ecol 101:28–36
46. Lu S et al (2019) Changes in Bemisia tabaci feeding behaviors caused directly and indirectly by cucurbit chlorotic yellows virus. Virology Journal 16(1):1–14
47. Maiti R et al (2020) Advances in cotton science: botany, production, and crop improvement. CRC Press
48. Marian M et al (2020) A novel strain of endophytic Streptomyces for the biocontrol of strawberry anthracnose caused by Glomerella cingulata. Microbiol Res 234:126428
49. Mengistie BT, Mol APJ, Oosterveer P (2017) Pesticide use practices among smallholder vegetable farmers in Ethiopian Central Rift Valley. Environ Dev Sustain 19:301–324
50. Noureen N, a. (2016) Cotton mealybug management: a review. J Entomol Zool Stud 4(4):657–663
51. Pan L-L et al (2018) Cotton leaf curl disease: which whitefly is the vector? Phytopathology 108(10):1172–1183
52. Patel Y, Patel P (2014) Evaluation of cotton (Gossyipium hirsutum L.) genotypes for their reaction to Thrips tabaci Lindemann. Res Environ Life Sci 7(4):267–270
53. Price T et al (2020) First report of cotton leafroll dwarf virus in Louisiana. Am Phytopath Society
54. Rajendran TP, Birah A, Burange PS (2018) Insect pests of cotton. In: Pests and their management. Springer, pp 361–411
55. Reay-Jones FPF et al (2017) Within-plant distribution and dynamics of thrips species (Thysanoptera: Thripidae) in cotton. J Econ Entomol 110(4):1563–1575
56. Reed B (2000) Screening for resistance to Aphis gossypii Glover, the cotton aphid, in cotton genotypes. Texas Tech University
57. Rehman A et al (2019) Economic perspectives of cotton crop in Pakistan: a time series analysis (1970–2015)(Part 1). J Saudi Soc Agric Sci 18(1):49–54
58. Rodríguez E, Téllez MM, Janssen D (2019) Whitefly control strategies against tomato leaf curl New Delhi virus in greenhouse zucchini. Int J Environ Res Public Health 16(15):2673
59. Rojo-Gutiérrez E et al (2020) Biotechnological potential of cottonseed, a by-product of cotton production. In: Valorisation of agro-industrial residues–volume II: non-biological approaches. Springer, pp 63–82
60. Sanjta S, Chauhan U (2015) Survey of thrips (Thysanoptera) and their natural enemies in vegetables from mid hills of Himachal Pradesh. The Ecoscan 9(3&4):713–715
61. Savva AP, Frenken K (2002) Agronomic Aspects of Irrigated Crop Production. Irrig Manual Module 3
62. Seboka N (2020) Mapping Cotton value chain of Ethiopia. Ethiop J Text Apparel 1(2)
63. Singh B et al (2020) Studies on Alternaria leaf spot of cotton with special reference to disease management by use of chemicals and botanicals. J Pharmacog Phytochem 9(5):1857–1863
64. Steyn LAI et al (2020) A review of leaf-mining insects and control options for their management, with special reference to Holocacista capensis (Lepidoptera: Heliozelidae) in vineyards in South Africa. S Afr J Enol Vitic 41(2):218–232
65. Suganthy M, Baskaran RKM, Mani M (2022) Pests and their management in medicinal plants: (Muskdana, Aonla, Ashwagandha, Black Datura, Black Nightshade, Coleus forskohlii, Gymnema sylvestre, Isabgol, Glory Lily, Long Pepper, Noni, Opium Poppy, Periwinkle, Phyllanthus amarus and Phyllanthus niruri. Trends Horticult Entomol 1255–1303

66. Tabassum A et al (2019) First report of cotton leafroll dwarf virus infecting cotton in Georgia, USA. Plant Dis 103(7):1803
67. Tena E et al (2020) Review on agronomy research in increasing the yield and resource use efficiency of cotton crop in Ethiopia
68. Tena E, Dejene M (2012) Quantitative and qualitative determination of weeds in cotton-growing areas of Humera and Metema, northwestern Ethiopia. Ethiop J Appl Sci Technol 3(1):57–69
69. Terefe G, Shonga E (2006) Cotton protection handbook
70. Taye W (2021) Effect of Weed Management on Insect Pest Population on Cotton (Gossypium Hirsutum L.) At Werer, Ethiopia. Int J Res Stud Agricult Sci (IJRSAS) 7(1):24–31
71. Yankova V et al (2014) Plant oils for control of cotton aphid (Aphis gossypii Glov.) in greenhouse cucumbers. Türk Tarım ve Doğa Bilimleri Dergisi 1(Özel Sayı-2):1565–1568
72. Yesuf NS, Alemu Z, Getahun S (2022) Comparative efficacy and economic efficiency of different insecticides against cotton thrips (Thrips tabaci L.)(Thysanoptera: Thripidae) on cotton in the Middle Awash, Ethiopia. Afr J Agric Res 18(10):850–859
73. Zeleke M et al (2019) Cotton production and marketing trend in Ethiopia: a review. Cogent Food Agricult 5(1):1691812
74. Zhang A et al (2004) Sex pheromone of the pink hibiscus mealybug, Maconellicoccus hirsutus, contains an unusual cyclobutanoid monoterpene. Proc Natl Acad Sci 101(26):9601–9606
75. Zhang J et al (2014) Genetics, breeding, and marker-assisted selection for Verticillium wilt resistance in cotton. Crop Sci 54(4):1289–1303
76. Zhang Y et al (2022) A review of the pathogenicity mechanism of Verticillium dahliae in cotton. J Cotton Res 5(1):3
77. Zhu Y et al (2019) Etiology of Alternaria leaf spot of cotton in Southern New Mexico. Plant Dis 103(7):1595–1604

Cotton Biotechnology

Dawit Tesfaye Degefu and Zerihun Desalegn Gebregiorgis

Abstract The cotton plant, *Gossypium hirsutum L.*, is one of the four cultivated cotton varieties grown in tropical environments primarily for its natural fiber. Cotton is cultivated by many farmers in Ethiopia as a source of income in both rainfed and irrigated areas. It can also create job opportunities for thousands of individuals. In addition to its fiber, the by-products are utilized for various purposes. Despite these advantages, the production of cotton faces challenges from both biotic and abiotic constraints. Among these, insect pests such as *Helicoverpa armigera* and *Pectinophora gossypiella* are significant threats to cotton in Ethiopia, and managing these pests through insecticide spraying is difficult due to their concealed feeding habits. The extensive use of different types of insecticides poses numerous challenges to humans, animals, and the environment. Moreover, these pests have developed resistance to multiple insecticides. Therefore, finding environmentally friendly alternatives for pest control is crucial. However, the limited genetic diversity among cotton germplasms presents a breeding challenge for developing cultivars that can overcome various production constraints. In plant breeding, molecular markers are valuable tools for measuring and identifying economically important traits that are otherwise difficult to assess visually. Molecular markers are essential for plant breeders in the development of cotton cultivars that meet market demands. Additionally, advancements in recombinant DNA technology offer the potential to improve crops with limited genetic diversity, such as cotton. Through genetic engineering, desirable traits such as insect resistance, herbicide tolerance, increased lint strength, length, and fineness can be introduced to cotton, adding value to the crop. This technology can help farmers achieve lower production costs, protect them from hazardous chemical exposure, ensure environmental safety, and maximize potential profits by reducing pest infestation at the early boll formation stage and providing high-quality cotton lint.

D. T. Degefu (✉)
Ethiopian Institute of Agricultural Research, Melkassa Agricultural Research Center, Shewa, Ethiopia
e-mail: dawit.tesfaye@office.eiar.gov.et

Z. D. Gebregiorgis
Solidaridad Ethiopia, Yeka, Ethiopia

K. Murugesh Babu et al. (eds.), *Cotton Sector Development in Ethiopia*, Textile Science and Clothing Technology, https://doi.org/10.1007/978-981-99-9149-5_4

Keywords Biotechnology · *Bt* Cotton · Cotton variety · Genetic diversity · Gene of interest

1 Introduction

Cotton, in the *Gossypium* genus, belongs to tribe *Gossypiae*, in a member of the Malvaceae family and the genus emerged as a separate evolutionary lineage around 11–14 million years ago [98, 99]. The *Gossypium* genus contains about 54 recognized species globally, out of this only 4 are cultivated for their spinnable fiber and the remaining 48 species are distributed throughout the tropics and subtropics in wild forms [6, 7, 31]. Cotton has four cultivated species with two diploids ($2n = 26$), namely *G. arboreum L.* and *G. herbaceous L.* which belong to the old-world cotton, while the other two species *G. barbadense* and *G. hirsutum* are tetraploid ($2n = 52$) known as new-world cotton [62]. Among those species, *G. hirsutum L.* species is one of the most commonly cultivated species in the world followed by *G. barbadense.* More than 95% of the world's cotton area is covered by tetraploids, which plays a vital role in cotton agriculture across the world [29]. On the other hand, diploid cotton is mostly cultivated in Asia and the Middle East. India is the only country where all the cultivated species and some of their hybrid combinations are commercially grown. Each of these species has a unique domestication history, diversification, and use that have exhibited their genetic structure through management and artificial selection of the variation caused by evolutionary processes over millions of years [24].

Cotton is cultivated by numerous farmers in Ethiopia as a means of generating income in both areas with rainfall and areas with irrigation. It can also create employment opportunities for a large number of individuals. In addition to its fibers, the by-products of cotton are used for various purposes. Despite these advantages, the production and efficiency of cotton have not yet reached the level needed to meet the national demand of the textile and garment industries in the country. The production of cotton in Ethiopia has been hindered by both biotic and abiotic constraints. As a result, the sector has become inefficient in terms of contributing to the national economy, considering the available resources in the country. Furthermore, the limited genetic resources pose a challenge for cotton breeding in order to improve the cultivars for different agronomic characteristics. Over the past few decades, the national cotton research program has conducted extensive research aimed at enhancing yield and quality-related traits, resulting in the release of several varieties suitable for both irrigated and rainfed agriculture [15] However, the adoption of these released cultivars by producers has been limited in terms of success.

Conventional plant breeding has played a significant role in the development of new cultivars, but it has been limited by various factors such as multiple stages of crossing, selection, testing, and crossing barriers [11]. Moreover, the current global climate change poses an additional challenge to agricultural production systems, as it brings about emerging and invasive insect pests and diseases that have a significant impact on the quality and yield of various crops [119]. Additionally, the long-term domestication of cotton has resulted in a narrowing of the genetic diversity in the available germplasm for breeding purposes. As a result, relying solely on traditional plant breeding and pest management practices is insufficient to meet the increasing demand for cotton lint in the textile and garment industries. Therefore, it is crucial to explore environmentally friendly alternative technologies and management strategies to enhance the cotton production system.

The recent advancement in biotechnology has brought about a revolution in the cotton breeding program, allowing for more efficient utilization of available genetic resources (79). The implementation of modern biotechnology techniques, such as genomic selection, high-throughput phenotyping, molecular markers, and genetic engineering, has played a crucial role in the development of cotton varieties with desirable characteristics [76]. In this context, the use of biotechnology through molecular breeding and recombinant DNA technology has the potential to enhance important economic traits of cotton. Currently, the adoption of genetically modified cotton, specifically designed to address various production challenges, particularly insect and herbicide tolerance, is increasing worldwide. Since the initial commercialization of *Bt* cotton in the United States in 1996 [49], the global acceptance of *Bt* cotton, primarily used as an insect-resistant measure, has significantly boosted economic benefits and revolutionized cotton productivity globally. Specifically, in the case of cotton, these two traits, *Bt* cotton (named after *Bacillus thuringiensis*) and herbicide-tolerant cotton (HT-cotton), serve as effective alternatives for controlling pest and weed issues that often plague cotton production. As a result, this technology keeps farmers safe from harmful chemical exposure, ensures environmental safety, and maximizes potential economic gains by reducing pest infestation during the early boll formation stage and minimizing pesticide expenses.

2 Biotechnology in the Cotton Sector

Cotton is one of the most cultivated crops primarily used as a source of natural fibers globally. The Gossypium genus consists of 54 species, with 47 being diploid ($2n = 2x = 26$) [31] and 7 being tetraploid ($2n = 4x = 52$) [6, 7, 31], each possessing distinct characteristics and utilized in various breeding programs. Genetic diversity with diverse characteristic traits is crucial for successful plant breeding initiatives. Breeders must possess knowledge and a thorough understanding of the gene pool of the plant materials, which serves as a guide for effective utilization of the germplasm resources. In Ethiopia, significant efforts have been made in the past few decades to enhance cotton through traditional breeding methods, employing research techniques such as classical breeding and artificial selection, albeit with limited success.

Conventional breeding has many limitations unless supplemented with modern molecular breeding. These limitations include challenges related to crossing barriers [87] and reliance on morphological markers that can be influenced by environmental fluctuations. Additionally, conventional breeding requires repetitive backcrossing, selfing, and testing, which is time-consuming, less precise, and expensive compared to DNA markers. The use of molecular markers is a faster and more cost-effective technique that allows plant breeders to select individual plants based on their genotypic marker patterns at an early stage of plant growth [63], thus accelerating the development of new varieties. Considering the limited gene pool of cotton [47, 75, 117, 123], the use of molecular markers facilitates breeding programs by providing valuable insights into the genetic diversity of available germplasm, as diversity is crucial for successful breeding.

The progress of biotechnology, both in the private and public sectors worldwide, has generated novel tools for molecular applications. These tools have the potential to enhance cotton yield, quality, and address biotic and abiotic constraints [27]. Molecular marker technologies can be categorized into hybridization-based, polymerase chain reaction (PCR)-based, and sequence-based markers, based on their mechanisms [76]. Various classes of molecular markers have been used for cotton, targeting alleles associated with biotic and abiotic constraints [9, 89]. PCR-based markers, such as random amplified polymorphic DNA (RAPD) [100, 113], amplified fragment length polymorphism (AFLP) [67, 90], inter simple sequence repeats (ISSR) [101], and simple sequence repeat (SSR) [89], are the most commonly used marker classes in cotton genomic research. Another frequently employed marker class in cotton improvement research is single nucleotide polymorphism (SNP), a sequence-based marker that enables the detection of genes based on nucleotide sequence variations [9]. This high-throughput technology is a valuable resource for discovering economically important genes in cotton [76].

Very recently, several studies have been conducted on cotton using SSR molecular markers to identify genetic markers associated with various phenotypic traits, including fiber quality traits [4, 26, 82, 126], yield traits [5, 114, 120], (Zhang et al. 2013), [55], agronomic traits [61, 121, 71], early-maturing traits [69]), and drought and salt tolerance [56, 94]. The discovery of such closely linked markers to the gene of interest helps breeders select plant materials that carry the desired gene for introgression and the stacking of major effect genes. Furthermore, the recent development of a SNP chip containing around 70 K markers is a valuable resource that can be utilized by scientists and researchers worldwide to enhance cotton research programs [76].

Thanks to the progress made in biotechnology, the limited range of genes in cotton, which has been a hindrance to its improvement, has been overcome through the use of genetic engineering. This has allowed for the incorporation of potential candidate genes for various beneficial traits that could not be achieved through traditional methods of crop improvement. The application of genetic engineering in cotton began in 1987 with the development of the first genetically engineered cotton plant for resistance against bollworm [44, 112]. Since then, genetic engineering tools have been further utilized to address challenges posed by both biotic and abiotic constraints in cotton. These efforts have resulted in the development of various cotton technologies, such as resistance against fungal pathogens, tolerance to herbicides and insects, salt tolerance, protection against high light intensity, water stress tolerance, resistance to leaf curl disease, reduced gossypol level, increased fiber length and strength, and higher fiber and biomass production [57, 12, 78, 86, 84, 85, 13, 64,

70]. Therefore, adopting an impact-oriented approach by prioritizing key areas of intervention can bring about significant changes in the cotton sector in Ethiopia.

2.1 Introduced Cotton Biotechnology and Environment

Agriculture is a crucial productive sector that ensures the largest portion of global food supply, but it also has a significant impact on the environment unless its activities are managed in an eco-friendly manner. However, agriculture is heavily influenced by weather and climatic conditions and is susceptible to various living factors such as emerging pests and diseases, as well as non-living factors like drought, salinity, and higher temperatures. Over the past few decades, agricultural practices worldwide have placed immense pressure on the environment. These include the use of harmful chemicals for crop protection, changes in the water cycle due to extensive irrigation, habitat destruction due to the expansion of agricultural lands, the introduction of invasive species, soil degradation caused by deforestation and excessive plowing, and more [103, 105, 59, 115]. Adding to this, the global population is expected to reach almost 10 billion by 2050, further straining natural resources for food production [43]. The reality is that all forms of agriculture have an impact on the environment, whether they use conventional or high-tech methods. Therefore, it is crucial to find a balance between agricultural practices and their environmental consequences in order to effectively utilize and improve natural resources. Additionally, in order to maintain a financially viable agricultural industry, the sector must embrace innovative technologies driven by continuous research and monitoring.

Cotton is the most consumed natural fiber that represents approximately a quarter of all fibers processed by the textile industry worldwide. According to the [40], cotton planted across the globe covers 32.9 million ha, of which 76% was genetically modified. This means that the remaining 24% of conventional cotton required only a small amount of pesticides to protect the planted cotton. Insect-resistant genetically modified (GM) cotton plants contain one or more genes from the bacteria *Bacillus thuringiensis*, which enable them to protect themselves from one or more groups of insects. Since the first commercialization of GM crops in 1996, this technology has made significant contributions to the natural ecosystem by conserving biodiversity, preventing deforestation, and protecting biodiversity sanctuaries. The widespread adoption of biotech crops worldwide reflects the substantial benefits derived from increased productivity and a cleaner environment through reduced use of conventional pesticides [58]. Biotech crops have contributed to a better environment by saving approximately 671 million kg of active ingredient pesticides from 1996 to 2016, and a further 48.5 million kg in 2016 alone, which represents an 8.2% reduction in pesticide use [23]. Due to the cultivation of GM cotton alone, 21.3 million kg of insecticides were saved from being released into the environment in 2013, resulting in a 48.3% reduction in pesticide use [21]. Since 1996, the global area of biotech crops has increased by 112-fold, from 1.7 million hectares to 190.4 million hectares in 2019. Currently, biotech crops are grown in 29 countries, with 24 of them being developing countries, and an additional 43 countries importing biotech crops [58].

2.2 Cotton Biotechnology and Quality of Life

In Ethiopia, cotton plays a crucial role in enhancing the country's economy due to the vast amount of cultivable land, which exceeds three million hectares. Cotton accounts for over 70% of the lint utilized in textile factories, garment production, and small-scale industries. There are numerous textile factories, ginning facilities, and fabric manufacturers, highlighting its significance as an economic resource for impoverished farmers, cultivators, and processors in the nation. With a population of over 100 million, a significant portion of which consists of unemployed young individuals, there is ample opportunity for them to participate in various stages of the cotton value chain, from farming to fabric production. However, despite the growing demand for natural cotton fibers from the textile industry in Ethiopia, cotton cultivation faces numerous challenges such as the emergence of insect pests, climate change, the rapid development of insecticide resistance, and other production limitations.

Several studies have demonstrated that the rapid growth in the adoption of *Bt* cotton brings tangible advantages for *Bt* farmers worldwide. While *Bt* cotton does not directly increase crop yield, the indirect increase in yield comes from the added value of the crop, which reduces pest damage. This reduction in pesticide use has also led to a decrease in the risk of pesticide poisonings. From a human health perspective, working on the farm becomes safer due to reduced pesticide exposure, ultimately improving the quality of life [1]. Additionally, the decrease in pesticide expenses has lowered production costs for *Bt* cotton and increased net revenue [45, 52]. Similarly, several other reports have shown that even though the cost of *Bt* cotton seed is higher, this is offset by the reduction in pesticide expenses and the significant increase in yield compared to non-*Bt* cotton [14, 15, 50, 18].

The impact of adopting *Bt* cotton on the welfare and distribution at the village level was reported and documented by Subramanian and Qaim [104]. The study revealed that the advantages gained from *Bt* cotton are not solely due to increased yield and reduced pesticide costs, but also include improved employment opportunities for female agricultural workers and increased household income for growers. The study also reported that growers experienced an 82% increase in aggregate income for every additional hectare of *Bt* cotton. This global experience of *Bt* cotton adoption holds great significance for a country like Ethiopia, which has over 3 million hectares of potentially available land for cotton production, with many of these areas located near river basins for irrigation. However, the country has only utilized 3% of this available land for cotton production, and the existing production system is hindered by the use of poor quality seeds due to the lack of an appropriate seed system in the country.

2.3 Cotton Biotechnology and Economic Profitability

The application of genetic manipulation offers numerous advantages in agricultural production systems. This technology allows crops to possess genetically altered characteristics through the artificial manipulation, modification, and recombination of DNA, making them more resistant to biotic and abiotic constraints. Since mid-1996, there has been a significant introduction of genetically modified (GM) crops in various countries worldwide, and they have been extensively cultivated. In the field of agriculture, the adoption of GM crops, particularly soybeans, corn, cotton, and rapeseed, encompasses approximately 190.4 million hectares globally. According to the ISAAA report, GM cotton accounted for 79% of the total cotton cultivation area in 2019, followed by 74% for soybeans, 31% for corn, and 27% for rapeseed. Among the 26 countries that have adopted GM crops, 21 of them are developing nations, which account for 54% of the GM crops planted in 2018, with Africa contributing 11% of this total. Over the past 22 years, from 1996 to 2018, the global area of biotech crops has shown a consistent increase in developing countries [58].

The socio-economic benefits of biotech crops, documented over the past 23/24 years, demonstrate their contribution to food security, sustainability, and solutions for climate change. The global economic gains, in terms of increased crop productivity by 822 million tons valued at US$224.9 billion, have benefited more than 16–17 million farmers, 95% of whom are from developing countries. Additionally, biotech crops have played a significant role in conserving biodiversity by preserving 231 million hectares of land. Furthermore, these crops have contributed to a safer environment by reducing pesticide usage by 776 million kg of active ingredients and lowering CO_2 emissions by 23 billion kg in 2018 alone. Moreover, more than 60 million individuals, primarily those living in poverty, have experienced economic benefits through the adoption of biotech crops [23].

According to the income benefit document analyzed by [22], farmers cultivating genetically modified cotton over the past 21 years have achieved an economic advantage totaling US$59.9 billion, with US$3.8 billion attributed to the year 2016 alone. India is included in the top 10 countries that grow genetically modified crops, and they have been cultivating insect-resistant cotton since 2002, as stated by [20]. Farmers in India have increased their income by US$27.3 billion over an 18-year period from 2002 to 2020 and earned US$1.5 billion in 2020 alone. These significant benefits have been enjoyed by over 7.5 million farmers, leading to an improvement in the economic status of the community [58]. Insect-resistant genetically modified cotton has also benefited cotton growers in South Africa and Burkina Faso, resulting in a combined economic gain of US$74.9 million and US$204.6 million, respectively, from 1996 to 2020. Therefore, biotech crops, particularly cotton, hold great value for farmers due to their significant contributions in terms of increasing yield, reducing pesticide expenses, cutting production costs, and preserving the environment.

3 Biotechnology in Ethiopian Agricultural Development

In the beginning, contemporary biotechnological research applications such as the utilization of Molecular markers, recombinant DNA, and genome editing techniques are in their early stages in Africa. Additionally, in Ethiopia, the strictness of the former biosafety regulation and concerns about the contamination of recombinant DNA technology have influenced the establishment of state-of-the-art biotechnology platforms. The previous limitation of the country's biosafety regulations could potentially impact and slow down the progress of capacity building in terms of both human capital and infrastructure. However, the current situation has significantly changed, and these application tools have made tremendous advancements, aiding us in developing the necessary human capital and infrastructure. This is a valuable asset in the research environment, enabling us to effectively provide demand-driven technologies to farmers. From this perspective, due to the previous circumstances, the application of modern biotechnological research on cotton in the country and its resulting outputs have been minimal.

Plant biotechnology encompasses two fundamental sciences in the realm of agriculture, specifically plant tissue culture and molecular biology. Plant tissue culture is a biological technique that can be achieved by utilizing small fragments of plant components or organs and transferring them onto synthetic culture media in order to cultivate them into complete plant organs. Tissue culture is a component of biotechnology that can be effectively utilized to clonally propagate commercially valuable crops, preserve endangered plant species and germplasm, introduce new genetic variations into breeding lines, generate transgenic plants, produce secondary metabolites, and expedite the breeding cycle. In Ethiopia, the application of plant tissue culture began at both Addis Ababa University (AAU) and the Ethiopian Institute of Agricultural Research (EIAR) in 1980 and 2000, respectively. Since then, tissue culture research has primarily focused on optimizing protocols for mass propagation, disease elimination, and in vitro conservation of economically significant crop species. Notably, significant achievements have been made in the mass propagation of hybrid coffee, pineapple, potato, and banana for both small-scale and commercial production [8]. However, there have been no reported studies on tissue culture in the field of cotton. In the realm of molecular biology, research has been conducted on various agricultural crops, excluding cotton, to analyze their genetic diversity using molecular markers such as Restriction Fragment Length Polymorphisms (RFLPs), Random Amplified Polymorphic DNA (RAPD), Amplified Fragment Length Polymorphism (AFLP), and Single Sequence Repeats (SSR) [8].

The introduction of recombinant DNA technology offers promising tools that could be utilized to enhance the agricultural sector worldwide. The competitive advantage of this technology lies in its ability to facilitate the transfer of desired genes from various sources, thereby expanding the genetic diversity [96]. Cotton, in particular, was the first genetically modified plant to be widely accepted and has been a remarkable success in agriculture, leading to significant economic benefits for farmers through reduced production costs, increased yields, and positive environmental impacts [65]. In Ethiopia, the approval of the biosafety proclamation on May 19, 2015, by the House of Peoples has paved the way for the implementation of modern biotechnology, including recombinant DNA technology. This policy development provides an opportunity for the adoption of advanced genetically modified technologies, such as *Bt* cotton and Bt-Dt maize, in the country. Currently, triple gene effects of Bt-Gt cotton are being tested under controlled conditions in Confined Field Trials (CFT). This creates a favorable environment for research and development in cotton biotechnology, with the ongoing Ethiopian *Bt* cotton project aiming to protect against the major cotton bollworm, *Helicoverpa armigera.*

4 Biotechnology in the Ethiopian Cotton Sector

For the past few decades, conducting research at both the molecular level and in controlled environments like greenhouses has been incredibly challenging. As a result, the country has missed out on numerous opportunities to utilize technology to improve the national economy through increasing agricultural productivity, creating jobs, ensuring food security, and advancing technological advancements. However, the current situation has changed, and this field of science has made significant progress. This progress has allowed us to develop a highly skilled human capital, which is invaluable in a research and development setting. This opportunity will enable both public and private organizations in the agricultural sector to develop and provide sustainable, demand-driven, and climate-smart technologies that are preferred by farmers. Additionally, advanced research in the field of biotechnology will help us meet the growing need for expertise in this modern-day science, particularly in the life sciences sector in Ethiopia. The country's interest and political willingness to utilize biotechnology as a tool for research and development in agriculture, medicine, and the environment led to the amendment of the Ethiopian biosafety laws through new proclamations and directives. This amendment resulted in the establishment of institutes like the Bio and Emerging Technology Institute, formerly known as the Ethiopian Biotechnology Institute. The amendment signifies the expectations the country has for the biotech institute. From this perspective, there is a timely opportunity to address the gap in plant biotechnology research and development.

Cotton is one of the largest crops in the agricultural production system that utilizes up to 25% of the insecticide produced globally [37]. In Ethiopia, cotton accounts for approximately 65% of the total pesticide purchase [2]. This chemical is primarily used for the protection of major insect pests of cotton. Among the numerous insect pests attacking cotton, Hemiptera and Lepidoptera insect pests are of utmost concern due to the damage they cause to the plant [38]. Lepidopteran insect pests, specifically *Helicoverpa armigera* and *Pectinophora gossypiella*, are globally recognized pests of cotton and economically significant insect pests in Ethiopia [109, 92]. Their concealed feeding behavior prevents them from coming into contact with sprayed insecticides. As a result of this challenging feeding habit, the extensive use of various types of insecticides poses numerous challenges to producers, including increased production costs, as well as negative impacts on beneficial non-target organisms and groundwater pollution. Additionally, these insect pests have rapidly developed resistance to nearly all classes of insecticides worldwide, including Ethiopia [107, 121]. This raises concerns about the future production of cotton under the current circumstances, as similar cases of resistance are likely to occur. Therefore, exploring alternative options and investing in demand-driven, high-tech cotton technologies can help Ethiopian farmers lead sustainable lives and secure their basic needs. In this regard, *Bt* cotton, named after *Bacillus thuringiensis*, which provides resistance to *H. armigera* in cotton, is considered as a viable alternative to address the current challenges in cotton production caused by this pest. In the years 2015/16, two *Bt* cotton varieties, namely JKCH-1947 and JKCH-1050, underwent 2 years of confined field trials and received official approval for commercial release in 2018. These hybrids demonstrated superior yield performance and reduced yield losses from bollworm damage. Furthermore, the cry protein gene, which is effectively expressed and protects against bollworms, provides an additional advantage by significantly reducing the need for chemical sprays compared to non-Bt cotton varieties. This suggests that the utilization of these two *Bt* cotton varieties improves efficiency and excellence in the cotton industry of Ethiopia.

4.1 Cotton Varieties in Ethiopia

Cotton is a versatile cash crop cultivated worldwide. In Ethiopia, the crop is grown on both large and small farms using irrigation and rain-fed agriculture. The cotton technology employed in the production system, such as the type of varieties, sources of seeds, and management practices, varies significantly. However, the available technologies do not adequately meet the national demand. The cotton breeding program in Ethiopia is primarily led by the Werer Agricultural Research Center (WARC), which is responsible for developing cotton varieties through selection and various breeding methods. To enhance the competitiveness of the cotton sector in the global market, efforts should focus on genetic improvements of the crop. These improvements rely heavily on the availability of diverse genetic materials. To date, WARC has released 42 cotton varieties, with 22 of them being released in the past 13 years (Table 1). Five of these varieties are recommended for rain-fed areas (Table 2), while 35 are recommended for irrigated areas [16]. Additionally, two Bt cotton hybrids have been registered for irrigation (Table 3). The main objective of the cotton variety development program is to address the shortcomings in the production system and provide farmers with the best varieties that offer high yield, good fiber quality, and resistance to pests. However, only a few common varieties have been cultivated commercially so far. According to ICAC [54], upland cotton *G. hirsutum* accounts for over 96% of the Ethiopian cotton production area. Among the varieties, DP 90, which was released 30 years ago, covers 80% of the production area, followed by Stam 59A (15%), Acala SJ2 (0.5%), and land races (4%) (Table 4).

Table 1 Cotton cultivars released for irrigated agriculture for the last 13 years

Sl. no	Name	Year of release	Recommended
1	Stam59A	2007	Irrigated
2	Ionia	2008	Irrigated
3	YD-206	2011	Irrigated
4	YD-223	2011	Irrigated
5	YD-211	2011	Irrigated
6	YD-670	2013	Irrigated
7	YD-195	2013	Irrigated
8	VBCH 1203	2013	Irrigated
9	VBCH 1527	2013	Irrigated
10	STG-14	2014	Irrigated
11	Candia	2014	Irrigated
12	Claudia	2014	Irrigated
13	Gloria	2014	Irrigated
14	Sisikuk-02	2015	Irrigated
15	Werer-50	2015	Irrigated
16	Weyto-7	2015	Irrigated
17	Werer-12	2019	Irrigated
18	Malkasadi	2019	Irrigated
19	WARC-LS1	2019	Irrigated
20	WARC-LS2	2019	Irrigated
21	Werer-13	2020	Irrigated
22	Sille-13	2020	Irrigated

Table 2 Cotton cultivars released for rainfed agriculture from 1960 to 1989

Sl. no	Name	Year of release	Recommended
1	A-333–57	1960s	Rainfed
2	Albar 637	1960s	Rainfed
3	Reba B-50	1960s	Rainfed
4	Arba	1987	Rainfed
5	Bulk-202	1989	Rainfed

Table 3 Genetically modified cotton hybrids registered for irrigated agriculture

Sl. no	Name	Year of release	Recommended
1	JKCH 1947 (*BT*)	2018	Irrigated
2	JKCH 1050 (*BT*)	2018	Irrigated

Table 4 Upland cotton cultivars, *G. hirsutum*, area coverage in Ethiopia

S/N	Name	Year of release	Recommended for	% area coverage
1	Deltapine-90	1989	Irrigated	80
2	Acala SJ2	1986	Irrigated	0.5
3	Stam59A	2007	Irrigated	15
4	Germplasm	–	–	4

At present, the production and efficiency of cotton are insufficient to meet the increasing demand from the textile and clothing industry. Despite numerous successful initiatives since 1994 when WARC was established as a national center for cotton improvement, there are still many limitations in meeting the demand for high-quality cotton fibers and production. Various factors restrict the production and efficiency of the cotton sector, including the absence of an extension system that leads to improved technologies being ignored for years and a lack of trust from producers in newly developed technologies [80]. In this context, the country lacks any regulations or laws to eliminate old varieties from the production system. Additionally, the limited availability of germplasm in the country, combined with the lengthy conventional method, poses significant challenges in developing technologies that address specific biotic and abiotic constraints such as pests, moisture stress, drought, frost, salinity, nutrient-deficient soils, and other factors.

Currently, a different situation exists, with the adoption of *Bt* cotton as a new pest management strategy gaining acceptance for enhancing the cotton sector by reducing production costs and improving cotton quality. The release of two genetically modified (GM) cotton varieties for commercial production, following approval by the Environmental Protection Authority, is an important step toward improving the cotton sector. However, due to the limitations of the released *Bt* cotton varieties, which only have bollworm protection traits, farmers have shifted to cotton varieties with two stacked traits, namely *Bt* and herbicide tolerance. This increased interest from farmers has led them to obtain illegal cotton varieties with stacked *Bt* and herbicide tolerance genes from unknown sources (Personal communication). This indicates the need to combine available efforts to improve and sustainably and economically compete the country's cotton sector, taking into account the adaptability of technologies to the country's climatic and environmental conditions, as well as emerging pest challenges in cotton production. These efforts should include high-yielding and high-quality cotton varieties, innovative approaches to cotton management practices such as post-harvest quality control, integrated pest management, and high-quality seeds.

4.2 Biotic and Abiotic Stress-Resistant Varieties in Ethiopia

Cotton is vulnerable to a wide range of insect pests in addition to various environmental factors. Biotic and abiotic factors negatively impact the yield and quality of cotton lint [30]. Therefore, it is crucial to develop cotton varieties that are resistant to these constraints. The success of developing varieties that can withstand pest and environmental constraints relies heavily on the available germplasm for hybridization and selection for host plant resistance. However, the intensive breeding process conducted over several decades with limited genotypes for yield and quality parameters has narrowed the gene pool available for further improvement. Additionally, the limited availability of germplasm makes it challenging to advance host plant-resistant breeding for insects, pests, diseases, drought, salinity, heat tolerance, and other environmental factors.

In Ethiopia, efforts have been made to identify cotton varieties that are resistant to major insect pests such as African bollworm [36], whitefly [17, 53, 33], and jassids [53]. However, there have been very limited studies on host plant resistance using available cotton germplasms. Despite these limitations, some promising genotypes, such as Frego-bract, Del. SL.DSR, La okra leaf-2 (which showed resistance to whiteflies), and Albar 637 (which showed resistance to jassids), could be used as sources of resistance in cotton breeding for the future [38]. Currently, there are no technologies specifically developed for addressing environmental challenges such as moisture stress, drought, frost, salinity, nutrient deficit soils, pests, and other factors. The increasing variability in climate conditions favors the pressure of pests and diseases on cotton, which becomes a key challenge in the crop's production system. Therefore, studies on host plant resistance should focus on thoroughly exploring the available germplasm to develop high-yielding cultivars that are resistant to economically important insects, diseases, and environmental factors. The availability of cotton germplasm is crucial for varietal development. Therefore, it is important to establish a national collection of germplasm and establish strong collaborations with international partners to enrich the germplasm resources at the national level.

In recent years, the utilization of innovative technologies, specifically characteristics beyond the cotton genetic pool, for pest management and other factors such as temperature, drought, salinity, seed oil quality, and fiber properties through the adoption of modern biotechnological approaches enhance the agronomic performance of cotton varieties. The cotton cultivators worldwide have benefited from genetically modified cotton despite the notable accomplishments of traditional breeding in the cotton industry. Due to certain drawbacks that currently exist in the country like the limited cotton genetic pool, restricted germplasm, and compatibility constraints through traditional breeding, future research efforts should focus on and prioritize the use of modern molecular techniques that significantly enhance the cotton industry.

4.3 *Identified Genes of Interest from the Wild and Cultivated Varieties of Cotton in Ethiopia*

The domestication of crop plants in general is recognized to lead to reduced genetic variation, thereby limiting their potential for developing new varieties with improved traits. In such situations, wild relatives are believed to experience less selection pressure, adapt to a wide range of environmental conditions, and preserve many valuable genetic variants. As a result, they become a valuable source of novel genes for breeding programs [32, 77, 127]. Among the four cultivated cotton species, G. *hirsutum*, which accounts for 98–99% of the total cotton-growing area [68], has experienced intense selection pressure during domestication, resulting in reduced genetic diversity and a narrow germplasm base [10, 42, 48, 111, 73]. Similarly, diversity studies have shown that the other domesticated cotton species, G. *barbadense*, known as "extra-long-staple" cotton, also exhibits low diversity [48, 51, 83]. Therefore, it is crucial to explore novel genes for selection and introgression of improved traits into elite cultivars in order to develop climate-resilient, high-yield, and high-quality cotton cultivars.

The advancements in biotechnology have greatly facilitated molecular genetic studies and created an enabling environment for cotton breeding and gene discovery [35, 72, 116, 124]. Molecular genetic studies provide unprecedented insights into the types of diversity within wild and cultivated varieties, which can assist breeders in accessing novel genes from wild germplasms to improve the highly domesticated upland cotton. Several studies have confirmed the presence of key genes that distinguish wild from domesticated cotton [28, 3, 122]. However, very little is known about gene discovery in the Ethiopian cotton research program. Therefore, based on the worldwide expertise and the progress of molecular technologies, it is necessary for the cotton research program of countries to focus on utilizing advanced technological tools to improve cotton productivity by identifying and incorporating significant genes for different agronomic characteristics.

4.4 *Genetic Diversity of Cotton Pests in Ethiopia*

Although cotton provides a habitat for a wide range of insect pests [95, 97, 125], only a few key species cause significant damage to the crop [74]. The abundance and distribution of these insect species depend heavily on factors such as temperature, surrounding vegetation cover, and pest management practices [118]. Based on their feeding habits, cotton insects can be grouped into two categories: chewing insects, which damage the plant by chewing on various tender parts, and sucking insects, which extract sap from the plant. Sucking insects in the orders Thysanoptera and Homoptera are also known for transmitting plant diseases.

A study conducted in Ethiopia documented over 68 species of insects and mites associated with cotton, belonging to the orders Coleoptera, Lepidoptera, Homoptera,

Thysanoptera, Heteroptera, Isoptera, Acari, and Orthoptera [38]. Sixteen insect species from various families were identified as major pests of economic importance, including cotton flea beetles (Chrysomelidae), cotton red mites (Tetranychidae), cotton seed bugs and tobacco whiteflies (Lygaeidae), cotton aphids (Aphididae), cotton jassids (Cicadellidae), pink bollworms (Gelechiidae), Sudan bollworms, African bollworms, lesser armyworms, and cotton leafworms (Noctuidae), as well as cotton thrips and bean thrips (Thripidae) (Table 5).

Insects due to the expansion of their host range, geographical distribution, and novel host association may induce the selection of adaptive traits and genetic differentiation in population [34, 93]. Moreover, the ability of the pest to develop resistance to multiple synthetic insecticides might result in heterogeneity in population due to the overexpression of resistant gene [41, 102]. Several studies using different molecular markers confirmed that there is significant diversity and species complex of cotton bollworm, aphid, whitefly, and thrip, worldwide [25, 39, 66, 91, 106, 108, 117]. However, in Ethiopia thus far no studies have been conducted to characterize the population genetic structure of cotton insect and the presence of host-associated species complex. A better understanding of the genetic variation of cotton insect pests occurring on diverse host plants and different geographical populations can be very helpful in gaining insights into the structure and population dynamics, their behavior, and response to various selection pressures [39]. Therefore, clarifying their genetic diversity in different geographical populations can assist in designing the most suitable tools for their management, including determining the economic threshold level [88], monitoring insecticide resistance [19], selecting biological control [81], identifying endosymbiotic relationships [2], and studying virus–vector relationships [46].

5 Challenges and Recommendations

Cotton production has been challenged by many issues that directly impact its productivity and quality across its growing regions worldwide. Both living organisms (insect pests, weeds, and pathogens) and non-living factors (drought and salinity) greatly limit the economic gain generated from cotton. Among the primary obstacles, damage caused by insect pests is the most prominent. They can have a significant impact on reducing both the yield and quality of cotton products. Overall, the economic importance of cotton is greatly overshadowed, hindered, and endangered by various living and non-living stress factors.

Another challenge in the cotton sector is the limited dissemination and popularization of cotton technologies and their management to end users. Over the past few years, improved cotton technologies such as released varieties, agronomic practices, pest management, and soil and water management have been developed but have not been effectively and sufficiently distributed and promoted to producers. This weak extension service that targets both small and large-scale cotton producers has had a negative impact on the cotton sector. Additionally, the limited sources of germplasm

Table 5 Major types of insect pest species recorded associated with cotton in Ethiopia

Order	Scientific name	Common name	Economic importance
Acari			
	Tetranychidae		
	Tetranychus cinnabarinus (Boised.)	Cotton red mite	Major
Coleoptera			
	Chrysomelidae		
	Podagrica pallidicolor Pic	Cotton flea beetle	Major
	Podagrica puncticollis Wse	Cotton flea beetle	Major
	Podagrica uniformis Jac	Cotton flea beetle	Major
Hetroptera			
	Lygaeidae		
	Oxycarenus hyalinipennis (Costa)	Cotton seed bug	Major
	Bemisia tabaci (Genn.)	Tobacco whitefly	Major
	Aphididae		
	Aphis gossypii Glov	Cotton aphid	Major
	Cicadellidae		
	Empoasca fascialis (Jac.)	Cotton jassid	Major
	Empoasca lybica (De Berg.)	Cotton jassid	Major
Lepidotera			
	Gelechiidae		
	Pectinophora gossipiella (Saund.)	Pink bollworm	Major
	Noctuidae		
	Diparopsis sp.	Sudan bollworm	Major
	Helicverpa armigera (Hb.)	African bollworm	Major
	Spodeptera exigua (Bb.)	Lesser armyworm	Major
	Podeptera littoralis (Boisd.)	Cotton leafworm	Major
Thysanoptera			
	Thripidae		
	Caliothrips lupirus (Pries)	Cotton thrips	Major
	Caliothrips sudanensis (Bagn. And Cam.)	Bean thrips	Major

(*Source* Modified from [38])

in both in situ and ex situ conservation, as well as narrow genetic variability, make future cotton improvement very difficult. Even currently, the technology we have is far from meeting the national demand. Similarly, the inadequate cotton seed system, grading, and the shortage and lack of high-quality seeds are major obstacles that hinder the production and productivity of cotton in Ethiopia. The lack of modern application tools for research and development, such as the use of genomic selection and marker-assisted molecular breeding, has left the research effort dependent on conventional research approaches, which have many limitations as mentioned in other sections of this article. The limited financial support for cotton research and development is the main obstacle preventing the establishment of state-of-the-art laboratories for cotton research and quality testing facilities. Such modern facilities could help accelerate the development of high-yielding, quality cotton cultivars with greater precision.

Currently, there is a strong local demand for cotton fibers and cotton by-products in Ethiopia due to the recent growth in textile and garment manufacturing industries, in addition to the local handloom industry. Moreover, the country's growing population presents an opportunity to utilize the young workforce in the large-scale industries established in the country. Therefore, in order to meet the industry's demand, there is a need to enhance the production of this crop using advanced technologies within the country. Additionally, the development of new cotton-based products and improved living standards in developing countries will further increase the demand for cotton. Consequently, Ethiopia should strive to increase the production and productivity of cotton through a climate-smart approach, in order to benefit from the growing domestic and international markets, while also strengthening the capabilities of farmers, textile and garment manufacturers, oil mills, and other processing industries within the country. In this regard, both public and private institutions have a responsibility to organize and lead advanced research at the national level, and to take the initiative in modernizing the research culture to a more competitive level, providing alternative technologies to farmers, processors, consumers, and ultimately, to the national economy.

6 Summary

Cotton is one of the world's leading natural fiber crops with various species under the genus Gossypium, which belongs to the family Malvaceae. Cotton is utilized for all types of fabrics, home furnishings, and industrial purposes, and the by-products are used as animal feed and edible oil. It is the most crucial agricultural commodity across all the value chains from production to processing industries. It is one of the strategically important commodities for both the textile and garment industries in Ethiopia. A commodity like cotton holds significant importance for countries with a large productive workforce and potential resources such as production land and river basin for irrigation. From this perspective, Ethiopia has enormous potential for intensifying cotton production and the associated textile and garment industries. Currently, cotton is cultivated by both large private farms and small-scale farmers. The production and efficiency of cotton have not yet reached the level required to meet the national demand of the textile and garment industries in the country. Several factors contribute to the sectors' inefficiency in enhancing the national economy of the country relative to the available potential resources that the country possesses.

It is an evident truth that cotton production is greatly influenced by both biotic and abiotic constraints. In addition, the insufficiency of adequate germplasms, which is a limitation for breeding, and the limited choices of cultivated cultivars and their control of pests should be appropriately addressed. These restrictions can be overcome by enhancing the national germplasm collection and utilizing modern biotechnology tools. For example, modern molecular breeding can assist in increasing the genetic improvement of profitable traits by introducing new alleles from wild species. Moreover, for cotton, which has a limited genetic range, the use of recombinant DNA technologies to develop cultivars that are resistant to economically significant insect pests and weeds is of utmost importance. Therefore, in order to make the cotton industry more dynamic and productive, all stakeholders at various levels along the value chains need to be in agreement and more focused on establishing advanced research facilities, both privately and publicly, and facilitating access to high-yielding, quality cotton cultivars with diverse options.

References

1. ASSOCHAM—Association of Chambers of Commerce and Industry of India (2007) Economic benefits of Bt cotton cultivation in India. In: Bt cotton farming in India. New Delhi, India
2. Abate T (1997) Pesticide use in Ethiopian agriculture. Pest Manag J Ethiop 1:35–39
3. Abate T (1982) Cotton pest problems and their control in Ethiopia. In: Proceedings of the symposium on cotton production under irrigation in Ethiopia. IAR, Melka Werer (Ethiopia), pp 111–127
4. Abdurakhmonov I, Saha S, Jenkins J, Buriev Z, Shermatov S, Scheffler B et al (2009) Linkagedisequilibrium based association mapping of fibre quality traits in G. hirsutum L. variety germplasm. Genetica 136:401–417
5. Abdurakhmonov IY (2007) Exploiting genetic diversity. In: Proceedings of world cotton research conference-4. Lubbock, Texas, p 2153
6. Adams KL, Cronn R, Percifield R, Wendel JF (2003) Genes duplicated by polyploidy show unequal contributions to the transcriptome and organ-specific reciprocal silencing. Proc Natl Acad Sci U S A. 100:4649–4654. https://doi.org/10.1073/pnas.0630618100
7. Adams KL, Wendel JF (2005) Allele-specific bidirectional silencing of an alcohol dehydrogenase gene in different organs of interspecific diploid cotton hybrids. Genetics 171:2139–2142. 10.1534/ genetics.105.047357
8. Adane A (2009) Agricultural biotechnology research and development in Ethiopia. Afr J Biotech 8:7196–7204
9. Agarwal M, Shrivastava N, Padh H (2008) Advances in molecular marker techniques and their applications in plant sciences. Plant Cell Rep 27:617–631
10. Ahmad MQ, Khan SH, Azhar FM (2012) Decreasing level of genetic diversity in germplasm and cultivars of upland cotton (Gossypium hirsutum) in Pakistan. J Agric Soc Sci. 8:92–96
11. Ahmar S, Gill RA, Jung K-H, Faheem A, Qasim MU, Mubeen M, Zhou W (2020) Conventional and molecular techniques from simple breeding to speed breeding in crop plants: recent advances and future outlook. Int J Mol Sci 21:2590
12. Amudha J, Balasubramani G, Malathi VG, Monga D, Kranthi KR (2011) Cotton leaf curl virus resistance transgenics with antisense coat protein gene (AV1). Curr Sci 101(3):300–307
13. Asad S, Mukhtar Z, Nazir F, Hashmi JA, Mansoor S, Zafar Y, Arshad M (2008) Silicon carbide whisker-mediated embryogenic callus transformation of cotton (Gossypium Hirsutum L.) and

regeneration of salt tolerant plants. Mol Biotechnol 40:161–169. https://doi.org/10.1007/s12033-008-9072-5
14. Bakhsh, A., Anayol, E., O¨ zcan, S. F., Hussain, T., Aasim, M., Khawar, K. M., O¨ zcan, S. (2015) An insight into cotton genetic engineering (Gossypium hirsutum L.): current endeavors and prospects. Acta Physiol Plant 37:171
15. Bambawale OM, Singh A, Sharma OP, Bhosle BB, Lavekar RC, Dhandapani A, Kanwar V, Tanwar RK, Rathod KS, Patange NR, Pawar VM (2004) Performance of Bt cotton (MECH162) under Integrated Pest Management in farmers' participatory field trial in Nanded district. Central India. Current Sci. 86:1628–1633
16. Barwale RB, Gadwal VR, Zehr U, Zehr B (2004) Prospects for Bt cotton technology in India. AgBioforum 7(1&2):23–26
17. Bennett RM (2004) Economic impact of genetically modified cotton in India. AgBioforum 7:16
18. Bennett R, Kambhampati U, Morse S, Ismael Y (2006) Farm-level economic performance of genetically modified cotton in Maharashtra, India. Appl Econ Perspect Policy 28:59–71. https://doi.org/10.1111/j.1467-9353.2006.00273.x
19. Berhane G/Yohannes (1987) An integrated approach in the management of cotton whitefly, Bemisia tabaci in the Middle Awash valley. In: Proceedings of the 7th CEE Annual Meeting. Addis Ababa, Ethiopia, pp 53–65
20. Bielza P, Quinto V, Contreras J, Torné M, Martín A, Espinosa PJ (2007) Resistance to spinosad in the western flower thrips, Frankliniella occidentalis (Pergande), in greenhouses of south-eastern Spain. Pest Manag Sci 63:682–687. https://doi.org/10.1002/ps.1388
21. Brookes G (2022) Farm income and production impacts from the use of genetically modified (GM) crop technology 1996–2020. GM Crops Food 13:171–195. https://doi.org/10.1080/21645698.2022.2105626
22. Brookes G, Barfoot P (2015) Environmental impacts of genetically modified (GM) crop use 1996–2013: impacts on pesticide use and carbon emissions. GM Crop Food: Biotechnol Agric Food Chain 6:103–133. https://doi.org/10.1080/21645698.2015.1025193
23. Brookes G, Barfoot P (2018) Environmental impacts of genetically modified (GM) crop use 1996–2016: Impacts on pesticide use and carbon emissions. GM Crops Food 9:109–139. https://doi.org/10.1080/21645698.2018.1476792
24. Brookes G, Barfoot P (2020) GM crop technology use 1996–2018: farm income and production impacts. GM Crops Food 11:242–261. https://doi.org/10.1080/21645698.2020.1779574
25. Brubaker CL, Bourland FM, Wendel JF (1999) The origin and domestication of cotton. Cotton; Origin, History. Technology and Production. John Wiley, New York, pp 3–31
26. Brunner PC, Chatzivassiliou EK, Katis NI, Frey JE (2004) Host-associated genetic differentiation in Thrips tabaci (Insecta; Thysanoptera), as determined from mtDNA sequence data. Heredity 93:364–370. https://doi.org/10.1038/sj.hdy.6800512
27. Cai C, Ye W, Zhang T, Guo W (2014) Association analysis of fibre quality traits and exploration of elite alleles in Upland cotton cultivars/accessions (Gossypium hirsutum L.). J Integr Plant Biol 56:51–62. https://doi.org/10.1111/jipb.12124
28. Campbell BT et al (2010) Status of the global cotton germplasm resources. Crop Sci 50(4):1161–1179. https://doi.org/10.2135/cropsci2009.09.0551
29. Chaudhary B, Flagel L, Stupar RM, Udall JA, Verma N, Springer NM, Wendel JF (2009) Reciprocal silencing, transcriptional bias and functional divergence of homeologs in polyploid cotton (Gossypium). Genetics 182:503–517. https://doi.org/10.1534/genetics.109.102608
30. Chen ZJ et al (2007) Toward sequencing cotton (Gossypium) Genomes. Plant Physiol 145:1303–1310. https://doi.org/10.1104/pp.107.107672
31. Constable GA, Bange MP (2015) The yield potential of cotton (Gossypium hirsutum L.). Field Crop Res 182:98–106. https://doi.org/10.1016/j.fcr.2015.07.017
32. Cronn RC, Small RL, Haselkorn T, Wendel JF (2002) Rapid diversification of the cotton genus (Gossypium: Malvaceae) revealed by analysis of sixteen nuclear and chloroplast genes. Am J Bot 89:707–725. https://doi.org/10.3732/ajb.89.4.707

33. Demissie A, Refera A (1989) Screening of cotton varieties for resistance to whitefly, Bemisia tabaci Gennadius. In: Proceedings of the 9th annual meeting of the committee of Ethiopian entomologists. Addis Ababa Ethiopia, pp 57–67
34. Dempewolf H, Baute G, Anderson J, Kilian B, Smith C, Guarino L (2017) Past and future use of wild relatives in crop breeding. Crop Sci 57:1070–1082. https://doi.org/10.2135/cropsci2016.10.0885
35. Diehl SR, Bush GL (1989) The role of habitat preference in adaptation and speciation. Speciation and its consequences. Sinauer Associates, Sinauer, Sunderland, pp 345–365
36. Du X et al (2018) Resequencing of 243 diploid cotton accessions based on an updated A Genome identifies the genetic basis of key agronomic traits. Nat Genet 50:796–802. https://doi.org/10.1038/s41588-018-0116-x
37. EARO, Ethiopian Agricultural Research Organization (2004) Progress report of national cotton research project for the period 2003–2004. Ethiopian Agricultural Research Organization, Werer Agricultural Research Center (WARC)
38. EJF (Environmental Justice Foundation) (2007) The deadly chemicals in cotton (Environmental Justice Foundation in collaboration with Pesticide Action Network UK No. ISBN No. 1–904523 10–2). London, UK
39. Ermias Shonga, Geremew Terefe, Zeray Mehari (2009) Review of Research on Insect Pests of Fibre Crops in Ethiopia. In: Abraham Tadesse (ed) Increasing crop production through improved plant protection—volume II. Plant Protection Society of Ethiopia (PPSE). PPSE and EIAR, Addis Ababa, Ethiopia, p 542
40. FAO (Food and Agriculture Organization of the United Nations) (2017) The future of food and agriculture: Trends and challenges. FAO, Rome
41. FAOSTAT (2018) Food and agriculture data. http://www.fao.org/faostat/en/#home
42. Fakrudin B, Prakash S, Krishnareddy K, Prasad PB, Patil B, Kuruvinashetti M (2004) Genetic variation of cotton bollworm, Helicoverpa armigera (Hübner) of South Indian cotton ecosystem using RAPD markers. Curr Sci 87:1654–1657
43. Fang DD, Hinze LL, Percy RG, Li P, Deng D, Thyssen G (2013) A microsatellite-based genome-wide analysis of genetic diversity and linkage disequilibrium in Upland cotton (Gossypium hirsutum L.) cultivars from major cotton-growing countries. Euphytica 191:391–401
44. Ffrench-Constant RH (2013) The molecular genetics of insecticide resistance. Genetics 194:807–815. https://doi.org/10.1534/genetics.112.141895
45. Firoozabady E, DeBoer DL, Merlo DJ, Halk EL, Amerson LN, Rashka KE, Murray EE (1987) Transformation of cotton (Gossypium hirsutum L.) by Agrobacterium tumefaciens and regeneration of transgenic plants. Plant Mol Biol 10:105–116
46. Gandhi VP, Namboodiri N (2006) The adoption and economics of Bt cotton in India: Preliminary results from a study. IIMA Working Papers WP2006–09–04, Indian Institute of Management Ahmedabad, India, Research and Publication Department
47. Ghosh A, Dey D, Timmanna Basavaraj Mandal B, Jain RK (2017) Thrips as the vectors oftospoviruses in Indian agriculture. A century of plant virology in India, pp 537–561
48. Gingle A, Yang H, Chee P, May O, Rong J, Bowman D, Lubbers E, Day J, Paterson A (2006) An integrated web resource for cotton. Crop Science—CROP SCI 46. https://doi.org/10.2135/cropsci2005.09.0328
49. Grover CE, Gallagher JP, Szadkowski EP, Page JT, Gore MA, Udall JA, Wendel JF (2017) Nucleotide diversity in the two co-resident genomes of allopolyploid cotton. Plant Syst Evol 303:1021–1042
50. Gudeta B, Arkebe G, Egziabher, (2019) Cotton production potential areas, production trends, research status, gaps and future directions of cotton improvement in Ethiopia. Greener J Agric Sci 9:163–170
51. Hardee DD, Herzog GA (1997) 50th Annual conference report on cotton insect research and control. In: Dugger P, Richter DA (eds) Proceedings of belt wide cotton conference. Natl Cotton Counc of Am., Memphis. pp 809–834

52. Hinze LL, Gazave E, Gore MA, Fang DD, Scheffler BE, Yu JZ, Jones DC, Frelichowski J, Percy RG (2016) Genetic diversity of the two commercial tetraploid cotton species in the Gossypium diversity reference set. J Hered 107:274–286
53. Huang J, Hu R, Rozelle S, Qiao F, Pray CE (2002) Transgenic varieties and productivity of smallholder cotton farmers in China. Aust J Agric Resour Econ 46:367–387
54. IAR (Institute of Agricultural Research) (1987) Department of field crops, fibre crops progress report for the period 1985/86, November 1987. IAR, Addis Ababa, Ethiopia
55. ICAC (International Cotton Advisory Committee) (2016) World cotton supply and demand. 1629 K Street NW, Suite 702, Washington, DC 20006 USA
56. ISAAA (International Service for the Acquisition of Agribiotech Applications) (2018). Pocket K No.16: Biotech Crop Highlights in 2017. Manila: International Service for the Acquisition of Agri-biotech Applications (ISAAA).
57. ISAAA (International Service for the Acquisition of Agribiotech Applications) (2019) Global Status of Commercialized Biotech/GM Crops in 2019 (No. ISAAA Brief no. 55)
58. Iqbal MA, Rahman M (2017) Identification of marker-trait associations for lint Traits in cotton. Front Plant Sci 8:86. https://doi.org/10.3389/fpls.2017.00086
59. Jia Y, Sun J, Wang X, Zhou Z, Pan Z, He S, Pang B, Wang L, Du X (2014) Molecular diversity and association analysis of drought and salt tolerance in Gossypium hirsutum L. germplasm. J Integr Agric 13:1845–1853
60. Jiang Y, Guo W, Zhu H, Ruan Y, Zhang T (2012) Overexpression of GhSusA1 increases plant biomass and improves cotton fibre yield and quality. Plant Biotechnol J 10:301–312
61. Jones CA, Sands RD (2013) Impact of agricultural productivity gains on greenhouse gas emissions: A global analysis. Am J Agric Econ, 95:1309–1316
62. Kalivas A, Xanthopoulos F, Kehagia O, Tsaftaris A (2011) Agronomic characterization, genetic diversity and association analysis of cotton cultivars using simple sequence repeat molecular markers. Genet Mol Res 10:208–217
63. Khadi BM, Santhy V, Yadav MS (2010) Cotton: An introduction. In: Zehr UB (ed) Biotechnology in agriculture and forestry. Springer, Heidelberg Dordrecht London New York, pp 1–14
64. Korir NK, Han J, Shangguan L, Wang C, Kayesh E, Zhang Y, Fang J (2013) Plant variety and cultivar identification: advances and prospects. Crit Rev Biotechnol 33:111–125. https://doi.org/10.3109/07388551.2012.675314
65. Kornyeyev D, Logan BA, Allen RD, Holaday AS (2003) Effect of chloroplastic overproduction of ascorbate peroxidase on photosynthesis and photoprotection in cotton leaves subjected to low temperature photoinhibition. Plant Sci 165:1033–1041
66. Kranthi KR (2018) Cotton production practices: snippets from global data 2017. The ICAC Recorder 36:4–14
67. Kranthi S, PRK, Kumar R, Suke R, Chawla S, Kranthi KR (2021) Mitochondrial COI based genetic diversity and phylogeographic structure of whitefly Bemisia tabaci (Gennadius) oncotton in India. Int J Trop Insect Sci, 41:1543–1554
68. Kranthi KR (2012) Bt cotton-questions and answers. Indian Society for Cotton Improvement (ISCI), Mumbai
69. Lacape J-M, Nguyen T-B, Thibivilliers S, Bojinov B, Courtois B, Cantrell RG, Burr B, Hau B (2003) A combined RFLP SSR AFLP map of tetraploid cotton based on a Gossypium hirsutum× Gossypium barbadense backcross population. Genome 46:612–626
70. Li F, Fan G, Lu C, Xiao G, Zou C, Kohel RJ, Ma Z, Shang H, Ma X, Wu J (2015) Genome sequence of cultivated Upland cotton (Gossypium hirsutum TM-1) provides insightsInto genome evolution. Nat Biotechnol 33:524–530
71. Li X, Wang X, Zhao X, Dutt Y (2004) Improvement of cotton fibre quality by transforming the acsA and acsB genes into Gossypium hirsutum L. by means of vacuum infiltration. Plant Cell Rep 22:691–697
72. Li C, Xu X, Dong N, Ai N, Wang Q (2016) Association mapping identifies markers related to major early-maturating traits in upland cotton (Gossypium hirsutum L.). Plant Breeding 135:483–491

73. Liang B, Fan S, Song M, Pang C, Wei H, Yu S (2014) Association analysis of agronomic traits in upland cotton using SSR markers. Cotton Science 26:387–395
74. Lubbers EL, Chee PW (2009) The worldwide gene pool of G. hirsutum and its improvement. In: Paterson AH (ed) Genetics and genomics of cotton. Springer US, New York, NY, pp 23–52
75. Luttrell R (1994) Cotton pest management: Part 2. A US perspective. Annu Rev Entomol 39:527–542
76. Ma Z et al (2018) Resequencing a core collection of upland cotton identifies genomic variationand loci influencing fibre quality and yield. Nat Genet 50:803–813. https://doi.org/10.1038/s41588-018-0119-7
77. Malik W, Ashraf J, Iqbal MZ, Ali Khan A, Qayyum A, Ali Abid M, Noor E, Qadir Ahmad M, Hasan Abbasi G (2014) Molecular markers and cotton genetic improvement: currentstatus and future prospects. Sci World J 2014:1–15
78. Mammadov J, Buyyarapu R, Guttikonda SK, Parliament K, Abdurakhmonov IY, Kumpatla SP (2018) Wild relatives of maize, rice, cotton, and soybean: treasure troves for tolerance to biotic and abiotic stresses. Front Plant Sci 9:886
79. Maqbool A, Abbas W, Rao AQ, Irfan M, Zahur M, Bakhsh A, Riazuddin S, Husnain T (2010) Gossypium arboreum GHSP26 enhances drought tolerance in Gossypium hirsutum. Biotechnol Prog 26:21–25
80. Mubarik MS, Ma C, Majeed S, Du X, Azhar MT (2020) Revamping of cotton breeding programs for efficient use of genetic resources under changing climate. Agronomy 10. https://doi.org/10.3390/agronomy10081190
81. NCCRS (National Cotton Commodity Research Strategy) (2017) National Cotton Commodity Research Strategy, for fifteen years (2016–2030)
82. Nguyen VH, Jonckheere W, Nguyen DT, de Moraes GJ, Van Leeuwen T, De Clercq P (2019) Phytoseiid mites prey effectively on thrips eggs: evidence from predation trials and molecular analyses. Biol Control 137:104012
83. Nie X, Huang C, You C, Li W, Zhao W, Shen C, Zhang B, Wang H, Yan Z, Dai B (2016) Genome-wide SSR-based association mapping for fibre quality in nation-wide upland cotton inbreed cultivars in China. BMC Genomics 17:1–16
84. Page JT, Liechty ZS, Alexander RH, Clemons K, Hulse-Kemp AM, Ashrafi H, Van Deynze A, Stelly DM, Udall JA (2016) DNA sequence evolution and rare homoeologous conversion in tetraploid cotton. PLoS Genet 12:e1006012
85. Parkhi V, Kumar V, Campbell LAM, Bell AA, Rathore KS (2010) Expression of arabidopsis NPR1 in transgenic cotton confers resistance to non-defoliating isolates of Verticilliumdahliae but not the defoliating isolates. J Phytopathol 158:822–825
86. Parkhi V, Kumar V, Campbell LM, Bell AA, Shah J, Rathore KS (2010) Resistance against various fungal pathogens and reniform nematode in transgenic cotton plants expressing Arabidopsis NPR1. Transgenic Res 19:959–975
87. Pasapula V, Shen G, Kuppu S, Paez-Valencia J, Mendoza M, Hou P, Chen J, Qiu X, Zhu L, Zhang X (2011) Expression of an Arabidopsis vacuolar H+-pyrophosphatase gene (AVP1) in cotton improves drought-and salt tolerance and increases fibre yield in the field conditions. Plant Biotechnol J 9:88–99
88. Pereira GS, Sousa RL, Araújo RL, Hoffmann LV, Silva EF, Barroso PAV (2012) Selective fertilization in interspecific crosses of allotetraploid species of Gossypium. Botany 90:159–166. https://doi.org/10.1139/b11-094
89. Prasanna Kumar N, Prasad BR, Bhat PS (2020) Distribution pattern and sequential sampling plan for chilli thrips, Scirtothrips dorsalis Hood (Thripidae: Thysanoptera). Int J Trop Insect Sci 40:131–139
90. Preetha S, Raveendren T (2008) Molecular marker technology in cotton. Biotechnol Mol Biol Rev 3:032–045
91. Rana M, Bhat K (2004) A comparison of AFLP and RAPD markers for genetic diversity and cultivar identification in cotton. J Plant Biochem Biotechnol 13:19–24
92. Rao N, Gaur RK, Singh S, Rachana R, Kumar R, Pandher S, Kaur R, Yadav S, Udikeri SS, Srinivas A (2023) Thrips composition in cotton agroecosystem: a single species or a species complex? Research Square, Preprint.

93. Refera A, Demissie A (1985) A review of research on insect pests of fibre crops in Ethiopia. In: Abate T (ed.) Proceedings of the first ethiopian crop protection symposium on review of crop protection research in Ethiopia. IAR, Institute of Agricultural Research, Addis Ababa, Ethiopia, pp 111–117
94. Rice WR (1987) Speciation via habitat specialization: the evolution of reproductive isolation as a correlated character. Evol Ecol 1:301–314
95. Saeed M, Wangzhen G, Tianzhen Z (2014) Association mapping for salinity tolerance in cotton ('Gossypium hirsutum'L.) germplasm from US and diverse regions of China. Aust J Crop Sci, 8(3):338–346
96. Sahito HA, Shah ZH, Kousar T, Mangrio WM, Mallah NA, Jatoi FA, Kubar WA (2017) Research article comparative efficacy of novel pesticides against Jassid, Amrasca Biguttula biguttula (Ishida) on cotton crop under field conditions at Khairpur, Sindh, Pakistan. Singap J Sci Res 1:1–8
97. Sakhanokho HF, Rajasekaran K (2016) Cotton regeneration in vitro. In: Fibre plants: biology, biotechnology and applications. Springer, pp 87–110
98. Sanjaya, Satyavathi VV, Prasad V, Kirthi N, Maiya SP, Savithri HS, Sita GL (2005) Development of cotton transgenics with antisense AV2 gene for resistance against cotton leaf curl virus (CLCuD) via Agrobacterium tumefaciens. Plant Cell Tiss Organ Cult 81: 55–63.
99. Sarwar M, Sattar M (2016) An analysis of comparative efficacies of various insecticides on the densities of important insect pests and the natural enemies of cotton, Gossypium hirsutum L. Pak J Zool 48(1):131–136
100. Seelanan T, Schnabel A, Wendel JF (1997) Congruence and consensus in the cotton tribe (Malvaceae). Syst Bot 22(2):259–290
101. Senchina DS, Alvarez I, Cronn RC, Liu B, Rong J, Noyes RD, Paterson AH, Wing RA, Wilkins TA, Wendel JF (2003) Rate variation among nuclear genes and the age of polyploidy in Gossypium. Mol Biol Evol 20(4):633–643
102. Shu B, Kong F, Zhou Y, Zhang G, Zhang Q, Wang X (2001) Genetic diversity analysis of representative elite cotton varieties in three main cotton regions in China by RAPD and its relation with agronomic characteristics. Scientia Agricultura Sinica 34:597–603
103. Sica M, Gamba G, Montieri S, Gaudio L, Aceto S (2005) ISSR markers show differentiation among Italian populations of Asparagus acutifolius L. BMC Genet 6:1–7
104. Siddiqui JA, Fan R, Naz H, Bamisile BS, Hafeez M, Ghani MI, Wei Y, Xu Y, Chen X (2022) Insights into insecticide-resistance mechanisms in invasive species: Challenges and control strategies. Front Physiol 13:1112278. https://doi.org/10.3389/fphys.2022.1112278
105. Stehle S, Schulz R (2015) Agricultural insecticides threaten surface waters at the global scale. Proc Natl Acad Sci 112:5750–5755
106. Subramanian A, Qaim M (2009) Village-wide effects of agricultural biotechnology: the case of Bt Cotton in India. World Dev 37:256–267
107. Tang W, Ao L, Zhang H, Shan B (2014) Accumulation and risk of heavy metals in relation to agricultural intensification in the river sediments of agricultural regions. Environ Earth Sci 71:3945–3951
108. Tenguri P, Kranthi S, Raghavendra K, Kumar R, Suke R, Chawla S, Kranthi K (2020) Mitochondrial COI based genetic diversity and phylogeographic structure of whitefly Bemisia tabaci (Gennadius) on cotton in India. Int J Trop Insect Sci 40:112. https://doi.org/10.1007/s42690-020-00354-x
109. Terefe G, Jamornmarn S, Hormchan P, Daorai A, Impithuksa S (2004) Evaluating four bioassay techniques for insecticide resistance monitoring in Helicoverpa armigera (Lepidoptera: Noctuidae). Agric Nat Resour 38:44–56
110. Toda S, Murai T (2007) Phylogenetic analysis based on mitochondrial COI gene sequences in Thrips tabaci Lindeman (Thysanoptera: Thripidae) in relation to reproductive forms and geographic distribution. Appl Entomol Zool 42:309–316
111. Tuttle JR, Nah G, Duke MV, Alexander DC, Guan X, Song Q, Chen ZJ, Scheffler BE, Haigler CH (2015) Metabolomic and transcriptomic insights into how cotton fibre transitions to secondary wall synthesis, represses lignification, and prolongs elongation. BMC Genomics 16:1–28

112. Tyagi P, Gore MA, Bowman DT, Campbell BT, Udall JA, Kuraparthy V (2014) Genetic diversity and population structure in the US Upland cotton (Gossypium hirsutum L.). Theor Appl Genet 127:283–295
113. Tyagi K, Tyagi I, Patidar A, Singha D, Kaczmarczyk-Ziemba A, Banerjee D, Kumar V (2022) Gut microbial composition in developmental stages of gall inducing thrips Gynaikothrips uzeli and associated plant pathogenesis. Saudi J Biol Sci 29:1439–1446
114. USGS (U.S Geological survey) (2007) Investigating the environmental effects of agriculture practices on natural resources (Scientific Contributions of the U.S. Geological Survey to Enhance the Management of Agricultural Landscapes No. Fact Sheet 2007–3001.). USA
115. Ulloa M, Cantrell R, Percy R, Zeiger E, Lu Z (2000) QTL analysis of stomatal conductance and relationship to lint yield in an interspecific cotton. J Cotton Sci 4:10–18
116. Umbeck P, Johnson G, Barton K, Swain W (1987) Genetically transformed cotton (Gossypium hirsutum L.) plants. Bio/Technology 5:263–266
117. Wang M, Tu L, Yuan D, Zhu D, Shen C, Li J, Liu F, Pei L, Wang P, Zhao G (2019) Reference genome sequences of two cultivated allotetraploid cottons, Gossypium hirsutum and Gossypium barbadense. Nat Genet 51:224–229
118. Wang X-Y, Yang X-M, Lu B, Zhou L-H, Wu K-M (2017) Genetic variation and phylogeographic structure of the cotton aphid, Aphis gossypii, based on mitochondrial DNA and microsatellite markers. Sci Rep 7:1920
119. Wang X, Yu Y, Li W, Guo H, Lin Z, Zhang X (2013) Association analysis of yield and fibre quality traits in Gossypium barbadense with SSRs and SRAPs. Genet Mol Res 12:335362
120. War AR, Taggar GK, War MY, Hussain B (2016) Impact of climate change on insect pests, plant chemical ecology, tritrophic interactions and food production. Int J Clin Biol Sci 1:16–29
121. Wilson L (2020) Estimating abundance, impact, and interactions among arthropods in cotton agroecosystems. In Handbook of sampling methods for arthropods in agriculture. CRC Press, pp 475–514
122. Wu J, Jenkins JN, McCarty JC, Saha S, Percy R (2008) Genetic association of lint yield with its components in cotton chromosome substitution lines. Euphytica 164:199–207
123. Xu SM, Brill E, Llewellyn DJ, Furbank RT, Ruan YL. 2012. Overexpression of a potato sucrose synthase gene in cotton accelerates leaf expansion, reduces seed abortion, and enhances fiber production. Mol Plant 5: 430–441.
124. Yang Y, Li Y, Wu Y (2013) Current status of insecticide resistance in Helicoverpa armigera after 15 years of Bt cotton planting in China. J Econ Entomol 106:375–438
125. Yang Z, Qanmber G, Wang Z, Yang Z, Li F (2020) Gossypium Genomics: Trends, Scope, and Utilization for Cotton Improvement. Trends Plant Sci 25:488–500. https://doi.org/10.1016/j.tplants.2019.12.011
126. Yoo M-J, Wendel JF (2014) Comparative evolutionary and developmental dynamics of the cotton (Gossypium hirsutum) fibre transcriptome. PLoS Genet 10:e1004073
127. Yuan D, Tang Z, Wang M, Gao W, Tu L, Jin X, Chen L, He Y, Zhang L, Zhu L (2015) The genome sequence of Sea-Island cotton (Gossypium barbadense) provides insights into the allopolyploidization and development of superior spinnable fibres. Sci Rep 5:17662
128. Yunus M, Yousuf M (1979) Insect and mite pests of cotton in Pakistan. Pak J Agric Sci 16:67–71
129. Zeng L, Meredith WR, Gutiérrez OA, Boykin DL (2009) Identification of associations between SSR markers and fibre traits in an exotic germplasm derived from multiple crosses among Gossypium tetraploid species. Theor Appl Genet 119:93–103
130. Zhang H, Mittal N, Leamy LJ, Barazani O, Song B (2017) Back into the wild—Apply untapped genetic diversity of wild relatives for crop improvement. Evol Appl 10:5–24

Irrigated and Mechanized Cotton Production

Getnet Belay Tesema

Abstract Cotton holds a unique position in Ethiopia's agrarian economy. Ethiopia has enormous potential for cotton production. Ethiopia has favorable weather and topography for cotton cultivation and production. Cotton (*Gossypium hirsutum* L.) is currently widely grown in irrigated lowlands on large-scale farms and in warmer mid-altitudes on small-scale farms under rainfed conditions. According to a recent Ministry of Agriculture study, there is 3,000,810 Ha of land suitable for cotton production, whereas it is currently limited to a maximum of 100,000 ha. Cotton demand is expected to remain strong in Ethiopia, and there is a clear need to improve productivity in future to meet this rising demand. Ethiopian cotton production is hampered by low productivity caused by 40% and 60% rainfed and irrigated cultivation, respectively. Following that was a small farm size, an increase in pest and disease, and a labor-intensive method of cotton cultivation. Because labor costs are rapidly rising in Ethiopia, mechanization in cotton cultivation will be critical in keeping costs under control. Furthermore, high-density planting will increase productivity. However, high-density planting in conjunction with mechanization necessitates the development of sympodial cotton hybrids as well as a complete transformation in agronomy practices.

Keywords Mechanization · Middle Awash · Tendaho · Setit-Humera · Irrigation · Drip irrigation

1 Introduction

Cotton (*Gossypium hirsutum* L.) grown in Ethiopia is also the primary raw material for the international textile industry, the most valuable natural fiber, and the world's second largest oil-seed crop [1, 2]. This cotton needs irrigation for the following main reasons:

G. B. Tesema (✉)
Ethiopian Institute of Textile and Fashion Technology (EiTEX), Bahir Dar University, Bahir Dar, Ethiopia
e-mail: getnet.belay@uni-bremen.de

K. Murugesh Babu et al. (eds.), *Cotton Sector Development in Ethiopia*, Textile Science and Clothing Technology, https://doi.org/10.1007/978-981-99-9149-5_5

(1) To provide a more consistent yield from year to year.
(2) To protect the crop's yield potential—being short 25.4 mm of water at the wrong time can easily result in the loss of 34 kg of seed and 23 kg of fiber.

Irrigation benefits both the producer and society by increasing yield per unit of land area and by providing a consistent and dependable source of food and fiber. Irrigation protects crops from poor crop performance and/or failure caused by insufficient and/ or untimely rainfall. In today's competitive markets, where significant investment has been committed at cotton planting time, protecting against rainfall uncertainties is highly desirable. Irrigation also makes agro-chemical management easier by allowing for the use of fertigation and chemigation practices.

It is estimated that approximately 70% of the world's freshwater consumption is for irrigation (all crops, not just cotton), and for good reason. Irrigation can increase yield as well as stabilize yield and quality by ensuring adequate soil water throughout the growing season, or at least during critical growth stages in areas where water resources are limited.

Reference [3] discovered that when dry conditions existed, cotton had a linear relationship with irrigation level. Furthermore, they reported a significant response even in a dry year,however, the relationship was quadratic rather than linear, indicating a decreased response at higher irrigation application rates. Reference [4] discovered a significant increase in cotton lint yields in response to irrigation. They reported that providing low levels of irrigation increased yields by nearly 91 kg per 4046.86 m^2 when compared to dry-land. However, they found no benefit from high levels of irrigation. As a result, properly managed irrigated mechanized cotton farming is a critical activity required for success in today's global competitive marketing system. Mechanization in cotton cultivation was nearly completed 20 years ago, from soil preparation for planting to the stage just before harvesting. Of course, there have been advancements since then as a result of new technology in this field.

Mechanical harvesting began experimentally with pickers and strippers in 1963, but there were numerous challenges to overcome, including very small agricultural holdings, land partitioning (fragmentation), high machinery costs, and so on. Cotton pickers were first used in a significant amount in 1972–73.

2 History of Mechanized Cotton Production in Ethiopia

The process of agricultural mechanization necessitates the gradual introduction of capital into the production function. Essentially, the agricultural production function combines land, labor, and capital under the supervision of the framer acting as manager. The nature of changes in the production function is determined by the intended function of the newly introduced capital. Capital invested in labor-saving equipment becomes a substitute for labor and is considered a labor-saving technological change. Conversely, agricultural equipment that improves cultivation without displacing labor becomes a substitute for land, resulting in a land-saving technological change. A change involving capital as a substitute for neither land nor labor is a neutral technological change. This latter type of change has the potential to boost the productivity of all factors.

Understanding the effects of mechanization and the possibilities for successfully implementing such changes necessitates a thorough description of the farming systems into which the changes of mechanization may be introduced.

Hand tool and draught animal power mechanization were the main methods of cotton production in Ethiopia during the Emperors' period (except for some mechanized farms such as the Middle Awash, Tendaho, and Setit-Humera). Cotton production of this type was widespread during the so-called "Derg," a socialist regime that ruled from 1974 to 1991. During this period, the socialist command economy was followed by the Land Proclamation of 1975, and acknowledging the importance of large-scale mechanization, the socialist regime established large mechanized (third category) farms, then known as "State farms," by confiscating private farms and establishing new ones [5].

2.1 The Middle Awash Mechanized Cotton Production

The Middle Awash mechanized cotton production scheme was based on irrigated cotton technology and was implemented prior to the "Derge" regime in 1969. The total area on the Amibara Plaines was 12,000 hectares. The scheme was intended to gradually introduce regime settlers to the complexities of mechanical power agricultural production to large-scale cotton production. The Middle Awash Settlement scheme's goal was primarily technical, with sociological implications. It was founded on the technology of irrigated cotton cultivation. The goal of growing cotton on land previously grazed by pastoralist herds was to encourage a number of these people to settle and become sedentary farmers [6].

There were several major factors that favored the settlement's success. The area is sparsely populated and undeveloped, aside from being a traditional grazing area for Afar herds and flocks. The scheme's goal is to settle nomadic people in a more prosperous way of life in an area where social pressures are increasingly hostile to nomadic traditional life. Economically, the scheme was based on cotton production in a relatively fertile region of a country with a high internal demand for cotton. There is also the huge benefit of being close to the Melka Werer Research Station. The station was established several years prior to the 1964 settlement and has experimental experience from which local farming enterprises can benefit.

2.2 *The Tendaho Mechanized Cotton Production*

The Tendaho Plantation Share Company was formed as a joint venture between Imperial Ethiopian Government agencies, Ethiopian shareholders, and Mitchell Cotts and Company with a share capital of $2,472,000 dollars. It is an example of a large-scale commercial plantation. The entire land concession in the Danakil Desert covers 10,000 hectares, and the plantation covered 5,500 hectares in 1968/69. During that cultivation stage, this case represents a highly mechanized operation, with hand labor reaching a peak of 7,000 workers at the peak of the raw cotton-picking season in November. For the 1968/69 season, 90,000 quintals of cotton were estimated, with ginned cotton sold in the domestic market and cotton seed primarily exported. In 1968/69, the estimated budget production costs of $366,830 would be paid out in day laborer wages. Export potential is critical, especially because Tendaho was producing high-quality cotton that competed in the world market at the time [7]. Tendaho plantation demonstrated intensive and efficient management capabilities and was one of the most successful private commercial enterprises at the time.

During its initial development and growth from 500 to 4,000 hectares in size, the Tendaho plantation required 55–60 hp. tractor for every 11 hectares under cultivation. Since expanding to over 5,000 hectares, the managers have been able to increase the amount cultivated per tractor to 150–160 hectares due to increased field efficiency, better operators, more level fields, fewer vet spots, and less breakage and downtime. At least half of this increase in efficiency can be attributed to improved training and experience among tractor operators who take on more responsibility for their equipment.

2.3 *Machine Field Efficiency*

Records showing work rates and average field capacities for various field operations from 1966 to 1968 are particularly revealing. Recognition of the importance of tractor operator training, recognition, and experience contributed to part of the indicated increase in efficiency. Table 1 displays data on the efficiency of field operations. Unless otherwise specified, all of the machines in the table are powered by a 58 hp. Massey Ferguson 165 tractor or equivalent.

The Massey Ferguson 165 (MF165) (Fig. 1) is a row-crop tractor that was manufactured in Massey Ferguson's Detroit, Michigan plant from 1964 until 1975. It, like the MF135, was a popular tractor on the market at the time.

The MF165 was available with one of four engines: the 3.3-L Perkins diesel engine, the 2.9-L Continental gasoline engine, the 3.5-L Perkins diesel engine, or the 3.5-L Perkins gasoline engine. Each of the four engines has four cylinders as well as liquid cooling. The bore/stroke of the 3.3-L Perkins is 91 × 127 mm, while the 2.9-L Continental is 91 × 111 mm, and both 3.5-L Perkins are 98 × 114 mm. Each of the four engines can produce 58.3 horsepower. It can hold 70 L of fuel. This

Table 1 Comparative efficiency of field operations: Tendaho Plantations Share Company 1966–1968

Field operation	Machine	Time period (hrs.)	Capacity (hectares)		
			1966	1997	1968
Plowing (2 cm)	4-furrow disk plow, mtd	10	2–3	4	4–5[a]
	5-furrow moldboard plow	10	-[d]	-[d]	6–8[a]
Harrowing	No. 52 Tandem, 3-m cut	10	8	20	20–25
Disking	M-36 wide -level, one way (24 disk)	10	10	25	30–35
Planting	4-row unit planter, (91 cm. rows, no fertilizing or spraying)	12	8	14	20–25
	6-raw planter (made from 1 1/2, 4-raw machines)	12	-[d]	-[d]	35[b]
Inter-row cultivation	3-row rear-mtd. rigid-tooth cultivators	8	6	16	16[c]
	4-row mid and rear-mtd. spring-tooth cultivators	8	8	20	20[c]
Slashing	182 cm. rotary single–blade	10	5	7–10	15
	3-m rotary double-blade	10	d	20	25
Leveling	3 m Eve-son land plane	10	-[e]	-[e]	-[e]
Hauling	2.5 × 3.0 × 2.5-m high bulk trailers (crop put in bags in field)	8	-[f]	-[f]	-[f]

[a] With 10 plows, 50 ha./day during the 1968 season were averaged. The HF-175 tractor was used with the 5-furrow moldboard plow
[b] With an experienced driver and an assistant to watch the planters, up to 48 ha. has been planted in one 12-h day
[c] Because of the good, straight, and even job of planting, 3-row cultivators in fields planted with 4-row planters, and also 4-row cultivators in fields planted with 6-row planters were used
[d] Not in use at that time
[e] Variable—depends on field condition and unevenness
[f] Variable—depends on distance and road conditions
Source Kline et al. [6]

tractor has a 4 × 2 two-wheel drive chassis. It features power steering, mechanical dry disc brakes, and an open operator station. In terms of transmission, the MF165 was available with one of three options: a 12-speed partial power shift, a 6-speed, or an 8-speed.

Reduced tillage systems will allow farmers to use smaller tractors for longer periods of time and with fewer repairs as cotton farms are mechanized. Aside from lower maintenance and repair costs, smaller horsepower and longer operating time of tractors imply lower total tractor cost per hour, and thus lower production costs. Estimating the relationship between various tractor cost variables (age, horsepower,

Fig. 1 Massey Ferguson 165 Tractor. *Source* https://gwtractors.com.au/massey-ferguson-165-tractor/#:~:text=Part%20of%20the%20100%20Series,the%20market%20during%20its%20time

repair cost, total cost per hour) is a critical activity that must be completed prior to the implementation of mechanized cotton farms.

2.4 The Setit-Humera Mechanized Cotton Production

In 1954, mechanized farming began in the northwest Ethiopia lowlands. It was started on the initiative of private entrepreneurs and has since been developed by them. The government's role has been limited to the allocation of land to farmers and the general administration of the area. Early development was slow, but between 1963 and 1968, there was rapid growth. By 1968, over 400 tractors were in use, and approximately 100,000 ha of land had been planted. Despite poor communications and a general lack of facilities, tractor dealers have either set up workshops or appointed dealers to stock fast-moving spares at Setit-Humera, and the service and spares situation for farm tractors was satisfactory.

The farming system was extensive, with low inputs and outputs per unit of land. Many farmers and their laborers come from cities and neighboring highland areas, and they only live on the farms during the crop season. Farmers use modern tractor-drawn equipment to cultivate and plant large areas at low cost, but weeding and harvesting are not automated. Fertilizers and pesticides were not used, despite the fact that cotton pests were becoming more of a problem as the area under cotton cultivation expanded. Economic responses to fertilizer were achieved as yields increased through the use of improved machinery, better weeding, and new varieties.

The estimated yield was 300 kg/ha. Cotton seed was ginned in Asmara (480 km from Setit-Humera). The Government owned the land in mechanized farming, and occupiers were assigned land by the Governor of the Woreda (District) through an annual permit system. However, no titles or leases have been issued, and the only condition of occupation is that the occupier agrees to pay education and health taxes. As a result, farmers' rights to the land were not defined, there was no incentive for them to invest in long-term improvements, and land could not be used as collateral for the credit most farmers require to carry out improvements. The government was aware of the issue and was informed during negotiations that satisfactory arrangements for the grant of long-term leases to farmers would be required for any lending for a second-stage project. (International Bank for Reconstruction and Development, International development association, Humera agricultural development project, Ethiopia, 1970).

3 Mechanization Status on Cotton Crop

Cotton is an important commodity in Ethiopian agriculture, and it has played an important role in Ethiopia's cultural and industrial development history. To comprehend Ethiopia's future role in domestic and international cotton markets, a better understanding of its cotton production system is required. Despite the fact that Ethiopia is Africa's second largest cotton consumer, cotton production productivity is very low when compared to the global average. It is a major concern for policymakers because the cotton sector is important in both social and economic aspects of Ethiopian society. The existence of lower-than-world-average yields in Ethiopia has been attributed to a variety of factors, including insufficient inputs, a lack of awareness about modern cotton cultivation practices among Ethiopian farmers, limited irrigation facilities, a lack of proper timing of field operations, and an overreliance on labor to cultivate cotton. Along with the aforementioned factors, a labor shortage in some large-scale farming areas of Ethiopia where picking and weeding are done by hand is reducing cotton crop profitability. In this context, a better understanding of Ethiopia's cotton sector and the impact of mechanization on cotton cultivation is required to assess Ethiopia's competitive position in the international market.

Cotton harvesting mechanization is dependent not only on the availability of suitable harvesters but also on the availability of appropriate cotton varieties, changes in agronomic practices such as seed rate, nutrient application, and so on, and finally on its economic feasibility in Ethiopia. No other African country has the same potential as Ethiopia to grow a diverse range of cotton varieties using rainfed or irrigated systems, cotton cultivation in various agro-climatic zones. Paradoxically, one single variety (DP 90) is currently grown on more than 90% of cotton cultivation areas, despite the fact that seed quality is poor and ginning outturn (GOT) is low by African standards, affecting profitability for both producers and ginners [4].

Farmers in many parts of Ethiopia still use human labor for many operations such as planting, weeding, and picking, as well as inefficient farm implements/machinery. Machinery adoption in farm operations is lagging due to a variety of factors such as a lack of credit to purchase expensive machinery, farmers' small land holdings (generally less than 1 ha), and a lack of technical knowledge and skills to operate complex farm machinery. The persistent low yields in Ethiopian cotton production are also attributed to a scarcity of disease-resistant and high-yielding cotton varieties.

Recently, in Ethiopia, a variety of cotton production systems coexist under irrigation (26% of total cotton cultivation area) or rainfed conditions, ranging from very large farms developed by investors (more than 200 ha of cotton) to very small family farms (generally less than 1 ha). The large-scale farms, not the small ones, are heavily mechanized (Fig. 2). Weeding and harvesting are still done by hand and require a lot of effort. In comparison to major African countries, average seed cotton yields are relatively high, at 2.5 tons/ha with irrigation and 1.5 tons/ha with rainfed conditions. The global average is 2.6 tons/ha (FAO production yearbook 2020). Large-scale farms are frequently established with the funds of investors whose goal is to make good and quick profits and returns. Prior investments were made with a view to short-term profitability (field preparation, machines, management), and the cropping system in place is not environmentally friendly or sustainable. Few temporary shelters or roads, wide open fields with rare trees, intensive plowing, and mono-cropping do not ensure the long-term sustainability of most of these farms (Fig. 3).

Fig. 2 Mechanized farm in the Afar region. *Source* Lançon and Woldu [8]

Fig. 3 Large-scale farms with open fields in the rainfed areas, Setit-Humera region. The crop has been broadcast. *Source* Lançon and Woldu [8]

3.1 Precision Agriculture Adaption to Mechanized Cotton Farms

Precision Agriculture (PA) is a management strategy that allows farmers to adjust input use and cultivation practices to match within-field variability in soil and crop conditions [9]. This ensures that the inputs are only applied at the required rates at various parts of the field with varying nutrient content and crop requirements. As a result, precision agriculture improves input use efficiency while lowering the potential negative environmental impact of agricultural chemical overuse [10]. In terms of input management, PA assists producers in making more informed management decisions, and site-specific knowledge allows producers to limit input use in accordance with the crop's spatial and temporal requirements [11].

Modern technological advances, such as Global Positioning Systems (GPS) and Geographical Information Systems (GIS), have added a new dimension to the practice of PA. Most researchers now regard PA as a system comprised of various components such as auto steering technology for tractors, automatic section control for sprayers and planters, geo-referenced soil sampling, various methods of soil variability analysis, and variable rate input application. However, in the traditional sense, PA entails collecting within-field variability data, processing this variability data to assess the extent and distribution of variability, and, if necessary, responding to this variability variable rate application of inputs to match the variability [12]. Common variability detection practices include using yield monitors, soil maps, geo-referenced soil grid and zone sampling, aerial photos, or satellite imagery to identify variability in soil fertility, pH of the soil, crop vigor, or moisture stress. Once the variability within the field has been detected and analyzed, the information is used to apply inputs such as fertilizers, lime, pix, or irrigation water in such a way that each portion of the field receives the input in the required quantities. Adopting PA strategies is important not only for increasing farm profitability and sustainability but also for protecting the environment because the inputs are not applied in excessive quantities, limiting the potential leaching of chemicals into water streams.

3.2 Irrigation Adaption to Mechanized Cotton Farms

Cotton (*Gossypium hirsutum L.)* is the primary raw material for the international textile industry, the most valuable natural fiber, and the world's second largest oilseed crop [1, 2]. Cotton yield is influenced by a variety of factors including edaphoclimatic constraints, genotypes, and crop management practices. Cotton cultivation is primarily rainfed in the majority of the world's producing regions, including Ethiopia. Rainfed cotton cultivation accounts for 74% of cotton cultivation in Ethiopia. Because of the high crop water demand (Fig. 4), the water deficit caused by semi-arid region droughts is the primary factor limiting high yields.

Irrigation is critical to ensuring the sustainability of production in water-stressed areas, particularly when combined with efficient water consumption and economic viability [9, 10, 26, 28]. Cotton, on the other hand, has a relatively long cycle and requires a lot of water when grown under full irrigation [13]. The average irrigation requirement for surface-irrigated cotton is reported to be 6000–7000 m3 ha^{-1}, depending on soil, weather conditions, and seasonal rainfall [14, 15].

Some parts of the world, such as cotton-producing areas in the United States of America, have low or no irrigation requirements due to an adequate distribution of rainfall during the growing season. However, several producers use Supplemental Irrigation (SI) to reduce drought stress, reduce risk, and improve yield stability across a wide range of environmental conditions [16]. The need for SI is obviously greater

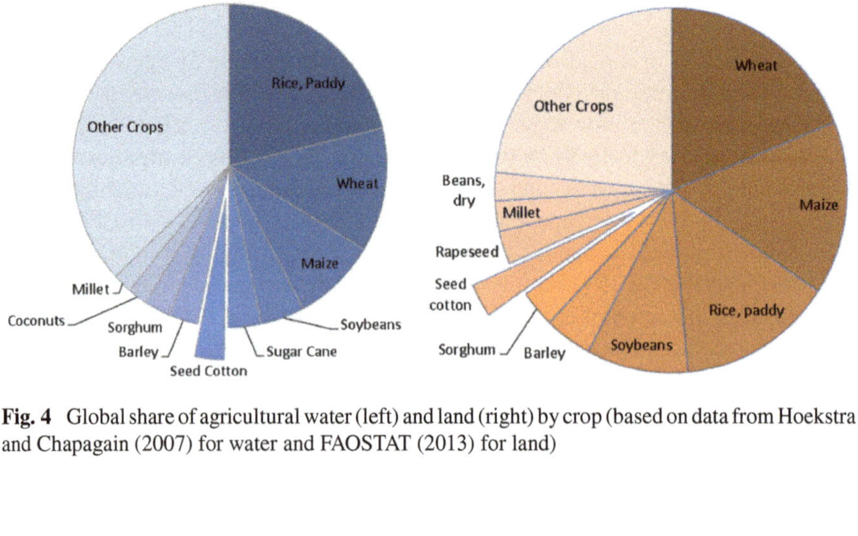

Fig. 4 Global share of agricultural water (left) and land (right) by crop (based on data from Hoekstra and Chapagain (2007) for water and FAOSTAT (2013) for land)

in semi-arid regions due to recurring droughts and long periods of dry spells, which tend to worsen as a result of global climate change [17].

Cotton crops are grown in six major agro-ecological zones in Ethiopia. The major irrigated cotton-producing agro-ecological zones are the hot to warm arid lowlands and the hot to warm semi-arid lowlands, which cover approximately 50% of the total cotton cultivation area. The remaining 50% of cotton coverage is covered by the major rainfed cotton producing agro-ecologies of hot to warm humid lowlands, hot to warm moist lowlands, hot to warm semi-humid lowlands, and Tepid to cool arid mid highlands. However, some rainfed areas, such as tepid to cool arid mid-highland, are marginal cotton production zones because the temperature in these zones occasionally falls below optimum, whereas hot to warm semi-humid lowlands zones are moderate production zones (Cotton research strategy 2016–2030, Ethiopian institute of agricultural research [14]. Implementing a supplement irrigation system in these zones may be beneficial in terms of reducing drought stress, lowering risk, and improving yield stability.

In Ethiopia, an irrigated large cotton farm is currently located in the Afar region. The farm was established in 2008. It now grows 250–350 ha of cotton and employs approximately 90 permanent employees, including a farm manager. Until the crop is in its early stages, mechanization is fully developed, but weeding and picking require 400–600 temporary workers. After plowing, ridging, and pre-planting irrigation, the cotton crop is machine planted in rows at 90 20 cm spacing in May–June. At 3 p.p. hole, the crop is tinned. Irrigation is done every 2 or 3 weeks, depending on the growth stage and water requirement, which is higher during the flowering period. Pest control necessitates 8–12 sprays. In 2019, the average yield of seed cotton per hectare was 3.3 tons.

The irrigation of cotton plants in this mechanized cotton farming region must be optimized through farm management schemes. The installation of modern, efficient irrigation systems should be discussed in light of recurring droughts, as well as a potential opportunity to increase cotton yields.

3.3 Irrigation Management Trends for Sustainable Cotton Production

Geospatial technology advancements have now provided global estimates of the level of water stress in the world. When viewing cotton through the lens of water scarcity indices, it is critical to consider the importance of irrigation to cotton in the region in question, as well as the sustainability of the irrigation water source if irrigation is required.

Irrigation can increase productivity even in humid areas when rainfall is delayed during the season or due to infrequent drought conditions [18]. Irrigation water management is an important tool for increasing land productivity and ensuring that no other inputs are wasted. New technologies have given producers several new tools to use in developing sustainable crop water management strategies. Companies that have taken advantage of widely distributed cellular networks to affordably transmit data from sensor networks monitoring water status in the field near-real time are one example (for example, [19] report recent work at the University of Zimbabwe). This provides farmers with easy access to the water status of the plants in their fields, allowing water to be applied only when necessary.

New technologies are also improving the precision of irrigation water delivery. Subsurface drip irrigation is one example that has been particularly successful for cotton in the southwestern United States, where it is used on over 100,000 ha of cotton [20]. Water is delivered through tubes buried 20–40 cm below the soil surface and beneath the planted row in this system. Water can be applied in small amounts as needed to the crop, and virtually no water is lost due to evaporation.

Another example is the use of global positioning system (GPS) technologies to map changes in the soil within a field and then control sections of an irrigation pivot to apply the volume of water required for the crop growing in that soil type [7]. The combination of sensors and irrigation controllers has resulted in systems that allow for completely automated control of irrigation water. Farmers can now purchase integrated control and wireless data handling systems from all of the major center pivot manufacturers.

3.4 Drip Irrigation and Conservation Tillage

Drip irrigation is a nearly 100% efficient water delivery system. The drip system's main feature is the direct application of water to plant roots via pipes on or below the soil surface (Figs. 5 and 6). This reduces evaporation and increases water use [21]. Conservation tillage is a crop production system that reduces field cultivation. This system frequently employs cover crops, which are planted prior to cotton, to reduce soil erosion and to conserve and trap rainfall within the field. Some cotton pests such as the cotton aphid, Aphis gossypii Glover (Homoptera: Aphidae), are more abundant in non-water stressed plants than in cotton plants grown in dryland [8], whereas the pests such as banded winged whitefly, Trialeurodes abutiloneus Haldeman (Homoptera: Aleyrodidae), are more abundant in water-stressed cotton [19].

The drip irrigation system was first developed in Israel in the 1960s, but it wasn't until the late 1970s that an Israeli company developed the integrated dripping pipe system that is still in use today in field crops (Fig. 7).

This dripping pipe is a continuous pipe with drippers molded every 60–100 cm that comes on 5 km rolls. Farmers install the pipes in the field at the start of the irrigation season and remove them before harvesting. Farmers can use the same pipe for about 20 seasons, so the system's annual cost is reasonable. Water savings range from 20 to 30% when compared to other methods.

Fig. 5 An irrigated California cotton field with hose pipes. *Source* photo of cotton drip irrigation system-Google Suche

Fig. 6 Drip irrigation successful on Pima cotton in El Paso Valley. *Source* photo of cotton drip irrigation system-Google Suche

Fig. 7 Drip irrigated cotton field. Source: M. Yogev, Irrigation management under cotton shortage. The Israel cotton production & marketing board Ltd. Netivot 3, Herzlia, Israel. *Source* 33rd International Cotton Conference: March 16–18, Bremen, Germany (2016)

Farmers are recently using a different method to save more water by burying the pipe at a depth of 40 cm, providing water directly to the root zone while leaving the soil surface dry and avoiding evaporative water loss.

4 Effect of Mechanization on the Cotton Industry

Cotton mechanization practices, cotton ginners, and the cotton industry are all interconnected. Ethiopia has developed agricultural mechanization based on cotton production and the textile industry. Ethiopia is the second largest consumer of cotton in Africa, after Egypt, and a net importer to meet the needs of the country's textile mills. Domestic spinning industries will drive the growth of cotton production. Ethiopia has a lot of potential to grow its spinning industry and become one of the biggest industrial users of cotton. Cotton production has a high potential in the country, and labor and electricity costs are low. Mechanization applications have already begun in some areas. It is necessary to introduce and/or increase the number of particularly powerful and new tractors, pneumatic planters, mounted and self-propelled sprayers for defoliant applications, self-propelled cotton pickers, and so on.

However, there may be some difficulties in mechanized cotton production. Production costs may rise sharply while cotton prices fall concurrently. This contradictory situation may have an impact on farmers and industrialists who invest in technological machines in the cotton industry, such as seeders and self-propelled cotton pickers. To address these issues, many development institutions, practitioners, and experts believe that investments in mechanization technologies such as tractors and cotton harvesting machines, such as tractors and cotton harvesting machines, are required to help Ethiopia emerge from agricultural stagnation [22].

The potential effects of mechanized cotton production provide an interesting field of speculation. It has also been established that cotton mechanization, particularly cotton harvesting, has a direct impact on the quality of cotton fiber, posing a challenge to the textile industries. Machine harvesting effects on neps have been studied, including the effects of harvest timing [23–25] and harvester setup [26], as well as comparisons between spindle and stripped harvested cotton [15, 22, 27] also demonstrated that the level of neps in spindle machine-harvested cotton was related to fiber maturity and linear density caused by differences in boll maturity at harvest. Reference [27] varied the timing of harvest aids to produce differences in boll maturity. The relationships between neps and fiber maturity, fiber linear density (fineness), or micronaire are well established [17, 27, 28] and show that fewer neps are associated with fiber that is more mature as measured by a higher maturity ratio, fiber linear density (fineness), or micronaire. After the ginning process, nep levels were measured in all of these relationships. Increased mechanical handling of lint during the ginning process also leads to an increase in neps, according to studies [23, 27, 29, 28], and differences in neps can be attributed to fiber maturity at harvest.

Lint contamination may also occur as a result of poor cotton production practices. Lint contamination has recently become a major problem for the global cotton industry, affecting both the textile and raw cotton industries. Cotton contamination has an immeasurable total cost. Textile mills around the world have spent an estimated $200 million on equipment that will detect some of the major contaminants in recent years. This is money only to help minimize the problem; it does not include revenue lost due to broken contracts, loss of business, or supplier loss of confidence, despite the fact that effective contamination devices are in place. Even though contamination detection devices are only partially effective, these investments were made.

Lint contamination can appear in a variety of ways in cotton. There are many suspects. Plastic twine, module cover materials (Fig. 8), plastic shopping bags, sewing twine, trash in irrigation ditch liners, sugar, and honeydew (Fig. 9), and colored clothing are all potential contaminants.

The fabric inspection line (Fig. 10, left) is the first place where contaminants are visible in the textile process after the yarns have been woven or knitted, scoured, bleached, and dyed. This is an expensive place to look for contaminants because one contaminated bale tainted several hundred thousand pounds of cotton. Contamination in the finished garment is even more expensive. In this case (Fig. 10, right), we're looking at men's shirts priced between $25 and $60. The garments are practically useless due to contaminants.

Fig. 8 Module cover materials (left) and trashes in cultivation field (right) as a source of lint contaminants. *Source* Andrew G. Jordan: Lint contamination a serious threat to U.S. cotton. 2004 Beltwide Cotton Conference, San Antonio, TX January 5–9

Fig. 9 Aphids and Whiteflies (left) as a source of cotton stickiness and baling twine (right) as a source of lint contaminants. *Source* Andrew G. Jordan: Lint contamination a serious threat to U.S. cotton. 2004 Beltwide Cotton Conference, San Antonio, TX January 5–9

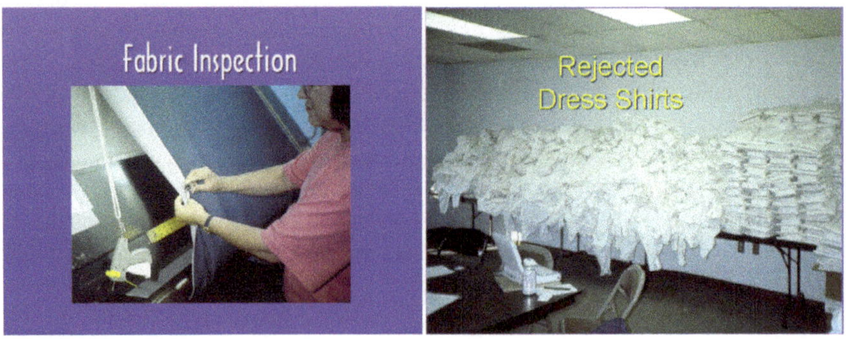

Fig. 10 Fabric inspection (left); Rejected dress shirts (right). *Source* Andrew G. Jordan: Lint contamination a serious threat to U.S. cotton. 2004 Beltwide Cotton Conference, San Antonio, TX January 5–9

5 Best Mechanization Practices for Cotton Industry

Every cotton farmer wants to increase his or her profit. To accomplish this, methods of producing cotton at the lowest possible cost must be developed. Of course, there are issues other than high production costs. Cotton production area reductions, for example, and lowered lint prices are significant. Individual farmers, however, have little control over such matters. Government programs determine cotton cultivation areas and, to a large extent, price subsidies. This leaves one critical factor—production costs, which are under the control of individual farmers. Currently, high labor requirements are the primary cause of high production costs. With current production and hand harvesting practices, 4046.86 m2 of cotton yielding 1 bale (226.8 kg) requires about 100 manhours. During the first and second pickings, a team of 32–35 laborers in Pakistan spent approximately 5–6 h picking one hectare of cotton. With labor becoming scarce and more expensive, reducing labor requirements provides the best opportunity to reduce production costs. It is well documented that the use of machinery can significantly reduce labor requirements for cotton production and harvesting. However, a reduction in labor does not always imply a reduction in production costs unless the machinery is used efficiently and the yields are high enough to justify the investment. Machinery is expensive, and its use must justify the investment. Many individually operated small farms cannot be completely and economically mechanized with the machines now available. This is especially true with regard to mechanical cotton harvesters. The majority of production equipment can be used to produce other crops that are typically included in a diversified farming program, but the cotton harvester can only be used for cotton. A farmer who wants to buy a picker must have enough cotton-producing land (at least 20 ha for smaller pickers) or arrange for custom work. In some cases, a farming program that allows multiple uses of machines can justify the purchase of equipment. A farmer with 5 ha of cotton, for example, might find it cost-effective to purchase a sprayer for applying chemicals for weed control in cotton if the sprayer was also used for applying cotton insecticides, spraying livestock and orchards, or custom spraying.

The latest recommended agronomic and insect control practices for producing high yields must be considered during cotton farm mechanization. High mechanical harvesting efficiency and economical machinery use are aided by good yields of sound bolls. Cotton production and harvesting with mechanical equipment require careful planning. From land selection to harvesting, every stage of mechanized cotton production has a direct impact on the success of the next operation. As a result, it is critical to get off to a good start by carefully selecting land and preparing the seedbed.

5.1 *Selection of Land for Mechanization*

In addition to selecting good soil capable of producing high yields, it is critical to select land that is suitable for efficient machine operation. Cotton ranks first in terms

Fig. 11 Seedbed preparation. *Source* Integrated crop management. National Cotton Council of America

of available land because the adjustment and operation of planting, cultivating, and harvesting equipment are more critical for cotton than for most other crops. Large fields with long rows are preferable. Fields can often be made larger by removing hedgerows and ditches, as well as changing fencing and road systems. Rocks and stumps that cause machinery breakdowns and obstruct planting, cultivating, and harvesting must be removed. Savings from fewer machine repairs and increased machinery efficiency will soon pay for the removal of rocks and stumps (Mechanized cotton production in Alabama).

5.2 *Seedbed Preparation for Mechanization*

The primary goals of seedbed preparation are to turn under plant residue, pulverize and farm the soil, and smooth the soil surface (Fig. 11). Seedbed preparation studies revealed that areas with the most thoroughly turned soil surface had the fewest weeds at harvest. To avoid pests, planting in a pest-free seedbed environment with a terminated cover crop containing no green matter is recommended. Shredding stalks, turning under residue, disking to firm soil and remove clods, and smoothing are all part of good seedbed preparation.

5.3 *Best-Performing Cotton Variety Selection for Mechanization*

The same varieties are recommended for machine harvesting as for hand harvesting. One of the most important characteristics to consider when selecting one of the recommended varieties for machine harvesting appears to be storm resistance. Cotton destined for machine harvesting is frequently left in the field until the majority or all of the bolls have fully opened. As a result, it is exposed to wind and rain for a longer period of time than handpicked cotton. Storm-resistant varieties have less weather or pre-harvest loss while maintaining high machine efficiency (Mechanized Cotton Production in Alabama, 1959). The main two commercial varieties in Ethiopia are still two US varieties bred for irrigated cotton farms, Deltabine 90, released in 1989, and Acala SJ2, released in 1986. Claudia, a more recent variety thought to be of Australian origin, is being bred. Before farms can be mechanized, these varieties must be tested for spindle picker (stripper) harvester tests. This will aid in calculating their yield per hectare. Before mechanization of cotton farms, their storm resistance properties (against wind and heavy rain) must be evaluated, as well as their overall machine efficiency.

6 Comparison Between Manual and Mechanical Cotton Harvesting

To use cotton pickers for harvesting, many changes in cotton cultivation agronomy must be made. The seed rate used for mechanical cotton harvesting is three times that of manual cotton picking. Because the height of the plants must be uniform and much lower than in the conventional method, the plant population must be much higher in order to achieve sufficient yields. Cotton plants grown traditionally will have many more branches and bolls per plant than cotton fields grown for mechanical harvesting.

Mechanical harvesting requires much less inter-row and inter-plant spacing to accommodate a greater number of plants. Cotton plants that will be mechanically harvested must also be sprayed with defoliant chemicals in order for the harvesting process to be clean and efficient. Cotton pickers gather more debris than manual picking, so it must be pre-cleaned before being sent to the cotton gin. All of the above changes in cultivation practices will increase costs, but the yields under this process are expected to be up to 35% higher than the conventional method of cotton cultivation.

7 Cost of Production of Raw and Lint Cotton

For the past 50 years, the Secretariat of the International Cotton Advisory Committee (ICAC) has conducted a cotton survey of the world's cotton producers. Every 3 years, reports on production costs are published. The most recent report, published in October 2016, is based on data from the fiscal year 2015/16. Ethiopia was 1 of 31 countries that participated in the survey, accounting for 87% of the world cotton area (ICAC 2016).

According to this survey, the net cost of cotton lint production decreased in 2015/2016 after years of continuous increases. In 2015/16, the net cost of producing a kilogram of lint, which excludes land rent and the value of seed after ginning, decreased by 23% to $1.16. The 34 cts/kg of lint produced decrease is due to two major factors: (1) There have been no increases in input costs. Insect control, weed control, ginning, and harvesting costs have all decreased. The cost of fertilizer per kilogram of lint produced remained unchanged from 2012/13; and (2) the value of commercial seed after ginning increased by approximately 50% in 2015/16 compared to 2012/13.

The average cost of producing seed cotton (before ginning) was $0.43/kg in 2009/10 and $0.52/kg in 2012/13, and it had been increasing since at least 2000/01. The cost of producing a kilogram of seed cotton doubled in 15 years, from $0.25 in 2000/01 to $0.52 in 2012/13. For the first time, this trend has reverted, with a net cost of production of seed cotton of $0.46/kg in 2015/16. The ICAC seed cotton cost calculations assumed that farmers are self-cultivators and do not pay rent for land use, as do small-scale cotton farmers in Ethiopia. The lower cost of production of seed cotton (before ginning) corresponds to the lower net cost of production in lint/kg (after ginning). The net cost of producing lint fell from $1.50/kg in 2012/13 to $1.16/kg in 2015/16.

The average net cost of production per hectare in 2015/16 was $1,006 under irrigated conditions and $776 under rainfed conditions, both lower than in 2012/13. Lint yielded 957 kg/ha under irrigated conditions and 647 kg/ha under rainfed conditions in 2015/16. Cotton is less expensive to produce under irrigated conditions due to higher yields than under rainfed conditions. The cost of a kilogram of lint produced under irrigated and rainfed conditions is $1.05 and $1.20, respectively. In 2015/16, irrigated conditions provided 69% of global production. In 2015/16, irrigation cost only 7 US cents per kilogram of lint produced. This data demonstrates that effective irrigation farming allows for the production of lint cotton at a lower cost than rainfed farming. The following chart (Fig. 12) depicts the cost structure of lint cotton (after ginning) production on a global scale.

In terms of the total cost of lint production, 18% (36 cents) of the total cost was spent on renting land to produce a kilogram of lint (ginned cotton). Fertilizers were

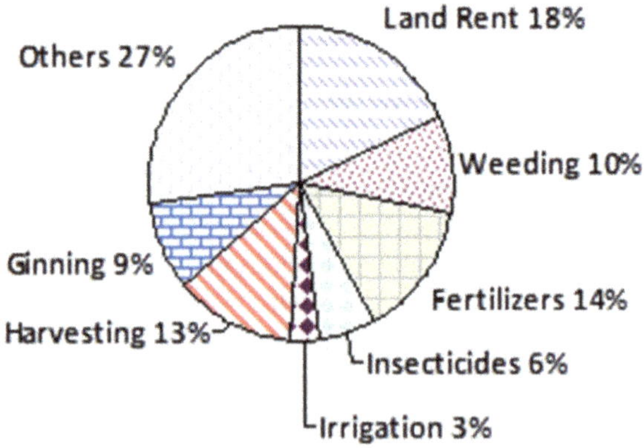

Fig. 12 Structure of cost of lint cotton production—world average. *Source* A report by the technical information section of the International Cotton Advisory Committee, October 2016 Washington DC USA

the most expensive input, accounting for 14% of the gross cost (27 cents/kg lint produced), followed by 13% (24 cts/kg) for harvesting/picking and 10% (21 cts/kg) for weeding. In 2015/16, the cost of insect control was only 6% of the total production cost, a decrease of over a decade. Irrigation costs only 3% of the total. The reported cost of irrigation, like all other inputs and operations, represents the average cost of irrigation per kilogram of lint produced in all participating countries during the study. In some cases, the 10 cts/kg lint spent by producers on planting seeds includes the technology fee for biotech trait(s). In addition to the operations and inputs not listed in the pie chart, the 'others' category includes economic costs, fixed costs, and fixed costs. The national average cost of producing one hectare of cotton in Ethiopia is shown in Table 2.

Table 2 shows that the harvesting cost for one hectare of cotton is $171.50 (3,720.00 Birr). According to the report, Ethiopia relies entirely on manual harvesting. In the same report (separate tables), the cost of harvesting one hectare of cotton spent by some world countries that use 100% manual harvesting is presented: Sudan $155.00/ha, Uganda $37.80/ha, Nigeria 86.67/ha, Mali $103.21/ha, Indonesia $92.59/ha, Pakistan $161.59/ha, Paraguay $144.23/ha.

Why the cost of manual harvesting in Ethiopia is higher than in the other countries mentioned needs to be investigated further. However, in Ethiopia's future plan to cultivate more land while keeping labor costs under control, mechanized harvesting will play a critical role.

8 Summary

Cotton occupies a unique position in Ethiopia's agrarian economy. Ethiopia has enormous potential for the production of cotton. Ethiopia has favorable weather and topography for cotton cultivation and production. Cotton (Gossypium hirsutum L.) is currently widely grown in irrigated lowlands on large-scale farms and in warmer mid-altitudes on small-scale farms under rainfed conditions. According to a recent Ministry of Agriculture study, there is 3,000,810 Ha of land suitable for cotton production, whereas it is currently limited to a maximum of 100,000 ha. Cotton demand is expected to remain strong in Ethiopia, and there is a clear need to improve productivity in future to meet this rising demand. Ethiopian cotton production is hampered by low productivity, which is caused by rainfed cultivation, small farm size, increasing pest and disease, and labor-intensive cultivation methods. Because labor costs in Ethiopia are rapidly rising, mechanization in cotton cultivation will be critical in keeping costs under control. Furthermore, high-density planting will result in increased productivity. However, high-density planting combined with mechanization necessitates the development of sympodial cotton hybrids as well as a complete transformation in agronomy practices.

Table 2 Cost of producing one hectare in Ethiopia (National Average)

Operation/Item	Unit	Quantity per ha	Cost or price per unit	Cost in Birr	Cost in US $
1. Pre-sowing					
Land rent for cotton	ha	1.00	150.00	150.00	6.92
Land revenue/tax	ha	1.00	60.00	60.00	2.77
Pre-soaking irrigation				210.00	9.68
Plowing	number	2.00	1500.00	3000.00	138.31
Planking					
Other					
Sub-total				**3420.00**	**157.68**
2. Sowing					
Soaking irrigation					
Land preparation				660.00	30.43
Seed				400.00	18.44
Seed treatment				95.00	4.38
Herbicides (Pre-sowing)					
Fertilizer					
Drilling					
Other				75.00	3.46
Sub-total				**1,230.00**	**56.71**
3. Growing					
Thinning				70.00	3.23
Weeding		3.00	162.00	630.00	29.05
Hoeing		3.00	162.00	630.00	29.05
Herbicides (Post-sowing)					
Fertilizer (Total)					
Irrigation		5.00	124.00	950.00	43.80
Insecticides		5.00	250.00	1,400	64.55
Defoliation					
Other					
Sub-total				**3680.00**	**169.68**
4. Harvesting					
Picking cost					
a. Hand picking (100%)		1.76	2,000.00	3,520.00	162.29

(continued)

Table 2 (continued)

Operation/Item	Unit	Quantity per ha	Cost or price per unit	Cost in Birr	Cost in US $
b. Machine picking (%)					
Stick cutting/ slashing				200.00	9.22
Other					
Sub-total				**3,720.00**	**171.51**
Seed cotton costs				**12,050.00**	**555.56**
5. Ginning					
Transportation to gin factory	ton	1.76	1,200.00	2,112.00	97.37
Ginning (Including bagging)	ton	1.76	1,300.00	2,288.00	105.49
Classing/grading charges					
Other					
Sub-total				**4,400.00**	**202.86**
Variable cash costs				**16,450.00**	**758.41**
6. Economic costs					
Management and administrative					
Interest on capital invested					
All repairs					
General farm overheads					
Other				148.05	6.83
Sub-total				**148.05**	**6.83**
7. Fixed costs					
Power supply					
Irrigation system at the farm					
Tractors					
Spray machinery					
Farm implements					
Other				213.85	9.86
Sub-total				**213.85**	**9.86**

(continued)

Table 2 (continued)

Operation/Item	Unit	Quantity per ha	Cost or price per unit	Cost in Birr	Cost in US $
8. Total cost				16,811.90	775.10
9. Value of seed cotton	**kg**	1,760.00	12.00	21,120.00	973.72
10. Net Value of lint	**kg**	650.00	33.00	21,450.00	988.93
11. Net Value of seed	**kg**	1,110.00	6.80	7,548.00	347.99

Source A report by the technical information section of the International Cotton Advisory Committee, October 2016 Washington DC USA

References

1. Abdelraheem A, Esmaeili N, O'Connell M, Zhang J (2019) Progress and Perspective on Drought and Salt Stress Tolerance in Cotton. Ind Crops Prod 130:118–129
2. Guo K, Tu L, He Y, Deng J, Wang M, Huang H, Li Z, Zhang X (2017) Interaction Between Calcium and Potassium Modulates Elongation Rate in Cotton Fibre Cells. J Exp Bot 68:5161–5175
3. Bronson KF, Onken AB, Keeling JW, Booker JD, Torbert HA (2001) Nitrogen response in cotton as affected by tillage system and irrigation level. Soil Sci Soc Am J 65:1153–1163
4. Raper TB, Pilon C, Singh V, Snider J, Stewart S, Byrd S (2020) Cotton Production in the United States of America: An overview. In Cotton Production; Jabram, K., Chauhan, B.S., Eds.; John Wiley & Sons, Inc.: Hoboken, NJ, USA:217–248
5. Kelemu F (2015) Agricultural Mechanization in Ethiopian: Experience, Status and Prospects. Ethiop. J. Agric. Sci. 25:45–60
6. Kline CK, Green DAG, Donahue RL, Stout BA (1969) Agricultural mechanization in equatorial Africa. East Lansing, Michigan: Institute of International Agriculture, Michigan State University
7. Kranz WL, Evans RG, Lamm FR, O'Shaughnessy SA, Peters RT (2012) A review of mechanical move sprinkler irrigation control and automation technologies. Appl Eng Agric 28(3):389–397
8. Pringle HC, Martin SW (2003) Cotton yield response and economic implications to in-row subsoil tillage and sprinkler irrigation. J. Cotton Sci. 7:185–193
9. Carvalho IR, Korcelski C, Pelissari G, Hanus AD, Rosa GM (2013) Demanda hídrica das culturas de interesse agronômico. Enciclopédia Biosf. 9:969–985
10. Cetin O, Bilgel L (2002) Effects of different irrigation methods on shedding and yield of cotton. Agric. Water Manag., 54:1–15
11. Bongiovanni R, Lowenberg-Deboer J (2004) Precision agriculture and sustainability. Precision Agric 5:359–387
12. Blackmore S, Godwin R, Fountas S (2003) The analysis of special and temporal trends in yield map data over six years. Biosys Eng 84(4):455–466
13. Multsch S, Krol MS, Pahlow M, Assunção ALC, Barretto AGOP (2020) Assessment of potential implications of agricultural irrigation policy on surface water scarcity in Brazil.Hydrol. Earth Syst. Sci. 24:307–324
14. Datta A, Ullah H, Ferdous Z, Santiago-Arenas R, Attia A (2020) Water Management in Cotton. In Cotton Production; Jabram, K., Chauhan, B.S., Eds.; JohnWiley & Sons, Inc.: Hoboken, NJ, USA:47–60
15. Uzen N, Cetin O, Unlu M (2016) Effects of domestic wastewater treated by anaerobic stabilizationon soil pollution, plant nutrition, and cotton crop yield. Environ. Monit. Assess. 188:664

16. Roth G, Harris G, Gillies M, Montgomery J, Wigginton D (2013) Water-use efficiency and productivity trends in Australian irrigated cotton: A review. Crop Pasture Sci., 64:1033–1048
17. Marengo JA, Torres RR, Alves LM (2017) Drought in northeast Brazil: Past, present and future. Theor. Appl. Climatol. 129:1189–1200
18. Whelan BM, McBratney AB (2000) The null hypothesis of precision agriculture management. Precision Agric 2:265–279
19. Mekonnen MM, Hoekstra AY (2011). The Green, Blue and Grey Water Footprint of Crops and Derived Crop Products. Hydrol. Earth Syst. Sci. 15:1577–1600
20. Bordovsky JP, Mustian JT (2012) Cotton response to crop row offset and orientation to subsurface drip irrigation laterals. Appl Eng Agric 28(3):367–376
21. Jacques LANÇON, Berhanu Woldu (2020) Cotton sector study in Ethiopia). Deutsche Gesellschaft für Internationale Zusammenarbeit (GIZ) GmbH
22. De Janvry A, Sadoulet E, Kyle E, Darr ML (2015) 'adoption des technologies agricoles: Quelles leçons tirer des expérimentations de terrain? RED, 4:129–153
23. Anthony WS, Meredith WR, Williford JR (1988) Neps in ginned lint: the effect o varieties, harvesting, and ginning practices. Textil Res J. 58:633–639
24. Columbus EP, Bel PD, Robert KQ (1990) Maintaining cotton quality. T ASAE 33:1057–1065
25. Mangialardi GJ, Lalor WF, Bassett DM et al (1987) Influence of growth period on neps in cotton. Textil Res J 57:421–427
26. Baker KD, Hughs E (2008) Cotton quality as affected by changes in the spindle picker. In: Proceedings of th beltwide cotton conferences. Nashville, TN: National Cotton Council of America.
27. Bange MP, Long RL, Constable GA et al (2010) Minimizing immature fibre and neps in Upland cotton. Agron J 102:781–789
28. Hebert JJ, Mangialardi GJ, Ramey HH (1986) Neps in cotton processing. Textil Res J. 56:108–111
29. Faulkner WB, Wanjura JD, Hequet EF et al (2011) (2011) Evaluation of modern cotton harvest Systems on irrigated cotton: fibre quality. Appl Eng Agric 27:507–513
30. Frydrych I, Matusiak M, Swiech T (2011) Cotton maturity and its influence on nep formation. Textil Res J. 71:595–604

Cotton Harvesting, Post-harvest Handling and Storage

Getnet Belay Tesema and Genet Nigatie Fetene

Abstract The percentage of world cotton production harvested by machines was around 30% in 2013–2014 and has remained stable over the last 15 years. With rising labour costs and a scarcity of labour, mechanization has gained traction in the Ethiopian cotton sector. The selection of appropriate mechanical harvesting systems is critical during machine harvesting. Cotton pickers and strippers are the two mechanical harvesting systems that are available. Strippers were developed as a cost-effective alternative to pickers for harvesting relatively low yielding cotton with closed or "storm resistant" boll types. Strippers use a non-selective harvesting mechanism that removes almost all of the material from the plants during harvest. Pickers employ a more selective harvesting mechanism that removes seed cotton only from well-opened bolls with minimal amounts of undesirable vegetative material. Ginning is the next step after harvesting. Saw gins produce about 85% of the world's cotton. Saw and roller ginning are available in Ethiopia, but custom ginning is not. Cotton can be harvested using a spindle picker or a brush-roll stripper. This chapter attempts to discuss the raw material, mechanical, and economic factors that influence a grower's decision to use picker or stripper harvesting machines. The evolution of each machine system will be discussed. A basic description of how the harvesting units on each machine work will be presented, as well as an operational description of the ancillary equipment used onboard the harvesters to convey, clean, and package seed cotton for infield storage.

Keywords Hand picking · Machine picking · Picker · Stripper · Storage · Grading · Marketing

G. B. Tesema (✉)
Ethiopian Institute of Textile and Fashion Technology (EiTEX), Bahir Dar University, Bahir Dar, Ethiopia
e-mail: getnet.belay@uni-bremen.de

G. N. Fetene
Ethiopian Textile Industry Development Institute (ETIDI), Addis Ababa, Ethiopia

K. Murugesh Babu et al. (eds.), *Cotton Sector Development in Ethiopia*, Textile Science and Clothing Technology, https://doi.org/10.1007/978-981-99-9149-5_6

1 Introduction

In Ethiopia, a plan to use machine harvesters to reduce cotton production costs was recently announced. Machine harvesting is a one-time operation that occurs when 85% of the cotton bolls are open and all of the leaves are desiccated. In general, the machines harvest 90% of the seed cotton from the plants [1]. Machine-harvested seed cotton, on the other hand, typically contains 10–30% more foreign matter than hand-harvested seed cotton [2, 3, 4].

Growers sell their seed cotton to a middleman, who buys cotton for himself or for ginneries on commission. When cotton is delivered to a ginner, it becomes his property, and he is responsible for the profit and loss on the sale of lint and seed. Estimating the cost of ginning under such conditions is difficult, and data on ginning costs is not available in Ethiopia. In any case, the ginning cost from countries where custom ginning is done is the grower's actual cost.

During cotton ginning, seed cotton cleaning primarily removes large foreign matter from seed cotton [5], whereas lint cleaning improves lint appearance by removing small foreign matter from lint, such as neps, small leaves, seed coats, small sticks, and funiculi [6, 7].

Lint cleaners reduce the availability of foreign-matter content in machine-harvested cotton while improving HVI colour grade and leaf grade but causing fibre damage [8, 9]. Many studies found that lint cleaners reduced fibre length and increased short-fibre content [9–13]. The effect of lint cleaners on fibre strength has received conflicting reports. Some researchers reported that lint cleaners reduce fibre strength [13, 14], whereas others reported that lint cleaners increased [15] or had no effect on fibre strength [10, 11]. Fibre maturity is closely related to fibre damage during lint cleaning. Mature cotton has less potential for fibre damage and is less affected by lint cleaning than immature cotton [11, 16]. Many studies have been conducted to assess the impact of lint cleaner amounts. In general, increased amounts of lint cleaners tended to improve HVI colour and leaf grade [8, 17], however, these consistently decreased net returns [18–20] and adversely affected fibre quality [8, 15, 21, 22].

The global cotton marketing practices, as well as the cotton marketing practices in Ethiopia, are discussed in this chapter. The cotton market information system in Ethiopia is generally deficient. The absence or inadequacy of such a system has denied producers and suppliers access to vital information such as alternative market outlets, levels of demand and price, and market standards.

Despite the absence of significant trade-distorting policies such as quotas or high tariffs, a number of cotton-producing countries use domestic measures to support their cotton industries. During the three seasons beginning in 1999/2000, support in the United States and the European Union averaged \$3.0 and \$0.6 billion, respectively. During the 2001/02 season, producer prices in the United States, Greece, and Spain were 91,144, 184%, and 91,144%, respectively, higher than world prices. China paid farmers \$330 per tonne of cotton harvested outside its top producing region in 2014, potentially slowing a drop in output in the world's top consumer of the fibre and further reducing demand for imports. Further investment in textile operations in the United States to help alleviate logistical issues in 2021/22 supports the increase. Furthermore, continued strong domestic demand is expected, as is a push to pre-pandemic exports of semi-processed textile and apparel products for finishing and shipping back to U.S. consumers. The current low cotton prices ($1.85/kg = 99.53 Birr/kg), which have undoubtedly been influenced by major players' support, have taken a toll on the rural sector of cotton-dependent households. To alleviate these issues, the government's support for the sector is required, which will result in continuous growth in Ethiopia's cotton mill industries.

Cotton is important in the Ethiopian economy because the country's textile industry is primarily based on cotton. The textile industry contributes significantly to the country's industrial output and is one of the most important sources of employment. However, the industry faces a number of challenges, including technological obsolescence, high input costs, and limited credit availability. To address these issues, the government must implement several subsidy schemes and is advised to develop a new textile policy for the sector's overall development.

2 Picking of Cotton

Cotton will be fully mature and ready for harvesting 160 days after it is planted. When the bolls have burst open, the farmers can prepare to remove the cotton bunch from the plant. Cotton picking/harvesting refers to the careful removal of the cotton bunch without causing significant damage to the fibres.

2.1 *Hand Picking of Cotton in the World*

Cotton can be picked by hand or by machine. Manual picking is time-consuming, but it preserves cotton fibre characteristics. Boll opening is the first action on the fibre that pushes fibres out of the place where they had been embedded for weeks before being exposed to external conditions. The gentle opening action of the boll has no effect on the fibre quality. A longer stay in the field, on the other hand, may change the colour and cause the fibres to shrink, affecting the three most important fibre properties, namely length, strength, and micronaire. If there is a lot of dew, one character may be affected more than the other. Such an effect cannot be avoided because not all bolls open at the same time and some open bolls must remain in the field for days, if not weeks. Hand picking allows you to pick open bolls at regular intervals, and weather effects on the fibre after the bolls have opened can be minimized.

This effect is minimized in China (Mainland) because land holdings are so small that the majority of growers who have planted cotton on about 1/10th of a hectare can return to the field several times to pick a few open bolls. Fibre quality is preserved in slightly larger plots by a number of pickings throughout the season. In many cotton-picking countries, 3–4 pickings per season are common [23]. In slightly bigger plots, fibre quality is preserved through a number of pickings during the season. 3–4 pickings/season are very common in many countries where cotton is hand-picked.

Hirsutum cottons can hold locks for a longer period of time before dropping the seed cotton to the ground. However, if cotton is not picked for weeks or months, it becomes loose and eventually falls to the ground. Burs in *G. arboreum* varieties lack sufficient holding force to keep seed cotton stuck in the burs. It is simple to pick such cotton, but it must be picked more frequently. In China (Mainland), India, Myanmar, and Pakistan, where such cottons are grown on a large scale, 8–10 picks are common. As a result, arboreum varieties are unsuitable for machine picking because the locks fall to the ground quickly. Upland cotton locks that fall to the ground are usually loose, whereas arboreum locks are mostly intact. *G. barbadense* cottons are similar to hirsutum cottons. Because of the position of the burs after opening, most *G. herbaceum* varieties are extremely difficult to pick. Bolls are smaller, and locks are positioned after opening so that each lock must be picked separately. Within all species, there are varietal differences.

2.2 *Machine Picking of Cotton in the World*

Machine picking accounts for approximately 30% of global output. Australia, Israel, and the United States are the only countries in the world where all cotton is picked by machine. A significant amount of cotton is picked by machines in Bulgaria, Greece, and Spain. Table 1 shows the percentages of machine and hand picking in the world's top ten cotton-producing countries.

Before the republics of the former Soviet Union became independent countries in 1991, a large percentage of cotton in all of them was machine picked. Machine picking has declined in almost all republics due to a lack of maintenance facilities for pickers. Kazakhstan and Kirghizstan have improved their maintenance facilities, and it is estimated that these two countries now handle 70–80% of total production. Hand picking has increased in Azerbaijan, Tajikistan, Turkmenistan, and Uzbekistan since 1991, and it is estimated that 60–70% of the total area is now hand-picked.

Argentina and Brazil are the only major cotton-producing countries where machine picking is expected to increase in the coming years. In Argentina, a prototype two-row stripper machine was designed, which is now commercially produced. The Sapucay 492 model is said to be less expensive, and its operational costs are also lower when compared to other machines on the market. The distance between rows can be adjusted between 0.85 and 1.0 m, and picking efficiency ranges from 85 to 90%. Last year, a number of these machines were imported into Brazil. Because of a growing labour shortage, Turkey may also implement machine picking in the coming years. By the end of the 1980s, labour shortages in Pakistan had become a problem as a result of doubling production in seven years. Machine picking would have been implemented by now if production had not been hampered by the leaf curl virus disease. Part of the labour problem was addressed by implementing various

Table 1 Percentage of hand and machine picking in major countries

Country	Hand picking (%)	Machine picking (%)
Argentina	25	75
Australia		100
Brazil	90–95	5–10
China (Mainland)	100	
Greece	8	92
India	100	
Pakistan	100	
Turkey	100	
United States		100
Bolivia	96	4
Bulgaria	15	85
Colombia	65	35
Ecuador	97	3
Paraguay	95	5
South Africa	80–85	15–20
Spain	5	95
Uruguay	30	70
Zimbabwe	95	5

social incentives for picking labour. Many issues must be addressed before machine picking can be implemented in any country.

However, some research on machine picking has already been conducted in Pakistan. Turkey is comparatively more prepared to implement machine picking. In addition to the cotton-producing countries listed in Table 1, all cotton is picked by hand in Cameroon, Chad, Côte d'Ivoire, Iran, Madagascar, Mali, Myanmar, the Philippines, Senegal, Sudan, Syria, Thailand, Togo, Uganda, and Vietnam. However, at least some cotton is picked by machines in the following countries (Table 1).

2.3 *Picker Versus Stripper Types of Cotton-Picking Machines*

Cotton strippers and pickers are the two mechanical harvesting systems available in the world. Strippers were developed as a cost-effective alternative to pickers for harvesting relatively low-yielding cotton with closed or "storm-resistant" boll types. Strippers use a non-selective harvesting mechanism that removes nearly all of the material from the plants during harvest (Fig. 1, right). Pickers use a more selective harvesting mechanism that removes seed cotton only from well-opened bolls with minimal amounts of undesirable vegetative material (Fig. 1, left). Cotton strippers require that the plants be dry and desiccated at the time of harvesting so that the material can be easily snapped off of the plants. Stripping is thus better suited to arid regions where relative humidity at harvest time is low enough to allow for dry crop conditions. Strippers have higher harvesting efficiency than pickers, which results in higher seed cotton and lint yields for stripper-harvested cotton. However, because cotton pickers typically leave a portion of seed cotton in the field, average fibre quality levels are generally improved for picker-harvested cotton. A study conducted in the United States' Southern High Plains region discovered that, despite productivity gains and improved fibre quality for picker-based harvesting systems, stripper-based harvesting systems had a higher net economic return per land area when all harvesting and ginning expenses were considered [24].

Stripper harvesters do have some advantages over picker harvesters, such as lower purchase prices, fewer moving parts in row units, lower fuel consumption and maintenance requirements, and faster ground speeds in low-yielding cotton. Picker harvesters, on the other hand, pick cleaner cotton, are thought to maintain fibre quality characteristics better than strippers, and can harvest cotton at higher speeds in high-yielding stands [25].

In the case of large-scale cotton farms in Ethiopia, it is recommended that appropriate machine technology be used. Because the world's textile mills have continued to raise their standards for fibre quality as cotton spinners are forced to compete with synthetic fibres that are not plagued by fibre contamination and degradation. Increased global cotton yields and higher quality demands are driving factors in the harvesting of mechanized large-scale cotton farms with appropriate harvesting technology (Picker or stripper).

Fig. 1 John Deere model 9930 Pro-12 VRS (variable row spacing) cotton picker (left). John Deere model 9960 brush-type cotton stripper with a view of the coupled field cleaner (right)

2.4 Picking of Cotton in Ethiopia

Cotton picking by hand is a common harvesting practice in Ethiopia. As previously stated, manual harvesting (hand picking) of seed cotton is time-consuming. A worker can typically pick 25 kg of seed cotton per day, depending on yield and plant density, but this can vary between 15 and 50 kg depending on variety, yield, and plant density. As a result, for larger fields and higher yields, manual harvesting becomes a bottleneck.

Picking practices in Ethiopia should also be improved so that impurities do not increase the weight of the crop. Instead, it is suggested that a reward scheme for the best cotton-picking practices be implemented.

In Ethiopia, large-scale farmers on irrigated fields pick twice, with the first picking done when 70% of the bolls are open and the second picking done three weeks later. In contrast to irrigated areas, large-scale and small-scale rain-fed farmers frequently harvest their cotton once when all of the bolls are open. All seed cotton handling is done manually and with various-sized bags. There is no bulk handling or compaction system in place for seed cotton. Although Ethiopia has an abundance of human resources for labour-intensive jobs such as cotton farming, the high cost of labour for harvesting makes cotton an unappealing business. Furthermore, the transportation cost of labour to farm locations, security concerns, and strong competition from sugarcane plantations would compel consideration of mechanical harvesting. Though no other method of harvesting can produce seed cotton of higher quality than good hand picking, seed cotton entering the market in Ethiopia contains a significant amount of trash, which can be attributed to poor collection practices, poor storage, and open transportation. The majority of trash, such as leaves, plant parts, immature bolls, small stones, dust, sand, and so on, is collected with seed cotton. Picking is done at regular intervals, and the cotton is piled on the ground without adequate protection from various types of trash, including sand. Sun, wind, dew, and rain all contribute to fibre deterioration. Weight being the primary criterion for both paying labour and selling seed cotton, little care is generally taken to remove insect-affected, immature, or otherwise defective bolls and impurities.

2.5 Cotton Hand Picking Guideline in Ethiopia

Cotton is hand-picked in Ethiopia by farm labour and, in the case of small-scale farms, by family members. Picking efficiency and quality are determined by picker training, boll weight, plant shape and size, and yield. Handpicked cotton is the cleanest cotton that can be harvested due to the intelligence built into the manual picking system, and it better preserves the quality characteristics.

It is documented that, while machine harvesting of cotton dramatically speeds up and reduces harvest costs, it decreases cotton fibre quality in general, particularly in terms of increases in nep content and lint foreign-matter level [8, 12, 26–29].

Manual picking, on the other hand, resulted in better fibre quality: low nep content and lint foreign-matter level, but it is a slow, tedious, and costly method of cotton harvesting. Ethiopian Institute of Textile and Fashion Technology (EiTEX) can intervene by training manual cotton-picking personnel trainers to reduce the disadvantages of manual cotton picking. The following guidelines are provided to help with this:

(a) Picking/harvesting stage

- Crop readiness should correspond to the "days of maturity" of the variety(s) in use.
- Procedures should check the cotton farm frequently for the status of the opened bolls present in the farm near harvest time.
- Picking begins when fluffed-out cotton that feels dry is ready for picking and 65% of the bolls are open, and the second/last picking begins two to three weeks later after the first pick.
- Weeds should be removed before causing harvesting irregularities.

(b) Picking/harvesting time

- Picking should be done early in the morning or late in the afternoon.
- Picking should be done after dews have evaporated in the morning, and seed cotton should be allowed to dry before picking begins. If moist cotton is picked, it should be allowed to dry in the shade before storing it.
- Picking should be avoided during hot noon for the sake of pickers' health and safety.

(c) Picker training

- Pickers should be trained to distinguish clean, undamaged cotton from trashy or stained cotton.
- Pickers should cover their heads and wear cotton clothing because artificial threads and human hairs are the most difficult to remove from ginning and textile factories.
- Pickers should avoid picking dried leaves, barks, weeds, and immature bolls.
- Pickers should be supervised to avoid potential contaminants in the picking, packing, and tying of material.
- Pickers should pick the bolls from either side of the row with both hands, picking the bolls closer to each hand and depositing the picked seed cotton into a sack or pouch.
- Pickers should be trained on how to handle sharp points of dried bracts, which can injure fingers and cause discomfort, as well as how to carry harvested cotton, which creates a moderate amount of drudgery in human pickers.

(d) Picking/harvesting step

- Cotton boll harvesting should begin with the lowest bolls and work its way up. This prevents contamination from upper dried leaves and bracts.
- Pickers should avoid collecting dried leaves, barks, weeds, or immature bolls.

(e) Picking/harvesting and storage materials

- Pickers should use cotton-made bags with twine ties suitable for handling seed cotton and hand-woven bamboo baskets for picking.
- Pickers should be given larger cotton-made bags in which to store the day's picking.
- Pickers should avoid using woven polypropylene materials at all stages of cotton handling to reduce contamination.

3 Post-harvest Seed Cotton Handing

Cotton is the most important industrial crop, providing 35% of the fibre to the world's textile industries, and approximately 350 million people are directly or indirectly employed in the cotton production and processing chain [30, 31].

As the world's fifth largest oil crop, it supplied enough seed to the oil industry while also becoming the world's second most important source of plant protein [32]. Cotton consumption is expected to rise further, as 33.56 million hectares of cotton were grown in 2018, with a total production of 123.78 million bales. Thus, proper post-harvest handling of lint cotton remains the most important requirement in the textile manufacturing process [33].

As a result, proper cottonseed handling cannot be overlooked. Because the production of high-quality seeds has always been a top priority for long-term agricultural growth [34]. Quality seed is required for uniform stand establishment, which creates a path to high yield, but a lack of good-quality cottonseed at planting is the primary cause of poor germination and crop stand failure in developing countries such as Ethiopia [35–37]. Every stage of seed production, from field selection to harvesting, processing, and storage, is critical for seed quality management [33].

Storage of seed cotton for any length of time prior to ginning frequently results in significant changes in the quality of the seed and lint. The nature and extent of these changes in any given case would be determined by a number of factors. The most important of these are (a) the amount of moisture present in the seed cotton heap; this moisture may be derived from rain, dew, or artificial humidification, or from the green leaves and unripe bolls, which are gathered along with the seed cotton at the time of picking; (b) the initial incidence of bacterial or fungoid infection, which is generally transferred to the cotton from sand, dirt, leaf-bits, or infected bolls; (c) the storage conditions, such as humidity, temperature, ventilation, sunlight, and heap density; (d) the use of mechanical or chemical treatments to combat bacterial activity, such as frequent stirring and sim-drying or the use of disinfectants; and finally (e) the cotton variety, which determines its inherent resistance to the attachment of cellulose-destroying microorganisms.

3.1 Effect of Post-harvest Storage on Cotton Fibre

The most significant and harmful effects of storage on cotton fibre are now widely acknowledged to be those caused by the activity of cellulose-destroying organisms. These organisms are primarily transferred to cotton from dirt, sand, or unhealthy, pest-infested bolls that are collected with it during harvest.

It is obvious that if seed cotton is laid out on damp ground, in the dew or rain, or is watered by spraying more contaminated foreign matter will adhere to it, and the damage done to the fibre will be proportionately greater. Microorganisms, bacteria, and fungi, usually enter the fibre from the broken end and multiply rapidly inside the lumen where nitrogenous food is available [38], but they can also attach to the fibre from the outside if its surface has been weakened by some mechanism injury [39].

The surface of the fibre is punctured wherever the pressure of the bacterial or fungoid moulds becomes excessive, its strength is greatly reduced, and the cotton becomes badly stained as a result of either mode of attack. Such cotton will produce more waste in the form of "fly," and the yarns spun from it will be comparatively weak and less amenable to dyeing and finishing processes.

When seed cotton is stored, the amount of food available, the moisture in the sample, and the humidity and temperature of the surrounding atmosphere all play a role in the deterioration caused by bacteria and fungi. Food is provided not only by the fibres themselves but also by stalk, boll, and leaf fragments. As a result, these impurities not only contaminate the material but also provide nutrition to the organisms. As a result, the disruptive effects of bacterial infection will be much more pronounced in the storage of a sample that contains a fair amount of trash due to poor and careless picking.

The presence of moisture in the sample, as well as a damp and warm environment, are other factors that contribute to the rapid multiplication of microorganisms and, as a result, the deterioration of cotton stored in seed. According to [40], serious deterioration occurs when the moisture content exceeds 11%. He is careful to point out, however, that even in a bulk sample with an average moisture content of only 7–8%, there may be many wet patches with localized overdamping that serve as suitable breeding grounds for bacteria and from which the infection spreads in all directions. This is especially true if the heap is left undistributed for an extended period of time, with no attempt at sun-drying or proper ventilation to reduce and equalize the moisture content.

It is thus discovered that if storage conditions are favourable to the growth and continuous activity of cellulose-destroying microorganisms, very serious injury to the fibres can occur, resulting in a corresponding reduction in the grade and quality of cotton. Burns [Burns, A. C. Ibid] discovered that after 44 days of storage in a very damp sample of seed cotton with no sun exposure or ventilation, 14.7% of the fibres were severely damaged and 29% were severely damaged. Even when the sample was occasionally ventilated, the percentage of infected fibres remained at 23.6 after 86 days of storage. He concluded that bacterial and fungoid infection should be sought and controlled in unginned seed cotton rather than ginned lint cotton as a source of cotton deterioration during storage.

3.2 *Effect of Post-harvest Storage upon Seed*

The most important economic effect of seed storage is the loss of germinating power, which is caused by oxidation or fermentation of the seed matter. Both of these processes are greatly aided by the presence of moisture in the material. Even if the moisture is not uniformly distributed in the bulk but is localized in wet patches, as is likely when the material is laid out on damp ground before storage or water is added to it by spraying. The oxidation process is accompanied by the evolution of heat, resulting in a significant increase in the temperature of the seeds.

Cottonseed accounts for 15–25% of crop value, and it is critical to maintain its quality after harvest and storage to avoid spoilage [41, 42]. Cotton (*Gossypium hirsutum* L.) is widely grown in Ethiopia, both in irrigated lowlands on large-scale farms and in warmer mid-altitudes on small-scale farms under rain-fed conditions. This species is widely used for its fibre, which is a raw material in the textile industry. The quality of the cotton plant's fibre is primarily determined by the mother crop's growth and reproductive efficiency. The mother crop's growth performance and productivity are naturally determined by the vigour of the initial seed used for sowing and the subsequently emerged seedling, which can only come from a healthy and viable seed. The mode of propagation is entirely through seeds, resulting in total planting seed replacement in every season of planting. [43] proposed that seed, as a living thing, acts as a carrier of genetic information that can be passed on to the next generation for the production of healthy seedlings to ultimate plants at various stages of growth.

Temperature and relative humidity of the storage atmosphere hasten the deterioration of planting seeds during storage [44], and it has been reported that cotton seeds stored at high temperatures have a lower germination percentage.

3.3 *Seed Cotton Grading in Ethiopia*

The proposal of seed cotton grading in Ethiopian cotton will help farmers get the best price for their produce. Most cotton-growing countries have standard-grade boxes for seed cotton that are based on visual grading. The grading in this standard is done using four main quality parameters: extraneous matter, colour grade, leaf grade, and ginning percentage. As shown in Table 2, each quality parameter has its own value out of 100%.

Table 2 Quality parameters of seed cotton grading

No.	Quality parameter	Assigned score (values 100%)
1.	Extraneous matter	45%
2.	Colour grade	40%
3.	Leaf grade	8%
4.	Ginning percentage	7%
5.	Total	100

3.4 Seed Cotton Transportation to the Local Market Places and Ginneries

The modern cotton-picking process has evolved from hand picking to machine picking, storing seed cotton in modules rather than trailers, storing modules on the turn-row, and transporting cotton modules to gins with module trucks (Fig. 2). Agricultural engineers have played critical roles in the development of the current system.

Cotton harvesting, transportation and scheduling, and ginning are three subsystems required for the fundamental process flow. Trampling at the field where the modules are stored shapes seed cotton into modules. For best results, modules should be formed on a high-elevation turn-row rather than in low spots throughout the field [45]. The round modules can be transported to the gin using conventional module trucks or semi-trailers. The modules must be picked up where they were dropped in the field, and staged together for pickup, four for a module truck and six for a semi-trailer.

Because the majority of Ethiopian farmers have small holdings and yields are generally low, they cannot afford to use modern methods of raw cotton transport from the farm to the nearest market or ginnery. Transport systems are quite primitive, and raw cotton is subject to the vagaries of poor handling and storage practices by farmers. Following transportation to the market yards, seed cotton is frequently

Fig. 2 John Deere round modules stored for transportation to the gin. The yellow plastic wrap forms a lip on the ends of the modules

unloaded on the ground in the market yards, where the flooring and surroundings are not kept clean. In the absence of covered sheds and cemented clean flooring, seed cotton destined for auction becomes contaminated/deteriorates. The transportation from market yards to ginning factories is also done in open vehicles, which affects the quality of the product. Cotton should be collected in fully covered cloth bags to reduce contamination and ensure safe transportation to avoid the addition of any contaminant/deterioration during transportation.

To avoid the addition of any contaminants, the following safe transportation procedures are recommended:

(a) Avoid transporting non-cotton materials alongside seed cotton
(b) Transport different lots separately (different varieties, first and second harvest)
(c) Avoid transporting seed cotton from farms to ginning factories in open vehicles, which may reduce exposure to contaminants that affect cotton quality
(d) Transportation of seed cotton should also be fully covered to avoid exposure to dust, rain, and wind.

4 Cotton Handling Practices During Ginning

Cotton becomes the property of the ginner once it is delivered to it, and he is responsible for the profit and loss on the sale of lint and seed.

4.1 Seed Cotton Storing and Ginning

Delays in the cotton ginning industry could expose modules to inclement weather for longer periods of time. If not properly protected, exposing seed cotton to the elements can reduce both producer and ginner profits. Rain damages cotton by increasing moisture content, lowering colour grade, causing rot, and heating the module. These avoidable quality changes affect the price received by the producer while also increasing the cost of ginning due to increased drying energy and decreased ginning rate. Wind can blow cotton from exposed modules, resulting in less salable lint and seed [46].

Prerequisites for Seed Cotton Storage in Ginning

- To prevent cotton colour distortion, seed cotton should be stored in a separate package rather than in a heap.
- Cotton material shall be used for wrapping and packing materials.
- In the ginnery compound, the packed seed cotton shall be placed on pallets that keep the cotton free of contaminants and not subjected to flooding.
- The packed seed cotton shall contain a tag that reflects the variety field lot number, load number, and any other necessary information.

Seed Cotton Ginning

The primary function of a cotton gin is to separate lint from seed, but the gin must also be equipped to remove a large percentage of foreign matter from the cotton, which would significantly reduce the value of the ginned lint. A ginner must have two objectives:

1. To produce lint of satisfactory quality for the grower's market and
2. To gin the cotton with the least amount of reduction in fibre spinning quality, so that the cotton meets the needs of its ultimate users, the spinner and the consumer.

As a result, quality preservation during ginning necessitates the proper selection and operation of each machine in a spinning system. Cotton's natural quality characteristics can be altered by mechanical handling and drying. At best, a ginner can preserve the quality characteristics inherent in the cotton when it enters the gin. Ginning can have an impact on lint quality. The gin stand, whether saw or roller, pulls the fibre from the seed and is the heart of the ginning system. The capacity of the system, as well as the quality and potential spinning performance of the lint, are determined by the operating condition and gin stand adjustment.

Cotton quality may be reduced if gin stands are operated beyond their design capacity. Short fibre content increases as saw speed increases, resulting in yarn imperfections. While most Ethiopian cotton is saw ginned, a few roller ginning facilities have begun to appear. Roller ginned lint is longer and contains fewer short fibre, seed coat fragments, and neps. As a result, upland cottons command a few cents per pound premium over sawn cottons. A detailed cost-benefit analysis should be performed to determine whether roller ginning is a viable alternative to saw ginning. Upgrading ginning technologies with cleaning facilities and timely maintenance of renewable parts to reduce contamination and maintain the inherent characteristics of fibre are critical.

4.2 Lint Cotton Grading in Ginning

Ginners should voluntarily adopt the grading and classification system. The Ethiopian Institute of Textile and Fashion Technology (EiTEX) should send technical field staff to ginneries to carry out seed cotton grading and lint classification in the future. Ginneries could also serve as information management hubs for raw and lint cotton sampling, grading, and fibre quality.

The ideal ginning factory mainly should consist of the following:

1. Appropriate weighing facilities upon arrival of cotton and removal of ginned cotton/bales.
2. Appropriate raw cotton storage arrangements for moisture-controlled contamination-free storage as well as neat and clean surroundings.
3. Properly set ginning machines and well-maintained conveying systems at all stages for maximum output and contamination-free cotton.

4. Appropriate handling arrangements for raw cotton from storage to gin machine and lint from gin machine to lint storage places and baling to avoid contamination/deterioration of cotton.
5. A well-maintained baling press and the desired packing materials for cotton bales result in a contamination-free suitable bale.
6. Adequate testing equipment and moisture control arrangements as needed during ginning and baling. Aside from the machinery and baling arrangements, it is critical that the people working in the ginning factory are serious about controlling contamination in cotton and maintaining cotton quality. The ginning machine's settings and overall maintenance of all the equipment are also critical. However, in Ethiopia, many of the above requirements are not being met, resulting in lower/deteriorating cotton quality.

Cotton fibre quality is affected by every stage of production, including variety selection, harvesting, and ginning. Certain quality characteristics are heavily influenced by genetics, whereas others are primarily determined by environmental factors or harvesting and ginning practices. Problems at any stage of production or processing can result in irreversible damage to fibre quality and lower profits for both the producer and the textile manufacturer. The day a cotton boll opens, the fibre quality is at its peak. Weathering, mechanical harvesting, handling, ginning, and manufacturing all have the potential to degrade natural quality. Many factors influence the overall quality of cotton fibre. The most important ones are strength, fibre length, short-fibre content (fibres shorter than 12.7 mm), length uniformity, maturity, fineness, trash content, colour, seedcoat fragment and nep content, and stickiness. Although not all of these factors are measured on each bale, the market generally recognizes them. The ginning process has a significant impact on fibre length, uniformity, and the content of seed coat fragments, trash, short fibres, and neps. The two ginning practices that have the greatest impact on quality are the regulation of fibre moisture during ginning and cleaning and the degree of saw-type lint cleaning used. The recommended lint moisture range for ginning is 6–7%. Gin cleaners remove more trash at low moisture but cause more fibre damage. Higher fibre moisture preserves fibre length but causes ginning issues and poor cleaning. When drying is increased to improve trash removal, yarn quality suffers. Although yarn appearance improves with drying to a point due to increased foreign-matter removal, the effect of increased short-fibre content outweighs the benefits of foreign-matter removal.

Cleaning has little effect on the true colour of the fibre, but combing and removing trash changes the perceived colour. Lint cleaning can sometimes blend fibre so that fewer bales are classified as spotted or light spotted. Ginning has no effect on fineness and maturity. Each mechanical or pneumatic device used during cleaning and ginning increases the nep content, but lint cleaners have the greatest impact. The number of seedcoat fragments in ginned lint is influenced by seed condition and ginning action. Lint cleaners reduce fragment size but not fragment number. Yarn strength, yarn appearance, and spinning-end breakage are three important aspects of spinning quality. All are affected by length uniformity and, as a result, the proportion of short or broken fibres. These three elements are typically best preserved when cotton is ginned with minimal drying and cleaning machinery. Recommendations for the sequence and amount of gin machinery used to dry and clean spindle-harvested cotton were developed to achieve a satisfactory bale value while preserving the inherent quality of cotton. The recommendations take into account marketing-system premiums and discounts, as well as cleaning efficiency and fibre damage caused by different gin machines. Some deviation from these recommendations is required for special harvesting conditions. When gin machinery is used in the recommended sequence, 75–85% of the foreign matter is usually removed from cotton. Unfortunately, in the process of removing foreign matter, this machinery also removes small amounts of good-quality cotton, reducing the amount of marketable cotton during cleaning. Cleaning cotton is thus a trade-off between foreign-matter level and fibre loss and damage.

Optimal ginning is the ultimate preservation of inherent quality characteristics and can be achieved by:

- Proper selection and operation of each machine in the ginning system.
- Tracking the impact of each machine on weight loss and fibre quality.
- Current knowledge of raw cotton fibre testing technologies.
- Careful evaluation of each cotton lot and adaptation of appropriate ginning practices.
- Controlling ginning rates and gin stand speeds that tend to reduce quality.
- Strike a balance between increased grade, reduced length parameters, and reduced turnout.
- Moisture content monitoring and adaptation of ginning to the optimum fibre moisture content for any given gin process.

Requirements during lint cotton storage in ginning:

- Because cotton bales can catch fire, they must be stored separately.
- Make sure there is no open light in the go-down.
- Because moisture causes the colour of cotton to change, the commodity should not be exposed to humid conditions.
- Because cotton can heat up during storage, the first-in, first-out rule may be followed when the commodity is delivered.
- It is critical that each consignment be stored separately, with stack cards displaying the date of arrival, lot number, weight, staple length, and grade, among other information.
- During the rainy season, all windows and ventilation should be closed. On sunny days, ventilation should be provided.
- The godown must be rodent-proof.
- Automatic fire hydrants should be installed in the godown where cotton bales are to be stored, as it is difficult to transport other firefighting equipment inside the godown when a fire occurs and bales begin to break due to heat expansion.

4.3 Cotton-Packaging (Wrapping) and Labelling Bales

A. ***Packaging (Wrapping)***

- The bales must be completely wrapped in protective cotton fabric (woven or knitted).
- The cotton fabric's mass per unit area must be at least 270 g/m^2.
- Inside the protective wrapping, the bales must be strapped. Rustproof steel or synthetic material hoops can be used. The strapping must be strong enough to withstand the stress of stand handling and transportation.
- The strapping (bands/wires) arrangement inside the protective wrapping must allow for automatic opening of the bales by machine, i.e. the hoops must be parallel to each other and all the locks must be aligned on one side of the bale; additionally, a free space of 300–350 mm may be left in the middle for sampling, if required by the buyer.
- A bale length of 1060 mm requires six hoops, while a bale length of 1400 mm requires eight hoops.

B. ***Labelling***

- Each bale must bear a mark identifying the shipping lot. The mark must be the same as those on the bill of lading, delivery order, and other shipping documents. The marking colour/ink must not penetrate the protective wrapping. All bales must be labelled in the same place.
- Each bale must have a label that includes the following information: bale number (in figures and/or bar code) and gin number and/or name.
- The labels could also include the following information: compressor number and/or name
  a. Shipper's name
  b. gross mass
  c. tare
  d. cotton name or any other information as requested by the buyer.

5 Cotton Marketing

Cotton is ready for sale once the quality parameters for each bale have been determined by instrument classing. Cotton marketing is a complex operation that includes all transactions that involve buying, selling, or reselling cotton from the time it is ginned until it reaches the textile mill. Growers usually sell their cotton to a local buyer or merchant after it has been ginned and baled, but they can store it and borrow money against it if they decide not to sell it right away. Cotton stored in a government-approved warehouse provides a secure basis for a monetary loan because it is a nonperishable crop. Cotton is grown primarily for its fibre, which is used to make yarn. As a result, the value of spun yarn is directly related to the quality of the lint cotton delivered to the spinning mill (the cost of the cotton can account for up to 65% of a spinning mill's total operating costs). As a result, advances in spinning technology are increasing the pressure on cotton farmers to produce cotton that is longer, stronger, finer, more uniform, and free of contaminants. Because these cotton characteristics are particularly important to spinning mills, as they are critical to maximizing the speed and efficiency with which the mills operate. As a result, the quality of the fibre produced by the farmer is critical to its marketability and value, which is primarily influenced by the level of trash within and contamination of the seed cotton. In this regard, three broad characteristics of cotton are important: the first is the inherent characteristics of the fibre, the second is the level of trash, and the third is the level of contamination. As a result, seed cotton delivered to gins should be as clean as possible, free of contaminants, and not too wet or dry. As a result, sustainable cotton farmers must consider the needs and requirements of the users of the cotton they produce. In general, the higher the quality of the cotton, the higher its value, which should result in a higher price for the farmer.

5.1 Global Cotton Market Price

Prices for raw cotton fell sharply in the spring of 2020 during the first wave of lockdowns since the start of the COVID-19 pandemic. The onset of the pandemic drove consumers away from stores, particularly apparel retailers, and it hampered yarn spinning in many countries. This price drop, however, did not have the same impact on planting decisions as it would have had it occurred a few months earlier. Nonetheless, the decrease in cotton production in the 2020/2021 marketing year was the highest since 2016 for a variety of reasons, including the fact that corn and soybean prices had gained some ground on cotton prices prior to the COVID shock, putting downward pressure on cotton area in several countries (Fig. 3).

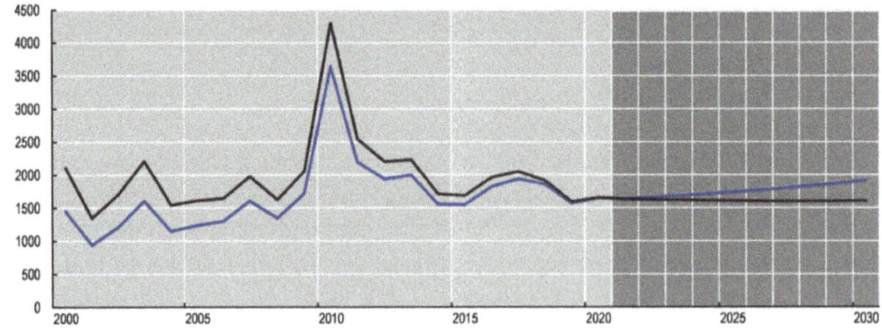

Fig. 3 World cotton prices (USD/t) with reference to cotton type: length 28.58 mm. *Source* OECD/FAO (2021), 'OECD-FAO Agricultural Outlook OECD Agriculture statistics (database)', http://dx.doi.org/10.1787/agr-outl-data-en

Cotton is an important export crop for Sub-Saharan Africa, which currently accounts for 15% of global exports. Overall, cotton production in the region has increased in recent years, owing primarily to increased acreage, though a drop in cotton prices in 2020 resulted in a drop in acreage and production, particularly in Mali, a key producing country. Spinning mill consumption remains limited throughout Sub-Saharan Africa, with many countries exporting nearly all of their output. However, the textile and apparel industry is expanding in some countries, particularly Ethiopia, because the region offers some attractive conditions for direct foreign investment (FDI), which has been significant in recent years. In the long run, this could change Sub-Saharan Africa's net export situation. Sub-Saharan African exports are expected to grow at a 2.7% annual rate in the coming decade, with the region's market share remaining at around 15% in the base period; South and Southeast Asia are the major export destinations (OECD-FAO Agricultural Outlook 2021–2030).

5.2 *Cotton Marketing Practices in Ethiopia*

In general, Ethiopia's cotton market information system is deficient. The absence or inadequacy of such a system has denied producers and suppliers access to critical information such as alternative market outlets, levels of demand and price, and the standards required by the various market players. As a result, production and marketing decisions have been made arbitrarily or based on incomplete information. Needless to say, producers require market information to help them decide what, how much, and at what quality to produce, and traders require market information to decide at what price to buy and sell. Farmers' and local suppliers' competitiveness and bargaining power have been reduced as a result of a lack of market information, forcing them to be price takers.

Cotton marketing in Ethiopia is of the open type. The following are the primary cotton marketing agencies:

1. Private Sector: Private traders, ginning and pressing industry owners who operate as sole proprietors, partnership firms, and private limited companies.
2. Agencies in the Public Sector: Ethiopian Industry Input Development Enterprise (EIIDE).

   Producers, traders or collectors, wholesalers, and consumers (spinners from the textile industry and the local handloom sector) are key actors.
3. Seed cotton business model in which a company (trader, ginner, cooperative, commercial farm) supplies inputs to smallholder cotton growers, purchases seed cotton from them, gins it (commission ginning), and sells lint and cottonseed.
4. Lint business model: a large commercial farm gins on commission and sells the lint.
5. A fully integrated model (from farm to garment) has recently emerged as a third model in Ethiopia's cotton sector.

There is no proper marketing mechanism to regulate the market, such as weighing, sale process, grading method, payment process constraints, and future prospects. Low cotton prices, combined with a lack of domestic policies, may have a negative impact on the amount of acreage dedicated to cotton production.

The government's support for the sector is required to ensure Ethiopia's cotton mill industries continue to grow. Cotton is important in the Ethiopian economy because the country's textile industry is predominantly cotton-based. The textile industry is an important component of the country's industrial production and one of the largest sources of employment. However, the industry faces a number of challenges, including technological obsolescence, high input costs, and limited access to credit. To address these issues, the government must implement several subsidy schemes, and it is recommended that a new textile policy be developed for the sector's overall development.

References

1. Hughs SE, Valco TD, Williford JR (2008) 100 years of cotton production, harvesting, and ginning systems engineering: 1907–2007. Trans ASABE 51:1187–1198
2. Faulkner WB, Wanjura JD, Boman RK, Shaw BW, Parnell CB Jr (2011) Evaluation of modern cotton harvest systems on irrigated cotton: harvester performance. Appl Eng Agric 27:497–506
3. Hughs SE, Gillum MN (1991) Quality effects of current roller-gin lint cleaning. Appl Eng Agric 7:673–676
4. Kerby TA, Carter LM, Hughs SE, Bragg CK (1986) Alternate harvesting systems and cotton quality. Trans ASABE 29:407–412
5. Anthony WS (1990) Performance characteristics of cotton ginning machinery. Trans ASAE 33:1089–1098
6. Boykin JC, Armijo CB, Whitelock DP, Buser MD, Holt GA, Valco TD, Findley DS, Barnes EM, Watson MD (2009) Fractionation of foreign matter in ginned lint before and after lint cleaning. Trans ASABE 52:419–426
7. Mangialardi GJ Jr (1992) Lint cleaning effect on seed-cotton fragment size distribution in cotton. Text Res J 62:335–340
8. Baker RV, Brashears AD (1999) Effects of multiple lint cleaning on the value and quality of stripper harvested cotton. In: Proceedings beltwide cotton conferences. Memphis, TN, pp 1391–1393
9. Li C, Thibodeaux D, Knowlton AR, Foulk J (2012) Effect of cleaning treatment and cotton cultivar on cotton fibre and textile yarn quality. Appl Eng Agric 28:833–840
10. Dever JK, Gannaway JR (1988) Influence of cotton fibre strength and fineness of fibre damage during cleaning. Text Res J 58:433–438
11. Krifa M, Holt G (2013) Impacts of gin and mill cleaning on medium-long staple stripper harvested cotton. Trans ASABE 56:203–215
12. Xu H, Cao JQ, Ye W, Xie ZL (2014) Influence of saw type lint cleaning on performance of machine stripped cotton. J Text Res 35:35–39
13. Zurek W, Greszta M, Frydrych I, Balcar G (1999) Cotton fibre length changes in the spinning process on the basis of AFIS measurements. Text Res J 69:804–810
14. Tesema GB, Drieling A (2020) Effect of ginning on tenacity, Elongation and other fibre quality properties. J Text Inst 111(3):326–333. https://doi.org/10.1080/00405000.2020.1865502
15. Ethridge DE, Barker GL, Bergan DL (1995) Maximizing net returns to gin lint cleaning of stripper-harvested cotton. Appl Eng Agric 11:7–11

16. Wei H, Chen M-L, Zhao W, Chen B, Wang Y-H, Wang S-S, Meng YL, Zhou Z-L (2017) The effects of sowing date on cottonseed properties at different fruiting-branch positions. J Integr Agric 16(6):1322–1330
17. Gillum MN, Armijo CB (1997) Pima seed cotton cleaning for maximum profit. Trans ASABE 40:513–518
18. Bennett B, Misra S, Barker G (1997) Lint cleaning stripper harvested cotton for maximizing producer net returns. Appl Eng Agric 13:459–463
19. Mangialardi GJ Jr (1993) Effect of lint cleaning at gins on market value and quality. Appl Eng Agric 9:365–371
20. Nelson J, Misra S, Bennett B, Barker G (1999) Gin lint cleaning to maximize producer net returns revisited. Appl Eng Agric 15:621–626
21. Hughs SE, Armijo CB, Foulk JA (2013) Upland fibre changes due to ginning and lint cleaning. J Cotton Sci 17:115–124
22. Willcutt MHE, Columbus TD, Valco P Gerard (2002) Cotton lint qualities as affected by harvester type in 10 and 30-inch production systems. In: Proceedings of the Beltwide Cotton Conference 2002 CDROM. Memphis, Tenn.: National Cotton Council of America
23. Chaudhry MR (1997) Harvesting and ginning of cotton in the world. In: Proceedings of the Beltwide Cotton Conference, vol 2. National Cotton Council, Memphis TN, pp 1617–1619
24. Wanjura JD (2016) Cotton harvesting, USDA agricultural research service. International Cotton Conference, Cotton Production and Processing Research: Lubbock, TX
25. Faulkner WB, Wanjura JD, Shaw BW (2009) Picker versus stripper harvesters on the high plains of texas. In: Beltwide Cotton Conferences, San Antonio, Texas, pp 5–8
26. Baker KD, Hughs SE (2008) Cotton quality as affected by changes in the spindle picker. In: Proceedings of the world cotton research conference, Lubbock, Texas. 2007 CDROM
27. Faulkner WB, Shaw BW, Hequet E (2008) Effects of harvesting method on foreign matter content, fibre quality, and yarn quality from irrigated cotton on the high plains. In: Proceedings of the beltwide cotton conference. Memphis, National Cotton Council of America, pp 612–619
28. Hughs SE, Bragg CK, Owen C (2000) Where neps in Pima cotton are made. In: Proceedings of the Beltwide Cotton Conference, Memphis, National Cotton Council of America, pp 1593–1595
29. Sui RX, Thomasson JA, Byler RK, Boykin JC, Barnes EM (2010) Effect of machine-fibre interaction on cotton fibre quality and foreign-matter particle attachment to fibre. J Cotton Sci 14:148–153
30. Afzal I, Kamran M, Basra SMA, Khan SHU, Mahmood A, Farooq M, Tan DKY (2020) Harvesting and post-harvest management approaches for preserving cottonseed quality. Ind Crops Prod 155:112842.
31. USDA Economic Research Service, U.S. Department of Agriculture (2019). Annual report
32. Whitelock DP, Armijo CB, Boykin JC, Buser MD, Holt GA, Barnes EM, Valco TD, Findley DS, Watson MD (2011) Beltwide cotton quality before and after lint cleaning. J Cotton Sci 15:282–291
33. Afzal et al. (2020) USDA and economics research service, 2019
34. Sharma KK, Singh US, Sharma P, Kumar A, Sharma L (2015) Seed treatments for sustainable agriculture-A review. J Appl Nat Sci 7(1):521–539
35. Biabani A, Boggs LC, Katozi M, Sabouri H (2011) Effects of seed deterioration and inoculation with Mesorhizobium ciceri on yield and plant performance of chickpea. Aust J Crop Sci
36. Kamran M, Afzal I, Basra SMA, Mahmood A (2016) Improvement of cotton (Gossypium hirsutum L.) crop performance by estimating optimum sowing and picking time. Int J Agric Biol 19:241–247
37. Snider J, Collins GD, Whitaker JR, Grey TL (2014) Seed size and oil content are key determinants of seedling Vigor in Gossypium Hirsutum. J Cotton Sci
38. Denham HJ (1922) Action of micro-organisms on cotton hairs. J Text Inst XIII:T240
39. Wanjura JD, Fanlkner WB, Holt GA, Pelletier MG (2012) Influence of harvesting and gin cleaning practices on Southern high plains cotton quality. Appl Eng Agric 28:631–641

40. Burns AC (1927) Deterioration of cotton during damp storage. Bulletin No. 71, Technical and Scientific Service, Ministry of Agriculture, Egypt
41. Afzal I, Rehman HU, Naveed M, Basar SMA (2016) Recent advances in seed enhancements. In: New challenges in seed biology—Basic and translational research driving seed technology. InTech. https://doi.org/10.5772/64791
42. Ashley H, Thomas J, Holt G, Valco T (2018) Cottonseed air-handling and storage requirements. Eng Ginning. https://doi.org/10.56454/FZZP5205
43. Heydecker W (1969) The vigour of seeds—A review. Proc Int Seed Test Ass 34(2):201–219
44. Abdel Magid AS, Osman AM (1975) Influence of storage period and temperature on viability and chemical composition of cotton seeds—Ann. Bot 39:237–248
45. Simpson SL, Wanjura JD, Capareda SC, Shaw BW, Parnell C (2006) Seed cotton transport analyses using GIS. In: Beltwide cotton conferences, San Antonio, Texas
46. Searcy SW, Wanjura JD, Willcutt MH, Brashears AD, Buschermohle MJ, Barnes EM (2010) Seed cotton handling and storage. In: Cotton incorporated, America's cotton producers and importers, pp 1–30
47. Brashears AD, Baker RV (2000) Comparison of finger strippers, Brush roll strippers and spindle pickers on the texas high plains. In: Proceedings of the beltwide cotton conference, vol 1, pp 452–453
48. Griffin AC (1979) High-capacity ginning and fibre breakage. Text Res J 49:123–126
49. Thaysen AC, Bunker HJ (1924) Bacterial decomposition of cellulose. Rev App Mycol 3:37

Cotton Fibre Technology in Ethiopia

Cotton Fibre Demand and Supply in Ethiopia

Haymanot Enawgaw and Belete Baye Gelaw

Abstract Ethiopia has a long tradition of cotton cultivation and has over 2.6 million hectares of suitable land for cotton cultivation. The country has important export markets in Africa, Asia and Europe, with Asia alone accounting for 67% of total cotton exports. Domestic cotton consumption in the country shows increasing trends due to the increased demand of the existing and newly built spinning mills, industrial parks and the hand-weaving sector. This chapter attempts to explain two basic concepts of cotton fibre in Ethiopia: the historical background in the first section and the domestic demand for cotton fibre in the second section. The first section focuses on the historical trends of annual cotton production capacity in the country, small cottage industry and the development of modern textile mills in ancient times, and finally tries to trace the import–export culture of cotton and domestic consumption trends. Challenges and opportunities of the history of cotton production in the country. The second section explains the current scenario of cotton fibre demand in the country, the spinning industry and its capacity, the cotton demand for the hand-weaving sector and the modern fibre demand trends in spinning, and the cotton fibre demand–supply balance. Finally, the main factors affecting the balance between supply and demand of cotton in the country are explained in detail.

1 Introduction

Cotton is one of the most outstanding and inexpensive agricultural commodities and is mainly grown in tropical and semitropical areas of the world. Even though this annual crop consists of various important parts that provide very important input materials for various agro-processing industries such as oil milling and livestock farming, the fibres are the most important raw material source for the textile industry [3, 11, 16].

H. Enawgaw · B. Baye Gelaw (✉)
Department of Textile Chemical Process Engineering, Ethiopian Institute of Textile and Fashion Technology, Bahir Dar University, P.O. Box 79, Bahir Dar, Ethiopia
e-mail: astawl49@gmail.com

K. Murugesh Babu et al. (eds.), *Cotton Sector Development in Ethiopia*, Textile Science and Clothing Technology, https://doi.org/10.1007/978-981-99-9149-5_7

Cotton is produced and grown in more than 100 countries in different parts of the world. China, India, USA, Pakistan, Brazil, Australia, Turkey, Uzbekistan, Turkmenistan and Burkina Faso are the top ten cotton-producing countries [9, 22], accounting for more than 85% of global cotton production [16, 17, 21, 22]. The world produces about 25 million tonnes of cotton annually, worth about $12 billion [22], and the cotton production trend has increased by 75.86% over the past three decades.

In Ethiopia, cotton is an indispensable source of income for small farmers and poor rural families are heavily dependent on cotton cultivation. The country has huge land potential, a favourable climate, cheap labour and market opportunities in the textile and clothing industry. Cotton is a single raw material in Ethiopia for traditional and modern textile industries, local handicraft and ginning factories [3, 7]. Because of this, cotton cultivation is deeply rooted in the country's history, dating back to AD 350 and the era of the Axumite king, King Aizana, and began on a commercial scale in 1935–1941 [4, 18] and make cotton the country of origin [4, 13, 18]. In general, this chapter explains the domestic demand for cotton fibre, past and current trends in cotton production, the spinning industry, the balance between cotton supply and demand, and the factors affecting cotton production, demand and supply trends and other related factors.

2 History of Cotton in Ethiopia: (Cotton Production and Cotton Industry)

Cotton is one of the preeminent economically valuable agricultural commodities in the world. It is grown as an annual crop which is mostly grown in tropical and semitropical areas of the world. Among parts of cotton, the fibre is the most significant source of raw materials for textile industries [16]. It has a yearly economic impact of no less than $600 billion in the biggest textile industries [10, 16, 22]. Cultivated cotton belongs to the genus Gossypium which has 54 species which commonly grow in the arid and semi-arid regions of tropical areas. In which only *G. hirsutum*, *G. barbadense*, *G. herbaceum*, and *G. arboretum* are cultivated [16, 27]. Cotton crop can provide an input for different agro-processing industries like textile, oil mills and the livestock sectors [3, 11].

More than 100 countries produce cotton in different part of the world. Among these cotton-producing countries, the top ten are China, India, the USA, Pakistan, Brazil, Australia, Turkey, Uzbekistan, Turkmenistan, and Burkina Faso [9, 22]. China, the USA, India, Pakistan, Uzbekistan, Turkey, and Brazil contribute 83% of the world's cotton production [16, 17, 21], (Shakeel et al. 2020).

Nearly 25 million tonnes of cotton is produced worldwide per year and it is worth about 12 billion dollars (Shakeel et al. 2020). Worldwide cotton production showed 75.86% increase during the last 23 years. During the last 50 years, the cotton cultivation area in Africa has been increased from 4.0 to ~4.5 mha [6]. In the year

2016, the global cotton cultivation area was 30.2 mha, out of which 4.5mha was contributed by Africa. Globally, purchases of cotton for the period of the year 2017–2018 were US$49.9 billion in which Africa covered 7.8% (Shakeel et al. 2020). Cotton production in sub-Saharan Africa upraised from 200,000 tonnes per year to over 1,700,000 by a factor of 8.5 in 2004/05. However, over the past decade, cotton production has stagnated approximately by half due to lack of irrigation and inconsistency in the provision of inputs and advice across the region.

In Africa, cotton is an essential cash crop for smallholder farmers and more than 2 million poor rural families depend on cotton cultivation to earn cash [6, 11]. The cost of cotton farming is lower in Africa when compared to other countries [28]. It is cultivated around the world either as rain-fed or irrigated, while rain-fed cotton production is prominent in Africa [6, 28], where the use of purchased inputs such as chemicals and fertilizer is minimal [28]. Cotton in Africa is grown by the traditional crop rotation system. Cotton produced in Africa is measured of high quality due to the hand-picking method used during harvesting. However, this is a labour-intensive system as compared to the highly mechanized system practiced by developed countries [6].

Ethiopia is part of sub-Saharan African countries that produce and export cotton. The country has a favourable climate, cheap labour and market opportunities that make the country have comparative advantages in agro-industries. Ethiopia has the huge potential of over 2.6 million hectares that are suitable for cotton production but only about less than 3% is being utilized [1, 7, 14, 21, 28]. Ethiopia contributes only 5% of total production in Africa [12]. Despite its available land and cheap labour to produce plentiful cotton, its production and productivity are negligible and our contribution to the world cotton commercial exchange is so small [21, 28].

Ethiopia has a long tradition of cotton cultivation and it is one of the main cash crops in the country. Cotton is the only crop fibre in Ethiopia that is used as a single raw material for the traditional and modern textile industries, local handcraft and ginning factories [3, 7]. Cotton cultivation is deeply rooted in the history of the country's agricultural activity and it is believed that Ethiopia is one of the countries of origin of cotton [4, 13, 18]. Cotton which is 'tet' in Amharic word, is dated back to 350 A.D., including in the era of the Axumite King, King Aizana [4, 18]. Traditionally, cotton fibre was supplied by small-hold cotton farmers, and homespun using age-old spinning drop wheel and then converted into fabrics using handlooms [21, 28]. However, production of cotton at a commercial level was started in 1935–1941 which was initiated and then stopped soon after by the Italians.

The next phase in cotton expansion began in 1948 with the first request of the Government of Ethiopia for technical assistance from the United Nations and other agencies for further surveys, planning, research and training of Ethiopian personnel. In 1960s, the commercial scale cotton production re-initiated, and the entire cotton and textile sector was nationalized in the 1970s. Starting from the year 2000, state cotton farms and ginneries were privatized, and state-owned textile factories were either leased or sold to investors [4, 28].

Ethiopia was importing raw cotton during the period of 1940s to 1970s, to fulfil the domestic demand of the textile factories. After the establishment of state farms and large-scale private farms in 1970s, the country started exporting cotton which discontinued in the 1980s due to the drought, and then continued in MY 1994/95 [28]. Cotton in Ethiopia is widely grown in the lowlands on large-scale and small-scale farms under both rain-fed systems and some of which are irrigated [14, 24]. Most of the cotton grown in the country is around the Awash Valley, with some smaller amounts being grown in Gambella, the Omo Valley, Humera and Metema [24].

3 Annual Cotton Production Capacity of Ethiopia

Ethiopia possesses over 2.6 million hectares of land suitable for growing cotton; an area that is equivalent to the cotton land in Pakistan which is the world's fourth largest producer [8, 28]. 65% is found in 38 high-potential cotton-producing areas while the remaining 35% is in 75 medium-potential areas [7, 28]. Studies indicate that cotton fibre contributes about 70% of the total raw materials required for textile industries in Ethiopia [3]. Although Ethiopia has more than 2.6 million hectares suitable for cotton cultivation, only about 80,000 hm^2 are currently cultivated by cotton plant in Ethiopia with productivity of 230,000 tonnes of raw cotton per year [4] and lint cotton production is about 51,400 tonnes per year [3]. The annual production of seed-cotton was nearly 120,000 tonnes in the year 2011 to 2013 with an overall productivity of 1.42 tonnes/hectare [28]. According to Ethiopian development research institute (EDRI), there were about 52,754 smallholder farmers, 408 mechanized rain-fed farms, and 107 mechanized irrigated farms engaged in cotton farming in the production year of 2015 [28].

In the market year (MY) of 2018/19, the total cotton cultivated area was 80,000 hm^2 and about 154,000 tonnes of raw cotton was harvested. According to the Ethiopian Textile Industry Development Institute (ETIDI) and Foreign Agricultural Service (FAS) report, the current Ethiopian ginning out turn (GOT) is 37% with a cotton yield of about 642 kg/hm^2. Therefore, in MY 2018/19, lint cotton after ginning will be about 57,000 tones [3]. With this production level, the country faces a gap of 70,000 tonnes per year of lint cotton to meet the local demand of the evolving textile sector [4]. Unlike other countries, the Ethiopian textile mills do not hold a significant volume of stocks due to shortage of working capital and capacity constraints [25].

4 Emerging Cotton Industry in Ethiopia

Most textile industries exist in Ethiopia are cotton-based. Cotton produced by private commercial farms and state-owned farms is mostly used in the modern textile manufacturing sector and to some extent exported to foreign countries, while cotton produced by peasant farms is mainly used by the handloom sector [28]. The handloom industry which showed vigorous growth commonly consumes cotton produced by smallholder farmers [25]. The textile, garments and apparel sector is growing much faster than cotton production in the country [9]. According to TIDI and USDA report in 2015, there are 11 ginning companies with a collective capacity of more than 60,000 metric tonnes (276,000 bales, 9 spinners, 35 garment manufactures, and about 15 integrated textile factories (TIDI; [24]).

As part of country's five-year Growth and Transformation Plan (GTP II) economic plan (2015–2020), the government made considerable investments to develop the textile and apparel sector. The GOE, in its industrial development policy, has given priority to build textile and industrial parks around the country to accelerate economic transformation by investing considerable resources [4]. At the same time, it works to attract potential international investors some of whom have vertically integrated operations of cotton growing, ginning, spinning, and using it for their local textile and apparel manufacturing business [4, 25]. There is a fairly better FDI movement in the textile and garment sector, especially since many Turkish textile firms are moving to Ethiopia. Hence, the demand for raw cotton and fabric continues to expand [28]. Foreign investors from China, the USA, India, Korea, Sri Lanka, Bangladesh, and others are setting up shop in these parks [24, 25]. The GOE will provide a variety of incentives, such as reasonable land and electricity rates, tax dispensations, duty exemptions, and capital financing to help these investors [4, 24]. For example, Hawassa industrial park (HIP) was planned and established as a leading textile and apparel eco-industrial park in 2015 and is one of the first Ethiopian and Africa's most significant export-oriented textile and garment industrial parks [19]. Industrial Parks Development Corporation (IPDC) has been developing additional parks in Adama, Jimma, Dire Dawa, Kombolcha, Bahir Dar and Mekelle [4].

According to industry sources, cotton demand is around 460,000 bales (100,000 metric tonnes) in the country. However, due to insufficient amount of cotton production, and the challenges of importing cotton from abroad, a large portion of this demand is unsatisfied and textile and apparel manufacturers that use cotton are forced to reduce consumption and/or to stand idle until they get sufficient cotton [24]. This will affect the productivity and competitiveness of the global market of the textile industries in the country, because cotton shortage can reduce their productivity and overall efficiency.

In the upcoming years, cotton consumption is expected to rise especially as the GOE works to attract foreign investment in the textile and apparel sector. More than a dozen spinning mills have been planned to be opened in the country in the short run (ETIDI). After the installation of the planned spinning mills, the country's annual processing capacity of lint cotton will reach 200,000 metric tonnes [25]. Therefore, due to the growing demand from the local textile and apparel sector and limited local production, cotton is expected to be imported in the future. Some of which are expected to come from India and the USA [24]. The Ethiopian Industrial Input Development Enterprise (EIIDE) is a state-owned organization that plays a great role in market stabilization. It has an obligation to bridge the supply and demand gaps of industrial raw materials, including cotton, sugar, and other raw materials.

5 Cotton Production Trend and Gaps in Ethiopia

Of the total land under cotton farming, 33% is cultivated by small holders, 45% by private farms and 22% are state-owned farms [5, 7]. Currently, majority of the cotton cultivation takes place in the Awash Valley, with some cultivation also taking place in Gambella, Humera and Metema. Ethiopia can also produce irrigated cotton. Out of 84,000 hectares of land that is under cotton cultivation, only 35,000 hectares are irrigated. The major potential cotton-growing areas include Omo, Ghibe, Wabi-Shebelle, Awash, Baro-Akobo, Blue Nile and Tekeze river basins [15].

The share of Ethiopia in terms of international production and marketing of cotton was only 0.13% of the total cultivated land and 0.1% of the produced cotton for the year 1998–2000 [17]. The average yearly national production of lint cotton during the period 1996/97–2000/01 was about 29,849.7 metric tonnes in Ethiopia, of which about nearly 83% was locally consumed. The respective share of textile mills and handlooms, and handcrafts was 86% and 14% of the annual domestic sales of lint cotton, respectively [28].

Figure 1 shows the cotton production trend of Ethiopia during the market year (MY) 2000–2018. An average of 33,842.10 metric tonnes of cotton was produced within indicated year durations. The country produced a minimum of 14,000 metric tonnes in the year 2001 and a maximum of 62,000 metric tonnes in the year 2012. Ethiopia's cotton production was estimated at 40,000 metric tonnes (184,000 bales) in MY14/15. While production seems to be on a rising path, cotton output in previous years has been relatively flat, thus keeping the country from reaching its production targets [24].

As seen in Fig. 2, the area harvested and the production of lint cotton in Ethiopia showed a great fluctuation within the years between 2010 and 2020 in which the maximum and minimum areas harvested were about 99,000 and 57,000 ha in the MY 2010/11 and 2013/14, respectively. The yearly cotton harvested area showed a gradual decrease in the production period of 2010/11 to 2013/2014 to 57,000 ha which considerably increased to 98,000 ha by the next year and then declined for the next five years. An average area for cotton cultivation in these ten years of cotton production was about 79,600 ha, which is only about 2.7% of the country's available suitable land of more than 3 million hectares. The cotton harvested area was 82,000 hectares in the MY 2016/17, which fallen by 26.83% to 60,000 hectares in MY 2017/18. Total cotton harvested area moved up by 28.33% to 77,000 hectares in MY 2018/19 over the previous year and was forecasted to reach 92,380 hectares in MY 2020/21 with a CAGR of 9.53% from MY 2018/19. In MY 2019/2020, Ethiopia accounts only 0.24% of the world area coverage of cotton harvest [7].

In terms of productivity, the maximum and minimum amounts of lint cotton produced were about 62,000 and 28,000 metric tonnes during the harvest year of 2011/12 and 2013/14, respectively. One great factor for the minimum amount of lint cotton during the production year of 2013/14 would be clearly the decrease in the area of cotton harvested. The total lint cotton production was 45,000 tonnes in marketing year (MY) 2016/17, which increased by 17.78% to 53,000 tonnes in MY 2018/19.

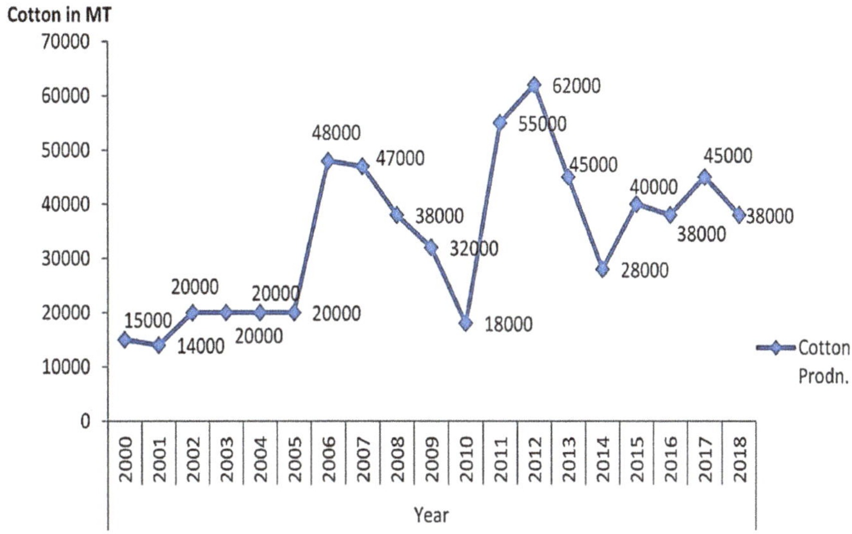

Fig. 1 Lint cotton production trend in Ethiopia (2000–2018). *Source* FAOSTAT (2011) and USDA (2018)

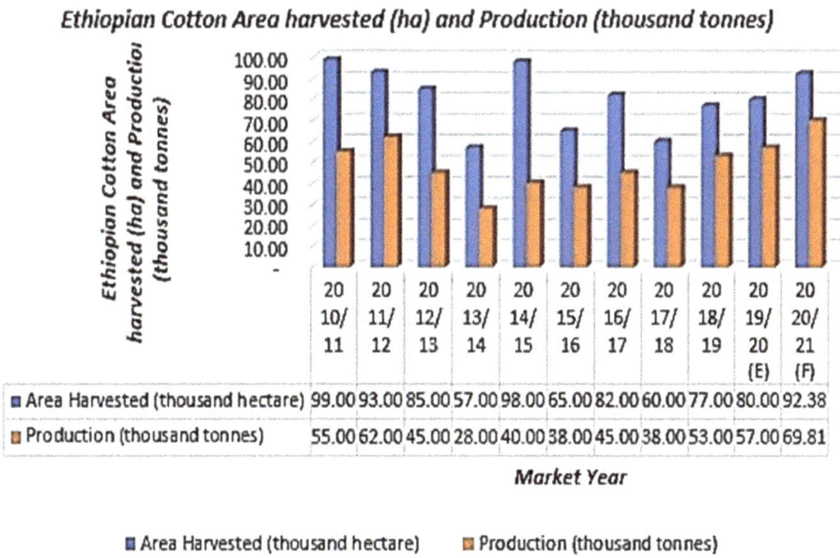

Fig. 2 Cotton production figures in Ethiopia during market year of 2010/11–2020/21. *Source* TIDI and FAS Addis Ababa Forecast Production Figures (2014/15–2019/20) are FAS Addis Ababa estimates

The lint cotton production of the country showed an increment by 39.47% in MY 2018/19 over the previous year and was expected to reach 69.81 thousand tonnes in MY 2020/21 with a CAGR of 14.76% from MY 2018/19. The average production was about 46,000 metric tonnes within this cotton harvesting periods. MY 2019/20 lint cotton production forecast was 57,000 metric tonnes (262,000 bales), up by 8% from the previous year's production estimate. The forecast was established mainly based on the predicted increase of cotton area harvested from 77,000 hectares to 80,000 hectares.

There are several reasons resulted to the weak efficiency of cotton productivity in Ethiopia. For instance, in Amhara and Southern regions, cotton plantation in MY 2018/19 had got delayed due to late arrival of rain and in the same year, the harvests also faced the challenge of the shortage of labour. The cotton harvests, approximately 40%, in some of the large cotton farms in Amhara, Gambela, Benshagul Gumuz and Afar regions were delayed due severe labour problem [25]. Besides labour issues, the cotton farming got affected by land right issues, shortage of finance, improved seeds, agrochemicals, farm machinery, spare parts, and out-of-date ginneries among others.

The Ethiopian government has launched the NCDS. Under NCDS, the government has planned to make Ethiopia one of the world's top cotton producers with annual cotton lint production of 1.1 (Table 1) million metric tonnes during the market year 2017–2032 [25].

6 Cotton Import, Export and Domestic Consumption Trend in Ethiopia

As seen in Fig. 3, Burkina Faso, Mali and Cote d'Ivoire are the leader countries among the top ten cotton exporting countries in Africa.

Domestic Consumption of Cotton in Ethiopia: The total local consumption was 56,608 tonnes in MY 2016/17 (Fig. 4), which decreased by 7.31% to 52,471 tonnes in the next market year and then moved up by 14.11 per cent to 59,874 tonnes in MY 2018/19. During MY 2019/20, the total consumption increased by over 4,000 tonnes compared to the previous year. Total cotton consumption of the country reached 68,899 tonnes in MY 2020/21 with a CAGR of 7.27% from MY 2018/19. The increase in the country's local cotton consumption was predicted mainly due to increasing demand from existing and newly installed spinning mills and the newly opened and planned textile and apparel industrial parks [25].

Import Export Trend of Cotton in Ethiopia: Ethiopian cotton import was 13,716 tonnes in MY 2016/17 (Fig. 5), which increased by 1,742 tonnes to 15,458 tonnes in MY 2017/18. Cotton import was 6,531 tonnes in MY 2018/19 which moved down by 57.75% over the previous year and is expected to reach at 10,479 tonnes in MY 2020/21 with a CAGR of 26.67% from MY 2018/19. The country imported minimum

Table 1 Targets of Ethiopia's national cotton development strategy (2020–2032)

	Market year		
	2020	2025	2032
Cultivated land (ha)			
Smallholder rain-fed	48,000	260,000	290,000
Smallholder irrigated	2,000	5,000	10,000
Large farms rain-fed	100,000	200,000	350,000
Large farms irrigated	100,000	200,000	350,000
Total	**250,000**	**665,000**	**1,000,000**
Productivity (kg seed cotton/ha)			
Smallholder rain-fed	1,600	2,000	2,200
Smallholder irrigated	2,600	3,000	3,300
Large farms rain-fed	1,600	2,000	2,2000
Large farms irrigated	2,600	3,000	3,300
Average	**2,008.00**	**2,308.00**	**2,596.00**
Production (tonnes seed cotton)			
Smallholder rain-fed	76,800	520,000	638,000
Smallholder irrigated	5,200	15,000	33,000
Large farms rain-fed	160,000	400,000	770,000
Large farms irrigated	260,000	600,000	1,155,000
Total	**502,000**	**1,535,000**	**2,596,000**
Ginning outturn (%)	39	42	43
Lint cotton production (tonnes)	195,780	644,700	1,116,280
Yield (kg lint cotton/ha)	783	969	1,116
Domestic mill use (tonnes lint cotton)	100,000	350,000	600,000
Exports (tonnes lint cotton)	95,780	294,700	516,280

Source ETIDI

cotton (4,354 tonnes) in the MY 2013/14 and maximum cotton (15,458 tonnes) in MY 20117/18. MY2019/20 cotton import was forecast at 38,000 bales (8,000 MT). Persistent and severe foreign exchange shortage has significant downward pressure on the ability to import. Hence efficient utilization of available resources to produce cotton with sufficient amount and quality is mandatory to overcome the challenges of cotton import [25].

Ethiopia exported 1,088 tonnes of cotton in MY 2016/17, which moved up by 200% to 3,265 tonnes in MY 2017/18. However, the total cotton export of the country moved down by 46.67% in MY 2018/19 compared to the previous year and is expected to reach at 2,721 tonnes in MY 2020/21 with a CAGR of 25% from MY 2018/19. MY2019/20 cotton exports are estimated to fall down slightly from the previous year to 2,000 tonnes. The major markets for Ethiopian cotton are Africa, Europe, and Asia (67% of the total exports) [27]. In Ethiopia, income gained from

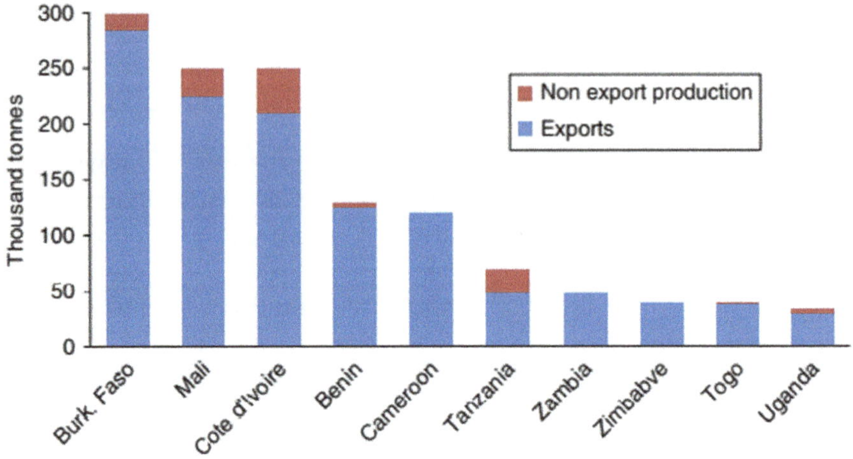

Fig. 3 Share of cotton exports of top African countries (2015/2016). *Source* Amanet et al. [6]

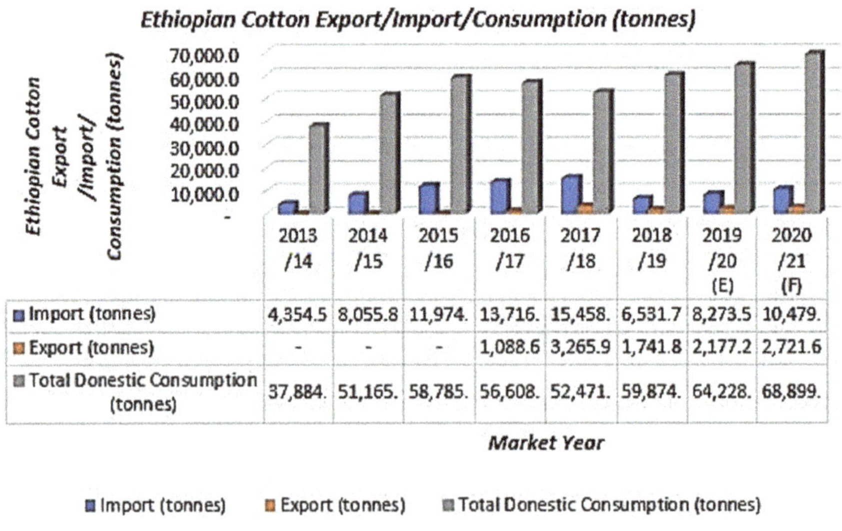

	2013 /14	2014 /15	2015 /16	2016 /17	2017 /18	2018 /19	2019 /20 (E)	2020 /21 (F)
Import (tonnes)	4,354.5	8,055.8	11,974.	13,716.	15,458.	6,531.7	8,273.5	10,479.
Export (tonnes)	-	-	-	1,088.6	3,265.9	1,741.8	2,177.2	2,721.6
Total Donestic Consumption (tonnes)	37,884.	51,165.	58,785.	56,608.	52,471.	59,874.	64,228.	68,899.

Fig. 4 Cotton import, export and domestic consumption trend in Ethiopia. *Source FAS Addis Ababa*

the export of cotton and textile products is low when compared to other commodities [11]. One of the reasons would be the government ban on cotton export, which was effective for almost seven years [25]. However, currently, the GOE targets making Ethiopia the textile and apparel manufacturing hub of Africa with annual exports of $30 billion by 2025 [25].

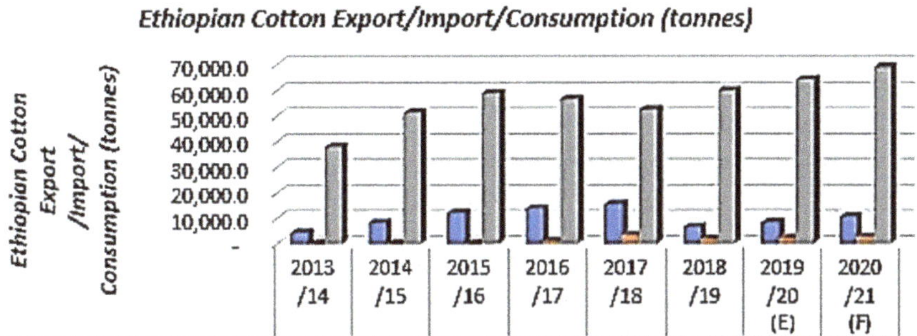

	2013 /14	2014 /15	2015 /16	2016 /17	2017 /18	2018 /19	2019 /20 (E)	2020 /21 (F)
Import (tonnes)	4,354.5	8,055.8	11,974.	13,716.	15,458.	6,531.7	8,273.5	10,479.
Export (tonnes)	-	-	-	1,088.6	3,265.9	1,741.8	2,177.2	2,721.6
Total Donestic Consumption (tonnes)	37,884.	51,165.	58,785.	56,608.	52,471.	59,874.	64,228.	68,899.

Fig. 5 Cotton import–export trends in Ethiopia (2013–2021)

7 Challenges and Opportunities of Cotton Production in Ethiopia

Though Ethiopia has a huge potential capacity to produce both rain-fed and irrigated cotton, the country performed low productivity. In addition, the amount and quality of cotton produced are very low compared to other major cotton-producing countries [4]. Different production and marketing constraints such as erratic weather conditions, difficulty in accessing inputs like fertilizer and improved quality seeds, absence of extension services on irrigation practices, shortage of seasonal labour, pest challenges, financial constraints and increasing production costs are recognized as a great constraint [28].

The most important challenges being faced by cotton production in Africa are low emergence rate of cotton seeds, poor storage facilities for seeds, late seed distribution, pest attack, ineffective pesticides and changing climate situations [6]. Cotton farming in Ethiopia faces a serious shortage of labour for planting, weeding and harvesting. This led to delays in some farms harvesting cotton which also affects the quality of the cotton [28]. Cotton can either be picked by hand or by machines. About 30–35% of world cotton harvest is machine-picked and the rest by hands [3]. Cotton production in Sub-Sahara Africa including Ethiopia, no machinery is available for the harvest. So the harvesting of cotton is done by hand picking, with the help of seasonal workers [14]. Therefore, if sufficient labour is not found at the time of cotton harvesting, then a considerable amount of cotton may remain un-harvested on the land. Meanwhile the farmer will not be able to sell the cotton and get the required profit within the market period. A large-scale private cotton farm alone requires 2,000–3,000 daily labourers to collect cotton during the main harvest season [25].

The other constraint for poor productivity of cotton in the country was that the GOE was emphasized on ramping up sugar production with the intent of making the country one of the world's top ten sugar producers. For example, a former largest cotton farm in the country (Tendaho cotton farm), and several other cotton operations were transformed to sugar production several years ago [25, 28]. Moreover, cotton farmers are switching to other crops, due to limited access of inputs, increasingly erratic weather conditions, pest challenges, lack of credit, increasing production costs and lack of a well-organized marketing system [28]. Natural disasters, for instance, floods, particularly along the Awash River have also negatively affected cotton production [24]. Different sources indicated that the GOE declared restriction on exporting surplus cotton during 2010–12 with the fact that other crops, like sesame, were more profitable to grow. So farmers planted less cotton and more of other cash crops. This in fact could diminish the cotton production in the country [25, 26, 28].

Moreover, for the last several decades, one major cotton seed variety Deltapine (DP 90) was primarily been used in Ethiopia [23, 24] which is susceptible to pests and diseases. The DP 90 cotton seed variety currently covers more than 90% of cotton production areas [4]. During the most recent crop season, farmers reported substantial losses resulting from bacterial blight, Pink Bollworm, Flea Beetle and Mealy bug. In the meantime, the cotton bollworm is one of the pest challenges farmers struggle to manage. Ethiopia recently adapted *Bt* cotton (a genetically modified pest-resistant plant cotton variety) to overcome the constraints of bollworms in cotton production and two genetically modified cotton hybrids were recommended for commercial production [14]. In general, factors that constrain the production of cotton in Ethiopia are shortage of improved seed varieties, fertilizer, shortage of technical inputs including labour, insect pests, absence of extension service and limited irrigation practices, and climatic constraints [16, 24, 28].

The current attractive domestic market prices, approval of *Bt* cotton seed varieties for commercial cultivation and the growing textile and apparel industry will encourage existing and new commercial farms to boost cotton production in Ethiopia [25].

Ethiopia has a long practice of cotton cultivation with an estimated area of above 3 million hectares suitable for cultivation. The major markets for Ethiopian cotton are Africa, Asia and Europe, with Asia alone accounting for 67% of the total exports [7, 28].

In 2017, the domestic demand for cotton in Ethiopia was 45 thousand tonnes. The demand has increased from time to time. The sector has increased by 35.9% in 2021 [28].

However the country shares only 5% of total production in Africa [12]. Going through these reports, despite its available land to produce abundant cotton, Ethiopia has scored very weak performance in the production and marketing of cotton products [21].

The domestic consumption of the country has been increasing due to growing demand from existing and newly installed spinning mills and increased number of textile industry parks. The handloom industry has shown vigorous growth, it mostly consumes cotton produced by smallholder farmers. The sharp increase in the country's cotton consumption was anticipated mainly due to demand from the newly opened and planned textile and apparel industrial parks [21].

Currently, even though there are some difficulties with finance and forex, power outages and logistical challenges, there have been more spinning mills that have been on their installation of facilities that can increase the country's annual processing capacity of lint cotton 200,000 metric tonnes mark (ETIDI) [21, 28].

8 Cotton Spinning Industries in Ethiopia

Ethiopia has the earliest legacy in the utilization and production of textiles in weaving, before 3,500 BC [2, 18]. However the industrialized textile and apparel sector in the country has a history of less than a century that was started in 1939 with the establishment of the Dire Dawa Textile Factory during the Italian occupation (1935 to 1940) [2]. From this historical Benchmark and earlier experiences, the Imperial government of Ethiopia signed an agreement with the Indian government in the late 1960s Akaki Textile Factory (former name Lazaridis Cotton Mills Ltd, S.C) was established. In the meantime, the Bahir Dar Textile Company was founded in 1961 by the Italian government as a war compensation for Ethiopia [2, 18, 21].

In recent years, the Federal Government has given a priority to the textile sector especially for cotton processing and manufacturing industry [2]. The main reason for this demand has been mostly related to the presence of cheap manpower for cotton processing industries, and suitable environmental conditions for the cultivation of cotton. Even though the prevailing few textile processing industries in Ethiopia are cotton-based, the domestic demand for cotton fibre is almost null which can't even satisfy the demand of these limited textile industries. Because cotton farming and cultivation in the country is not only traditional but there is also no availability of cotton and even poor in its quality fibre supply. Due to other factors like political and security problems, the cotton spinning industry has not advanced as well. Even though the history of cotton spinning companies was started there was no growth and development in the sector. Nowadays, the country has more than 15 spinning industries [21].

Even though there was an increased number of spinning companies, Ethiopia has produced less than an average of 34,000 metric tonnes of lint cotton for the last two decades with a minimum production capacity of 14,000 metric tonnes of lint cotton (2001 production year) and a maximum production capacity of 62,000 metric tonnes in the year 2012 [2]. This shows that the country has not utilized its potential for cotton spinning and processing industries for different reasons of single-minded seed variety which is extremely exposed and vulnerable to peptides and diseases like bacterial blight, Flea Beetle, Pink Bollworm and Mealy bug, a serious shortage of labour in commercial farms, increasingly unpredictable weather conditions, difficulty accessing inputs, lack of credit, rising production costs and an inefficient marketing system [2, 28].

The seed varieties are prolonged and outdated varieties sourced from the USA before 20 years ago [23, 28]. This led to the country's inefficiency in cotton production and got the cotton spinning companies in trouble with raw materials in domestic markets and exposed them to import from foreign which cost high hard currency. As a result, most of the spinning companies were out of production and even some others were limited to some specific spinning products limited by time and raw material [21, 28].

9 Cotton Demand for Hand Loom Sectors

Ethiopia was importing raw cotton starting from the early 1940s to 1970s in order to fulfil the domestic demand of the country's textile factories. But later there was the development and establishment of state-owned and large-scale commercial farms. Even though the country has started to participate in cotton exporting markets in the world, but not sustained due to the drought in the early 1980s. Further, the country has tried to resume exporting lint cotton 1994/95 (MoARD 2005) though interrupted by different conditions [21]

In general, the cotton mill sector in Ethiopia no longer occupied the demand of the textile industries in the country. This interrupted cotton mill sector also affects the handloom mill sector which has played a vital role in the share of the total cloth output of the country. This handloom sector nowadays has been falling rapidly. Some of these mills are incurable and on the track of scrapings and have been incurring cash losses. Even though these sectors have generated high cash flows, they are in trouble due to the fact that the producers do not incorporate any stock adjustment factor to take durability into, consideration. On the other hand, these handloom sectors have been time-consuming, costly and single-minded product-oriented sectors and perform less productivity as compared to the power loom sector and the pure spinning units. As a result, there is a decline in the market share of the handloom sector.

The handloom sector has not given an attention-getting into troubles due to low supply and costly raw materials from the domestic lint cotton production of the country (only took 14%, 1996–2001).

In contrast, the textile, garments, apparel sector and handloom products demand in Ethiopia has been growing at much faster rate than cotton production. The increased demand in the textile and apparel sector has also created a profound effect on the demand of cotton for the local handloom sector. This has also created different problems on the handloom sector and refrained the stakeholders of the sector from cotton utilization and handloom manufacturing operations. These handloom sectors have only cotton demand problems but also they are in trouble with adequate knowledge about market standards, market information, system for production and marketing arrangements, support through service cooperatives and finance [2].

On the contrary, the handloom sector has a competitive advantage over the power loom mills that can produce the lowest counts so that the overhead cost of yarns as a percentage of total cost is highest. As the cloth quality increases, weaving costs as a percentage of total costs increase correspondingly with the greater difficulty of the operation so that at the highest counts handloom products may be considered skill- as well as labour intensive.

Furthermore, the technical backwardness of Ethiopia has been histrionically clarified through the miscarriage of the Ethiopian handloom industry and the house industries as a whole. The textile experts and the society as a whole are still unable to adopt and develop a system with domestic and indigenous knowledge rather than engaged on automatic looms or automatic attachments on any large scale.

10 Capacity of Spinning Mills in Ethiopia

The textile Mill was the first integrated mill in the country's history that was started in 1939 by the establishment of the Dire Dawa Textile mill by the Italian government. Furthermore, the capacity of the sector was enhanced in the course of 1960s with the development and establishment of more Five large-scale integrated textile companies by different agreements of the imperial government with foreign governments like India and Italy. There after starting from the late 1970s to the late 1990s, the Capacity of the Spinning Mills in Ethiopia has enhanced due to the establishment of additional four integrated textile mills (Hawassa, Kombolcha, Arba-Minch and Adey Abeba textile mills) and other state-owned private textile and apparel companies by the socialist regime, to satisfy the domestic demand of the textile and apparel products by substituting imported products (National cotton development strategy (2018–2032) and Road map, 2017) [9].

Different platforms have been applied to attract and motivate different investors from the USA, China, Korea, India, Sri Lanka, Bangladesh, and others, by investing considerable resources and building textile and industrial parks all over the country. Due to these favourable conditions, Nowadays the capacity of Spinning Mills in the country has increased. There are about more than 21 large-scale textile mills integrated with spinning sections in the country with three of them having modern spinning mills. In general, these spinning mills have a spinning capacity of about 29,000 ring spindles and about 14,000 rotors [21] (National cotton development strategy (2018–2032) and Road map, 2017). There are also newly installed spinning mills and an increased number of textile industry parks.

There are other different sources that show the growth of Spinning Capacity of cotton in Ethiopia (295,000 bales (64,000 metric tonnes), which is more than 4,000 metric tonnes than the previous years in 2020).

As per the Ethiopian Textile Industry Development Institute (TIDI), there was a short-run plan to instal more 12 spinning mills with integrated facilities and increased the country's processing capacity of lint cotton about 200,000 metric tonnes.

11 Trends of Cotton Demand in Ethiopia

When we see the cotton demand trends in Ethiopia, it varies from time to time. The cotton demand from the 1940s to 1970s, the domestic demand of the country was managed to satisfy the high demand of the textile factories by importing raw cotton from different countries. Thereafter, starting from the late 1970s, the country has started exporting cotton by the establishment of state-owned commercial farms and large-scale private farms in different parts of the country. But later by the late 1980s, the exporting market of cotton was disrupted due to the drought in the country. In 1995, the country has tried to start exporting lint cotton again following the rehabilitation of the state-owned farms by the transitional government (MoARD, Citation 2005) [9, 21].

In recent years, from domestic production of lint cotton, about 83% was consumed by the domestic textile mills and handloom sector manufacturers. Most of the demand was consumed by the spinning mills (about 86%), and the rest of the demand was consumed (14%) by handlooms and handcraft sectors each year from 1996 to 2001.

12 Cotton Production and Marketing in Ethiopia

The development of a cotton textile industry in Ethiopia has increased based on the national production of raw cotton was gradually taking place with multiple interruptions on the sustainable production trends. Cotton has been scowled and used in Ethiopia since the ancient times, and hand spinning and weaving are still a well-established and widespread craft technology in the country. Ethiopia is one of the centres of variability and domestication of several cultivated plants, and it is feasible that cotton has been also domesticated in the different regions of the country. Ethiopia has been found as the important world centres of domestication of plants (more than 5.2 billion domesticated plants) and Cotton has been found one of the more valuable and broadly grown crops of the country for a very long period of time [18].

Seed cotton has been sold in all the Ethiopian region markets. The cotton has been classified into different qualities, each of which was sold at a different price. Spinners have examined these qualities very carefully before buying small amounts of each. During that time all of the produced raw cotton in very small quantities by peasant farmers all over the country was consumed by the cottage industry [18]. These cottage industries produced yarn and cloth, but much of the yarn was utilized by the hand-weaving sector. Further, there was an attempt for the development of larger scale production of cotton with several introduced upland varieties grown at about 4,000 feet on deep, heavy, black soils in some special places of the country like Eritrea for the consumption of the textile mills.

However, the capacity of the small house (cottage) industries and the developed textile mills was not sufficient to meet the demands of the country, as a result, the country was obligated to import manufactured cotton from abroad and made it the most important Ethiopian import (about 1/3 of the total value of imports in the country). The major products of these textile mills were coarse yarns and rough clothes [18].

Different researchers reported that the American upland cottons have been found the most suitable species for Ethiopian conditions, particularly in upland areas having 4,000–5,000 ft above sea levels; where cotton has been grown both under rainfall and under a combination of rainfall and supplementary irrigation. In the irrigated lowlands of the country like the Awash Corridor Region finer, and long staple Egyptian types (*G. barbadense*), of both North American and African origin have been grown. There was also a possibility of cultivating the longer staple American Upland cotton (1 inch to $1^3/_2$ inches) which has been used in the Ethiopian textile mills [18]. Generally, Ethiopia is the land of high uplands having more than 4500 ft suitable for the production variety cotton species like 'Delta pine', 'Empire', 'Wilds', 'Acala' and 'Stoneville' varieties under a combination of rainfall and irrigation [18].

Ethiopia's textile and apparel sector was highly confident that the deliveries to the USA, which was the domain leading market for apparel marketing, was flood following the 10-year renewal of the African Growth and Opportunity Act (AGOA) [2, 7, 21, 28]. This system was governed under the US General System of Preferences signed on June 11, 2015, which allows the country a duty-free import of a wide range of textiles and apparel products [2]. Even though AGOA agreement (18th of May 2000) has created a driving force for different countries' investors in Ethiopia to enhance economic development, help the country to fill the rising production and labour costs at home, and avail of the duty-free exports, but for the last two years, it was disrupted and affected the production and marketing scheme of the sector in the country [28].

Furthermore, the Generalized System of Preference (GSP) allowed the country's textile and apparel sector products export to different European countries like Finland, Canada, Sweden, Austria, Japan and Norway. Even Everything but Arms (EBA) has given a privileged access to Ethiopian export goods to the EU Markets. In all aspects, we found that the cotton-based textile and apparel products have been given the first priority by the government and even by investors. The Government has imposed different packages by applying fundamental principles for this sector including private sector-led industrialization, export-oriented industries policy and capacity enhancement and strengthening of existing industries and promoting in international markets, emphasizing the foreign investment and utilizing the labour-intensive technologies in order to alleviate poverty in the country. Due to these and other opportunities in the country, the cotton production and marketing sectors have been in its era of exploration [2, 7, 18, 28].

Nowadays, Ethiopia has created a huge potential for the production of cotton and marketing business sector's investors. The country has suitable land for cotton production which has been estimated to be more than 3 million Hectares (Table 2), which is greater than Pakistanis suitable land (2.9 million Hectares) for cotton production [2, 21, 28]. It is obvious that Pakistan is the 4th largest cotton producer country in the world. But cotton area cultivation in Ethiopia is only about 3.6% of the total area. Even the country has a multi potential resources for Cotton production in both

Table 2 Availability of land for cotton production in different parts of the country [21]

Cotton production region	Production capacity (Hectares)
Tigray	269,130
Amhara	678,710
SNNP	600,930
Oromia	407,420
Gambella	316,450
Benshangul	303,170
Afar	200,000
Somalia	225,000

rain-fed and irrigated environments with almost all cotton production in regions in the country (National cotton development strategy (2018–2032) and Road map, 2017) [21].

The cotton production in the country has increased due to the increasing trends in harvested cotton area from year to year with some fluctuating circumstances. For instance, in 2016/17 production year, the cotton harvested area in the country was about 82,000 hectares, and there was a decreasing trend (6%) in the next year 2018/19 about 77,000 hectares. Furthermore, in the following production years, there was an increasing trend (28.33%) for that the cotton harvested area was found about 93,000 hectares in 2020/21. There was also another factor that increased the production of cotton in the local markets. These factors include lower local prices, growing demand in the textile and apparel sector and the approval of *BT* cotton seed varieties (JKCH1050 and JKCH1947) by Ethiopian Ministry of Environment, Forest and Climate Change (MEFCC) in 2018 for commercial cultivation [9, 20].

13 Cotton Import–Export

The Domestic consumption of cotton in Ethiopia was about 57,000 tonnes in 2016/17, which increased by 5.8% almost about 60,000 tonnes in 2018/19. Total cotton consumption of the country was increased by 14.11% in 2018/19 over the previous years and was grown up by 7.3% of that of 2018/19 which was about 69,000 tonnes in 2020/21.

Ethiopian cotton export was 1,088.60 tonnes in 2017, which was then increased by 60% to 1,741.80 tonnes in 2019. Total cotton export of the country moved down by 46.67% in 2019 over the previous year and was grown by 2,721.60 tonnes in 2021. Latter on the Ethiopian export market was declined due to some reasons like the disruption of AGOA agreements led to the country's small export margins, and there was an attractive local market prices for the producers due to the increasing capacity of the country's spinning industries and expanded industrial parks.

Table 3 Cotton demand and supply balance in Ethiopia [21]

Market year	Cotton demand (in Million tonnes)	Cotton supply (in Million tonnes)
2014	23.48	26.30
2015	24.05	26.34
2016	24.47	23.89

For the last two to three years, cotton producers have been trying to increase the export rate of cotton after the government removed the export ban on cotton. Despite attractive local prices, some commercial growers were demanding to export cotton in order to earn foreign currency.

On the other hand, in 2017, the Ethiopian cotton import was 13,716.60 tonnes, and 6,531.70 tonnes in 2019, which was decreased by 52.38% from the previous times. Total cotton import of the country has decreased by 57.75% by the year 2019 over the previous year and again increase by 26.67% in 20,221.

In general, the cotton import–export trend has a fluctuated scheme due to various reasons like Persistent and severe foreign exchange shortage, government policy and cotton market unpredictability from time to time have significant downward pressure on the ability of the country on the import and export of cotton market [2, 18, 28].

14 Cotton Demand and Supply Balance in Ethiopia

The cotton demand and supply balance in Ethiopia has a fluctuated trend from time to time. The demand–supply balance has oscillations highly dependent on different factors like cotton production, import–export scheme, international and domestic market price fluctuations and even the government policy has a great impact on the demand–supply balance. From the years 2013–2016 (Table 3), the cotton demand has some variations as shown in the table below. From these consecutive years, the cotton demand for the first two years was perfectly supplied by the domestic market but in 2016, the demand was over headed and more than the domestic supply of locally produced cotton due to the reduction in cotton production in the country [9].

15 Factors Affecting Cotton Supply in Ethiopia

The Ethiopian cotton supply has different factors that face the sector's many problems in meeting its expected goals. These factors can be considered as challenges that affect the cotton supply chain and can be dependent on the government policy, domestic and international market fluctuations and the manufacturing sectors themselves. The factors may source from different segments like government policy, working capital shortage, smuggle goods, loan and foreign currency cash delays, limited work order, power interruption, absence of nearby input raw materials and infrastructure. The factors may also be from the domestic market conditions of high sensitivity of market price, lack of sufficient and reliable spare parts, commodity price inflations and others [2, 21].

The cotton supply in the country has also different constraints in the international markets like lack of modern import/export market experience and low productivity, making them a low competitor in the market. Furthermore, there are other problems like lack of market information, lack of financial support, lack of contractual production, marketing and standards, lack of infrastructures, and limited research works in the sector.

There are also other positive factors related to the cotton supply in the country including the Priority of the government given to the textile sector, Government Assistance in capacity building, export, tax and investment, Availability of land and facility for cotton production, processing and expansion options of the cotton supply, and Availability of networking and infrastructures.

16 Summary

Cotton fibre is the main source of raw material for the textile industry and cotton was the sole raw material for the textile industry, local handicrafts and the ginning mills in Ethiopia. Various researchers have tried to organize and represent the cultivation, production and demand–supply trends of cotton in the country and have pointed out the different production trends and factors from different aspects of cotton. The researchers' unique approach is to take the surveys literally and ask about the demand for cotton fibre, which has a major impact on the country's GDP and textile sector. They suggest that encouraging cotton production and balancing domestic demand and supply could help reduce production costs.

The method used in this chapter was the analysis of available cross-sectional data from the survey from ETIDI, cotton sector organizations and research centres. The ratings were categorized based on domestic demand for cotton and its consumption over an average annual market period. In addition, in the different years of production, the country either shows an import–export of cotton that shows a balanced demand–supply trend or not. The review was analysed to verify whether there is a relationship between cotton cultivation, production, consumption and several dependent variables: import–export trends, spinning industry capacity, domestic demand and other important factors.

Although Ethiopia offers great potential for cotton cultivation and cotton production with high domestic and international demand, the country cannot afford this demand by exploiting its great opportunities due to various factors being somewhat avoided. There is an important link between cotton consumption and production trends. However, it was found that cotton production was slightly more likely to have been further developed through the use of modern and commercial agricultural mechanization to meet the increasing domestic demand for cotton fibre. Based on the various literatures, cotton cultivation does not maintain the supply balance. These suggest that this finding could contain suggestions for reducing the cost of cotton cultivation and cotton production, considering the high annual cost fluctuations of the cotton market and foreign exchange, as well as the country's low-cost land for cotton cultivation. However, the review also notes several limitations in different literatures: Most importantly, cotton studies in the country likely have limited sources of information. In order to establish a fundamental relationship between cotton cultivation, production, consumption and import–export supply balance, we recommend investigative studies and further research.

References

1. Abebe EA, Bayeh M, Tebkew T, Mulatu W (2021) Susceptibility of several cotton varieties to the cotton flea beetle, Podagrica puncticollis Weise (Coleoptera: Chrysomelidae), in a hot dry tropical environment of Ethiopia. Entomologia Hellenica 30(1):1–19
2. Abdella A (2016) Assessment of challenges, opportunities and prospects of textile industry in Ethiopia: the case of Yirgalem Addis Textile Factory PLC
3. Abuhay A, Mengie W, Tesfaye T, Gebino G, Ayele M, Haile A, Yillie D (2021) Opportunities for new biorefinery products from Ethiopian ginning industry by-products: current status and prospects. J Bioresour Bioproducts 6(3):195–214
4. Addis T, Kachi A, Wang J (2021) A review of current state and future directions of cotton production in Ethiopia. Cogent Food Agric 7(1):1880533
5. Admassie A, Seid N, May JF, Megquier S, Moreland S (2015) The demographic dividend: an opportunity for Ethiopia's transformation. Population Reference Bureau and Ethiopian Economics Association, Washington, DC
6. Amanet K, Chiamaka EO, Quansah GW, Mubeen M, Farid HU, Akram R, Nasim W (2019) Cotton production in Africa. In: Cotton production, pp 359–369
7. Balcha M, Bekele Y, Gurmessa D, Dejen T, Damtew S, Kebede M, Tegegn F, Seid N, Getahun S, Alemu Z, Arega M (2022) Cotton production guideline in Ethiopia, Doctoral dissertation, Wcrer Agricultural Rcscarch Center, Ethiopian Institute of Agricultural Research
8. Bayrau A, Bekele F, Assefa B, Mihiretu M (2014) Ethiopian Development Research Institute
9. Belay G, Mussema R (2020) Production and marketing constraints of cotton under smallholders: the case of Arbaminch Zuria District, Southern Ethiopia
10. Blaise D, Kranthi KR (2019) Cotton production in India. In: Cotton production, pp 193–215
11. Bosena T, Bekabil F, Gebremedhin B, Hoekstra D (2011) Structure-conduct-performance of cotton market: the case of Metema District, Ethiopia. J Agric, Biotechnol Ecol 4(1):1–12
12. EIA (Ethiopian Investment Agency) (2012) Investment opportunity profile for cotton production and ginning in Ethiopia. Updated Report on Cotton Ginning. Addis Ababa, Ethiopia. https://doi.org/10.1094/PDIS-11-11-0999-PDN
13. Feyso A (2018) Cotton value chain analysis: the case of Smallholder Farmers Arbaminch Zuria District, Gamo Gofa Zone, Ethiopia. Asian J Plant Sci Res 8(2):19–30
14. Gudeta B, Egziabher AG (2019) Cotton production potential areas, production trends, research status, gaps and future directions of cotton improvement in Ethiopia. Greener J Agric Sci 9(2):163–170
15. International cotton advisory committee (ICAC) (2014) Production and Trade Policies Affecting the Cotton Industry. A Report by the Secretariat. International Cotton Advisory Committee, Washington, DC
16. Jabran K, Ul-Allah S, Chauhan BS, Bakhsh A (2019) An introduction to global production trends and uses, history and evolution, and genetic and biotechnological improvements in cotton. In: Cotton production, pp 1–22, Accessed 28 Nov 2015

17. Merima A, Ayele G (2008) Agri-chain analysis of cotton sub-sector in Ethiopia
18. Nicholson GE (1960) The production, history, uses and relationships of cotton (Gossypium spp.) in Ethiopia. Econ Botany 14(1):3–36
19. Negesa D, Cong W, Cheng L, Shi L (2022) Development of eco-industrial parks in Ethiopia: the case of Hawassa Industrial Park. J Ind Ecol 26(1)
20. Partzsch L, Kemper L (2019) Cotton certification in Ethiopia: can an increasing demand for certified textiles create a 'fashion revolution'? Geoforum 99:111–119
21. Seboka N (2020) Mapping cotton value chain of Ethiopia. Ethiopian J Textile Apparel 1(2):1–21
22. Shakeel A, Hasanuzzaman M (eds) (2020) Cotton production and uses: agronomy, crop protection, and postharvest technologies. Springer
23. TIDI (2015) Textile Industry Development Institute (TIDI): supervises the performance of both the cotton production and textile manufacturing industries
24. United States Department of Agriculture (USDA) (2015) Annual Cotton GAIN Report number: ET1512
25. United States Department of Agriculture (USDA) (2019) Ethiopia Cotton Production Annual GAIN Report number: ET1906.
26. Wagaye BT, Walle GA (2018) Overview of Ethiopian textile industry. J Textiles Polymers 6(2):117–120
27. Walelgn G (2020) Effect of inter-and intra-row spacing on lint yield, yield related traits and fibre quality of cotton (*Gossypium hirsutum L.*) in Middle Awash, Ethiopia, MSc Thesis, Haramaya University
28. Zeleke M, Adem M, Aynalem M, Mossie H (2019) Cotton production and marketing trend in Ethiopia: a review. Cogent Food Agric 5(1):1691812

Ginning Industry in Ethiopia

Gashaw Ashagre Walle, Abera Kechi Kabish, Gizachew Zeleke Mindaye, Mesay Dubale Tigabu, and Zeine Nessre Selamo

Abstract Ginning is a prerequisite for the processing of cotton in textile factories. Ginning is a mechanical process that separates the cotton fibres attached to the seed. The main objectives of this chapter are to provide an overview of ginning technology, the existing ginning industry and its practices in Ethiopia, and the plethora of challenges and opportunities of the sector. Generally, secondary data are compiled from books, company documents, research institutes, and internet sources. The data are analysed with a simple statistical tool (Excel) and the information is logically arranged. The total ginning capacity is 1,362 tons/day; where roller ginning is becoming more popular in the country than traditional saw ginning for upland cotton. However, the capacity utilization of the ginners is less than 50% due to outdated technology and lack of input materials. The quality of the fluff is of paramount importance, but it can be reduced if sorting of the cotton is carried out and everyone involved works together. The general trend indicates that the ginners have not been able to meet the textile industry's demand for lint cotton in the local market. In this regard, there is a huge negative trade balance. Despite the availability of excess land for cotton cultivation, a trainable workforce, and supportive policy frameworks for textile industry expansion, factories are found to be importing cotton due to a lack of raw materials. Finally, there is limited research to solve the multiple problems faced by the sector.

Keywords Ginning · Saw ginning · Roller ginning · Bale · Productivity · Ginning outturn

---

G. A. Walle (✉) · A. K. Kabish
Ethiopian Institute of Textile and Fashion Technology, Bahir Dar, Ethiopia
e-mail: gashaw15@gmail.com

G. Z. Mindaye · M. D. Tigabu · Z. N. Selamo
Textile and Garment Industry Research Development Center, Addis Ababa, Ethiopia

K. Murugesh Babu et al. (eds.), *Cotton Sector Development in Ethiopia*, Textile Science and Clothing Technology, https://doi.org/10.1007/978-981-99-9149-5_8

1 Introduction

Cotton is one of the most important natural fibres grown worldwide. Cotton is a soft, fluffy staple fibre that grows in a boll around the seeds of cotton plants, as shown in Fig. 1. There are three main products of cotton production: cotton lint, linters and cotton seed. Cotton fluff is used in clothing, shoelaces, pillowcases, jeans, towels and dollar notes. Linters are used in plastics, paper products, films, yarns and cosmetics. Cottonseed is ground into three separate products: petroleum, flour and hulls. The oil is the most valuable by-product of the cottonseed. It is obtained by crushing the kernel of cotton seed. The oil is used in cooking oil, salad dressings, soaps, cosmetics and in the preparation of snack foods such as chips, crackers and cookies. The trays are used for animal feed, fertilizer, fuel and packaging materials. Flour is the second most valuable by-product of cottonseed. Flour, made by grinding the cottonseed, is used as fodder for livestock and poultry, and as a natural fertilizer for lawns, gardens and flower beds [1].

To make effective use of seed cotton, it must first be ginned. Ginning is the mechanical separation of cotton fibres clinging to the seed that allows cotton to be converted into a continuous thread [42]. Figure 1 depicts a mature and opened seed cotton boll. The ginning process is simplified in the diagram below (Fig. 2). It also shows the percentages of lint and cotton seed.

Lint is the mass of soft cotton fibres extracted from the seeds during ginning and pressed into bales in the cotton ginning plant. Lint cotton arriving at the spinning mill contains varying amounts of foreign matter such as seed fragments, dust and particles, all of which are referred to as trash1.

The main objectives of ginning are:

1. To separate the cotton fibres from the seed.

Fig. 1 Opened seed cotton boll

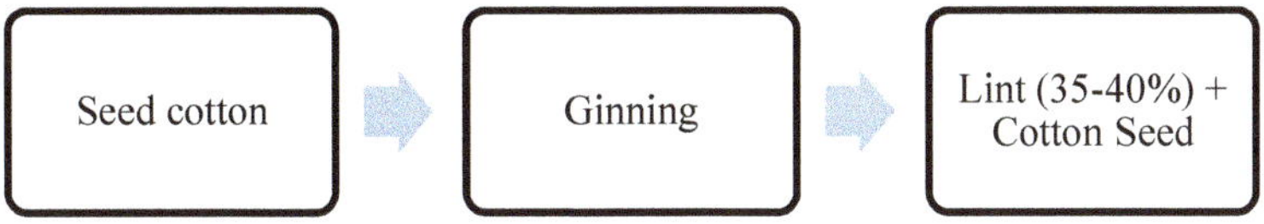

Fig. 2 Simplified ginning process. *Source* Authors

2. To eliminate foreign matter that lowers the value of the cotton bale. Under the current marketplace conditions, foreign matter removal offers the maximum total monetary return for the resultant lint, seeds, and so on.
3. Moisture Control.
4. Gin the cotton with the least amount of loss in fibre spinning quality, so that the cotton meets the needs of its final users, the spinner and the customer. A perfect ginning procedure would be one in which the fibres and seeds were separated without any damage to the fibres or seeds.

Before going into the ginning practices in Ethiopia, the next section provides a general overview of the ginning machines to understand the process and its impact on lint quality.

2 Seed Cotton Storage

2.1 Seed Cotton Storage Technology and Developments

Trailers were the major means of transporting harvested cotton from the field to the cotton gin for many years, as seen in Fig. 3. Cotton was generally manually picked until the mid-1940s, and trailer capacity had a minimal influence on harvesting timeliness. With the introduction of automated harvesting, harvesting capacity grew substantially, making trailers the limiting element in the harvesting-ginning system. Larger and more trailers, as well as steady increases in ginning capacity, were unable to keep up with harvesting capacity. Harvesters would therefore be idle until cotton was ginned from trailers and delivered to the producer. This prolonged the harvest season, exposing remaining cotton in the field to adverse weather conditions, resulting in production and quality losses [40, 41].

The new approach is now known as the module system of seed cotton storage. Since 1972, there has been a continuous growth in the usage of modules for storing harvested cotton [30]. A cotton module is a freestanding stack of cotton that is created by pouring collected material into a shape known as a module builder.

The Module Builder: A module builder is fitted with a mechanism that compacts the harvested material to a density of around 12 lb/ft^3, allowing the stack to remain freestanding when the module builder is removed. The entire modules are transported using specially designed self-loading trucks and trailers. The modules' function is

Fig. 3 Cotton seed trailers machine. *Source* [41, 40]

to offer reservoirs where harvested cotton can be held when it cannot be ginned immediately, allowing the harvesting procedure to operate independently of ginning. When harvesting is not feasible due to inclement weather, gins can draw from the reservoir. Having a supply of cotton results in a more predictable, managed and cost-effective business [30].

Module is a system where cotton is dumped from the harvester into the module builder and then compacted into a trapezoidal shape. Originally, the modules were built on pallets made of wood or metal. Further studies found that the cotton modules could be formed and stored directly on the ground with no appreciable loss in cotton quantity or quality, as long as the soil surface was well drained. Viewed as a system, the module builder, transporter and feeder represented a revolutionary change in the way cotton was handled between the field and the ginning. The modular system was quickly adopted wherever cotton was mechanically harvested as it resulted in significant operational efficiencies in both harvesting and ginning [39, 40].

The original modular system concept has been further developed with many changes and improvements. The module builder deployed in the United States has retained the standard dimensions of 7.5 feet wide and 32 feet base. The height of the modular assembly can vary from 9 to 11 feet, with the more common height being 11 feet. The choice of heights depends on other equipment used in the harvesting system. Other improvements have included hydraulic systems, controls, extended wire sides to allow air to escape during deflation and operator cabs. A modern module builder is shown in Fig. 4.

The biggest change since the early development of the modular builder was larger capacity cotton harvesters. The working width of the harvester was gradually increased from 2 to 4 rows and then to 6 rows and 8 rows. The wider multi-row units cannot be dropped directly into module builders due to the header width. To accommodate the wider machinery and to keep the expensive harvesters running, a system of carts called Boll Buggies was developed to transport the harvested cotton

Fig. 4 A modern module builder with cab for operator comfort and hydraulic drive system. A tractor provides PTO power to drive the hydraulic pump and drawbar power to move the builder (Photo courtesy of KBH, Inc.)

from the field to the modular growers, usually located at the edge of the field. The boll buggy (Fig. 5) pulled by a tractor follows the harvester through the field until the harvester bucket is full. The boll buggy pulls along the side of the harvester, picks up the cotton that has been unloaded from the harvester basket and transports the load to the module builder [40, 41].

In the late 1970s and early 1980s, engineering efforts to develop a cotton picker that would also form modules began, but it was not until 2007/09 that these systems became commercially available. The Case IH machine makes a 16-foot-long unit (half a module) that needs to be covered after unloading. The John Deere cotton picker forms seed cotton into a round module weighing about a quarter the weight of a conventional module and encloses this module with a plastic film forming the cover (Fig. 6).

Fig. 5 The boll buggy used to transport harvested cotton between the harvester operating in the field and the module builder located at the edge of the field (Photo courtesy of KBH, Inc.)

Fig. 6 Conventional, half-length and round modules stored in a gin yard. Photo was taken after rainfall and some modules have water collected on the cover surface. *Source* [40, 41]

These machines represent another attempt to increase the efficiency of cotton harvesting. By performing harvesting and module building operations, these machines can eliminate the need for boll buggies, module builders, tractors and manpower. Achieving this level of efficiency, however, required trade-offs. These machines form smaller modules and have significantly higher acquisition costs than conventional cotton pickers. While these new machines produce different module shapes and sizes compared to traditional modules, there remains a need to maintain the quality of the seed cotton in the package [40, 41]. Three types of modules are currently in use: conventional modules, Case IH half-length modules and round John Deere modules.

The conventional module practised in forming module forming with the help of a module builder making 32-foot-long modules with a trapezoidal cross-section from a mechanical harvester may have different dimensions of the forming chamber depending on the country. After forming, the modules must be protected from the weather. Protective covers for the conventional module are available in various designs, materials and prices. The design differences include the material used for the top, the skirt design (shape, depth and material), the presence of fastening loops on the side skirts and the type of fastener (rope, belt, etc.).

The modules formed by the Case IH Module Express cotton picker are 16 feet long, rectangular and trapezoidal in cross-section. These modules are protected by a cover similar to traditional modules but shorter. The number of manufacturers who sold covers for these half-length modules was initially limited. But as the number of Module Express cotton pickers in use increases, the choice of makes and models is expected to increase as well. Some people have tried placing two 16-foot modules end-to-end and using a 32-foot cover over both, but this practice is discouraged. Figure 7 shows two half-length modules joined together, one of which is loaded into a module trolley.

Fig. 7 Two case IH half-length modules staged together and being loaded into a module truck. *Source* [40, 41]

The John Deere 7760 Cotton Picker forms round modules that are completely encased around the perimeter by a specially developed polyethylene film that protects the cotton while providing a compressive force to maintain module density. John Deere also sells the upholstery materials. The sheet materials are sized to overlap a few inches at the ends of the circular module, preventing water flowing on the soil surface from penetrating the seed cotton. Figure 8 shows round modules with plastic film wrapped at the end. Figure 9 shows how a round module with a three-point linkage is provided for transport to the ginning site.

Various types of trucks, semi-trailers and agricultural semi-trailers are available for loading and transporting modules on public roads. Special movers are used on gin

Fig. 8 Deere round modules stored for transportation to the gin. Note the yellow plastic wrap forms a lip on the ends of the modules. *Source* [40, 41]

Fig. 9 Carrying a round module with a three-point hitch attachment. The module is being staged for transport to the gin. *Source* [40, 41]

farms to transport modules from the warehouse to the gin. The highway transporters and gin yard movers are designed for easy transfer of modules onto the feeder platform. Another special charger, a so-called portal loader, is able to pick up a module, lift it and place it on a flatbed truck or trailer. Straddle loaders are also used to unload modules from trucks or trailers and to transport modules from their storage location in the gin yard to the module feeder platform [30]. A safe storage location for modules should be well-drained, free of gravel (stalks and debris such as long grass), smooth (firm and nearly constant), accessible in wet weather, away from busy roads and other possible sources from fire and vandalism, and free from obstacles such as power lines.

Experiences across the Cotton Belt have shown that storing seed cotton in modules can benefit farmers and giners when the crop volume for the ginning community is greater than the ginning volume and when there is sufficient volume to justify the investment. Trailers are an efficient and inexpensive way to deliver cotton to the giner but can become expensive when used to store seed cotton. Farmers and ginners must decide together whether they want to use the modular system or not. The gin must be equipped for module storage and handling. Roads and bridges to be used by modular trucks must be able to handle gross loads of up to 50,000 pounds.

2.2 *Quality Changes During Storage*

Several factors influence seed and fibre quality in seed cotton storage. The most critical factor is moisture content. Other variables include storage length, amount of high moisture, foreign matter, variation in moisture content throughout the stored mass, seed cotton initial temperature, seed cotton temperature during storage, weather factors during storage (temperature, relative humidity, rainfall), and protection of the cotton from rain and wet ground. The study's findings give important guidance for understanding how storage conditions impact quality. These findings clearly demonstrate that, while it is hard to forecast exactly how storage factors would impact quality, suggestions for safe, effective storage are beneficial [30].

Lint Quality: According to studies, lint can withstand higher levels of moisture than seed before losing quality. Previous research on the subject concluded that drying or aerating seed cotton to retain quality during storage was impossible. The advised technique was to utilize excellent harvesting practices and preserve only low-moisture seed cotton. Furthermore, study results demonstrate that seed cotton may be securely kept in high-density bales (25 lb/ft^3) if specific measures, particularly regarding moisture, are taken. Lint yellowing was seen in lots with seed cotton moisture levels over 10.5%, however, the amount of yellowing was not statistically significant. There was no heating in bales with initial moisture levels of 12% or below. In addition, research conducted in central and south Texas (USA) suggests that some minor spotting occurs in seed cotton kept at moisture levels ranging from 13 to 15%.

A quality comparison between cotton stored in modules and trailers shows no significant differences at 11–13% humidity. A study conducted in California (USA) shows that changes in both the temperature at harvest and the average ambient temperature during storage have a moderate impact on yellowness. On the other hand, the moisture content at values above 13–14% led to a sharp increase in yellowing, especially when the storage time exceeded 45 days. For longer storage, the humidity should be below about 12% [30].

Yellowing is enhanced at high temperatures, therefore both temperature rise and maximum temperature are critical. Temperature rise is most likely caused by heat created by biological activity rather than heat absorbed from the environment. Seed cotton with a moisture content of 12% or less will allow for safe, long-term storage if cultivation, harvesting, and storage procedures are followed. Higher seed cotton moisture levels can be tolerated for brief storage periods. The rate of lint yellowing, on the other hand, begins to accelerate significantly at moisture levels over 13% and can continue to increase long after a module's temperature declines.

Seed Quality: When storing seed cotton, the length of storage time is important to maintain seed quality and should be based on the moisture content of the seed cotton. If the relationship between moisture content and storage time is not understood, seed quality will be lost (germination is reduced and free fatty acid and aflatoxin levels are increased). Table 1 shows recommendations from studies by the US Agricultural Research Service (ARS) for safe storage. The studies were conducted on small quantities of seed cotton with densities up to 12 lb/ft^3.

The moisture content of seed cotton during storage is the most important variable affecting seed germination and oil quality. A general, conservative recommendation is that the moisture content of cottonseed should not exceed 10% in module storage

Table 1 The moisture content versus storage time of cotton seed

Moisture content of seed, wet basis (%)	Maximum safe storage (Days)
8–10	30
10–12	20
12–14	10
14–15	<3

Source Adapted from [30]

when seed is held for sowing. The quality of the oil, on the other hand, appears to be less sensitive and can be preserved at a moisture content of 12% during storage.

Module Moisture Monitoring: If seed cotton is to be stored safely without lint or compromised seed quality, it should be harvested below 12% moisture content and contain a minimum of green waste. Water added at picking can also be a major factor in creating excess moisture. Ideally, farmers should monitor seed cotton moisture before and during harvest so they can prevent the formation of moisture modules by either delaying harvest or ginning the cotton just before fibre quality is lost. In a module, a rise in temperature is a retrospective indicator of excessive humidity and possible avoidable degradation. Regular moisture readings with a meter are important, especially early and late in the day when moisture levels are at their highest.

Portable moisture meters should be calibrated periodically according to manufacturer's procedures to ensure they are providing accurate readings. If measuring devices are used, the samples should be representative of the machine (rather than hand) harvested material. The meter should be clean and dry and the samples should be hand mixed (preferably wearing latex gloves) to mix any surface moisture into the seed cotton. Each sample should be filled firmly into the measuring cup.

Moisture readings should be taken two or three times per sample and the readings averaged. Between measurements, the sample should be removed from the meter and mixed. Latex gloves should be worn when handling seed cotton as moisture from hands can get onto the fibre and cause false readings. When manufacturing modules, it is possible to automatically monitor and record the moisture content of each individual module using a meter mounted on the module builder's hitchhiker's foot. Each time the ram reaches the bottom of its stroke, a measurement is taken in the compacted cotton. The readings are stored in a computer, which determines the average and maximum moisture content of the module [30].

Module Temperature Monitoring: Portable moisture meters should be calibrated periodically according to manufacturer's procedures to ensure they are providing accurate readings. If measuring devices are used, the samples should be representative of the machine (rather than hand) harvested material. The meter should be clean and dry and the samples should be hand mixed (preferably wearing latex gloves) to mix any surface moisture into the seed cotton. Each sample should be filled firmly into the measuring cup. During storage, the module temperature should be checked daily at at least six locations for the first 57 days. In addition, the probing can be done every 34 days or depending on the temperature. The temperature probe should extend at least 2 1/2 feet into the module. Figure 10 shows a typical temperature diagram of modules stored at two humidity levels. For modules harvested in safe storage, humidity generally will not rise more than 1015°F for the first 57 days and will then moderate and even cool over time in storage. A rapid and sustained rise in temperature of 1520°F or more during the first few days generally indicates a moisture problem. If a temperature of 120°F is reached or the temperature rises more than 20°F, the module should be gutted immediately to avoid the possibility of a major loss. High moisture modules (particularly those harvested late in the season when ambient temperatures are low) may continue to increase in temperature slowly over several weeks if not ginned immediately. All modules should be inspected on a weekly basis and promptly after a storm to discover damaged or missing tarps, water pools in depressions, or rainfall leaking through the tarps. Water should be drained from tarp depressions regardless of tarp material. If water has entered the module, it should be ginned as quickly as possible. Check for excessive surface water or moisture around the base of the modules, as well as for any theft or vandalism.

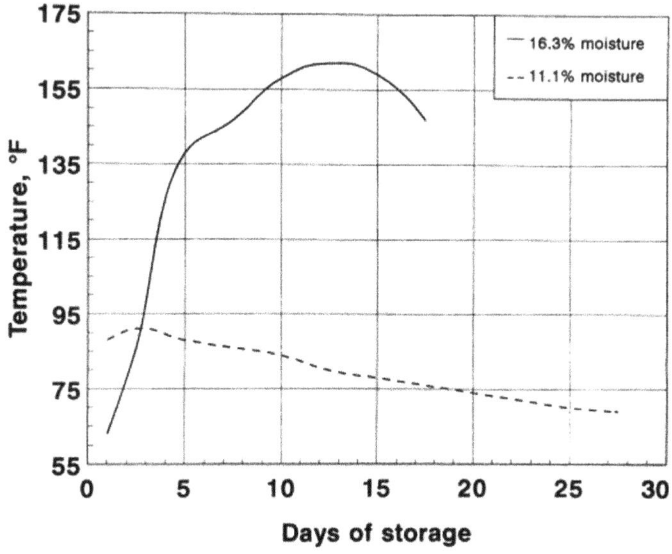

Fig. 10 Typical temperature rise as a function of seed cotton moisture content and days of storage. *Source* [30]

2.3 Seed Cotton Storage Experiences in Ethiopia

The transaction of seed cotton from small-scale cotton farmers who produce 30% of the country's seed cotton involves many actors until it reaches the ginning factory. The way seed cotton is handled and stored by each of the actors in the line does not allow for the preservation of the quality of lint cotton and cotton seed as harvested due to a lack of knowledge and opportunities. On the other hand, the transaction of seed cotton between commercial farmers and the ginning plant may not be complex. Commercial farmers are also expected to have better seed handling, storage and transportation capabilities than smallholders. In general, the storage and transportation of cotton seed to the ginning facility varies for small farmers and tradespeople [43, 44].

Seed Cotton Storage by Smallholder Farmers: Unlike other crops, seed cotton is mostly not stored at home by small-scale cotton producers because it is inherently voluminous and bulky and requires a lot of storage space. Additionally, seed cotton storage requires industry standards to maintain fluff and seed quality when harvested. Compliance with seed cotton storage requirements is difficult for small farmers due to lack of knowledge, technical ability, finance, lack of space, etc. Proper seed cotton storage is a critical first step in preserving the spinnability, colour and purity of cotton seed from foreign cotton impurities. Not only fluffy cotton but also cotton seeds are affected by poor storage conditions.

Research shows that small-scale cotton farmers in Ethiopia employ different practices in post-harvest handling and storage of seed cotton. Some farmers sell immediately after harvest, while most of them store seed cotton for up to six months and expect the price to increase in the future [41]. Small-scale cotton farmers in remote areas who lack access to paved roads and transportation, especially during the rainy season, must face major problems in transporting seed cotton to markets. Therefore, they either sell the cotton to farmers living in their area at a relatively low price, or are forced to store the seed cotton until weather conditions allow it to be transported to the local market or ginning facility [41]. In return, local farmers sell seed cotton to craft workshops, retailers, primary cooperatives or cooperative associations [Feyso, 2018].

One of the challenges is the unavailability of storage facilities for small cotton farmers [Feyso, 2018]. The small-scale cotton farmers store cotton in simple huts built of unwalled covered grass, or store it around the homestead in very poor conditions. After filling, the seed cotton is stacked in sisal or polyethylene bags [41]. These storage practices potentially affect the quality of seed cotton as it is sensitive to environmental and weather conditions. The two types of sacks (bags) differ in cost and quality. Sisal bags are the most used by farmers to pack cotton because they are sturdier and can hold a larger volume. Farmers who have access to wagons and tractors tend to use sisal sacks for packaging, while those far from local markets tend to use polyethylene sacks (bags) to facilitate transport by pack animals (Fig. 11). Polyethylene bags are cheaper and of higher quality than sisal bags.

Fig. 11 Seed cotton transport to ginnery using pack animals [Feyso, 2018]

Seed Cotton Storage by Commercial Farmers: Commercial cotton farmers typically store seed cotton in the field until harvest is complete. The duration of storage in the field depends on the time of harvest. Some commercial farmers simultaneously harvest and transport the seed cotton to the ginning facility. After harvesting, the seed cotton is usually filled into polyethylene bags and transported by truck to the ginning facility, as shown in Fig. 12. The details of Ethiopian practices of storing and transporting seed cotton to the ginners are discussed in the next section.

Fig. 12 Seed cotton transportation to ginnery using trucks [Authors, 2023]

When storing seed cotton in the field, any delay in harvesting means a quality risk and a loss of quantity due to bad weather. Therefore, there is a strong motive to pick the crop faster than it can be ginned and storage is required. The form in which cotton is stored depends on many factors. Special bearings offer maximum protection. However, field storage is also practised. Seed cotton can be stored successfully if certain precautions are observed. The most important factors that determine the safety of seed cotton storage are moisture and foreign objects. Clean and dry cotton can be stored indefinitely without any loss of quality. Cotton must be picked when the relative humidity of the environment is below 50%. Damp cotton should be ginned immediately. Portable moisture meters used at ginning plants do not provide accurate readings, but there is a high correlation between their readings and the relative humidity of the environment. Other variables affecting fibre quality include length of storage, variation in moisture content throughout the stored mass, initial temperature of the seed cotton during storage, weather during storage (temperature, relative humidity, rainfall), and protection of the cotton from rain and wet ground. Seed cotton should be stored on clean and well-compacted ground, since when storing it on loose soil or on dried grass, garbage will accumulate along the field. In general, seed cotton should be stored in compliance with the following requirements:

- The storage should be dry, moisture-controlled, contamination-free, neat and have clean surroundings;
- Picked seed cotton should be allowed to dry in shade as excessive sun exposure can cause yellowing of cotton, lowering the grade;
- Grading should also begin in the field during picking to minimize foreign matter in the seed cotton;
- Seed cotton should be stored at the farm and/or warehouse for a short time.

Seed Cotton Storage at Ginning Factory: Seed cotton is delivered to ginners packaged in either sisal bags or polythene bags and remains in the packages until ready for ginning. At the ginning site, the packaged seed cotton must be placed on pallets to avoid contamination and sometimes flooding. During storage, seed cotton must be packed in separate packaging and not in a heap to avoid colour distortion of the cotton. Wrapping and packaging materials must be cotton material. The packaged seed cotton must have a label stating the variety, field, lot number, load number and any other required information. A batch of seed cotton may not exceed 30 tons.

3 Seed Cotton Transportation from Farm to Ginnery

During transport to the ginners, the seed cotton should be placed in clean trucks. When loading the cotton into the trucks, immature, rotten and damaged bolls should be picked out. To ensure purity, only one variety should be loaded into a truck. Seeds picked individually in the trucks should be loaded at once. Cotton from the last harvest should be kept separately. Cotton should be covered with cotton cloth during transport (USDA 1994).

The practice in Ethiopia is that after harvest, seed cotton is generally stored in an open area on the dust (dirt) of the field. On the farm, different plots are unnecessarily mixed up (Fig. 13). It is then trucked to the ginners in either jute or polypropylene sacks. Sometimes seed cotton is stored on the farm in sacks with the mouth open and exposed to dust (Fig. 14). After loading, the trucks are not covered with traps to prevent rain and dust. Certain amounts of seed cotton are lost from sacks by the wind during transport, as shown in Fig. 15.

The transportation of seed cotton from the farm to the ginning factory is one of the areas where the quality of the cotton fibre should be preserved. However,

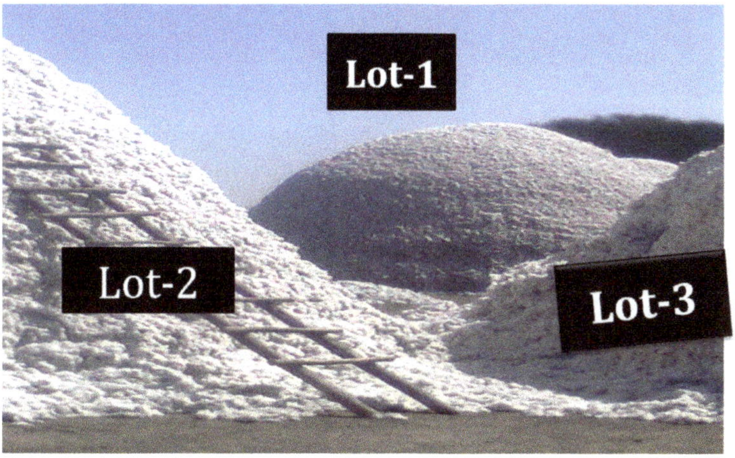

Fig. 13 Lots mixed unnecessarily at farm area

Fig. 14 Storage at farm on full of dust floor

(a)

(b)

Fig. 15 Seed cotton transportation with worn-out bags: **a** using PP bags **b** jute bags

there are poor seed cotton handling practices during transportation, such as: the use of uncleaned trucks before loading, the transport of mixed batches of different seed cotton varieties together without separation, and the use of old, worn and torn polypropylene and jute sacks. The causes of these practices are poor surveillance during transport, lack of awareness of the severity of cotton contamination, insufficient supply of cotton bags and the increasing preference for polypropylene and jute bags. In addition, cotton bags are expensive and cannot be used for more than a year due to their lower strength. The effects of these practices are far-reaching and require additional efforts in the cleaning process, which unnecessarily increase costs, reduce the quality of the end product, which can lead to outright rejection of the shipment, reduced machine efficiency in spinning and weaving processes, damaged machine parts and fires and waste caused by colouring material.

4 Ginning Process

4.1 Seed Cotton Feeding to the Gins

Hand feeding could never provide smooth and consistent feeding, which usually results in up to 20% efficiency loss and has a significant impact on production costs. Consistent feeding of the seed cotton to the drying, cleaning and ginning machines is very important to achieve proper efficiency of these machines. In recent years, some machines such as cotton dispenser and cotton feeding control box have been successfully introduced to achieve smooth feeding, which has benefited cotton ginning immensely. The introduction of these devices has improved the efficiency of processing machines by over 20% compared to manual feeding [37].

Ginning machines work more efficiently when the seed cotton feed rate is constant. In early ginning, cotton flow varied during feeding. The automatic feed control was developed to solve the problems by providing a smooth flow of cotton to the ginning drying and cleaning system. A mechanical module feeder also performs a similar function and can be used to feed seed cotton directly from the module (http://www.cotton.org/).

As discussed in Sect. 2.1, modular trucks and trailers transport cotton from the field to the ginning. A pneumatic system removes the cotton from the trailers and either a pneumatic system or a module feeder removes the cotton from the modules. A combination of conveyor belt and pneumatic system transports the cotton to a separator and feed control unit. Prior to this first separation point, some giners use a stone and greencap trap for preliminary trash removal. The screen assembly in the separator allows air to escape but collects the cotton and allows it to fall into the feed controller. The conveying air then flows from the separator to a cyclone system where it is cleaned and vented to atmosphere.

The green boll trap is important for removing green bolls, stones and other heavy foreign objects from coarse cotton. These large, heavy materials should be removed early in the ginning system to avoid machine damage and preserve fibre quality. Green boll traps use sudden changes in flow direction and/or reduced air velocities to separate heavy foreign matter from seed cotton. A typical green boll trap is shown in Fig. 16.

The most important factor in quality assurance in ginning is the moisture content of the fibres. At higher humidities, cotton fibres are stronger, but dirt is harder to remove and cleaning machines are less efficient. Consequently, the choice of moisture content for ginning is a compromise between good dirt removal and quality preservation. Under most conditions, cotton should be ginned with a lint moisture content of 67.5%. The tower dryer is the most widely used gin dryer. Tower dryers typically have 16–24 shelves arranged in such a way that the cotton must slow down as it moves through the machine (Fig. 17). Heated air transports the cotton through the shelves in 10–15 s. In order to control the degree of drying, the temperature of the conveying air is regulated. To avoid fibre damage, the maximum temperature in the drying system should be kept below 350°F. The temperature control sensor should

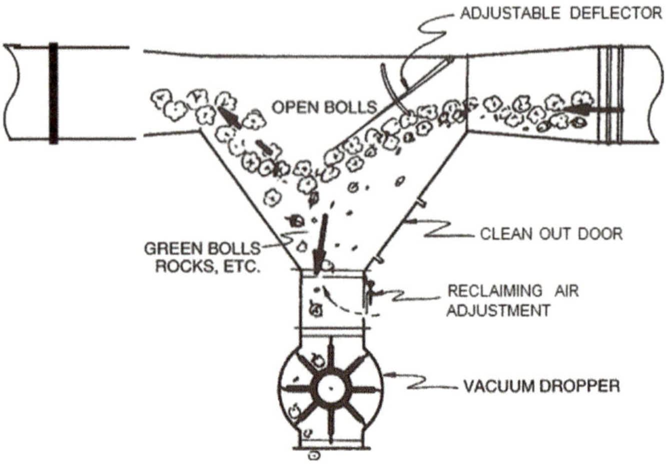

Fig. 16 The green boll trap removes heavy objects such as green bolls and rocks from the seed cotton [36, 44]

be located near the dryer inlet and a maximum temperature limit switch should be placed between the burner and the mix point to keep the temperature below 350°F. If the temperature control sensor in the gin is near the bottom of the dryer, the reading can be 200–40° lower than the temperature at the mix point.

Various types of mechanical, pneumatic and electromechanical conveying systems are used to feed seed cotton to individual twin-roller (DR) gins. For the feeding of each individual gin, an overhead distributor conveyor system with automatic controls was most preferred for each line (Fig. 18). However, this also required additional capital

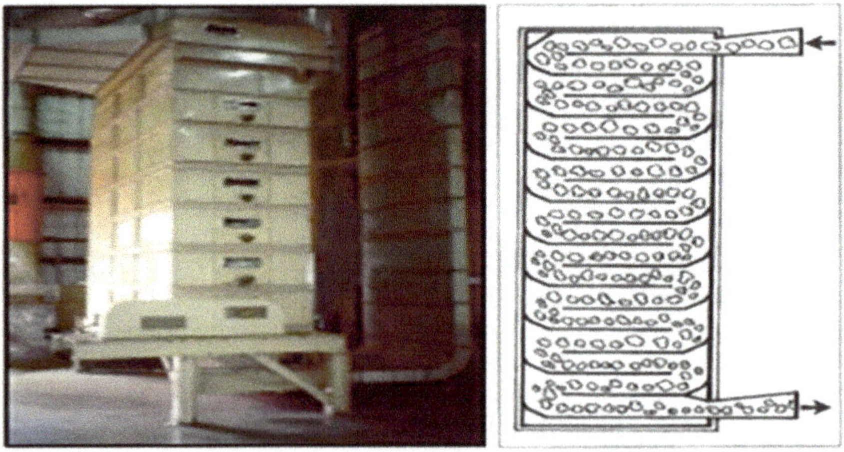

Fig. 17 The tower drier reduces the seed cotton moisture content for smooth ginning. *Source* [36, 37, 44]

Fig. 18 Twin auto regulator individual gin feeding conveyor system. *Source* [40]

costs and additional power consumption. In 2012, a double-line overhead distribution system with advanced automatic regulation for simultaneous feeding of two gins on two parallel lines was introduced, which has revolutionized individual ginning feeding for twin-roller gins, reducing investment costs and power consumption. This system has also introduced a slide where one can control the escaping of short fibres coming out of seed cotton during the feeding system and can be covered up from the feeding point to the Auto Feeder, resulting in the control of spreading of small fibres in the ginning area and is environmentally friendly.

4.2 Pre-cleaning

Seed cotton can be picked by hand or by machine. There are currently two main types of machines used for seed cotton harvesting. These are cotton pickers and cotton scrapers. Seed cotton picked with all existing picking methods requires different treatments to clean the seed cotton.

4.2.1 Seed Cotton Picking

The picking methods of seed cotton, i.e. hand picking, manual stripping, machine stripping and machine stripping, etc., require different treatments to clean seed cotton. Seed cotton is mainly hand-picked in Ethiopia and should normally not contain any waste. However, for various reasons, Ethiopian cotton is considered trashy. In addition to plant parts such as stems, twigs, unopened seed pods, leaves, burrs and waste, it also contains foreign substances. International experience shows that when hand-picking, workers sometimes pick cotton with large amounts of waste such as leaves, stalks, bracts, and immature and unopened bolls. The waste content in the hand-picked seed cotton is 46% [37].

The contaminants commonly found in hand-picked and traditionally handled, packaged, shipped and stored cottonseed are identified and classified as described in Table 2. Loose sisal sacks are used to transport the cotton to market and from there to the ginning, and the sacks are considered the source of the sisal fibres. Cotton canvas bags are currently used in some places to avoid contamination. At the ginning facility, cotton is often dumped on unpaved ground, opened and laid out to dry in the sun for further processing. At each of these stages, it picks up sand, dust and other foreign matter. Therefore, seed cotton can naturally contain impurities and foreign substances during harvesting, transport and storage. With modern harvesting, packaging and storage, contamination does not occur.

There are currently two main types of machines used in cotton harvesting: cotton pickers and cotton strippers, with or without field cleaners. These different types of machines package seed cotton with different amounts of burrs, rods, granules and leaves.

The amount of burrs, sticks, dust grains and leaves in the seed cotton varies depending on the type of cotton harvest by pickers and strippers, as shown in Table 3.

The trash content of seed cotton varies greatly because of the various harvesting methods used and the year-to-year variability in the weather throughout the cropping season. Trashy cotton, in principle, requires pre-cleaning so that only pure seed cotton is sent into the ginning machinery.

4.2.2 Seed Cotton Pre-cleaning Technology and Developments

If no pre-cleaning is done, trash particles stick to the fibres during the high-pressure baling process. Trash cleanup in the blow room in spinning mills becomes difficult, costly, and harmful to fibre quality. Pre-cleaning is important to optimize gin stand function and lint quality. The phrase "seed cotton cleaning" is frequently used interchangeably by the ginning industry to refer to the process done by either the complete cleaning or extracting system of a gin or by certain types of machinery within the system. In a more limited sense, "cleaning" refers to the use of several types of cylinder cleaners designed primarily for the removal of dirt and small fragments of leaves, bracts, and other vegetative matter. "Extracting," on the other hand, strictly refers to techniques meant to remove large trash, such as burs and sticks, from seed cotton. Bur machines, stick machines, extractor-feeders and combination bur and stick machines are examples of extracting-type machinery [30].

The cleaning and extraction system of a modern gin performs two functions. First, large trash components such as burs, limbs and branches must be separated from the seed cotton so that the gin stand may run at full efficiency and without undue downtime. Second, seed cotton cleaning is frequently required to get optimal grades and market values, particularly when ginning high-trash-content seed cotton. Cleaners and extractors also assist open the seed cotton for more effective drying, which is normally done concurrently with cleaning. The quantity of cleaning and

Table 2 Contaminants in Ethiopian cotton. *Source* Authors

Plant contaminants	Inorganic contaminants	Handling and storage contaminants		
		Fabrics/Yarns/ Threads	Gin house origin	Others
Leaf	Sand, dust	Plastics and polyethylene films	Grease	Bird feathers
Stems, twigs	Soil particles	Sisal fibres, yarns, and threads	Oil	Papers
Cotton of other varieties	Stones	Cotton threads	Stamping ink	Leather pieces
Seed cot fragments	Metal pieces	Plastic ribbons	–	–
Cut seeds	–	–	–	–

Table 3 Typical trash levels for machine picked and stripped seed cotton

Type of trash	Trash level (lb/bale)	
	Machine picked	Machine stripped
Burrs	34	450
Sticks	9	115
Fine trash	26	110
Motes	30	25
Total	**99**	**470**

Adapted from Patil and Arude [37]

extraction apparatus required to clean cotton properly varies with the trash content of the seed cotton, which is heavily influenced by the harvesting procedure [30].

Most gins nowadays process either picked or stripped cotton and are normally equipped with simply the amount and kind of cleaning and extracting technology necessary for the most harsh circumstances expected in their trading region. For less severe circumstances, a portion of the system should be skipped to minimize significant weight losses and to limit the potential of over-machining the cotton. Cleaning of seed cotton should be limited to what is required to promote smooth, trouble-free ginning and optimum bale values.

Cylinder Cleaners: Cylinder cleaners are used to remove finely split particles as well as to open and prepare seed cotton for drying and extraction [35]. The cylinder cleaner is made up of a series of spiked cylinders, often four to seven in number, that agitate and carry the seed cotton over cleaning surfaces with small apertures or slots (Fig. 19). Cleaning surfaces might be concave screen or grid rod portions, or serrated disks like those used in impact cleaners. Foreign matter displaced from the seed cotton by the cylinders is collected and disposed of through the screen, grid rod, or disk apertures. A typical screen is made of 2-mesh woven galvanized wire cloth.

Although screen cylinder cleaners have largely been superseded by the more durable mesh cylinder cleaners, screen cleaners remain popular in some areas as they are the last inclined cleaner in the cleaning order. Grid sections usually consist of 3/8 inch diameter rods spaced about 3/8 inch apart (Fig. 20). Grid spacing greater than 3/8 inch provides additional cleaning but also results in more cotton being lost [29]. However, the Trash Master Cleaner (Lummus Industries) overcomes this problem by using a cotton seed recovery device to recover the cotton from the trash (Fig. 21). The Trash Master Cleaner uses a 5/8" grid spacing.

In another type of machine, the impact cleaner (Continental Eagle Corp.), cotton is transported over a series of rotating serrated discs that form a unique rotating lattice system. Cylinder cleaners can be further classified in terms of their use in gin. In this regard, they are either airline, air-fed or gravity-fed cleaners [24].

Extractors: There are two types of extraction machines: drill and rod machines. The drill machine was developed in Texas, USA, in the 1920s in response to the manual cracking and mechanical stripping of cotton. Once upon a time, the ginning

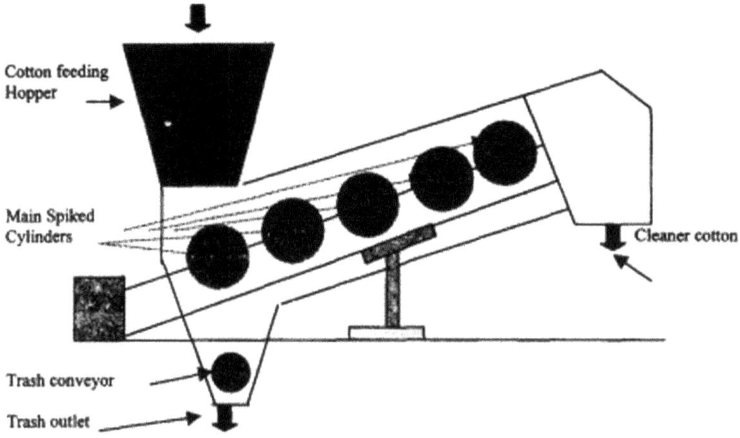

Fig. 19 Five-cylinder, air-fed inclined cylinder cleaner equipped with screen sections [37]

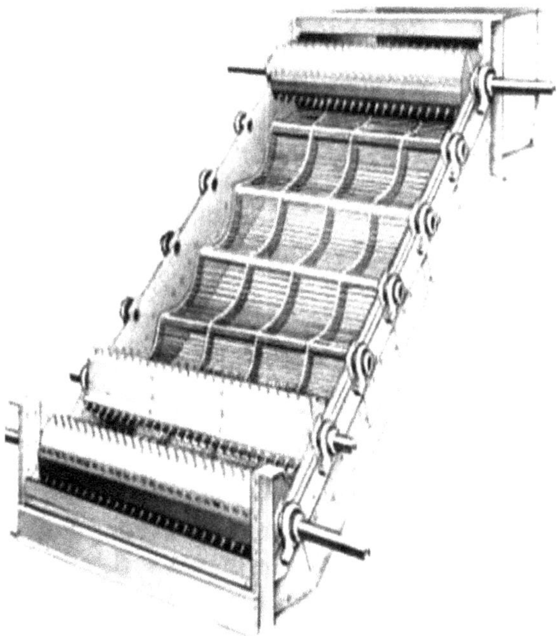

Fig. 20 Inclined cylinder cleaner equipped with grid-rod sections. *Source* Courtesy of Continental Eagle Corporation

machine was the most popular bulk extraction machine used in the ginning industry, particularly in areas where seed cotton was harvested. Due to its limited operating capacity and low stick removal efficiency, the drill machine has been largely replaced by other, more efficient extractors in most modern cotton ginning machines. The stick machines use the centrifugal action of high-speed saw cylinders to extract burs and

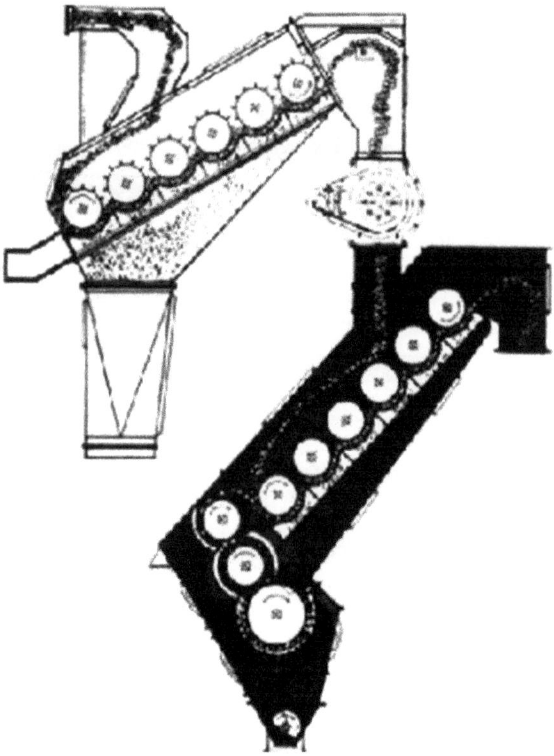

Fig. 21 Air-fed inclined cylinder cleaner feeding a trash-master cylinder cleaner equipped with a reclaiming saw cylinder. *Source* Courtesy of Lummis Corporation

sticks from seed cotton by centrifugal force [22]. Seed cotton is fed to the primary desling saw cylinder and wiped across the saw teeth by one or more stationary brushes (Fig. 22). Foreign matter and some seed cotton are thrown off the saw cylinders by the centrifugal force 2550 times that of gravity. Grid bars are strategically placed around the perimeter of the saw cylinder to control seed cotton loss and aid in the extraction process. However, some loss of seed cotton is inevitable for satisfactory cleaning. Additional saw cylinders are used to reclaim the seed cotton obtained with the drills and sticks. Reclaimer saw cylinders are similar to primary saws, but typically operate at slower speeds and have more grid bars.

Commercial stick machines vary widely in terms of the number of saw cylinders, the type and spacing of the bars, the sawing speed and size, and the position of the stationary brushes. However, all models have at least one main loop saw and one reclamation saw. In general, bar machines are classified as either two-saw or three-saw machines. In addition, a five-saw bar machine was developed as shown in Fig. 23 [10]. This machine has outperformed conventional extraction equipment in laboratory and field experiments. Dubbed a multi-stage extractor, the machine offers three stages of extraction in a single compact machine and shows promise for commercial adaptation.

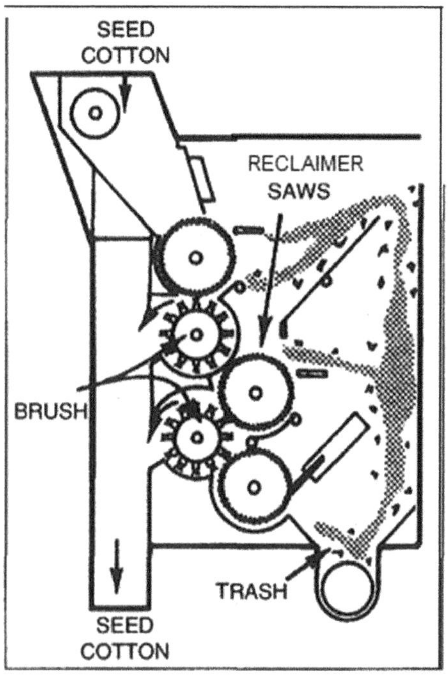

Fig. 22 The three-saw stick machine primarily removes plant parts 1–4 inches long. *Source* [36, 44]

Combination Bur and Stick Machine: The upper part of a combined drill-handle machine (CBS) resembles a drill machine in that it is equipped with a screw feed and waste extraction system and a large-diameter saw cylinder (Fig. 24). However, the CBS machine differs from a drill machine in several important ways. The CBS machine is not as wide, although its rated capacity is much higher. Seed cotton is generally fed across the width of a CBS machine as opposed to the end-feed method used on Klette machines. The CBS machine is fitted with a stationary stripper plate while the drill machine is fitted with a steel stripper roller.

The upper part of the CBS machine has a foreign object removal function that is not found on traditional drill machines. The foreign matter and about half of the seed cotton are discharged into the lower part of the CBS machine by throwing them off the large diameter saw cylinder. The discharged cotton is cleaned in the lower section, which, with minor modifications, consists of a standard two- or three-saw bar machine (Figs. 24 and 25). Thus, the upper part of the machine serves as the primary cleaner feeding the lower stick machine assembly [8].

The CBS machines have grown in popularity since their introduction, particularly in gins processing stripper-harvested seed cotton.

Extractor-Feeders: Gin stand feeders with extraction capabilities have been utilized since the early 1900s [11]. Early extractor-feeders employed the same stripping or dislodging mechanism as bur machines. However, with the invention of the stick machine, most manufacturers abandoned the stripping idea in favour of the stick

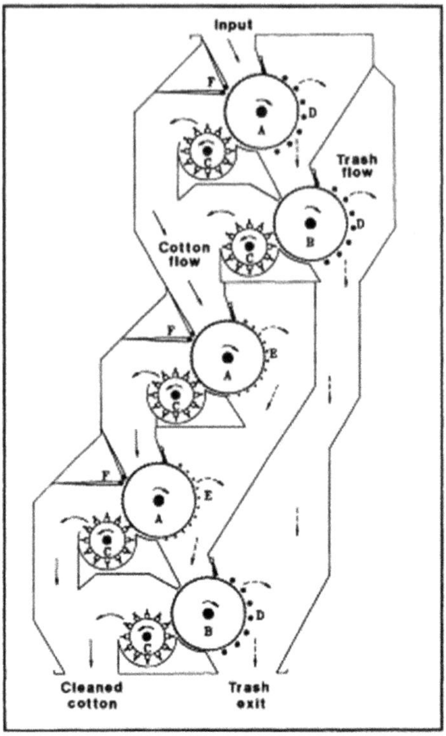

Fig. 23 Multistage bur and stick extractor: **a** Sling-off saw cylinders; **b** Reclaimer saw cylinders, **c** Doffing brushes; **d** Large-diameter grid bar sets; **e** Small-diameter grid bar sets; **f** Cotton bypass valves. *Source* [10]

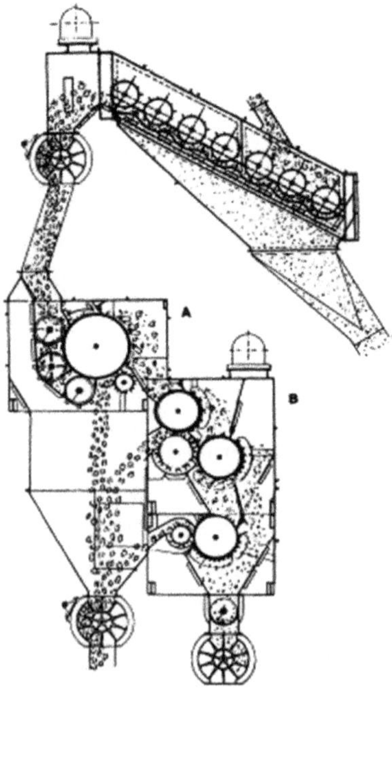

Fig. 24 Air-fed inclined cylinder cleaner feeding a combination bur and stick machine. **a** Upper feeding and extracting section; **b** lower three-saw stick machine unit (Courtesy of Consolidated Cotton Gin Co., Inc.)

Fig. 25 Combination bur and stick machine equipped with lower, two saw stick machine. *Source* Courtesy of Lummus Corporation

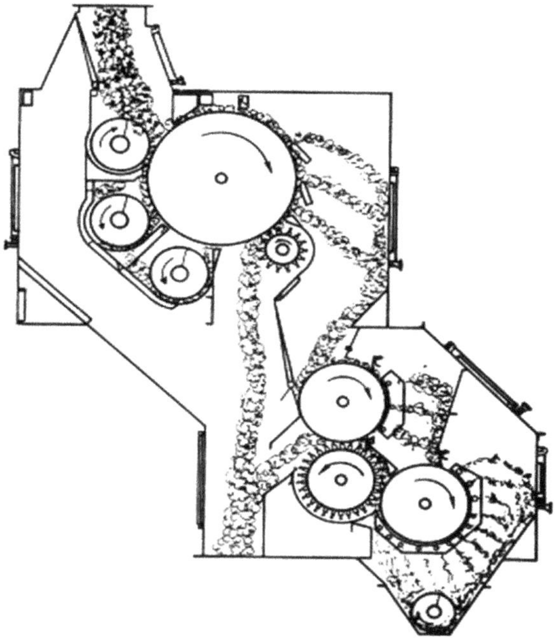

machine's more efficient sling-off function. In addition, the development of high-capacity gin stands in the late 1950s compelled manufacturers to simplify and streamline feeders in order to attain required capacities. This was done by utilizing the sling-off function for extractor-feeders (Fig. 26).

In addition, many modern extraction feeders improved fine waste removal through the use of cleaning cylinders. The primary function of a modern, high-capacity extractor feeder is to feed seed cotton to the ginning stand evenly and at controllable rates, with extraction and cleaning being a secondary function. The feed rate of the seed cotton is controlled by the speed of two star-shaped feed rollers located on top of the feeder just below the distribution hopper. These feed rollers are driven by variable speed hydraulic or electric motors and are controlled manually or automatically by various interlocking systems within the gantry. The drive can be designed to start and stop automatically when the ginning breast is activated or deactivated. The system can also be designed to stop or start the supply of seed cotton in the event of overload or underload of the ginning stand. Many of the systems are designed to maintain a constant seed roll density. This is typically accomplished by regulating the speed of the feed rollers in response to feedback control signals from the ginning station. The signals are based on monitoring the power consumption of the electric motor driving the ginning tower, measuring the displacements of the cove board in the seed roller box, or monitoring the pressure required to drive the hydraulically driven seed roller agitator.

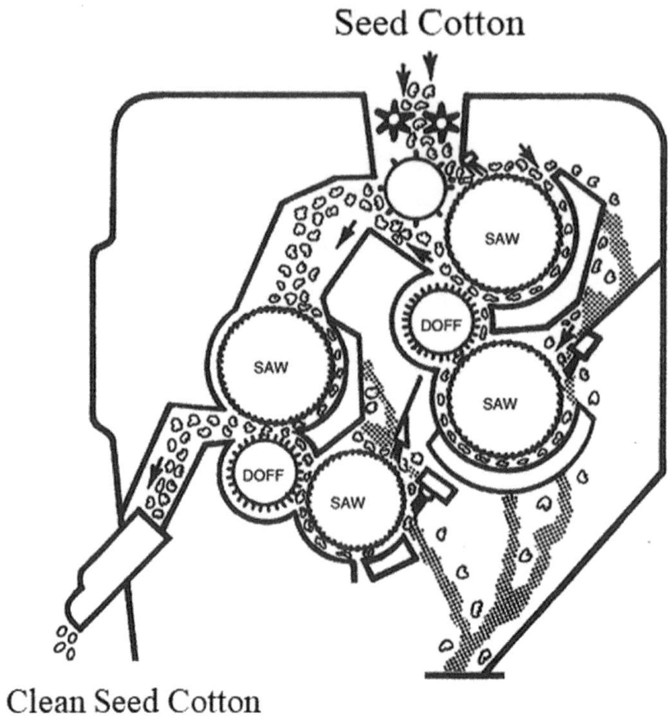

Fig. 26 Extractor feeder equipped with extracting and reclaiming cylinders feeding a saw gin stand. *Source* [36, 44]

4.2.3 Cleaning and Extracting Efficiency

The efficiency of a cleaner or extractor depends on many factors, including machine design; cotton moisture content; processing speed; settings, speed and condition of the machine; the amount and type of trash in the cotton; distribution of the cotton over the machine; and the cotton variety. At the beginning of the cleaning process, a high percentage of the dirt in cottonseed consists of large, loosely attached particles that are relatively easy to remove. However, later in the process, the remaining dirt particles tend to be smaller and more entangled in the fibre than the initial particles. This makes cleaning more difficult later in processing.

The overall trash removal efficiency of cylinder cleaners is generally low. As a rule, however, they are not used alone, but in combination with other machines. Cylinder cleaners perform an extremely useful function in opening the cotton and removing fine debris. Studies using both machine-picked and machine-cleaned cotton have shown that the overall trash removal efficiency of a six-barrel, six-barrel inclined cleaner is generally between 10 and 40% [4, 8, 12, 39]. However, these efficiencies were based on the total trash content of the test cotton, including burrs and chopsticks, which were not removed to any appreciable extent by the inclined cleaners. The performance of an inclined cleaner is much more impressive when the efficiency figures are based solely on fine trash levels before and after cleaning. Fine trash removal efficiency of 50–55% has been observed with both grid-rod and screened inclined cleaners when processing stripped and picked cotton [4, 29].

Trash weight data from testing at the United States Department of Agriculture (USDA) Ginning Laboratories indicates that the Airline Cleaner performs similarly efficiently to a regular sloped cleaner. Some specific types of cylinder cleaners feature grille or disk openings that are larger than the 3/8" openings used in standard angle cleaners. Consequently, these special cleaners should be somewhat more efficient. The cleaning performance of a grinder largely depends on the amount of trash in the seed cotton. Efficiency is highest for cotton with a high trash content. When extracting machine-stripped cotton, the bur machine can be assumed to have a 45–65% efficiency in removing burs and an efficiency of 30–35% in removing sticks [6–8, 22]. Bur machines usually only remove small amounts of fine trash. In many cases, seed cotton contains more fine trash after extraction by the bur machine than before extraction because the bur machine tends to pulverize large trash particles. This tendency is particularly noticeable when processing seed cotton, which contains very dry and brittle burrs and sticks.

The cleaning efficiency of the bur machine is lower for machine-picked cotton than it is for stripped cotton. The bur machine is projected to remove 7–12% of the total waste from machine-picked cotton [39]. Fine trash removal from machine-picked cottons is also poor.

The cleaning efficiency of stick machines varies greatly depending on the condition of the seed cotton and machine design factors. For machine stripped cotton, a contemporary commercial stick machine may be anticipated to remove around 65% of the burs, 50% of the sticks, and 10–35% of the fine trash [6, 8]. The total cleaning efficiency for stripped cotton is generally in the 60–65% range for the most recent models. Because of the huge fluctuations in the number of burs and sticks present in this type of cotton, the efficiency for machine-picked cotton is very variable [4, 39].

The CBS machine is slightly more efficient than a stick machine at processing machine-stripped cotton [7]. While there is usually little difference between the two machines when it comes to removing sticky or fine trash, the CBS machine is more efficient at removing burrs. Overall, the CBS machine is expected to remove 47% more waste from seed cotton overall than a stick machine. Extractor-feeders are efficient cleaners. Seed cotton is usually well distributed when it enters an extractor-feeder and the feed rate through this machine is often lower than the feed rate of other seed cotton cleaning machines. Studies in which all seed cotton cleaners prior to the extractor-feeder were bypassed revealed that the extractor-feeder removes 70% of the hulls, 15% of the motes, and 40% of the remaining trash components, with an overall cleaning efficiency of about 40% for machine picked cotton [3].

Cleaning efficiencies for sequences of four seed cotton machines consisting of a cylinder cleaner, a stick machine, a second cylinder cleaner, and an extractor-feeder range from 40 to 80%, depending on the conditions stated earlier. Normal machine sequences for cleaning selected and stripped seed cotton were created not only to provide the necessary cleaning and extraction but also to mesh the machines conveniently. The number and types of machinery required under typical ginning circumstances were identified by research investigations done at the USDA ginning laboratory [9]. The precise configuration of the machinery, on the other hand, evolved gradually over the years, primarily in response to the functional requirements for managing seed cotton in the dryers and between the various machines. Typically, machine-picked seed cotton is processed by the following equipment in the following order: a dryer, cylinder cleaner, stick machine, dryer, cylinder cleaner, and extractor-feeder. The standard machine-stripped cotton sequence is an aircraft cleaner, dryer, cylinder cleaner, and combined bur and stick extractor, drier, cylinder cleaner, stick machine and extractor-feeder.

Cleaning and extraction machines are subject to considerable wear and tear due to the large amounts of dirt and soil particles contained in seed cotton. In addition, seed cotton occasionally contains stones, scrap metal, large pieces of wood, or other foreign objects that can damage various machine components. Accordingly, regular repair and maintenance of purification and extraction machines is required. Therefore, all machines should be checked regularly for excessive wear or damage. Worn or damaged machine parts should be repaired or replaced in a timely manner and adjusted according to manufacturer's specifications to maintain maximum operational efficiency. Worn machines or improper machine settings can also have a negative impact on fibre quality. Poor doffer settings or worn brushes, for example, can cause cotton to tumble around the saw cylinder. Recirculation contributes to low grades and overprocessing. Any machine issue that affects smooth and consistent feeding can also negatively impact quality. It is imperative that the seed cotton is evenly distributed across the width of the cleaning machines to avoid poor cleaning and excessive cotton wastage. Common wear and damage problems include:

1. Worn or bent grid rods, screens, or drum spikes in inclined cleaners;
2. Excessively worn, bent, or missing channel saws and stationary brushes in extractors;
3. Worn or damaged doffing brush cylinders in extractors or reclaimers;
4. Rough or corroded grid bars in extractors;
5. Damaged or ill-fitting doors, lids, or access panels and;
6. Worn or improperly tensioned belt and chain-drive assemblies.

Paying close attention to these and other related issues will significantly increase gin operational efficiency and avoid expensive downtime. Many clogs in cleaning and extraction gear are caused by malfunctioning rubber flights in seed cotton droppers and waste vacuum droppers. Vacuum droppers must be carefully maintained for smooth, trouble-free operation. Other operational issues can be traced back to seed cotton or garbage that accumulates inside machines over time and interferes with the smooth flow of material. To avoid these issues, all machinery should be examined and cleaned on a daily basis.

When cleaning and maintaining gin machinery, extreme caution is required. Many revolving cylinders in cleaners and extractors may seriously harm or kill negligent personnel. One should not attempt to unclog, clean, or service a machine while it is in use. Before opening access doors, lids, or panels, or removing guards on belts, chains, or gears, a machine should be totally stopped and electronically shut out for safety. After servicing machines, access doors, panels, and equipment guards should be reinstalled before restarting the machine.

4.2.4 Ginning

The fibre-seed attachment force varies depending on variety, field degradation, moisture content, and other conditions, but it is normally around 55% of the breaking force [5], implying that the fibres might be withdrawn from the seed without breaking. The gin stand, whether saw or roller, separates (pulls) the fibre from the seed and is the core of the ginning operation. The capacity of the system, as well as the quality and potential spinning performance of the lint, are determined by the operating circumstances and gin stand adjustment. Gin stands must be correctly adjusted, maintained, and operated at or below design capacity. Cotton quality may suffer if gin stands are overcrowded. The short fibre content increases as the ginning rate increases beyond the manufacturer's recommendation. With increasing sawing speed, the short fibre also increases. As the ginning rate increases, yarn defects also increase. Also, increasing the ginning speed can cause seed damage, especially if the seeds are dry. High ginning rate and low seed moisture cause seed damage ranging from 2 to 8% in gin stands. Thus, it is critical to keep the gin stand in excellent mechanical condition, gin at optimum moisture levels, and not exceed the capacity of the gin stand or other system components.

Roller gins were the first mechanically assisted way of separating lint from seed. Churka, reciprocating knife, and rotary-knife roller gins are examples of roller gins. The rotary-knife gin has a ginning rate of roughly 20% of the saw gin, which is 225–350 kg/hr of extra-long staple lint and 100–175 kg/hr of upland lint. Although saw gin has a higher production capacity than roller gin, it has the following disadvantages: fibre damage (increased SFC and reduced uniformity), increased neps, lower mill efficiency and inferior yarn quality [25].

Roller gins have a larger ginning outturn than saw gins, resulting in an increase in lint output of 3.5–6%, depending on the kind and type of roller gin. For example, for the same seed cotton, the ginning outturn is 1.5–2% greater with double roller gins than with saw gins. Furthermore, roller ginned cotton gets a premium of up to 4.4 cents/kg of lint above saw ginned cotton because it is longer and has higher uniformity with fewer short fibres, seed coat pieces and neps [25].

The seed cotton conditioning equipment used in roller gins is the same as that used in saw gins. Lint cleaning in modern reciprocating knife roller gins is commonly done with cylinder and impact cleaners similar to those used for seed cotton, as well as air-jet cleaners [36].

4.3 Post Cleaning

Lint cleaners remove leaf particles, motes, grass, and bark that remain in cotton after seed cotton cleaning, extracting, and ginning. They can improve the grade of cotton by removing foreign matter if the cotton has the necessary colour and preparation characteristics. Most gins that process machine-harvested cotton have one or more stages of lint cleaning. Lint cleaners are of two types: flow-through air type and controlled-batt saw type.

The flow-through air lint cleaner, commercially known as the Air Jet/Super Jet®, Centrifugal Cleaner®, or Super Mote Lint Cleaner®, is commonly used in both saw and roller gins. It is usually installed immediately behind the gin stand. It has no saws, brushes, or moving parts. Air and cotton moving through the duct change direction abruptly as they pass across a narrow trash-ejection slot. Foreign matter that is heavier than the cotton fibres and not too securely bound by the fibres is ejected through the slot by inertial force. It is less effective in improving the grade of cotton than saw lint cleaners because it does not comb the fibres. However, its fibre loss is less. Fibre length, fibre strength, and neps are unaffected by the air lint cleaner [44]. Lint from the gin stand or another lint cleaner is formed into a batt on a condenser screen drum in controlled-batt saw lint cleaner (Fig. 27). The batt is then fed via one or more compression roller sets, passing between a very closely fitting feed roller and feed plate or bar, and fed onto a saw cylinder. Lint cleaners comb and combine cotton to give a smooth look in addition to eliminating rubbish. However, they also increase SFC, which degrades several favourable mill properties, particularly at low moistures [44].

Lint fed to the cleaning machinery at high rates results in poorer cleaning efficiency, extra fibre damage and lint loss, and perhaps lower bale values. Feed rates should average around 750 kg/h/m of saw-cylinder length for efficient cleaning and minimal damage. Lint cleaners can process 1,119–1,492 kg/h/m of saw cylinder length without obvious operating issues; this rate corresponds to around 1,591 kg/h for a 1.7 m wide lint cleaner (40.6 cm saw cylinder) and approximately 7,500 kg/h/m for a 2.4 m wide cleaner. These feed rates are also proportional to saw diameter and speed. Increased saw speeds also increase fibre damage and fibre loss. Larger diameter saws (61 cm) have higher feed rates than smaller diameter saws (30.5 or 40.6 cm).

The number of grid bars in a modern lint cleaner can range from four to eight depending on the type. Clearance gauges are used to position the grid bars in relation to the saw cylinder. Cotton gin factories use grid-bar air wash to enhance lint cleaner performance, pick up and eliminate trash, and minimize air pollution within the gin plant. Air movement throughout the grid bar region should be at least 33 m^3/min/m of grid bar length.

Increasing the number of saw lint cleaners at the gin reduces waste during spinning but frequently has the unintended consequence of increasing neps in the card web and reducing yarn strength and appearance [44].

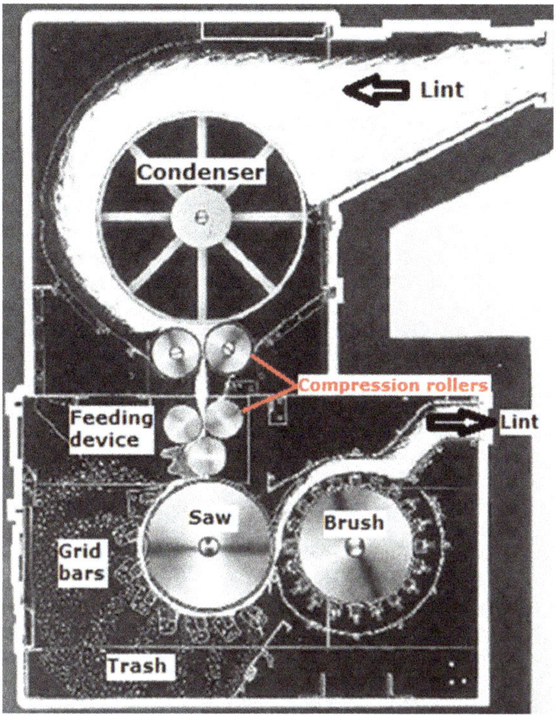

Fig. 27 Controlled-batt saw lint cleaner. *Source* [26]

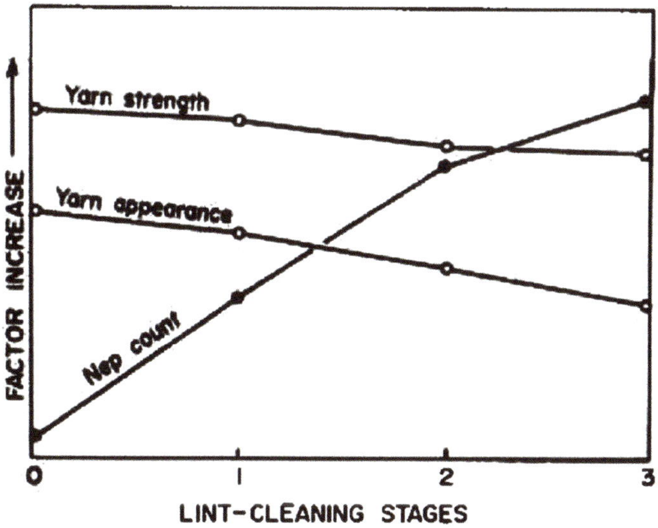

Fig. 28 Effect of saw lint cleaners on nep count, yarn strength and appearance. *Source* [44]

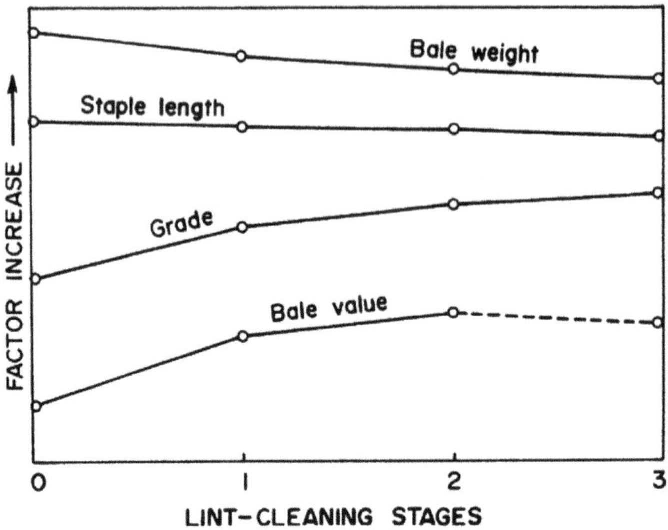

Fig. 29 Effect of lint cleaners on classer's lint grade and staple length, and bale weight and value. *Source* [44]

Lint cleaning generally improves the grade classification (colour, leaf, and smoothness) of the lint. As the number of lint cleaners increases, grade tends to increase. However, each succeeding cleaner gives less grade improvement than the preceding one. In addition, lint cleaners blend Light-Spotted cottons so that some of these pass into the White grades [31–33]. Lint cleaners can also decrease the number of bales that are reduced in grade because of grass and bark content. But when grades are improved, bale weights are reduced and staple length may decrease, and these opposing factors affect bale value as shown in Fig. 29 [44]. In some cases, the net effect of multiple stages of lint cleaning is a loss in bale sales value as well as an increase in neps and SFC which decreases its spinning value [34].

4.3.1 Lint Cleaning for Roller Ginned Cotton

Cylinder cleaners, textile-type beaters, impact cleaners and air jet cleaners can be used to remove motes, broken seed, fibre entanglements, and small trash that was not removed during seed cotton washing. There is no standard machinery sequence for cleaning lint from roller-ginned cotton. Lint cleaning is often done in the United States by a sequence of inclined cleaners, impact cleaners, and one air jet cleaner. The controlled-batt saw cleaner is not utilized in roller gins because it alters the typical look of roller ginned cotton.

4.4 Moisture Restoration

Adding moisture to lint that has previously been ginned and cleaned will not improve fibre length. Rather, the benefits of moisture restoration include lowering the static electricity level of the cotton, reducing the amount of cotton necessary to attain a particular bale size, and lowering the force required to crush the bale [26, 44].

Lint moisture in the bale must be consistent and not exceed 7.5% at any point in the bale to minimize fibre discolouration and weight loss during storage. One method is to utilize humid air to wet cotton. The air is first heated to high temperatures before being subjected to atomized water droplets, which evaporate into the air. The evaporation process decreases the air temperature while increasing the "dew point" temperature. This humid air is then blown through the cotton, lowering the air temperature below its dew point and causing small water droplets to collect on the cotton fibres throughout the cotton batt. The quantity of moisture restored by this technique is restricted, especially at greater ginning rates [26].

Another technique for restoring moisture is to atomize water and spray it straight onto the cotton. A wetting agent is sometimes added to the water to speed its diffusion through the cotton. Most gins that employ this technique spray water on the cotton at the lint slide as the cotton moves by gravity from the battery condenser to the bale press. However, care must be taken to avoid moist patches in the bale, which promote bacterial and fungal development and cause fibre deterioration; hence, this approach is not advised [26].

4.5 Bale Management

Bale wrapping is the final step in cotton processing at the ginning facility. Cotton must be baled and packaged to protect it from contamination during transportation and storage. The packaging system consists of a battery capacitor, lint chute, lint feeder, tramper, baler, bale cover and bale tying system.

The basic tramping and pressing system can be supplemented with bale conveying, weighing and wrapping systems. The bale press is made up of a frame, one or more hydraulic rams, and a hydraulic power system. Tying subsystems might be completely manual, semi-automatic, or totally automated. Restraining ties are often steel wire or flat, steel straps, although they can also be plastic straps. Six to ten ties are normally placed throughout the bale, however a spirally wrapped continuous tie is occasionally employed. The tension on the ties when the bale is released from the press is determined by the regularity of the lint distribution, bale weight, bale dimensions, density to which the bale was pressed, moisture content, tie length, and other parameters. Bale tie strength must be properly matched to the bale press mechanism to avoid tie breakage and associated contamination and handling challenges. Ties for gin universal density bales (227.3 kg) contained in a 53.3 cm wide × 137.2 cm long

× 76.2 cm thick package in the United States must have minimum breaking strengths at the junction ranging from 9.3 kN to 17.8 kN depending on the kind of tie.

Bales must be completely covered (including any holes generated by sampling). All bale covering materials should be clean, in good shape, and strong enough to protect the cotton. Natural fibres such as cotton (preferable), burlap, and jute, as well as synthetics such as polypropylene and polyethylene, are used to cover bales. The substance must not contain salt or other corrosive compounds, as well as sisal or other harsh fibres or any other components that will taint or harm cotton. UV inhibitors must be included in bale covers for outdoor storage in proportion to the estimated storage time.

4.6 Lint Cotton Grading

Cotton grading/classification refers to the use of official standards and defined techniques to measure the physical characteristics of lint cotton that impact the end product's quality and/or manufacturing efficiency. Classification technique is based on both grade and instrument standards, which are utilized in conjunction with cutting-edge procedures and technology to offer the cotton industry with the highest possible quality information for marketing and processing. Cotton classification encompasses the cotton quality criteria of colour grade, leaf grade, preparation, fibre length, length uniformity index, fibre strength, micronaire, colour Rd, colour + b, trash content and extraneous matter identification. As classification systems evolve across the world, dependence on human senses decreases as instrument classification increases. Australia, Brazil, China, Uzbekistan and the United States have all implemented or are on their way to adopting instrument classification on 100% of their cotton crops [28].

The United States Department of Agriculture prepares and maintains International Cotton Standards. Every 3 years, a group of international signatories (merchants and textile mill representatives who buy or use US cotton) and the US Cotton Advisory Committee examine, modify, and approve these requirements. They are then utilized as global reference materials for manual classification as well as reference samples for the HVI system [27].

4.6.1 Manual Classification—Upland Cotton

There are 25 different colour grades, 15 physical examples, and 10 description samples. Colour grades are classified into five colour categories, as follows: White, Light Spotted, Spotted, Tinged and Yellow Stained are all options. Six samples arranged opposite to one other in a standards box show the range of each colour grade for which a physical standard exists. Colour and leaf grade requirements are kept in the same package for practical reasons. For example, the Strict Low Middling colour grade criteria box also includes the size and number of leaf that would be classified as leaf grade 4. Each descriptive standard describes cotton that is above, below, or in between specified physical requirements.

Weather and the period of exposure to weather conditions after the bolls open have a large impact on the colour. Varietal traits, as well as harvesting and ginning procedures, may have an impact. Upland cotton has a brilliant, white colour when it opens naturally. A change in colour implies a drop-in quality. Continued exposure to the elements and the activities of microbes can cause white cotton to lose its luster and darken in colour. When plant development is halted prematurely due to cold, dryness, or other climatic circumstances, the cotton may exhibit varying degrees of yellow colouration. Cotton can also get discoloured or speckled as a result of insects or fungus.

Leaf grade refers to the quantity of leaf content in cotton. There are seven leaf grades, and they are all represented by physical criteria. In the spinning industry, leaf material is regarded as trash, and its removal decreases fibre quality and increases production costs. Cotton variety, as well as various harvesting methods and circumstances, influence leaf content. The amount of leaf material left in the lint after ginning is determined by the amount present in the seed cotton before to ginning, as well as the type and quantity of cleaning and drying equipment employed during ginning. Even with the most meticulous harvesting and ginning processes, a little quantity of leaf will remain in the cotton lint.

The degree of "roughness" or "smoothness" of the ginned lint cotton is measured by preparation. Smooth cotton, on average, has less spinning waste and creates a more consistent yarn than rough cotton. However, as previously stated, the desirable smooth appearance may induce overprocessing at the gin.

Any item in a cotton sample that is neither cotton fibre or leaf material that is not included in the official prepared cotton standards is referred to as extraneous matter. Extraneous matter includes things like bark, grass, spindle twists, dust and grease.

4.6.2 High Volume Instrument (HVI) Determinations—Upland Cotton

Traditional classifying systems use human classifiers to evaluate the colour, leaf and length of cotton, whereas tools evaluate the micronaire. The use of the HVI classification system is mandatory in the United States for cotton to be eligible for the Commodity Credit Corporation loan program. The HVI system is supplemented by humans especially in assessing leaf and discount factors such as bark, grass and preparation.

Fibre length is measured by passing a small tuft of parallel fibres through a sensing point of the HVI system. It may also be determined manually. Fibre length is a varietal characteristic that is influenced strongly by environmental conditions; it may be reduced by processing equipment, especially at low moistures. Length uniformity, which is typically 80–82 for upland cotton, is the ratio of the average or mean length of the fibres to their upper half mean length and is expressed as a percentage. If all fibres in a sample were the same length, the length-uniformity index would be 100.

Fibre strength is determined by the HVI system on the same fibre tuft used for length measurement. Results are reported in grams per tex. A Tex corresponds to the weight in grams of 1,000 m of fibre. The stated strength is the force in grams required to break a one-Tex fibre bundle.

Micronaire refers to an air flow measurement that indicates the fineness and maturity of the fibres. Micronaire is determined by passing air compressed to a standard pressure and temperature through a cotton sample of standard weight and volume. The volume of airflow through the sample is expressed in micronaire, which may also be referred to as the mike value or simply mike. Some fibres are extremely fine simply because they are immature. These fibres cause dyeing irregularities, increase manufacturing waste during picking and carding, and detract from the product's appearance. A micronaire below the optimal range may indicate immaturity; A value above the optimal range may indicate that the fibre is too coarse for many high-quality products.

Fibre fineness is a varietal characteristic estimated using micronaire and is also influenced by growing conditions in the later stages of fibre development. Favourable growing conditions result in fully mature fibres and higher micronaire values. Unfavourable conditions such as lack of moisture, early frosts or other conditions that disrupt plant processes result in immature fibres and lower readings.

Currently, the USDA classification system includes colour determination using HVI. The HVI colour determinations are made in the form of greyness (measured in Rd) and yellowness (measured in + b). The gravy level (or percent reflectance) indicates the lightness or darkness of the sample. Cotton Rd values typically range from 48 to 82. Yellowness indicates the amount of yellowing in the sample and typically ranges from 5 to 17. Typically, opened cotton has an Rd of 70 or greater and a + b of 9.0 or lower. The various combinations of gravy and yellow tones can be converted into colour values using the Rd and + b values using a colour measurement card.

The content assessed by the HVI system is determined by scanning the sample surface and recording the particles present. The proportion of the sample surface covered by nonlint particles is provided as a result. The nonlint particle content of the fibre should be less than 5.0%.

American Pima grade standards are also presented in physical form. There are six American Pima grades, numbered 1 through 6. American Pima and upland grade standards vary widely. The leaf content of American Pima standards is peculiar to this cotton and does not equal that of upland standards. Because American Pima cotton is often ginned on roller gins and is more stringy and lumpy, the preparation differs greatly from that for upland standards. Upland cotton is often cleaned with saw-type lint cleaners, which provide a smooth, blended, combed sample. Roller ginned cotton is often cleaned with air or cylinder-type cleaners and has a rough look.

5 Balancing Fibre Quality and Cleaning

Table 4 shows the usual influence of gin equipment on several quality characteristics such as reflectance, yellowness, leaf grade, HVI trash, uniformity, HVI length, short fibre percentage, neps, seedcoat fragment weight, number of seedcoat fragments, trash and dust.

Whether performed in a ginning plant or in a textile mill, cleaning generally reduces most important fibre quality attributes except colour, foreign matter and appearance, and reduces the amount of usable fibre [13]. Ginners must make compromise between trash removal and fibre damage when choosing their cleaning machines. For machine-picked cotton, ginners should use 12-to-14-cylinder cotton cleaning cylinders along with a stick machine and one or two lint cleaners, depending on the cotton's trash content and colour potential. For stripped cotton, a second extractor and an airline cleaner or cleaning separator should be included; A second saw lint cleaner may be required. To deliver products of the absolute highest quality for spinning performance, growers and ginners must be careful during production, harvesting, ginning and textile manufacture to avoid practices that could compromise fibre quality.

5.1 *Impacts of Ginning*

As explained above, the first mechanical operation in the processing of cotton is ginning. A ginning system essentially consists of pre-cleaning, ginning and post-cleaning machines. When ginning, fibres (lint) are separated from the seed. It also removes foreign objects from the cotton. Finally, it prepares bales suitable for transportation to and storage in textile mills. In conclusion, however, the key effects of ginning are discussed below:

1. It easily forms neps (i.e. small balls of entangled fibres) which are an undesirable part of the cotton pulp. There is a high correlation between the neps in the ginned cotton and the neps in the resulting yarn. Neps in yarns lead to stains on dyed or printed fabrics, lowering the market value of the end product. Therefore, not only must neps be removed during yarn manufacture, which increases waste, but they should also be prevented from being formed by these processes. The relationship between neps potential and fibre properties shows that fine and immature fibres tend to form neps. The neppiness of the yarns correlated significantly with the amount of immature fibres in the fluff. The recommended method for measuring neps is the Zellweger Uster Advanced Fibre Information System (AFIS). The AFIS neps tester indicates the average number of neps per gram in the fibre mass and the average nep's size in millimetres.
2. It increases SFC (Short Fibre Content) (fibres shorter than 12.7 mm) due to fibre breakage, especially in saw gins. SFC is around 10% in bales. Yarn strength,

Table 4 Various quality parameters of ginned cotton between 2011 and 2017 [21]

Year	Moisture	Micronaire	Maturity	UHML	UI%	SF%	Strength	Elongation	Reflectance	Yellowness	Colour	Trash	Sticky
2011	10.79	3.76	0.85	28.41	79.84	11.6	31.53	6.92	79.69	8.69	33.09	8.25	**6.11**
2012	8.75	4.41	0.87	28.14	80.86	11.79	28.98	6.35	76.05	9.09	31.06	6.27	11.3
2013	8.49	4.36	0.85	29.05	78.95	13.2	27.39	8.85	75.08	9.15		4.71	5.44
2014	7.07	4.3	0.86	29.99	83.52	9.92	30.09	6.45	77.3	8.77	42.68	2.79	10.01
2015		4.44	0.85	29.21	81.96	11.14	28.01	7.17	76.03	9.73	25.42	2.8	14.13
2016		4.3	0.85	28.2	81.9	9.8	27.53					5.81	11.53
2017		4	0.84	27.7		8.71	28		81		34.67	4.79	5.5
	8.77	**4.22**	**0.85**	**28.67**	**81.17**	**10.88**	**28.79**	**7.14**	**76.83**	**9.08**	**33.38**	**5.06**	**9.14**

yarn appearance and spinning end breakage are three important spinning quality factors affected by SFC.
3. Intensive ginning produces a larger amount of fine waste fragments called pepper waste. This makes it difficult to remove these particles during spinning preparation.

6 Ginning Industry in Ethiopia

Ethiopia's textile sector is one of the fastest growing industries in Africa as the government has given it top priority by providing various initiatives to local and foreign investors. Besides the booming organized mill sector, Ethiopia also has the decentralized hand weaving sector that uses seed cotton [19]. Even when the installed ginning capacity is more than double the amount required in the spinning mill, textile mills operate below capacity and are forced to import lint due to a number of constraints in ginning plants [ETIDI, n.d]. This is a concern that must be addressed by the government and stakeholders such as cotton producers, ginners, exporters associations, Ethiopian Textile Industry Development Institute (ETIDI), higher education institutions, Ethiopian Textile and Garment Manufacturers Association (ETGAMA) and the like [16, 46].

The textile industry is flourishing in Ethiopia and is here to stay for the foreseeable future. Consequently, cotton cultivation and ginning are crucial issues that must be properly addressed. The focus of the next subsection is ginning in Ethiopia. What is the status of ginning industry? What are the challenges and opportunities of the ginning sector? What is the way forward? These questions are briefly dealt with.

The textile industry in Ethiopia is thriving and will continue to do so for the foreseeable future. Therefore, cotton growing and ginning are crucial issues that need to be addressed appropriately. The focus of the next subsection is on ginning in Ethiopia. How is the ginning industry doing? What are the ginning industry's difficulties and opportunities? What is the best course of action? These issues are addressed briefly.

6.1 Background

Ethiopia's ginning industry has existed since the 1950s. Tendaho was the first cotton ginning facility. Mitchell Cotts P.L.C. of the United Kingdom founded it in November 1960 in Dubti Woreda in the Afar region, 580 kms north-east of Addis Ababa. It was built with eight gin stands to supply lint cotton to the Dire Dawa Textile industry [47]. Other cotton ginning mills were developed in Addis Abeba (Abacot, Yerekesem, Shewa, and Ediget) and Gonder (Gonder ginning factory) throughout the 1960s. The ginning plants created in Addis Abeba (Yerekesem and Abacot) were transferred to cotton cultivating areas such as Middle Awash, Arbaminch and Abobo during the end of the 1970s and early 1980s.

Since the formulation of the Industrial Development Strategy by the Democratic Republic of Ethiopia, the first Growth and Transformation Plan (GTP I) of 2010–2015 has been and continues to prioritize the textile sector to attract US$1 billion in the textile sector. From this time onwards, the ginning industry rapidly developed throughout the country. The number of ginneries reached 27 in 2021 and their installed lint production capacity reached 1,362 tons per day. Of the 27 ginneries, 22 are operational, 4 factories are under construction and 1 factory is currently closed as shown in Table 5.

Although the total installed capacity of the ginneries is 1,362 tons per day, due to outdated technology and lack of input material, they use less than 50% of their capacity and produce 678 tons per day [17, 46]. There are three different types of ginning technologies in Ethiopia: saw gin, single roller gin and double roller gin. Of the 26 ginneries, 18 ginneries use saw gin, 4 ginneries use a single roller and the remaining 4 use double roller gin.

The saw gins available in the county are outdated technology and less productive compared to the latest state-of-the-art saw gins. The number of gin stands in a factory is between one and a maximum of four. The number of roller gins is relatively higher than that of saw gins, which is between 18 and 80 double rollers.

The ginning mills are geographically distributed across eight regions in the country, as shown in Fig. 30. Of these, 35% of the mills are installed near the cotton-growing areas, while 65% of the ginning plants are installed near the customers. The geographic distribution of ginneries is not economical; In particular, factories located far from agricultural areas face additional transportation costs.

The ginning result is a good indicator of the performance of the ginning mills. It is described as a percentage of the lint gain from seed cotton [25]. Ginning plants with high ginning scores are preferable. Most ginneries in the country do not use lint cleaners to increase or maintain GOT and deviate from best practices. Even if most of them do not have a systematic record of the ginning result, the average ginning result for both roller and saw ginning is estimated at 37%.

Most existing ginning mills have outdated technology that affects the quality of the cotton during processing, lack a moisture control system resulting in overheating of the seed cotton, and are inefficient in trash removal. In addition, ginneries use outdated ginning methods. In addition, ginneries have low productivity due to unskilled labour; outdated technology; shortage of raw cotton; lack of spare parts, workshop and infrastructure; power interruption; inappropriate layout; lack of a waste management system; and poor maintenance and labour practices.

Table 5 Ginneries in Ethiopia and their capacity

Serial number	Name of ginning factory and year of establishment	Location (City, Region)	Gin Type	Installed capacity (ton/day)		Number of gin stand
				Seed cotton	Lint	
1	Abdul Kedar Ahmed Ginning Factory (2014)	Genda Wuha, Amhara	Saw gin	100	37	3
2	Agricot Ginning Factory (2006)	Mojo, Oromia	Saw gin	80	30	2
3	Arba Minch Ginning Factory PLC (1989)	Arba Minch, SSNP	Saw gin	80	30	2
4	Awash Ginning Factory PLC (2011)	Awash 7, Afar	Saw gin	100	37	3
5	Chagni Ginning Factory (2019)	Chagni, Amhara	Saw gin	150	50	1
6	Des Ginning Factory (1998)	Gonder, Amhara	Saw gin	400	148	4
7	Dori Kebede Ginning PLC (2012)	Dukem, Oromia	Single roller	150	56	24
8	Else Addis Ginning Factory PLC (2011)	Adama, Oromia	Single roller	400	148	80
9	Gebreselam Ginning PLC (2005)	Dukem, Oromia	Saw gin	55	20	1
10	Haji Nur Hussien Ginning PLC (2007)	Gonder, Amhara	Saw gin	80	30	1
11	Hiwot Agricultural Mechanization PLC (1998)	Humera, Tigray	Saw gin	300	111	2
12	Genda Wuha Ginning Factory (formerly Loyal Tiret) (2014)	Genda Wuha, Amhara	Double roller	320	118	18

(continued)

Table 5 (continued)

Serial number	Name of ginning factory and year of establishment	Location (City, Region)	Gin Type	Installed capacity (ton/day)		Number of gin stand
				Seed cotton	Lint	
13	Middle Awash Ginning PLC (1989)	Amibara, Afar	Saw gin	150	56	2
14	Saudi Star Agricultural Business Group (2000)	Abobo, Gambela	Saw gin	84	31	3
15	Shewa Cotton Ginning Factory PLC (1970)	Akaki Kality, Addis Ababa	Saw gin	15	6	1
16	Studio 3D Ginning PLC (2010)	Mojo, Oromia	Saw gin	300	111	2
17	Lugudi Ginning Factory (2007)	Benishangul	Double roller	86	31.82	
18	Biruk Abebe Ginning (formerly Omo Shelko) (1994)	Adama, Oromia	Saw gin	70	25.9	4
19	MNS Ginning Factory (2017)	Legetafo, Oromia	Single roller	40	14.8	60
20	Genda Wuha Ginning Factory II (2020)	Chagni, Amhara	Double roller	48	17.76	
21	Mule/Addis Ginnery	Gambela	Double roller	160	57	
22	Hegayino Trading Pvt Ltd Company (2010)	Arba Minch, SNNPR	Single roller	11.2	4	2
23	Friel Ginnery (2021)	South Omo, SNNPR	Saw gin	420	151	3
24	Kalime Ginning Factory	Genda Wuha Amhara	Saw gin	10	3.7	
25	Nasa (2020)	South Omo, SNNPR				
26	Lucy Ginning Factory (2006)	Awash, Afar	Saw gin	103	37	2
SUM				**3,712.2**	**1,362**	**220**

Source [18, 19]

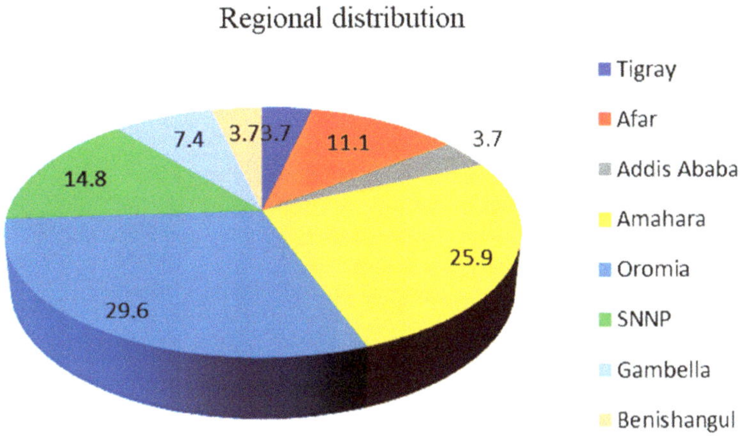

Fig. 30 Regional distribution of ginneries in Ethiopia

6.2 *Ginning Inputs*

The ginning mill uses raw cotton, electricity, tying material and labour as inputs. Raw cotton is the main input in the ginning process. As the ETIDI report 21 shows, there is a large gap between supply and demand for raw cotton. The supply of raw cotton accounts for the lion's share of the productivity of ginneries. The annual installed capacity of the ginneries reached 255,420 tons of lint in 2021. In order to meet this demand, however, the annual production of raw cotton would have to be 709,500 tons. However, the actual annual raw cotton production in 2014–2019 ranged from 86,194 to 182,744 tons, which means that it only covers 12.2–25.7% of the demand, as shown in Fig. 31.

The growth of cotton production should not be the only goal because the factories need high-quality raw materials to survive in the global competitive market and to succeed in the export strategy. Because there is no cotton grading and classification system in the country, lint cotton is marketed at more or less the same price for all types of cotton quality, making cotton farmers insensitive to quality. As a result, not only do factories suffer from a loss of production efficiency but growers and ginners also lose bargaining power. In addition, seed cotton is transported to the ginneries after being packaged in jute and polypropylene sacks. These fibres cause more impurities in the seed cotton and affect the quality of the yarn in spinning and textile processing. Cotton producers supply seed cotton by weight, therefore, they spray water or add unwanted materials to add weight to gain more profit.

One of the basic inputs for ginneries is labour. The ginning labour is divided into two large groups based on the nature of the cotton ginning season. Full-time staff are employed year-round and are responsible for the management, supervision and operation of the ginning during the active ginning season, as well as maintenance, repair and administration during the rest of the year. In contrast, seasonal workers

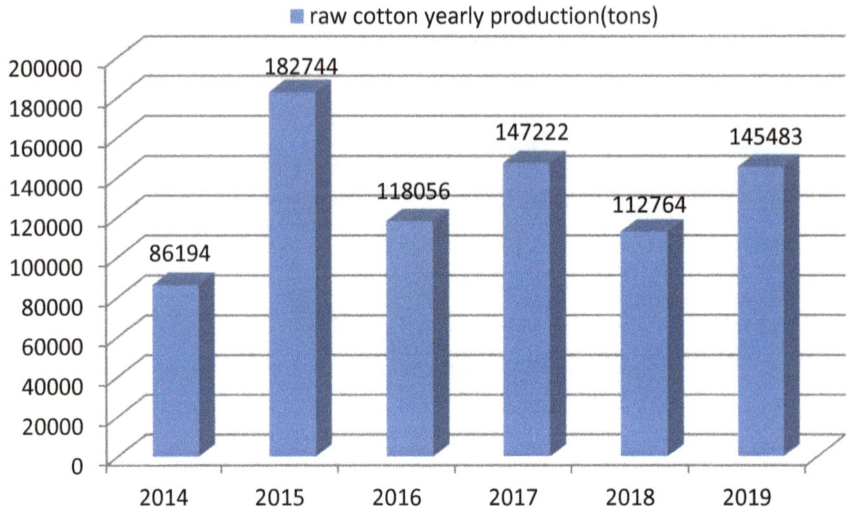

Fig. 31 Annual production of raw cotton (tons)

work during the active ginning season and are directly responsible for completing the work tasks required at the ginning facility [28]. In Ethiopia, the total number of full-time and seasonal workers is 1,649 and 3,767, respectively, according to ETIDI data.

Ginning is an energy-intensive process. The power supply is provided for production systems for drive supply as well as for sockets and light bulbs. The power consumption of different ginning plants correlated positively with the number of bales produced [27]. In Ethiopia, there are limited studies on energy use in ginning. In a 2013 survey, energy use accounted for about 25% of total variable costs, which included seasonal labour, packing and tying, and repairs and maintenance (Valco et al. 2015). Energy audit and monitoring studies from 2009 to 2011 [23] found that the average participating saw gin used 39.5 kWh to process a bale, while the average roller gin used 62.6 kWh [20]. Ethiopian ginneries are experiencing a lack of electricity and the ginneries around the farm in particular are victims of this problem. These power interruptions affect the production process, reducing productivity and increasing idle time and labour costs.

Bale wrapping is the final step in cotton processing at the ginning facility. The bales should be fully covered and the bale-covering material should be clean and sufficiently strong to adequately protect the cotton. Bale wrapping materials consist of steel wire or plastic straps to secure the bales and burlap or cotton fabric to cover the lint cotton.

6.3 Ginning Technology

The word technology is used to describe the application of systems, processes, techniques and tools that enable better productivity and quality of products in any manufacturing industry. It is well known that the ginning industry is one of the value-added steps in the textile and garment manufacturing process, which involves its own process from raw cotton storage to bale wrapping (the end of the process). For lint cotton quality and ginning productivity, it is imperative to use appropriate technology at each production step.

The general working conditions of the technologies in the Ethiopian ginning industries have been thoroughly assessed by UNIDO as part of their 2017 Competitiveness Improvement Study (Gerzi Textile Organization AG 2017). The ETIDI's (2017) ginneries assessment study ranked them by star level based on their infrastructure and management practices and various technical support reports. These studies and reports indicate that most machines are used machines that were dismantled in the United States and other countries due to obsolete technology, but installed as new in Ethiopia. These machines have a limited number of ginning stands that do not offer economies of scale, are subject to frequent breakdowns resulting in idle time, and a severe shortage of spare parts.

There are currently 26 ginning plants, of which 18 (70%) use saw ginning technology, while the remaining eight (30%) use roller ginning technology. Of the nationwide installed daily capacity of 1,418.24 tons, 950.42 tons (67%) are accounted for by saw-ginning technology and 468.58 tons (33%) by roller technology. Because the roller technology uses gentle force to loosen the fibres from the seed, its productivity is lower compared to saw ginning technology. Ethiopian Ginning Industry Performance Improvement Study [21] indicates that a roller ginning machine produces one bale (200–220 kg) per shift, while saw-ginning technology produces up to 14 bales (200–220 kg each) per shift, so saw-ginning machines are installed in fewer numbers. For example, the total number of saw gins in 20 factories is only 34 (with different numbers of saws), but the number of roller gins in 8 factories is 190 (mostly single roller gins).

Most of the installed saw-gins are brands of world-renowned manufacturers such as Lumus and Continental, although limited companies have also sourced from Chinese manufacturers. As mentioned above, the year of manufacture is a crucial parameter that determines the productivity of machines. However, in most factories except for two of them, it is not possible to determine the year of manufacture. Table 6 shows the year of installation in Ethiopia, although the machines are used machines imported from other countries due to outdated technology.

Table 6 Service years of ginning factories in Ethiopia

Number of factories	1	3	2	1	3	1	3	2	2	1	1	1	2	4
Age since installation in Ethiopia	50	31	23	18	15	14	11	10	6	4	3	2	1	Project

Roller gins are sourced from Bajaj and Jedaho of India and Summer of Turkey. Although double roller ginning is more productive than single roller ginning, most mills use the latter. As more rollers are required for this machine as spare parts, the giners face a serious threat due to the shortage, especially due to the foreign exchange restrictions. For example, in 2021 only 32 of the 80 installed machines in ELSE Addis industrial development ginning factory are working properly. At MNS Textil, of the 60 machines installed in 2020, only 30 are in operation. These companies cannibalize the decommissioned machines as spare parts to solve the shortage problem. The other problem observed in factories is that important machines are not installed in production lines. For example, seven factories have no lint cleaning machine at all, and some factories use outdated feed control machines that are manually operated.

6.4 *Quality Management and Marketing Trends in Ethiopian Ginning Industry*

Controlling cotton quality at the ginning stage will not only lead to an improvement in the living conditions of everyone associated with the ginning industry but will also bring yield benefits in both the forward and reverse linkage. Improving the quality of cotton fibres would mean providing higher quality raw materials for the textile and apparel industry, which would lead to an increase in the competitiveness of the higher value textile and apparel products. As a result, the industry will create more job opportunities and higher incomes for those associated with it. As for the backward links, quality control at the ginning stage will result in a demand for cleaner, higher quality cotton. Ultimately, farmers will be motivated to supply quality cotton in order to get better yields.

The production and ginning of high-quality cotton that is clean and meets the expectations of the textile industry has a distinct advantage, as quality depends on practicality or depends on the user or customer satisfaction. The need to care about the quality of cotton drives researchers to breed the best quality cotton, cotton farmers to get a fair price for the cotton they produce, ginners to optimize their ginning process, traders to lower the price of cotton set and satisfy the customers, etc. spinners to achieve required yarn quality and minimize raw material cost through properly running process. Therefore, general cotton quality parameters, the cause of poor cotton quality, its effects and remedies are discussed under this title.

As many publications indicate, cotton quality management begins with classification and grading, first established by the American Agriculture Research Center and commonly referred to as American Cotton Standards. This was first accepted in European countries and gradually spread to the rest of the world through the marketing of cotton. It has been referred to as the "Universal Standards of American Cotton". Eventually, many countries such as Australia, India, Europe and Egypt developed their own cotton quality grading and classification system for marketing purposes. The evaluation and classification system of cotton quality parameters differs from country to country. Since there was a close correlation between cotton price and quality, countries developed their own classification and grading systems. This led to high competitiveness between countries to produce their own premium cotton at the best prices, such as Egypt Giza Cotton. The globally recognized quality parameters for cotton are purity, fibre length, fibre strength, fibre count, colour, moisture content and impurities (foreign matter).

Similarly, in 2014, the Ethiopian Standardization Agency (ESA) had developed a standard entitled "Cotton Fibres The Classification of Lint" by a Technical Committee (TC 68) chaired by ETIDI and approved by the National Council. The standard was developed by collecting secondary data from ETIDI and textile industry test results. The approach followed was that of the USDA rating and classification system for Pima and Upland cotton. However, this standard was again improved as there was no quality improvement from stakeholders such as cotton farmers and ginneries.

The lint quality, which is important for the mill, is influenced by the operation of the ginning. Over the past decades, several issues related to the quality of cotton in ginning plants have been identified in the ETIDI report [20]. Firstly, good quality clean cotton is not supplied to the ginning factory. Cotton brought from cultivation areas contains a high proportion of dirt, dust and moisture. Farmers don't bother to keep cotton of different varieties separate; finally, individual characteristics of different types of cotton are lost. In addition, farmers must have appropriate skills in picking, stacking, packaging and transporting seed cotton. Secondly, among the various factors affecting cotton quality, ginning is one of the main factors. The cotton ginning industry in Ethiopia has significantly low-capacity utilization and the machinery used is outdated. There is no standardization of ginning machines.

In general, poor material handling and ginning practices are the main causes of lint cotton quality degradation when ginning. These are caused by a lack of awareness about cotton quality parameters; lack of quality control practices; same marketing prices for all types of cotton quality, regardless of the quality parameters, i.e. no premium price for contamination-free cotton and no discount for contaminated cotton; use of polypropylene and jute sacks for picking, transporting and wrapping bales; insufficient warehousing and inappropriate storage; poor supervision and ginning practices (e.g. not using lint cleaner); lack of organized data and.

Poor material handling and ginning practices lead to an increase in labour and maintenance costs (i.e. higher spare parts consumption and machine damage), a reduction in ginning yield (reduction in production), high levels of waste, and the production of inferior lint cotton. Compromised lint quality in turn results in efficiency reduction in the spinning mill and bad quality textile product, which eventually leads to the complete refusal of customer orders, the loss of income currency, and the creation of a bad image for the country.

6.4.1 Cotton Marketing

The national marketing of cotton consists of two segments. The first stage is before the ginning phase, known as seed cotton marketing. It is a weight-based marketing system conducted between small farmers and large investors or between small farmers and traders. The second segment is in the lint stage after ginning, which is called lint cotton marketing. In this segment, the quality parameters of the lint and the associated price are determined by government agencies, manufacturers and private associations concerned.

Normally seed cotton is transported to the ginning factory by cotton producers (farmers), gins or traders. Traders and producers ginning seed cotton by paying ginning fees. After ginning, the seller brings lint samples to the ETIDI lab for testing before the hearing. After that, the marketing of the lint takes place between the seller and the buyer according to the quality parameters and the established price. This is how the ginned lint cotton is delivered to the textile spinning industry. Consequently, the textile spinning industry purchases lint cotton from the ginning industry, cotton producers and/or merchants through negotiations. When domestic cotton demand is lower, producers do not find easy buyers, so the Ethiopian Industrial Inputs Development Enterprise (EIIDE) buys cotton from local producers at a minimum fixed price [17]. The national cotton marketing chain is shown in the diagram below (Fig. 32).

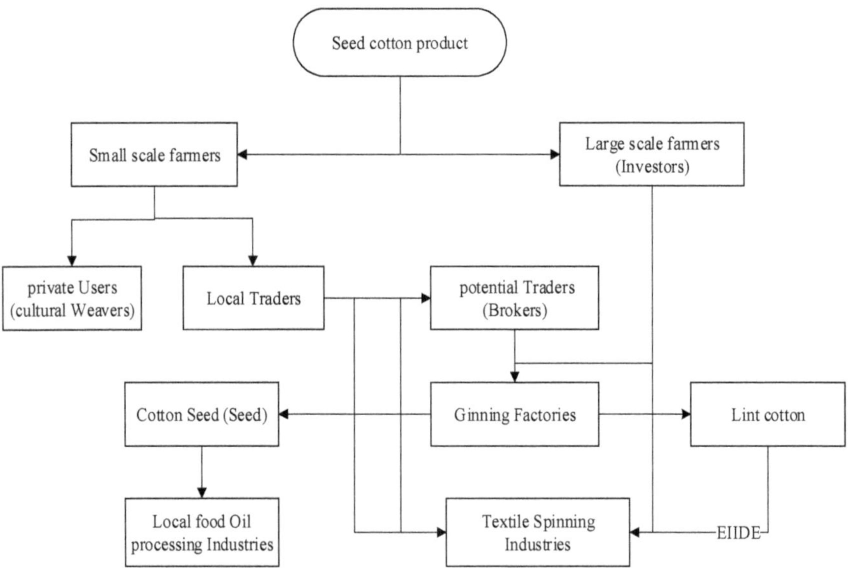

Fig. 32 Cotton marketing chain in Ethiopia. *Source* [17]

6.5 Maintenance Management in Ginning Industry

Currently, maintenance practices at the ginning facility are substandard. There is no documented maintenance schedule for machines. There is no daily maintenance checklist and no routine machine checks. Monitoring and maintenance tools are neither available nor used in mills. Suction fans and conveyor channels are not cleaned regularly. So, instead of preventive maintenance, reactive or breakdown maintenance is performed, resulting in lost production hours. Basically, the sector waits until the equipment fails before doing anything about it. Of course, not only is there a risk of losing valuable production time but also money due to unexpected downtime, which is no surprise. Obviously, unscheduled maintenance also tends to be more expensive.

The success of the ginning process depends on how efficiently the cotton is ginned. The ginning plant should be operated with maximum efficiency and minimum downtime. Therefore, a suitable maintenance plan should be drawn up. Successful maintenance programs require planning, execution, monitoring and tracking. Systematic documentation at the ginning facility would allow for accountability during operation of the ginning process, possible causes of inefficiency and remedial action. It also helps in performing seasonal repairs and maintenance. The off-season repair programs allow the machine to be kept in good condition and operational before the start of the ginning season. Preventative maintenance focuses on regular inspection, cleaning, lubrication, adjustment and repair to minimize machine failure and unplanned downtime.

Maintenance management must ensure that the ginning facility is equipped with machinery that maximizes the value of the cottonseed received. The training provided in the industry over the last two decades has instilled in gin managers and their staff the awareness that they have a basic understanding of maintenance principles as well as the ability to troubleshoot machines. Thus, the training brings some improvements in the ginning sector, albeit not yet significantly. The recommendations made by ETIDI to the sector in good time were not fully implemented in practice due to the lack of a systematic approach.

6.6 Sustainable Production in Ginning

Sustainable production is the creation of goods and services using processes and systems that are environmentally friendly and conserve energy and natural resources; economically feasible; safe and healthy for workers, communities and consumers. As such sustainable production is one of the concerns of the ginning industry, since the harvesting, storage and transport of raw cotton attracts dust to the raw cotton. This dust poses a serious threat to air quality, worker health, the environment and society if not properly disposed of.

Based on the diagnostic study by ETIDI [20], the sustainable production practices of the Ethiopian ginning industry have been thoroughly assessed. This study

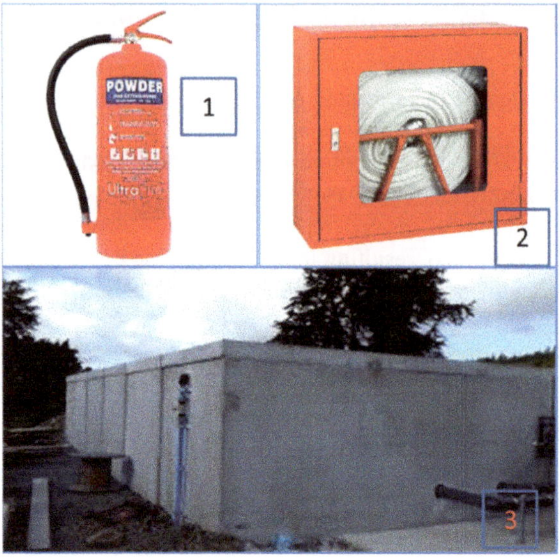

Fig. 33 Typical fire fighting system: (1) Fire extinguisher (2) Sprinkler

addressed issues of ensuring worker safety, installing a fire suppression system, dust collection and control practices, and involvement in community development. It has been noted that most workers are not provided with personal protective equipment such as coat, safety shoes, masks, goggles, etc., except for those working as mechanics. In addition, most ginning plants do not have a fire extinguishing system and lack the large infrastructure associated with it, as shown in Fig. 33, although cotton is very fire-prone.

In addition, control of the dust discharged from the raw cotton in the pre-cleaning, ginning and post-cleaning processes is imperative to protect workers' safety and environmental pollution by installing suction fans at appropriate places and constructing a dust collection room outside the production area. However, the study found that few factories have adequate exhaust fans and dust collection rooms. The following image (Fig. 34) is from one of the factories where polluting dust is emitted.

7 Research in Ginning

Research and development is a crucial factor for competitiveness in today's dynamic world. The ginning industry has several research areas that need to be addressed. Research in this sector would play an important role in bringing it to a globally competitive level. Table 7 lists the research work carried out and its results. However, as can be seen from the table, the research and development carried out in Ethiopia was very negligible.

Fig. 34 Inappropriate dust disposal

8 Challenges and Opportunities of the Ginning Sector in Ethiopia

The domestic ginning industry faces several challenges identified by the ETIDI. The main challenges are lack of quality control and assurance practices, poor maintenance practices, use of outdated technology or machinery, lack of infrastructure and skilled workforce, inappropriate organizational structure or management system, lack of adequate warehousing and warehousing practices, and limited concerted efforts by the researchers.

The lack of quality control and cotton grading systems results in seed cotton with high waste content being brought into ginners, particularly by merchants and ginners, who lack the knowledge to classify the cotton fibres. In addition, cotton traders are mostly concerned with quantity rather than quality. The regulator hasn't even set quality standards and price markups for fluff cotton. Almost all ginners do not have properly conditioned storage facilities for raw material (seed cotton) and finished goods (fluff cotton). Various lots of seed cotton are blended and delivered to ginners mainly by traders.

The ginning sector is in dire need of state-of-the-art technology. The existing factories use outdated technology or outdated machines that have no market access for spare parts. These were imported after prolonged use. The factories rely on old machines and outdated ginning methods.

Factories also lack skilled labour because they are located in remote areas and lack access to infrastructure. Low wages are also a big problem. Even for existing

Table 7 Research and their findings

Serial number	Research topic	Main objective	Findings	Condition of implementation
1	Cotton Quality Assessment in Ethiopia	To assess physical quality of Ethiopian cotton, which can be an input to prepare national grading and standardization system	The impact of ginning on the quality of cotton is critical. Cotton marketing, based on grading system, did not encourage quality orientation	The study output has been used for development of cotton grades that are being applied by industry inputs development enterprise of Ethiopia
2	Assessment of existing status and rating of Ethiopian ginning factories with support scheme recommendations	To avail a comprehensive document on the status of ginning factories rating them on star level (1 star lowest and 5 star maximum) based on infrastructure and management practices	Most ginning factories have poor infrastructural condition and management practices	Most companies were notified about their improvement areas based on gaps identified during assessment and technical support delivered. Additionally, some policy issues are on progress
3	Ginning industries' capacity improvement study	To prepare a study document that indicates need for policy initiative to be taken to improve capacity of ginning industry	Government should support ginning factories with finance at low interest rate and deliver technical support in collaboration with non-governmental organizations	In progress

employees, there are few standards to ensure their health and safety in the factories. The lack of infrastructure includes issues such as road access, internet, electricity, seed and lint ball storage, and workshop facilities. Any ginning, small or large, old or new, can be established without political constraints or economies of scale. These and related problems cause the existing ginning mills to underutilize their capacity or have a low ginning yield, produce lint with a high percentage of waste and the end product is completely sorted out, improving the quality of cotton and cotton products, creating employment opportunities and increasing the income of those with people associated with the ginning sector. Such interventions will essentially lead to improved livelihoods and poverty alleviation for all participants in the cotton value chain. The qualification gaps in the areas of production management, marketing management, personnel management, information technology, quality management and ergonomics could be closed through targeted and continuous training.

9 Summary

Ethiopia's existing textile sector includes cotton growing, ginning, spinning, weaving, knitting, dyeing and finishing and garment manufacturing. Although the country has 2.6 million hectares suitable for cotton cultivation, only a fraction of this (3%) is exploited [46]. Because cotton farmers are not incentivized, they cannot meet the demand of the ginneries for seed cotton, and the latter consequently have not been able to meet the cotton mills' demand for cotton fibre. Unfortunately, due to the lack of foreign exchange, importing lint is not easy.

Ginning companies face multiple problems to supply high-quality lint to the domestic and international markets. The impasse begins with receiving inferior seed cotton that is neither stored nor properly transported. The ginning industry mostly uses outdated technologies that degrade the quality of the cotton, especially by increasing the proportion of short fibres. In addition, the capacity utilization is less than 50%. In addition, ginners are unaware of the quality parameters of the cotton, which are crucial for the grading and classification of the cotton. Nonetheless, in recent years, roller gins have been installed and used for Upland cotton to reduce fibre length degradation that occurs when seed cotton is processed in conventional saw gins. While the idea is interesting, productivity needs less further investigation.

There are a number of issues that are compounding the problems facing the ginning industry. Shortages of seed cotton, lack of infrastructure, lack of spare parts, etc. require government intervention as well as concerted efforts of relevant stakeholders. Issues of sustainable production and safety should be given due consideration. Last but not least, the need for research and training should not be underestimated.

References

1. Agriculture in the Classroom (n.d) Cotton, USDA. http://www.agintheclassroom.org/TeacherResources/TerraNova/clr_cottonnews.pdf. Accessed on 15th Nov. 2021
2. Ambatkar RD, Sambhare RB, Dahake VB, Wakode RA (2020) Automatic Raw Cotton Feeding System. IOSR J Electron Commun Eng (IOSR JECE) 15(2):40–46
3. Anthony WS (1974) Development and evaluation of a small-scale cotton ginning system. U.S. Department of Agriculture, Agricultural Research Service, ARS-S-36:9
4. Anthony WS (1990) Performance characteristics of cotton ginning machinery. Transactions of the ASAE 33(4):1089–1098
5. Anthony WS, Griffin AC Jr, (2001) Fibre breakage at cotton gins: Causes and solutions. Proc Beltwide Cotton Conf National Cotton Council, Memphis, TN. 2001, 2:1347–1357
6. Baker RV (1971) Comparative performances of a stick machine and a bur machine on machine-stripped cotton. U.S. Department of Agriculture, Technical Bulletin 1437:16
7. Baker RV, Boving PA, Laird JW (1981) Comparative performances of bur and stick extractors. Cotton Ginners Journal and Yearbook 49:12–18
8. Baker RV, Boving PA, Laird JW (1982) Effects of processing rate on the performance of seed cotton cleaning equipment. Transactions of the ASAE 25:187–192
9. Baker RV, Columbus EP, Laird JW (1977) Cleaning machine-stripped cotton for efficient ginning and maximum bale values. U.S. Department of Agriculture, Technical Bulletin 1540:19

10. Baker RV, Lalor WF (1989) Multistage trash extractor for cotton gins. Transactions of the ASAE 33(5):1457–1463
11. Bennett CA (1962) Cotton ginning systems and auxiliary developments. Texas Cotton Ginners' Association, Dallas, TX
12. Cocke JB (1972) Effect of processing rates and speeds of cylinder-type cleaners on ginning performance and cotton quality. U.S. Department of Agriculture, Agricultural Research Service, ARS:42–199
13. Columbus EP, Bel PD, Robert KQ (1990) Maintaining cotton quality transactions of the ASAE. 33(4) : 1057–1065. https://doi.org/10.13031/2013.31438
14. Cotton Ginning Final Report for U. S. Environmental Protection Agency Office of Air Quality Planning and Standards Emission Factors and Inventory Group (1996) EPA Contract No. 68-D2–0159 Work Assignment No. 3–01 MRI Project No. 4603–01
15. Courtesy of Continental Eagle Corporation (2022) http://www.continentaleagle.com.au. Accessed on 03 May 2022
16. Embassy of Ethiopia (2008) Economy and Business Section, 3506 International Drive N.W., Washington, D.C.
17. ETIDI (2012) Cotton Cultivation and Marketing Strategic Plan
18. ETIDI (2014) Ethiopia Cotton Cultivated Land Areas and Regions
19. ETIDI (2016) Ethiopian Textile Sector Profile
20. ETIDI (2021b) Assessment of Existing Status and Rating of Ethiopian Ginning Factories with Support Scheme Recommendations
21. ETIDI (2021a) Ethiopian Ginning Industry Performance Improvement Study
22. Franks GN, Shaw OS (1959) Stick remover for cotton gins. U.S. Department of Agriculture, Marketing Research Report, 852:30
23. Funk P, Hardin R (2012) Cotton gin electrical energy use trends and 2009 audit results. Appl Eng Agric 28(4):503–510
24. Gamer WE, Baker RV (1977) Cleaning and extracting. In: Cotton Ginners Handbook, U.S. Department of Agriculture, Agricultural Handbook:18–29
25. Gérald Estur and Nicolas Gergely (2010) The Economics of Roller Ginning Technology and Implications for African Cotton Sectors. Africa Region Working Paper Series No. 129 (a):10–64
26. Gordon S, Hsieh Y-L (2007) Cotton: Science and technology. Woodhead Publishing Limited, Cambridge, England
27. International Cotton Advisory Committee (2001) Report of an Expert Panel on Ginning Methods
28. International Trade Center (2007) Cotton Exporters Guide
29. Laird JW, Baker RV, Childers RE (1984) Screen and grid dimensions and feeding method effects on performance of a cylinder cleaner at the gin. In: Proceedings of the Beltwide Cotton Production Research Conference, Atlanta, GA, Jan. 8–12, 1984, National Cotton Council, Memphis, TN:159–160
30. Lalor WF, Willcutt MH, Curley RG (1994) Cotton Ginners Handbook, United States Department of Agriculture, Agricultural Research Service
31. Mangialardi Jr GJ (1990) Performance of air-type lint cleaners in reducing motes and seed coat fragments. In: Proceedings-Beltwide Cotton Production Research Conferences (USA)
32. Mangialardi GJ (1993) Effect of lint cleaning at gins on market value and quality. Appl Eng Agric 9(4):365–371
33. Mangialardi GJ Jr (1996) Lint cleaning options to preserve fibre quality at gins. Appl Eng Agric 12(5):555–562
34. Mangialardi GJ, Anthony WS (1998) Field Evaluations of Air and Saw Lint Cleaning Systems. The Journal of Cotton Science, 2:53–61. The Cotton Foundation. https://www.cotton.org/journal/1998-02/1/upload/jcs02-053.pdf, Accessed on 23rd May 2022
35. Mayfield W, Baker RV, Hughs SE, Griffin AC (1983) Ginning cotton to preserve quality. U.S. Department of Agriculture, Extension Service, Program Aid 1323:11
36. National Cotton Ginners Association (n.d) Introduction to a Cotton Gin:3–12 https://www.cotton.org/ncga/techpubs/upload/Introduction-to-a-Cotton-Gin.pdf, Accessed on 27th April 2022

37. Patil PG, Arude VG (2015) Recent Advances in Cotton Ginning Technology in India. Central Institute for Research on Cotton Technology, (ICAR), Mumbai, India
38. Pendleton AM, Moore VP (1967) Ginning cotton to preserve quality. U.S. Department of Agriculture, Extension Service, ESC-560:19
39. Read KH (1972) Cylinder cleaner speed influences cleaner efficiency. Cotton Gin and Oil Mill Press 73:6–7
40. Searcy SW, Willcutt MH, Buschermohle MJ, Wanjura JD, Brashears AD, Barnes EM (2010) Seed Cotton Handling and Storage. Texas A&M University
41. Siddaiah M, Lieberman MA, Prasad NR, Hughs SE (2004) Automation in cotton ginning. Int J Intell Syst 19:111–129. Wiley Periodicals, Inc. https://doi.org/10.1002/int.10156
42. Tegegne B (2008) Analysis of Cotton Marketing Chains: The Case of Metema Woreda, North Gondar Zone, and Amhara National Regional State. Master Thesis. Haramaya University
43. Textile School (2019) Cotton Fibre Ginning. https://www.textileschool.com/366/cotton-fibreginning/, Accessed on 2nd April 2019
44. The Cotton Supply Chain in Ethiopia Project Report (2018) A country-focused commodity analysis in the context of the Bio-Macht project. Freiburg, Germany
45. United States Department of Agriculture (1977) Cotton Ginners Handbook. Agriculture Handbook No. 503, Agricultural Research Service
46. Vizia NC, Manojkumar TS, Arude VG, Patil PG (1999) Status of pre-cleaning and ginning technology in India. Agri Eng Today 23(5–6):39–55
47. Wagaye BT, Walle GA (2019) Overview of Ethiopian textile industry. J Textiles Polymers, ITAST 6(2):117–120. http://www.itast.ir/article_69035_8c7f2b99e674fe416145df0840d61726.pdf, Accessed on 12th Dec. 2021
48. Yiha O, Kumie A (2010) Assessment of occupational injuries in Tendaho Agricultural Development S.C, Afar Regional State. Ethiopian J Health Dev 24:167–174

Cotton Quality Testing and Characterization: Trends and Grading

Tiliksew Addis, Biruk Fentahun, Samual Genetu, and Belayneh Sisay

Abstract Cotton quality is an intricate characteristic that is influenced by various factors, such as the type of cotton, the growing conditions, and the processing techniques utilized. Different techniques can be utilized to examine and characterize cotton properties, including fibre length, fibre strength, elongation, micronaire, uniformity, colour, etc. There is a rising trend towards adopting more advanced methods to assess and characterize cotton quality. This is owing to the rising complexity of cotton processing and marketing, as well as the requirement for more accurate and dependable information regarding cotton quality. This chapter examines current trends in cotton testing and characterization, as well as the summary of the features of Ethiopian cotton. The classification and characterization of cotton in Ethiopia is in its preliminary stages. However, there is a growing curiosity in this field, and there are some research projects underway to develop better techniques for classifying and characterizing Ethiopian cotton.

1 Introduction

Ethiopia has a long history of cotton production. In the seventeenth century, Manuel de Almeida, a Portuguese explorer, wrote about the cultivation of cotton in Ethiopia. He observed that the local people were capable of creating thick fabric from native cotton and that cotton tents were utilized by the Abyssinian kings and their roaming courts, who did not have permanent dwellings at that time. Nicholson also mentioned that one of the types of cotton, *Gossypium herbaceum*, which is occasionally found in fields and gardens of Ethiopia, may be native to the country [1]. The cultivation and utilization of cotton in various regions of Ethiopia show a trend that has the potential to

T. Addis (✉) · B. Fentahun
Department of Textile Engineering, Ethiopian Institute of Textile and Fashion Technology, Bahir Dar University, P.O. Box 79, Bahir Dar, Ethiopia
e-mail: hulieaddis27@gmail.com

S. Genetu · B. Sisay
Cotton Quality Research and Inspection Directorate, Ethiopian Textile Industry Development Institute, Addis Ababa, Ethiopia

K. Murugesh Babu et al. (eds.), *Cotton Sector Development in Ethiopia*, Textile Science and Clothing Technology, https://doi.org/10.1007/978-981-99-9149-5_9

make a significant impact on the global cotton market. However, Ethiopia's practices for testing and classifying the quality of cotton are still in their early stages. There are several reasons why testing and classifying cotton quality are important. Firstly, it allows buyers to evaluate the quality of the cotton before making a purchase. This is crucial because the price of cotton can vary greatly depending on its quality. Secondly, testing and classifying cotton quality can help in identifying and using the appropriate quality of cotton for specific products [2].

Cotton research in Ethiopia has been ongoing since the mid-1960s. However, there has been limited progress in developing high-yielding and high-quality fibre cotton varieties. With the proper investment and support, Ethiopia can develop the high-yielding and high-quality fibre cotton varieties necessary to become a major player in the global cotton market.

Ethiopia has the potential to encourage the adoption of enhanced cotton types and agricultural techniques among farmers. This can be achieved through extension initiatives, on-farm demonstrations, and other forms of outreach. By providing farmers with knowledge on the advantages of improved types and techniques, Ethiopia can contribute to increased yields and productivity [3].

The cotton types that are commonly cultivated in Ethiopia primarily consist of Deltapine 90 and Acala SJ2, both of which originate from the United States. However, these types have been in use for over two decades, resulting in issues related to their ageing and decline in quality. To meet the changing needs of growers, domestic textile mills, and international markets, it is crucial to continually introduce and adapt high-yielding varieties with superior fibre quality.

2 Cotton Varieties in Ethiopia

Since the inherent characteristic of cotton relies on the variety, it is preferable to examine the quality traits of different varieties in Ethiopia, despite the fact that cotton is not classified in Ethiopia based on variety. Cotton is a significant cash crop in Ethiopia and is extensively cultivated in the lowlands through large-scale irrigation schemes. It is also grown on small-scale farms through rain-fed agriculture. The Werer agricultural research centre, since 1964, which is responsible for selecting and developing genotypes suitable for different agro-ecology in the country, has released 22 varieties and seven hybrids for irrigated areas, as well as five varieties for rain-fed areas [4], although only a few of them are commercially available. Currently, 96% of the cotton varieties grown in the country are of the upland type, with DP 90 occupying 80% of the acreage, followed by Stam 59A (15%) and Acala SJ2 (0.5%), while the remaining 4% is covered by landraces and local varieties. Commercially, Gossypium hirsutum cotton accounts for 98% of the annual cotton production in Ethiopia. This species offers superior quality due to its wide adaptability and high-yield potential [5]. Gossypium hirsutum species represents 98% of the annual commercial cotton production in the country. This species offers superior quality due to its wide adaptability and high-yield potential.

DP 90 was imported from the USA by the Ethiopian Government approximately 25 years ago. Despite the release of numerous varieties from research stations, only two prominent varieties are currently being cultivated. These two varieties (DP 90 and Stam 59A) perform well under both irrigated and rain-fed conditions. In addition to conventional varieties, several hybrid varieties have recently been registered for irrigated production [6].

DP 90, which was implemented in Ethiopia in 1989, is one unidentified local variant, and Stam 59A are cultivated in the Northern and Northeastern regions of Ethiopia. At present, there is a variety named Claudia being tested. In Ethiopia, the species Gossypium hirsutum is also more well-known and it accounts for 90% of the cotton produced in Ethiopia [7]. The primary variant in Ethiopia is DP 90, which accounts for 80% of the total cotton production. Stam 59A and the local cotton account for the remaining cotton types produced [8].

3 Overview of Cotton Quality

To produce high-quality yarn, it is crucial to select the appropriate cotton for the intended purpose, consider the quality of fibre properties, and efficiently utilize fibre blends. Fibre costs make up over 50% of the total cost of manufacturing cotton yarn, and the quality properties of the yarn vary depending on the fibre's quality [9].

The spinning performance is significantly influenced by the properties of cotton fibres, although they only explain a part of the variations in final yarn quality parameters. The connection between the quality of ring-spun yarn and fibre properties (measured using the High-Volume Instrument (HVI) and Advanced Fibre Information System (AFIS)) [10].

In the textile industry, quality is assessed through testing. Textile testing is a broad term that encompasses the various quality tests conducted on a textile product before it is introduced to the market. Rather than manual methods, cotton quality characteristics are increasingly measured using instruments. Increased industry competition, consumer demand for a wider range of products, and more stringent standards for product usage have all contributed to more extensive fibre quality testing [3, 11].

4 The Main Fibre Quality Aspects of Cotton

Cotton is one of the most widely used, largest textile fibre crops and comfortable fibres used for various applications including clothing for many years [12]. Two of the species known as *Gossypium arboreum* and *Gossypium herbaceum* are originated from Asiatic and *Gossypium barbadanse* are from American [13]. The majority of cotton produced for trade in the world belongs to the *Gossypium hirsutum* species, also known as upland cotton. Extra-long staple cotton, which is used to produce very fine yarns, belongs to another species that accounts for less than 5% of world cotton production [14].

The quality of a material can be defined in many ways. Many associations and scholars have defined it differently. According to ISO, quality is the extent to which a set of inherent characteristics fulfils the requirement. Accordingly, cotton quality is the one that meets the needs of the customer, specifically the spinners [15]. It is important to note that these are just some of the many perspectives on quality. There is no single definition of quality that is universally accepted. However, by understanding the different perspectives on quality, you can better understand how to improve the quality of your products and services. The main perspectives are Transcendent perspective, Product perspective, User perspective, Manufacturing perspective, and Value-based perspective [16].

As a spinner, the primary focus for spinning mills is the quality of fibre, which serves as the main raw material. To compete with synthetic fibres, cotton spinning mills must enhance the quality of their cotton fibre. The improvement of cotton fibre quality can be achieved through plant breeding based on scientific principles, in a suitable climate, and with appropriate management inputs [14].

The fundamental characteristics of cotton that describe its inherent quality are fibre length, strength, elongation, fineness, maturity, and colour. By comprehending the factors that impact the quality of cotton fibre, the textile industry can ensure the utilization of the finest cotton fibres to manufacture superior products [17]. The staple length of cotton fibre plays a significant role in determining its quality in the global textile industry [13].

The Fibre Quality Index (FQI), spinning consistency index (SCI), and modified fibre quality indexes (MFQI) combine various properties of cotton fibre and provide information about its spinnability [18].

1. **Length**: Fibre length is a genetic characteristic. This is typically understood to mean the Upper Half Mean Length (UHML). Uniformity Index (CI) and Short Fibre Index (SFI) are additional measures related to the fibre length distribution.

   Length is influenced by genetics, environmental conditions during the growing season, and the process of ginning [19, 20]. Length is one of the most important factors considered in all stages of the cotton production chain. Length is the most crucial attribute in the manufacturing of ring-spun yarn, as it affects the ability to spin the cotton and the number of twists per inch of yarn needed to achieve a desired level of strength. Length is also a critical factor in determining the drafting parameters within a textile mill. The distribution of length strongly impacts almost all yarn quality parameters. UHML influences yarn strength. Size consistency affects evenness, and SFI influences the level of hairiness.
2. **Strength**: The strength of fibre is determined by the type of seed and the conditions in which it is grown. It is measured as (g/tex) in brief. Excessive drying and the utilization of lint cleaners during ginning will diminish strength and result in more fibre breakage. Strength is the primary characteristic of Open End (Rotor) and Air-Jet spinning. The strength and length of the fibre influence yarn strength, which is vital for weaving yarns.

3. **Micronaire**: Micronaire is defined as a combination of fineness and maturity. The micronaire of a given sample of cotton is influenced by both genetics and environmental factors during the growing season. When comparing samples of cotton of the same growth, disparities in micronaire indicate disparities in maturity. However, when comparing samples of different growths but similar levels of maturity, disparities in micronaire indicate disparities in fineness. For producers, micronaire can assist in the comparison of seed varieties. For trading, it is used as a simple and dependable guide regarding the combination of fineness and maturity. For spinners, fineness is essential in predicting the spinnability of cotton and the fineness, uniformity, and strength of the yarn that might be produced from it. Micronaire is important to predict the dyeability, fibre neps, and appearance of yarn and fabric.
4. **Cleanness**: Trash comprises primarily of foliage from the cotton plant affected by the method of harvesting. Trash signifies the non-lint components of bales and consequently has an adverse effect on prices and textile processing.
5. **Colour**: Classers assign a single-colour grade to a sample. Instrument results are a combination of reflectivity (Rd) and yellowness (+b). Changes in colour indicate the history of a bale of cotton. Cotton can change in colour from white to grey or yellow, depending on its cultivation and harvesting methods, whether precipitation occurred during harvest, the moisture content of the cotton seeds, and the duration of storage prior to ginning. Grey or yellow cotton will typically possess less strength than white cotton. In processing, colour is vital for dyeing and the uniformity of dye absorption.
6. **Maturity**: Fibre maturity is one of the most important fibre properties that influences other fibre properties such as individual fibre strength and length [21]
7. **Other Fibre Measurements, Including Neps, Stickiness**: The development of neps (fibre entanglement) is affected by the maturity of the fibres and the level of processing that the cotton undergoes. If the cotton is processed slowly and with caution, from the initial ginning stage to spinning, the formation of neps is minimized. Neps have a negative impact on the appearance of the yarn. Any stickiness caused by infestation from white flies or aphids can disrupt the spinning process, especially during the drafting stage.
8. **Consistency Index**: It is an overview factor that is calculated based on the findings of micronaire, durability, length, length consistency, and colour in large-scale testing devices. Moisture content of the fibres impacts the processing. Dry fibres are more susceptible to increased breakage rates.
9. **Fibre Friction**: Frictional properties of cotton fibres differed greatly among types. For groups of fibres with identical weight and equivalent average length, the actual surface area increases as fibre immaturity and fineness decrease [22].

5 Testing of Cotton Fibre Properties

The quality of cotton fibre is determined by its most significant characteristics such as fibre strength, fibre fineness, staple length, maturity, uniformity, and grade. The dual-beard sampling technique can be utilized to assess the distribution of fibre lengths [23]. The staple length of cotton fibre plays a crucial role in determining its quality in the global textile industry [24]. Cottons with higher micronaire values have more twists compared to those with lower micronaire values, and longer fibres have greater frictional properties than shorter ones. The presence of twists in mature fibres enhances the spinnability of the fibre by increasing the cohesion between fibres. Longer fibres allow for the production of finer and stronger yarns, as the twist applied to longer fibres covers a greater length of yarn. Fibre length determines the settings of machines in a spinning mill. Longer fibres also require less twist, resulting in increased production speeds. Consistency in fibre length is crucial, as variations can lead to significant issues such as increased waste, decreased processing performance, and lower yarn quality [25, 26].

The characteristics of cotton fibre, including maturity and strength, as well as the impact of mechanical processes on the cotton, can affect the pattern of the distribution of fibre lengths [20].

Cotton fibre testing is carried out to assess the technological worth of cotton so that it can be priced accurately and utilize the appropriate fibre quality for the appropriate product [27]. The price of cotton lint is primarily associated not only with fibre characteristics but also with non-quality factors such as its international marketing methods [15, 28]. The instruments used to measure fibre quality parameters (HVI and AFIS) can be used to assess the inherent characteristics of cotton fibre such as fineness, length, maturity, and length distribution [24]. High Volume Instrument (HVI) measurement is a fundamental and regular method of providing fibre properties to cotton researchers [29]. The HVI instrument can also determine fibre elongation, but there is currently no standardized approach for it. Both extension and strength contribute to the work-to-break of fibres, which directly impacts fibre breakage and spinning performance [30].

Liu characterized cotton fibre using SITC and AFIS quality and also yarn skein strength of cottons harvested from different locations (Liu et al. 2015). The popularity of HVI can be confirmed by the fact that its results are being used to classify cotton in many countries. The fast testing speed of the HVI system allows the spinning mill to test each individual bale of cotton [31]. The Suter-Webb array method can be utilized to analyse fibre properties like AFIS and HVI [32]. Cotton breeders have used HVI as their primary instrument for accessing fibre quality data and also fibre data generated by Advanced Fibre Information System (AFIS) technology, which is now available to provide additional information on fibre length and fibre maturity characteristics [28].

There is a strong correlation between length and fineness, as well as length and maturity measurements in (HVI and AFIS), indicating that the information regarding fibre maturity and fineness is likely embedded in the distribution of fibre lengths [24].

The distribution of fibre lengths is heritable and can be modified through breeding efforts. There are significant interactions between the shape of the fibre length distribution and the strength, maturity properties, and mechanical processing of cotton [28].

There is a strong inverse relationship between fibre length measurements and the content of short and immature fibres [33]. The correlation between the appearance and physical properties of raw cotton is assessed. As the cotton samples become more yellow and contain more trash, they tend to have lower levels of short fibres and higher levels of length, length uniformity, and strength. The degree of yellowness or the amount of trash in raw cotton is positively associated with these properties [34].

The distribution of length is of utmost importance for excellent spinning performance and high yarn quality [10]. The properties of cotton fibres play a significant role in determining spinning performance, but only account for a portion of the variation in final yarn quality parameters. If the fibres are uniformly stronger, there is the potential to reduce the percentage of short fibres in cotton bales, resulting in stronger and more consistent yarns that can be processed at higher speeds [25]. The results from AFIS and SITC indicate that there is a moderate positive or negative correlation between fibre properties (Liu et al. 2015).

The HVI instrument can also measure fibre elongation. Both elongation and strength contribute to the work-to-break of fibres, which directly affects fibre breakage and spinning performance [30]. Descriptive statistics show that AFIS measurements of fibre length have a smaller range between the maximum and minimum values, but HVI measurements have a lower coefficient of variation (CV%) [35]. Previously, Stelometer testing was used to determine the tenacity and linear density of different types of cotton. Fibres with similar weights but varying fibre counts were classified as fine or coarse fibres [29]. The generation of variations in length parameters allows for the prediction of blend characteristics and the selection of optimal blends with greater accuracy than linear models [19].

Textile testing is a broad word that refers to a variety of tests conducted on textile materials to ensure their quality. These tests can be used to assess textiles' physical, chemical, and performance qualities. The findings of these tests can be used to: ensure textile quality, choose the appropriate fibres and yarns for the purpose, and adhere to industry standards [16]. Physical tests: These tests measure the physical properties of textiles, such as tensile strength, elongation, and abrasion resistance.

- Chemical tests: These evaluate the chemical characteristics of textiles such as colour fastness, flame resistance, and shrink resistance.
- Performance tests: These assess the performance of fabrics, such as moisture absorption, breathability, and wrinkle resistance. The specific tests carried out on a textile material will differ based on the type of cloth and the intended application.

5.1 *Methods and Equipment for Testing Cotton Fibre Properties*

Cotton fibre characteristics can be examined and analysed using various test methods and equipment. These could include either modern testing equipment such as HVI or AFIS or simple devices. Modern instruments are based on fibre bundle testing, which means that numerous fibre properties are tested at the same time and data is created using a central processing unit (CPU). Traditional testing methods such as Micronaire, Presseley, Stelometer, and fibrograph are intended to find the average value for a large number of fibres, often known as fibre bundle tests.

5.1.1 High Volume Instrument (HVI)

HVI is an instrument system with a high level of activity that automates the method of testing bundles. This means that the time required for testing is reduced and the number of samples that can be processed is significantly increased. The impact of the operator is also reduced. HVI testing is appealing for cotton classification and blending in textile spinning mills. It is suitable for extensive quality control of all bales processed in a spinning mill. Mills can determine their quality levels within a specific range of operations. Each sample takes 0.3 min to test, making it perfect for implementing optimal raw material conditions.

HVI strength testing is performed on the same bundle of fibres prepared for length measurement. The low-volume instrument Stelometer uses separate fibre bundles, the samples prepared after removing short fibres. For micronaire testing, a 10-g sample of cotton is used. Colour testing uses a random mass of fibres sufficient to cover the test window.

5.1.2 Advanced Fibre Information System (AFIS)

The Advanced Fibre Instrument System (AFIS) is a device that can analyse the physical characteristics of cotton fibre in individual fibre form. AFIS measures nep, which is a small, tangled cluster of fibres. The AFIS operates by initially inserting a fibre sample in the shape of a sliver weighing approximately 500 mg between the feed roller and the feed plate of the instrument. Opening rollers then separate and unravel the fibre assembly, distinguishing the fibres, neps, trash, and dust. The fibres and neps subsequently pass through an optical sensor, which determines the quantity and dimensions of the neps. The statistical information is then computed and printed through a printer. The procedure is managed by a central processing unit.

5.1.3 Contamination Testing

The fibre contamination test (FCT) system is used to test stickiness, neps, garbage, and seed coat fragments. Thermo detectors can also be used to measure the stickiness of cotton. Major contaminants such as stickiness, seed coat neps, and fibre neps are also taken into account while trading cotton. HVI, AFIS, and FCT are all fibre testing methodologies.

5.1.4 DNA Testing

This includes identifying genetically modified cotton and differentiating cotton varieties. The traceability of cotton varieties in various processing stages is becoming increasingly crucial, necessitating the development of acceptable and effective traceability technologies.

5.1.5 Trash and Micro Dust Testing

Trash can be measured using a variety of methods. The Shirley trash analyser utilizes a gravimetric measurement to detect the weight of trash in a cotton sample, whereas the HVI (High Volume Instrument) employs an optical measurement to determine the amount of garbage in a cotton sample. The AFIS (Advanced Fibre Instrument System) approach measures the quantity of garbage in a cotton sample using a combination of gravimetric and optical measurements.

MDTA (Micro Dust and Trash Analyzer) gives essential indications for the content of dust and fibre fragments, as well as the cleanability of raw cotton, which influences yarn processing on various spinning machines. Samples for the various segregated trash fractions of the examined sample are available upon request.

6 Standards for Textile Testing

Textile standards are sets of regulations that determine a textile's quality, performance, and safety. Organizations such as the International Organization for Standardization (ISO) and the American Society for Testing and Materials (ASTM) create them. A multinational group of cotton industry leaders urged in 1907 that standardized cotton standards be established in order to "eliminate price differences between markets and make farmers more aware of the value of their products." Cotton grade standards and cotton classification systems were developed and approved by the United States Department of Agriculture (USDA) in response to standardization requirements. Cotton classification is currently a system of established techniques for measuring raw cotton qualities that determine processing quality and product quality.

There are numerous conventional and HVI techniques available for characterization of cotton fibres.

Textile standards span a wide range of topics, and they are significant for a variety of reasons. Among the most important reasons are:

1. Consistency in quality and performance: Textile standards help to ensure that textiles are consistent in quality and performance. This is critical for both producers and customers. Manufacturers must be sure that their products will fulfil the expectations of their customers, and consumers must be confident that the textiles they purchase will be of high quality and operate as expected.
2. Safety: Textile regulations can help to safeguard customers from unsafe fabrics. Textile standards, for example, specify the flammability of fabrics. These guidelines aim to ensure that textiles do not easily catch fire and do not emit harmful fumes if they do catch fire.
3. Sustainability: Textile standards can help to increase the textile industry's sustainability. Textile standards, for example, require the use of recyclable materials as well as environmentally friendly dyes and finishes. These guidelines contribute to lowering the textile industry's environmental effects.

The USTER®STATISTICS is first and foremost a field guide to "good textile practices." The identification of specific flaws in overall yarn quality, which may be shown by utilizing statistics as a comparative standard, can lead to quick corrective action in the manufacturing process. There are numerous textile standards available, each of which can be utilized for a variety of reasons. Manufacturers utilize some criteria to ensure that their products satisfy a specific degree of quality. Consumers use other standards to assist them in buying safe and long-lasting fabrics. Governments often utilize various standards to regulate the textile sector.

The ISO and ASTM International textile committees collaborate to develop textile guidelines and standards for improving production processes. These committees' standards cover a wide range of topics, including fibre identification, yarn testing, fabric testing, and finishing. Guidelines for enhancing production processes are also provided by the ISO and ASTM International textile committees. Although membership in these organizations is not required, adopting their rules may assist textile manufacturers in ensuring consistent quality and developing a set of standard operating procedures.

Cotton is classified globally in accordance with United States Department of Agriculture (USDA) standards. Cotton classification is based on the following factors, according to the USDA:

1. Fibre length values, particularly the uniformity index UI (%) and upper half mean length
2. Micronaire and maturity value
3. Fibre strength (cN/Tex) and elongation E (%)
4. Whiteness Rt and yellowness degree.

SITRA (South India Textile Research Association) spinning mill standards are utilized as a benchmark in Ethiopia for cotton quality, blow room process control, card process control, and draw frame process control. According to SITRA, the following parameters are employed for cotton classification: flier frame process control, yarn quality control, and so on.

Cotton fibre classification takes into account span length, mean length, upper half mean length, micronaire and maturity, fibre strength, fibre quality index, fibre yarn relationships using HVI/FMT measured fibre properties, warp yarn properties, HVI system for cotton fibre testing, trash content, and nep content. Comparing Ethiopian cotton quality with other countries cotton quality is one method of levelling. Ethiopian cotton sector can acquire best practice from different cotton producing countries [36].

7 Cotton Fibre Testing in Ethiopia

Ethiopia has the potential to become a manufacturer and exporter of cotton yarn to the global market [37]. Because the cost of raw materials exceeds 50% of the cost of yarn, the quality of Ethiopian cotton must be examined using several standards such as Uster statistics, ISO USDA SITRA Standards, and so on.

There have been few studies on Ethiopian cotton fibre, particularly its quality in comparison with other nations [18, 31]. Werer Agricultural Research Centre (WARC) is an Ethiopian institution that has long worked in cotton breeding and genetics. Ethiopia has cotton fibre testing standards at the national level. Because the country is an ISO member, the preferred standard test techniques for cotton fibres are standardization for testing (ISO) standards.

Tesema, examined the three cotton cultivars and attempted to compare them to US and Egyptian cotton fibres using HVI, FAVIGRAPH, and AFIS instruments. According to their investigation, the FQI and MFQI of the three commercial cotton varieties Acala (SJ2), Arba, and Dp 90 are lower in comparison to US and Egyptian cotton varieties [18]. Addis, examined and contrasted the quality of Ethiopian cotton fibres with Texas cotton in their review article. The Ethiopian cotton fibres exhibit similar characteristics in terms of length, strength, and uniformity index, but they have a slightly higher content of short fibres and lower strength [8]. Conversely, Zerihun explored the relationship between Ethiopian cotton lint yield and its qualities. They discovered that there are strong negative genetic correlation coefficients between lint percentage and lint yield with fibre strength, while there is a positive correlation with fibre fineness [38]. Werer Agricultural Research Centre (WARC) has also conducted research on various Ethiopian cotton varieties, some of which are still in the experimental stage, while others have been commercialized.

7.1 Cotton Fibre Characterization and Grading Trends in Ethiopia

Knowledge of cotton quality is a crucial element of an effective cotton marketing system, and forecasting the appropriate raw material for the suitable product. Cotton is a natural fibre with a wide range of fibre characteristics, which can differ depending on the variety and farming practices used. Accurate description and measurement of these characteristics are vital for textile mills to determine the optimal blending levels and for cotton producers to receive premiums for qualities in the highest demand.

Textile mills utilize quality information to determine the optimal blending levels. By comprehending the quality of their cotton, both textile mills and cotton producers can make better choices regarding how to utilize and market their products. This can result in enhanced efficiency and profitability for both industries.

Ethiopia has a long-term objective of achieving self-sufficiency in lint cotton supply and exporting any surplus production. To accomplish this objective, the Ministry of Industry (MoI) has formulated a National Cotton Development Strategy (NCDS) for the period 2018–2032. The NCDS will be integrated into the Growth and Transformation Plan II (GTP II) and beyond once it is approved by the government.

For evaluation of cotton fibre characteristics various Cotton testing devices were utilized in Ethiopia. Werer agricultural research centre serves as the primary cotton research institute in Ethiopia. Previously, manual stapling and Stelometer equipment were employed to assess the staple length, strength, and elongation of the cotton fibre. At present, advanced instruments like the USTER® HVI are utilized for measuring cotton quality. Alongside the HVI, the Shirley Trash Analyser is utilized to quantify the amount of trash content present in the cotton, while the Thermodetector is used to gauge the stickiness of cotton caused by impurities or sugar. The USTER® HVI is a high-efficiency instrument that measures a broad range of cotton properties, including staple length, strength, elongation, micronaire, uniformity, and maturity. The Shirley Trash Analyser is a laboratory device that quantifies the quantity of trash content in cotton. The Thermodetector measures cotton stickiness. The use of advanced equipment has enabled the agricultural research centre to enhance its cotton quality evaluation capabilities.

The Ethiopian government has a strong commitment to boosting cotton production within the nation. They are supplying farmers with enhanced seeds, fertilizers, and irrigation systems. Additionally, efforts are being made to enhance the quality of cotton cultivated in Ethiopia [39].

The quality assurance of cotton seeds will be managed by the plant health regulatory directorate at the Ethiopian Ministry of Agriculture. This will take place at a national seed laboratory in Addis, which has received accreditation from the International Seed Testing Association (ISTA). The Ethiopian Textile and Industry Development Institute (ETIDI) is currently overseeing the Cotton Quality Research and Inspection Directorate. They are also working in collaboration with the Ministry of Agriculture to coordinate these efforts [40]. A national roadmap and strategic plan have been developed for cotton, with the goal of positioning Ethiopia as one of the leading global producers of high-quality cotton from 2017 to 2032 [39].

There have been few studies on Ethiopian cotton fibre, particularly its quality when compared to other countries. Tesema and Drieling [18] studied three cotton varieties and attempted to compare them with US and Egyptian cotton fibres using HVI, FAVIGRAPH, and AFIS instruments, as shown in Table 1. According to their findings, the FQI and MFQI of three commercial cotton types are lower than those of US and Egyptian cotton varieties (Table 1).

In their review study, Addis and others also analysed and compared the quality of Ethiopian cotton fibre variations to that of Texas cotton, as shown in Table 2. Ethiopian commercial cotton fibre qualities have a narrow quality range, limiting the sorts of yarns that can be produced from it. It has a higher uniformity index but lower quality in strength upper half mean length and high short fibre content [31].

Desalegn and colleagues explored the relationship between Ethiopian cotton lint output and quality. They discovered that lint percentage and lint yield have a strong negative genetic correlation coefficient with fibre strength, but a positive relationship with fibre fineness (micronaire).

Werer Agricultural Research Centre (WARC) is also conducting research on many Ethiopian cotton types. Cotton of different types is commonly mixed in textile mills to achieve a cost-quality balance [41]. In Ethiopia, the lack of multiple cotton cultivars with varying quality ranges forces textile mills to create a narrow yarn count range

Table 1 Comparison of the quality Index values of three Ethiopian commercial cotton cultivars with those of US and Egyptian cotton

Cotton varieties		Cotton fibre quality indexes		
		FQI	SCI	MFQI
Ethiopian cotton varieties	Acala (SJ2)	149.6	126.0	130.9
	Arba	150.2	121.4	113.7
	DP 90	138.6	110.0	106.3
US Upland, Pima	US PIMA	479.9	230.4	549.5
	US UPLAND	170.2	134.3	145.4
Egypt Giza 87	Giza 87	472.2	228.0	516.9

Abbreviations FQI, ibre quality index; SCI , spinning consistency index; MFQI, Modified cotton fibre quality index

Table 2 Ethiopian cotton versus Texas upland cotton fibre quality in 2017

No	Parameter	Ethiopian cotton	Texas cotton
1	UHML (mm)	27.86	28.52
2	Short fibre content (%)	11.30	9.75
3	Strength (g/Tex)	27.14	29.90
4	Uniformity (index) Micronaire	81.83	81.00

(FAS 2019) and [6]. Cotton growers and ginneries in Ethiopia, on the other hand, pay less attention to cotton quality [2] (Table 3).

The quality of yarn is determined by the inherent characteristics of the different fibre properties, which partially forecast the physical properties of yarn and the success of spinning [42]. Conversely, [10] asserted that the properties of cotton fibres have a significant impact on determining the spinning performance and final parameters of yarn quality to a certain degree.

The spinning performance and end-product quality of cotton are influenced by the different characteristics of the fibres and the spinning technology. It is important for the cotton industry to be aware of advancements in spinning technology and the quality requirements for the fibres. In order to assess the quality level, Ethiopian cotton is evaluated using the High-Volume Instrument (HVI) and the Cotton Quality Research and Inspection Directorate of the Ethiopian Textile Industry Development Institute is responsible for classifying the cotton and establishing the standards. Although the grading system set by the Cotton Quality Research and Inspection Directorate does not provide a clear distinction between grades, it is still used. Additionally, the stickiness of the fibres is measured using the HVI Thermodetector, the trash content is measured using the Shirley trash analyser, and the moisture content and maturity ratio are estimated using the HVI. Contamination is assessed through visual examination.

Table 3 Fibre properties of different cotton varieties

	Fibre quality				
Varieties	Micronaire	UHM (mm)	SIF	UI (%)	Strength (g/tex)
DP 90	5.09	26.5	82.3	13.7	24.1
Stam 59A	4.64	31.05	84.3	11.4	31
Ionia	3.8	31.2	85.0	10.5	30.9
YD-670	3.55	33.75	85.5	6.8	34.5
YD-195	3.46	31.06	82.7	12.8	35
YD-211	3.17	32.22	84.6	10.3	35.5
Stoneville	4.11	29.96	80.7	13.93	32.43
Arba	4.3	29.5	83.7	11.7	28.7
Carolina queen	3.7	28.1	82.8	12.4	26.6
Cucurona-1518	3.8	27.4	80.6	13.4	23.7
Cu-okra	3.7	26.4	80.5	14.6	23.1
Sille91	4.1	28.5	83.4	11.3	26.3
Claudia	4.36	30.9	84.6	13.64	32.4
Bulk 202	3.9	27.3	83.1	13.6	26.6

Source Werer Agricultural Research Centre (WARC), 2015; UI, fibre length uniformity (expressed as uniformity index UI (%); SIF, short fibre content (SF (%)); UHM, Fibre length (expressed as upper half mean UHM [mm]); Strength, fibre strength (as bundle strength STR (cN/tex))

To study the relationship between cotton fibre properties and yarn properties, the Fibre Quality Index (FQI), Spinning Consistency Index (SCI), and Premium-Discount Index (PDI) can be utilized. A new method based on a multiple-criteria decision-making technique has been proposed for this purpose [25].

Fibre Quality Index (FQI) integrates various cotton fibre qualities and employs formulae (Eq. 1), which are currently based primarily on HVI data.

$$FQI_{HVI} = \frac{FS \times UHML \times UI}{FF} \quad (1)$$

where: FQI_{HVI} is Fibre Quality Index of HVI, FS is the fibre bundle tensile strength (cN/tex), UHML is the upper half mean length (mm), UI is uniformity index (%), and FF is fibre fineness (micronaire) (Murthy and Samantha 2000). However, Eq. (1) does not take into account the elongation property of cotton fibre, which is a major contributor to the work-to-break of cotton fibres. To examine the cotton fibre elongation property, a modified fibre quality index (MFQI) is designed that takes into account both single fibre tenacity and elongation while taking into account both the HVI and FAVIGRAPH primary fibre qualities and assuming linear geometric features. To compare different cotton cultivars, MFQI is preferable. The MFQI formula is shown in Eq. (2):

$$MFQI = \frac{[UHML_H \times UI_H \times (1 - SFI_H) \times SFS_G \times (1 - EL_G)]}{MIC_H} \quad (2)$$

where $UHML_H$ is the HVI upper half mean length (mm), UI_H is the HVI uniformity index (%), SFI_H is the HVI short fibre index (%), MIC_H is the HVI fibre fineness (micronaire), SFS_G is Favigraph single fibre tenacity (cN/Tex), and EL_G is the Favigraph single fibre elongation (%).

7.2 Ethiopian Standards of Cotton Quality

Textile testing is the assessment of textiles using scientific test procedures and methodologies to ensure that the quality of the textiles fulfils the requirements of the set standards or the expectations of the consumers. A standard is a documented agreement by consensus that is authorized by a recognized entity. It is continually utilized as rules, guidelines, or definitions of qualities to ensure that materials, products, processes, and services are fit for their purpose. The Ethiopian cotton quality specifications and grading are presented in Table 4.

Standards are critical for removing trade barriers, saving money, and accelerating research. Cotton and related testing associations include the American Association of Textile Chemists and Colourists (AATCC), the American Society for Testing and Materials (ASTM), the International Organization for Standardization (ISO), and others.

Table 4 Ethiopian cotton quality specifications and grading [43]

No	Specification	Grade		
		A	B	C
1	Staple length, mm	≥28.5	27.0–28.4	25.0–26.9
2	Micronaire	3.5–4.2	4.3–4.9	3.2–3.4 and 5–5.2
3	Strength, gram/Tex	≥29	26.0–28.9	25.0–25.9
4	Average point of sticky point	0–10	11–20	21–32
5	Short fibre content, %	≤10	11–12	13–14
6	Trash content, %	<3.5	3.5–4.5	4.6–5.0
7	Moisture content, %	≤8	≤8	≤8
8	Maturity ratio, %	≥85	81– 84	75– 80
9	Length uniformity ratio, %	≥83	81–82	76–80
10	Colour	11–1 up to 21–4	31–1 up to 31–4	41–1 up to 51–4
11	Contamination, grams/ bale	≤5	5–10	10–15

Source ETIDI

The Quality and Standards body of Ethiopia (QSAE) was created in 1998 as Ethiopia's first body for quality and standardization of services, products, and materials. After twelve years, QSAE is restructured, and Ethiopia established an office for materials standards called "The Ethiopian Standards Agency (ESA)" in Addis Ababa, the country's capital city, in 2010, in accordance with regulation No. 193/ 2010. Ethiopian testing standards are among the world's national standards, and the Ethiopia Standards Agency (ESA) is a member of various international standard agencies such as ISO. Ethiopian standards, like any other national standards, have a national scope and applicability, and some standards are referred to as necessary by Ethiopian regulations. A list of acronyms for international standards is given in Table 5.

Ethiopian Standards are approved by the National Standardization Council and are continuously reviewed and updated after publication to reflect the most recent scientific and technological advances (ETHIOPIAN and S.E. 6586:2021, 2021).

Table 5 List of acronyms for international standards

	Acronym	Association full name
1	AATCC	American Association of Textile Chemists and Colourists
2	ASTM	American Society for Testing and Materials
3	ISO	International Organization for Standardization
4	ANSI	American National Standards Institute
5	ASABE	American Society of Agricultural and Biological Engineers

Table 6 Ethiopian lint and seed cotton testing standards

S. No	Standard name	Acronym
1	Standard test method for Neps in cotton fibres	ES 6698:2021
2	Seed cotton grading	ES 6676:2021
3	Standard test methods for measurement of physical properties of raw cotton by cotton classification instruments	ES 6286
4	Practice for sampling cotton fibres for testing	ES 148
5	Practice for conducting an inter laboratory study to determine the precision of a test method	ES ISO 24697
6	Terminology for cotton fibres	ES 124
7	Cotton fibres—Test method for sugar content—Spectrophotometry	ES ISO 18068:2017
8	Moisture content (max %)	ES 478
9	Guideline for bale management system	ES 6678:2021

Cottonseed standards are set by the Ethiopian Standard Agency Requirements in collaboration with the Ministry of Agriculture under the designation ES 441-2000 (Table 6). Ethiopia has national cotton fibre testing standards methods. Because the country is an ISO member, the preferred standard test procedures for cotton fibres are ISO standards.

8 Ethiopian Lint Cotton Quality Data Presentation and Dissemination

There is a requirement to store, examine, report, and distribute diverse cotton quality measurement data at various classing offices. This data can be utilized to enhance the quality of Ethiopian cotton, track the effectiveness of different cotton varieties, and recognize market trends. A nationwide database system would be a valuable instrument for gathering and managing this data. The system should be capable of storing data from all of the different classing offices in Ethiopia, and it should enable users to effortlessly search and examine the data. The system should also have the ability to generate reports and distribute the data to interested parties in the cotton industry.

The establishment of a nationwide database system and the formulation of a strategy for distributing cotton quality assessment information would represent a significant advancement for the Ethiopian cotton sector. The Ministry of Agriculture (MoA) of the Ethiopian government is entrusted with the task of maintaining the database on cotton seed production. A plan for the development of a nationwide database system to disseminate cotton quality measurement data would constitute a noteworthy step forward for the Ethiopian cotton industry. It would enhance the quality of Ethiopian cotton, optimize the textile industry, and attract additional investment. The advantages of a nationwide database system for cotton quality measurement data include a centralized repository of information, precision, and uniformity, monitoring performance, identifying market trends, and serving as a resource for government agencies. All in all, a nationwide database system for cotton quality measurement data would serve as a valuable instrument for enhancing the efficiency, transparency, and competitiveness of the Ethiopian cotton industry.

9 Summary

In order to remain competitive in the rapidly changing global market, it is essential to enhance the competitiveness of the Ethiopian cotton industry by offering extensive training on quality and implementing a quality management system. Cotton testing equipment such as the USTER® HVI, USTER® AFIS PRO2, Shirley Trash Analyser, and Thermodetector are utilized to examine the properties and impurities of cotton fibres. The primary type of cotton cultivated in Ethiopia is Dp 90, which has been utilized for an extended period of time. However, the lack of diverse varieties on a commercial scale and limited research are the main weaknesses of the cotton industry in terms of quality. The authors recommend that stakeholders establish a national programme and grading system for improving the quality of cotton and explore different varieties with a wide range of quality in order to produce various yarn counts from an industrial perspective.

10 Future Perspectives

Robust quality management systems must be developed and consistent monitoring must be implemented. The quality management system should consider the following aspects.

(i) Each sector in the value chain should have proper rapport with its customers. Improved communication and customer engagement, as well as addressing feedback from them, should be taken into account. Both short-term and long-term plans should be aimed at customer satisfaction, and everyone in the value chain should be responsible for ensuring quality. The quality of cotton should be defined and controlled from the moment it is planted.

(ii) The empowerment of high-quality cotton producers should be facilitated through various means, such as providing subsidies similar to other countries. Additionally, the cotton production sector should be motivated through ongoing training, subsidization of production inputs, and long-term credit options.

(iii) The current focus on inspection as the sole method for assessing quality is a problem. Another issue is the lack of laboratory equipment in the ginneries. The ginneries send their samples to the Cotton Quality Research and Inspection Directorate, which is located in the central part of the country, and they must wait for a considerable amount of time to receive the results.

(iv) The use of quality tools for documentation is essential. However, there is a lack of training on how to use these tools in farms and ginneries in some textile industries. Some types of checklists are used for assessment, but effective control of the process can be achieved through the use of quality tools. Therefore, training workers to use quality tools is crucial for improving quality.

(v) Benchmarking against best practices or standards is a significant measure for improving manufacturing processes. Benchmarking is a tool of total quality management that involves identifying and quantifying top-class or world-class performance (benchmarks) in a specific business or product category and comparing the data with the performance of one's own company or product.

(vi) To enhance the cotton quality testing and classification system, it is necessary to acquire modern laboratory equipment and provide training for cotton quality testers. Establishing a national cotton quality grading system and providing continuous training for all stakeholders in the value chain about the importance of producing quality cotton can lead to sustainable improvements in cotton quality. Overall, implementing best practices in cotton quality testing and classification will be beneficial for Ethiopia, benefiting farmers, buyers, and consumers.

References

1. Nicholson GE (1960) The production, history, uses and relationships of cotton (*Gossypium* spp.) in Ethiopia. Econ Bot 14:3–36
2. Hilbert I et al (2018) The cotton supply chain in Ethiopia. nfo@oekode, www.oeko.de
3. Gurmessa D et al (2022) Agronomic, yield and fibre quality performance of released introduced hybrid cotton varieties in irrigated agro ecologies of Ethiopia 10(2):53–58. https://doi.org/10.11648/j.ijgg.20221002.13
4. Bedane G, Egziabher AG (2019) Cotton production potential areas, production trends, research status, gaps and future directions of cotton improvement in Ethiopia. Greener J Agric Sci 9(2):163–170. https://doi.org/10.15580/gjas.2019.2.040619064
5. Global center for cotton and research (2018) 'Directory', ICA Bremen. ntertek.com/agriculture
6. NCDS (2017) National cotton development strategy (2018–2032) and road map
7. Zerihun (2016) Scoping study on sustainability issues. ADDIS ABEBA. http://www.worldbank.org/en/country/ethiopia/overview
8. Addis T, Kachi A, Wang J (2021) A review of current state and future directions of cotton production in Ethiopia. Cogent Food Agric 7(1):0–12. https://doi.org/10.1080/23311932.2021.1880533
9. Oner E, Topcuoglu S, Kutlu O (2019) The effect of cotton fibre characteristic on yarn properties. In: Aegean International Textile and Advanced Engineering Conference (AITAE 2018) IOP Conference Series: Materials Science and Engineering 459:012057 IOP Publishing

10. Faulkner WB et al (2012) Relationships of cotton fibre properties to ring-spun yarn quality on selected high plains cottons. Text Res J 82(4):400–414. https://doi.org/10.1177/0040517511426613
11. Collins KJ and Rogers CD (1981) Cotton quality evaluation. National Economics Division, Economic Research Service, U.S. Department of Agriculture. ERS-668
12. Adamu BF, Wagaye BT (2020) Cotton contamination. In: Wang H, Memon H (eds) Cotton science and processing technology: gene, ginning, garment and green recycling. Springer Singapore, Singapore, pp 121–141
13. Gazanfer S (2004) An overview of the Turkish cotton industry with special emphasis on the available tools and techniques of price risk management. Aegean Textiles and Raw Materials Exporters Unions, Izmir
14. Constable G et al (2015) Cotton breeding for fibre quality improvement. In: Cruz VMV, Dierig DA (eds) Industrial crops: breeding for bioenergy and bioproducts. Springer New York, New York, NY, pp 191–232
15. Estur G (2008) Quality and Marketing of Cotton Lint in Africa. (FAO Africa Region Working Paper Series No 121):82. gmpestur@aol.com
16. Saville BP (1999) Physical testing of textiles. Woodhead Publishing Ltd., Cambridge and CRC Press, Boca Raton, FL
17. Gordon S, Hsieh Y-l (2006) Cotton: science and technology. Woodhead Publishing
18. Tesema, GB and Drieling A (2019) Statistical analysis and instrumental characterization of commercial Ethiopian cotton varieties. J Textile Apparel, Technol Manag 11(1)
19. Azzouz B, Hassen M Ben, Sakli F (2007) Generation of length distribution, length diagram, fibrogram, and statistical characteristics by weight of cotton blends. Modell Simul Eng. https://doi.org/10.1155/2007/27521
20. Krifa M (2006) Fibre length distribution in cotton processing. Text Res J 76(5):426–435
21. Ayele A, Hequet E, Kelly B (2017) The impact of fibre maturity on estimating the number of cotton (Gossypium hirsutum L.) fibres per seed surface area. Ind Crops Prod 102:16–22. https://doi.org/10.1016/j.indcrop.2017.03.004
22. Hosseinali F, Thomasson JA (2018) Variability of fibre friction among cotton varieties: influence of salient fibre physical metrics. Tribol Int 127(January):433–445. https://doi.org/10.1016/j.triboint.2018.06.029
23. Wu HY and Wang FM (2014) Dual-beard sampling method for fibre length measurements 39(March):72–78
24. Paudel D (2012) Evaluating the potential of new testing methods for cotton (Gossypium hirsutum) breeding, Ph.D. Thesis, Texas Tech University, USA
25. Foulk J, Meredith W, McAlister (2009) Fibre and yarn properties improve with new cotton cultivar. J Cotton Sci 13(3):212–220
26. Klein W (1998) Technology of short staple spinning. Textile Institute, Manchester, UK
27. Militky J (1980) '1097 Comparison of Complex Indices for', Textile faculty, Technical university of Liberec, LIBEREC, Czech Republic:1–14. http://www.cottonic.com/CottonClassification
28. Kelly CM, Hequet EF, JKD (2012) Breeding and genetics. J Cotton Sci 16(1):1–16
29. Liu Y, Thibodeaux D, Rodgers J (2014) Preliminary study of linear density, tenacity, and crystallinity of cotton fibres. Fibres 2(3):211–220. https://doi.org/10.3390/fib2030211
30. McCormick KM et al (2019) Development of the correction procedure for high volume instrument elongation measurement. Text Res J 89(19–20):4095–4103. https://doi.org/10.1177/0040517519829002
31. Addis T et al (2022) Analysis of the cotton varieties under production in Ethiopia. J Nat Fibres 19(4):1522–1530. https://doi.org/10.1080/15440478.2020.1780181
32. Thibodeaux D et al (2008) Textile technology: a comparison of methods for measuring the short fibre content of cotton. J Cotton Sci 12(3):298–305
33. Kothari N et al (2017) Boll sampling protocols and their impact on measurements of cotton fibre quality. Ind Crops Prod 109:248–254. https://doi.org/10.1016/j.indcrop.2017.07.045

34. Cheng KPS, Cheng YSJ (2003) Relationship between appearance and physical properties of raw cotton. Textile Res J 73(3):245–250
35. Greveniotis V, Sioki E (2017) Correlations between cotton characteristics in Greek. J Agric Stud 5(4):170–178. https://doi.org/10.5296/jas.v5i4.12121
36. Incorporated C (2013) The classification of cotton. www.cottoninc.com
37. Adamu BF, Atalie D, Liyew EZ (2021) Quality evaluation of Ethiopian 100 % cotton carded ring Spun Yarn with respect to USTER standards
38. Zerihun et al (2009) Correlation and heritability for yield and fibre quality parameters of ethiopian cotton (Gossypium hirsutum L.) estimated from 15 (diallel) crosses. Kasetsart J—Nat Sci 43:1–11
39. Gian Nicolay (FIBL) GE, Conor Walsh (NRI), Paulos Desalegn (2021) Cotton Value chain analysis in Ethiopia
40. Lançon J, Woldu B (2020) Cottonseed sector study in Ethiopia. Lord E (1961) The characteristics of raw cotton, manual of cotton spinning, vol II, Part I. Textile Institute, Manchester and Butterworths, London
41. Sheikh RA, Lanjewar UA (2010) Decision support system for cotton bales blending using genetic Algorithm. Int J Comput Appl 28–34. https://doi.org/10.5120/1177-1608
42. Hallam D (2018) Revisiting Prebisch–Singer: what long–term trends in commodity prices tell us about the future of CDDCs, Food and Agriculture Organization of united nations. www.fao.org/publications
43. Service FA (2019) Ethiopia Cotton Production Annual in assessments of commodity and trade issues made by USDA staff and not necessarily statements of official U.S. government policy,Forein Agricaltural Service Addis Ababa
44. Liu et al (2015a) Variation and relationship of quality and near infrared spectral characteristics of cotton fibres collected from multi-location field performance trials. Text Res J 85(14):1474–1485. https://doi.org/10.1177/0040517514566106
45. Liu Y, Campbell BT and Delhom (2015b) Comparative relationship of fibre strength and yarn tenacity in four cotton cultivars. J Mater Sci Res 5(1):46. https://doi.org/10.5539/jmsr.v5n1p46

Cotton Quality Requirements for Spinning

Getnet Belay Tesema

Abstract A comprehensive understanding of the quality parameters of cotton and its influence on the spinning process and yarn quality is a very important aspect of cotton spinning. This chapter provides insight into the 'Cotton quality requirements for spinning' aspects mainly related to ring and rotor spinning systems. In the chapter, there are nine sections. These sections focus on detailed understanding of the quality parameters of the fibres and their influence on the spinning process and yarn quality, the hand and machine spinning sectors in Ethiopia, effects of saw and roller ginning on the cotton processability and quality of yarn, effect of cotton stickiness on the cotton processability and yarn quality, methods used to determine the technological value of cotton, optimization of cotton blending cost with respect to quality, mixing cost optimization by linear computer programing method, predicting ring and rotor spun yarn properties from fibre properties and machine parameters, and the future long-staple cotton requirements in Ethiopia.

Keywords Cotton quality · Ginning · Stickiness · Ring spinning · Rotor spinning · Yarn quality · Long staple cotton

1 Introduction

Cotton, being a natural product, varies widely in its fibre qualities, both physical and chemical, the former owing to genetic, environmental, harvesting, and ginning variables. Such variability impacts textile processing performance (including machine downtime and waste), cost, quality and utilization throughout the cotton pipeline, from farm to end product. The fibre alone contributes between 50 and 70% of the total yarn manufacturing costs. Ideally, the price of cotton should therefore be linked to the fibre properties. The connection between cotton fibre price and properties is shown in Fig. 1 [3, 8].

G. B. Tesema (✉)
Department of Textile Engineering, Ethiopian Institute of Textile and Fashion Technology, Selam Campus P.O. Box 1037, Bahir Dar University, Bahir Dar, Ethiopia
e-mail: getnet.belay@uni-bremen.de

K. Murugesh Babu et al. (eds.), *Cotton Sector Development in Ethiopia*, Textile Science and Clothing Technology, https://doi.org/10.1007/978-981-99-9149-5_10

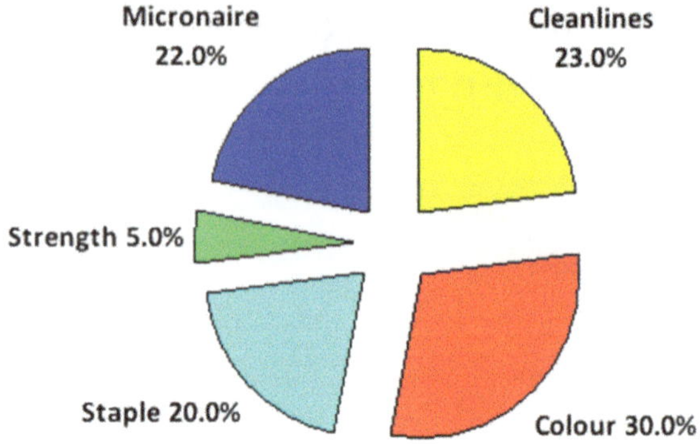

Fig. 1 Relationship between fibre price and properties. *Source* Chakraborty et al. [3] and Deussen and Neuhaus [8]

From spinning standpoint, length, strength, fineness (micronaire), fibre content, and colour are the most important fibre quality parameters and therefore the most important parameters for buying and selling cotton. However, there are other fibre properties such as neps, short fibre content, seed coat fragments, elongation, stickiness, maturity, and fineness or impurities that are also important [31].

Increasing quality and performance requirements are placed on the entire textile pipeline, from the raw material to the end product. For example, 15 irreparable faults per 100 m of cotton fabric were allowed about 20 years ago, today it is five and in the future possibly three [51]. The number of seconds has also dropped from 3 to 0.5%, in the future 0.3% is possible. Loom stops have also decreased by 50% over the same period, with about 20–30% of those stops being due to yarn defects and an end break costing about $0.70 to repair. It is known that thin places in yarns together with elongation and strength below certain minimum limits lead to end breaks in weaving, which are influenced by fibre properties and spinning conditions [23].

A spinner can hardly afford to use a flawless raw material, since it would normally be too expensive. Optimum conditions can only be achieved by mastering the raw materials. A good spinner achieves acceptable yarn quality with problem-free processing from a less expensive fibre, which significantly increases a spinner's profitability and competitiveness. Admittedly, even the best theoretical knowledge doesn't help much if the material is already at or beyond the spinnability limit. Excessive raw material savings usually do not lead to a reduction in costs, but to an increase in them due to the deterioration in processability in the spinning mill.

Therefore, it is very important to have a comprehensive understanding of the quality parameters of the fibres and their influence on the spinning process and yarn quality, so that the most suitable fibre blend can be decided. It is also important to know the fibre parameters for machine settings and the adjustment of all important process parameters. The relative importance of different fibre properties also depends on the type of spinning technology. The physical, chemical, and related properties of cotton lint, including the type and amount of nonfibrous matter present and fibre configuration (preparation, neps, etc.), determine its performance and behaviour in textile processing in terms of processing waste and efficiency (including machine stops and spinning breaks) and yarn and fabric quality (Figs. 2 and 3). Ultimately,

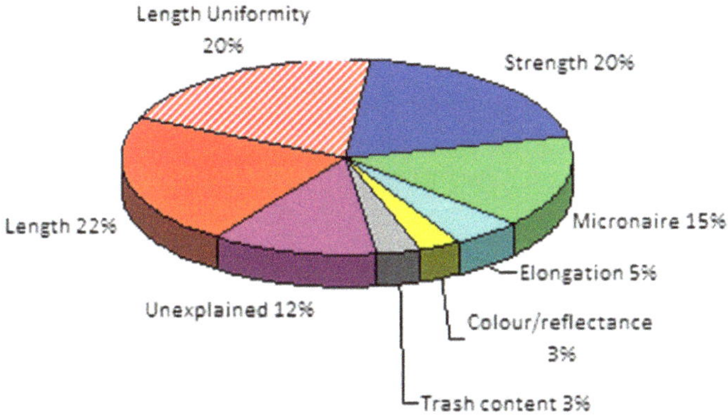

Fig. 2 The effect of cotton fibre properties on ring spun yarn strength

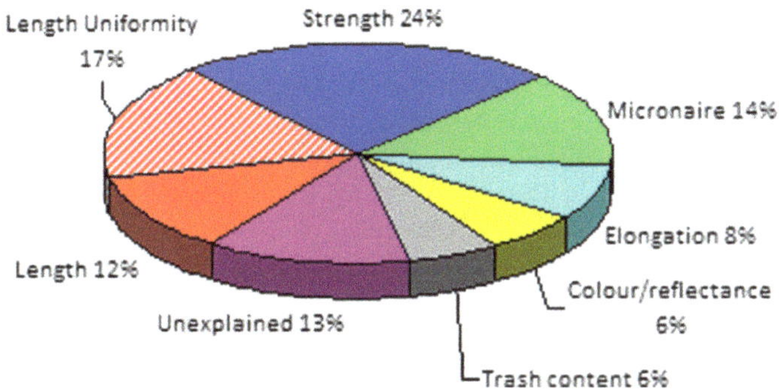

Fig. 3 The effect of cotton fibre properties on rotor spun yarn strength

these characteristics also determine both the conversion costs and the end use, price, and quality of the product.

In addition to ring spinning, compact spinning and rotor spinning systems are also used in Ethiopia. The fibre properties for the three main spinning technologies, including air-jet spinning, have been documented in order of importance and are presented in Table 1.

2 Introduction to Hand and Machine Spinning Sectors in Ethiopia

Ethiopia has a long history of traditional cottage textile sub-sector. Traditionally yarn from cotton fibre supplied by small-hold cotton farmers is home spun using age-old spinning drop spindle (Fig. 4). Before the spinner actually starts spinning with drop

Table 1 Order of importance of fibre properties for different spinning systems [23]

Type of spinning system				
Order of importance	Ring spinning	Rotor spinning	Compact spinning	Air-jet spinning
1	Fibre length	Fibre strength	Fibre length and length uniformity	Fibre fineness
2	Fibre strength	Fibre fineness	Fibre strength	Neps, trash, and dust content
3	Fibre fineness	Fibre length and length uniformity	Fibre fineness	Fibre strength
4	Neps, trash, and dust content	Neps, trash, and dust content	Neps, trash, and dust content	Fibre length and length uniformity

spindle, she/he should practice spinning without the fibre. The common procedures of spinning using drop spindle are:

1. **Making a Lead**: The spinner holds the cotton in her/his left hand and the spindle in her/his right hand. She/he will catch a bunch of fibres on the spindle hook, pinch them and draw them just slightly out of the Fibre mass. Rotate the spindle several turns clockwise. This will start her/his lead.

Fig. 4 Drop spindle spinning in Ethiopia. *Source* https://www.pinterest.com/pin/ethiopia-345229127663347829/

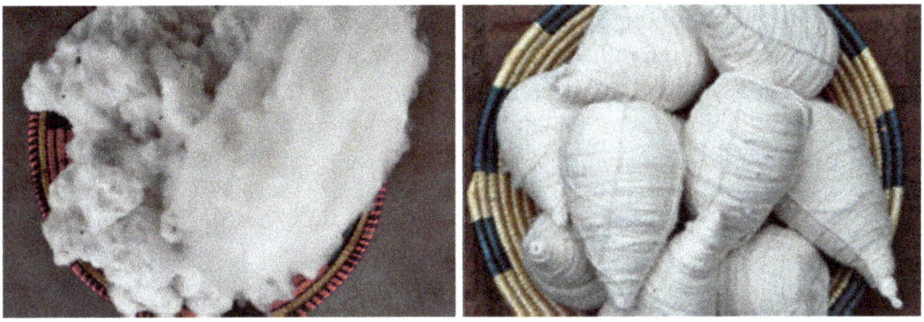

Fig. 5 Cone-shaped bobbins produced by local spinners. *Source* http://www.sabahar.com/store/Production/Cotton

2. **Drafting and Twisting**: The spinner will hold the bunch of fibres up left and draw the pinched fibres with the help of the hook to attenuate the required length fibre strand. Then she/he will spin the spindle to let the twist run up the lead.
3. **Winding**: The length of drafted strand is approximately 600 mm. During the drafting, any slubs are attenuated properly to distribute the twist evenly and produce a regular thread. After an element of thread is formed, the end will be detached from the hook and is wound clockwise around the base of the shaft, just below the whorl. In this method, a small conical shaped bobbin is formed (Fig. 5, right).

The yarn produced in this way is then converted into fabric using hand looms. The fabric thus produced is used for making traditional clothes like Netela (Kuta), Gabi, Yager Bahel Kemis, Tibeb, etc.

This traditional cottage industry is a tradition and continues to grow today. It makes an important contribution to meeting people's textile needs and provides numerous jobs to rural and urban households. The introduction of modern integrated mills is a recent phenomenon introduced by the Italians during World War II (1939). Dire Dawa Textile Mill (Ethiopia) was the first integrated textile mill founded in 1939 by foreign capital. This marked the beginning of the textile sector in Ethiopia. In the 1960s, five large integrated textile companies were established, mainly with private capital, and reached a spinning capacity of 175,000 spindles [6]. Bahir Dar Textile Mill was established in 1961 by the Italian government as war reparations for Ethiopia. The socialist regime that ruled from 1974 to 1991 nationalized private textile and clothing companies while simultaneously establishing four more integrated textile mills to expand the sector and meet domestic demand for regular textiles and replace imported products. The development of the modern textile sub-sector has historically helped meet the domestic textile needs, create employment opportunities, and boost the country's economic development, in addition to laying the fundamental foundation for the manufacturing industry in Ethiopia.

The command economy eventually took over as a tool in this sector. Due to neglect, lack of competition, and outdated technology, the sector has not been able to meet international market standards. As a result, cotton cultivation and the textile and apparel industry produced well below their capacity. After 1991, the previous planned economy partially changed to an economy based on the principles of the free market. There are currently more than 28 spinning mills and/or composite textile industries in Ethiopia, most of which use cotton as a raw material.

Most spinning mills in Ethiopia experienced frequent end breaks, which were relatively higher than expected by a spinning company, especially when processing local commercial genotypes/cultivars. This can be taken as an indicator to emphasize breeding new cultivars with improved cotton quality, which perform excellently during spinning processes. Cotton has value because it can be spun into yarn and woven into fabrics, which in turn can be made into clothing. High-quality lint cotton is the cotton used to make high-quality yarns, fabrics, and high-quality consumer products. High-quality cotton lint is characterized by a number of physical properties. Some of these properties are measurable, but others are not. High-quality lint cotton also means that it does not contain certain harmful substances.

3 Effect of Ginning on Cotton and Yarn Quality

Cotton is grown in about 80 countries, 70 of which are developing countries in the world. The cotton sector provides the livelihoods of 170 million workers in the developing world, 125 million of whom are directly dependent on cotton production. The value of global cotton production in 2018 was approximately US$50 billion [49]. The phenomenal increase in cotton production worldwide has been achieved through (i) the availability of more land for cotton cultivation and (ii) the increase in cotton yield/acreage thanks to the efforts of agronomists.

Among the various achievements in cotton production technology, the most important are (i) improved cultivation methods, (ii) the invention of higher-yielding cotton varieties with improved fibre quality, and (iii) the introduction of improved ginning methods to both increase productivity and preserve cotton quality.

To consider the ginning systems in different countries, it is necessary to consider both socio-economic and geographical factors, if necessary to understand and benefit from them. Among the major cotton-growing countries, the United States makes extensive use of Saw gins, while Egypt, India, and Pakistan use Roller gins (Table 2).

The ICAC (International Cotton Advisory Committee) conducted a survey into cotton production practices which was published in October 1996. The database contains information on 35 countries, which accounted for about 90% of world production in 1996/97. Information on ginning was also available for other countries. Based on 37 countries, only about 15% of the world's cotton production is ginned on roller gins. With the exception of a few countries, almost all cotton is saw ginned in most countries (Table 2).

*The data for the type of ginning systems in Ethiopia is compiled by the author.

The difference in ginning methods is caused by the different fibre properties, including the level of waste in each cotton variety, which in turn is a result of the type of harvest practiced in different countries.

Table 2 System of ginning cotton in some world cotton-producing countries (%)

Type of Ginning system		
Countries	Saw Ginning	Roller Ginning
Cameroon	100	
Egypt	–	100
Ethiopia*	67	33
India	40	60
Israel	85	15
Madagascar	90	10
Myanmar	20	80
Philippines	90	10
Spain	91	9
Sudan	–	100
Thailand	–	100
Togo	90–95	5–10
Turkey	25–30	70–75
Uganda	–	100
USA	98	2
Vietnam	90–95	5–10

Source Chaudhry [4]

3.1 Comparison of the Effect of Saw and Roller Gins on Cotton and Yarn Quality

Various researchers have studied the effects of the ginning processes in saw and roll gins on fibre and yarn quality. Cocke and co-workers [7] observed that roll-ginned cotton fibres generally have greater mean length and fewer short fibres and neps than saw-ginned cotton, but have more dust. Griffin and Columbus [14] reported in their studies that roller ginning produces more dust than saw ginning. [42], comparing the two types of ginning, explained that saw-ginned cotton is generally more regular but lumpier and also results in greater blowroom losses. [17] has carried out detailed studies on the properties of fibres and yarns obtained by various ginning. In his studies of four different types of cotton, he reported that roller gins had better effective length and short fibre percentage (Table 3).

The micronaire value or the fineness of saw-ginned cotton was observed to be low [43] (Table 4). 2.5% span length and uniformity ratio (UR) of fibre were also found to be better for roller-ginned cotton. Dever et al. [9] found that the upper quartile and mean length of roller-ginned cotton are higher than those or saw-ginned cotton. It was also found that the non-lint percentage was higher in roller-ginned cotton.

Harmancioglu [17] attributes the low micronaire value of saw-ginned cotton to the better cleaning effect of this type of ginning. The fact that cotton fineness was affected by saw gin was also observed by Dever et al. [9], however, they found no

Table 3 Effective length and percentage short fibres of cotton treated by saw gin and roller gin

Type of Cotton	Saw Gin				Roller Gin			
	Eff. length	SFC	2.5% SL	UR	Eff. length	SFC	2.5% SL	UR
A	28.7	22.0	29.9	0.45	29.0	20.0	31.3	0.50
B	27.7	28.5	28.6	0.45	28.8	24.0	30.1	0.50
C	26.1	31.0	26.3	0.41	27.0	25.5	27.8	0.45
D	27.5	29.1	28.4	0.41	28.2	24	29.2	0.41

Source Harmancioglu [17]

Table 4 Micronaire and neps of cotton fibres treated by saw gin and roller gin

Type of cotton	Saw Gin		Roller Gin	
	Micronaire, MIC	Neps/12'25 cm^2	Micronaire, MIC	Neps/12'25 cm^2
A	4.19	55.1	4.45	19.30
B	3.73	55.7	3.78	14.00
C	3.47	40.7	3.60	11.50
D	3.94	26.3	4.01	9.2

Source Harmancioglu [17]

significant difference in fibre strength values. It has been found that the trash content of roller-ginned cotton is higher than that of saw-ginned cotton. Saw gins produce more neps than roller gins [9, 17] and it has been observed that the percentage of seed coat (Prakash et al. 1991) is invariably higher in saw-ginned lint. Harmancioglu [17] concluded that saw-ginned cotton is better for spinning coarse yarns and roller-ginned cotton for finer yarns. The different ginning processes had little effect on the strength of the yarn, and the roller-ginned yarn had a better appearance. Prakash et al. [20] found that yarns made from saw-ginned lint were stronger and had fewer imperfections and less unevenness. The superior quality of yarns spun from saw-ginned lint has been attributed to the higher mean length of that lint.

Harmancioglu and Ercan [17] investigated the influence of the two ginning techniques on the open-end spinning process and yarn properties. It was observed that roller-ginned cotton produced more waste in the blow room and the purity of the carded sliver was better in roller-ginned cotton. Spinning finer, open-spun yarn with roller-ginned cotton experienced fewer end breaks because it had fewer neps and slightly longer fibres. Rotor contamination was lower with roller-ginned cotton.

The ginning methods had no statistically significant effect on the tenacity and elongation values. Yarn irregularities were less and appearance was better in roller-ginned cotton [40], which also resulted in yarns with fewer imperfections (thick places, thin places, and neps). Hughes et al. [30] found in their experiments that fabrics made from saw-ginned cotton had more neps (23.4 neps/58 cm^2) than fabrics made from roller-ginned cotton (17.1 neps/58 cm^2).

3.2 *Effect of Saw Ginning on Fibre Quality Properties of Ethiopian Commercial Cotton Varieties*

Cotton occupies a unique position in Ethiopia's agricultural economy. Ethiopia has tremendous potential for cotton cultivation and has favourable weather conditions and topography for cotton cultivation and production. Currently, cotton (*Gossypium hirsutum L.*) is commonly grown in irrigated lowland areas on large-scale farms and in warmer mid-elevations on small-scale farms under rainfield conditions [46].

A recent study by the Ministry of Agriculture says there are 3,000,810 hectares of land suitable for growing cotton, the equivalent of Pakistan, the world's fourth-largest cotton producer. Pakistan harvests about 2.3 metric tonnes of cotton annually from a total cotton area of 3.05 million hectares (in 2014/15) (Cotton Outlook, 2015). Of the total 3 million hectares of land suitable for cotton growing in Ethiopia, 1.9 million hectares or 63.3% are located in 38 high potential cotton-growing areas and the remaining 1.1 million hectares or 36.7% are located in 79 medium-potential districts. Despite this immense potential, Ethiopia has recently produced about 40,932.2 metric tonnes annually lint cotton on a total cotton area of 130,000 ha of land, which is only 4.33 hectares corresponds to % of the total area suitable for cotton cultivation [15].

The problems associated with the quality of raw cotton pose significant obstacles to the processing performance and efficiency of spinning machines. One of the factors causing the quality of lint cotton to deteriorate is poor ginning performance.

Tesema and Drieling [47] investigated the effect of saw ginning on lint cotton quality. For the study, they selected three commercial highland cotton (*Gossypium hirsutum L.)* varieties, namely Acala SJ-2, Arba, and Deltapine 90 (DP-90), from among the twelve cotton varieties found in Ethiopia. In their study, the first experiment to evaluate the fibre properties of seed cotton and the second experiment to evaluate the effect of ginning on cotton fibre properties were conducted by evaluating the fibre properties of lint cotton after ginning and comparing them with those obtained before ginning. For all experiments, samples were tested with single fibre testers (FAVIGRAPH, AFIS) and bundled fibre testers (HVI and CCS). The researchers came to the following conclusions:

(1) When comparing the quality characteristics of seed cotton (before ginning) with lint cotton (after ginning), a statistically significant decrease in FAVIGRAPH single fibre tenacity was observed for all varieties studied. Acala SJ-2, with higher mean tenacity and lower coefficient of variation in pre-ginning tenacity, was less affected than the other two cultivars studied. The mean FAVIGRAPH single fibre tenacity difference in cN/tex before and after ginning treatment was in the range (1.30, 1.35, 1.38) for Acala SJ-2, Arba, and DP-90, respectively. The results of the relative bundle strength of HVI and CCS also support the Acala SJ-2 variety.

(2) The mean difference of the absolute bundle strength of CCS in cN/tex before and after the ginning treatment was in the range (0.11, 0.16, 0.29) for Acala SJ-2, Arba and DP-90. The absolute tenacity with corresponding elongation values

are important parameters for predicting the spinnability of cotton for a desired yarn count. Higher tenacity with increased fibre elongation leads to higher yarn strength and elongation.

(3) Cotton with a higher elongation spin with fewer ends down than cotton with a lower elongation. In their study, Acala SJ-2 with higher mean FAVIGRAPH single fibre tenacity (28.81 cN/tex 164 millitex) and elongation (7.06%) experienced less fibre breakage during the ginning treatment than Arba (27.55 cN/tex 160 millitex) and elongation (6.95%) and DP-90 (27.11 cN/tex 156 millitex) and elongation (6.83%).

After ginning, the AFIS short fibre content by number and weight for Acala SJ-2 was ($SFC_n = 28.84\%$, $SFC_w = 12.45\%$) compared to Arba ($SFC_n = 29.13\%$, $SFC_w = 12.68$) and DP-90 ($SFC_n = 29.55\%$, $SFC_w = 12.76\%$). The length distribution changed significantly following the ginning process.

(4) Raw seed cotton of all varieties with few neps resulted in a high level of neps after ginning: Acala SJ-2 (75 count/g, 188 count/g), Arba (87 count/g, 203 count/g), and DP-90 (90 count/g, 221 count/g). Varietal variances were considerable, with the greatest difference being 17%.

4 Processing Sticky Cotton: Effect of Stickiness on Productivity and Yarn Quality

Stickiness is a form of cotton fibre contamination produced by too much sugar on the lint. Stickiness can be derived from the plant itself (physiological sugars) or from feeding insects such as the whitefly Bemisia tabaci (Gennadius) and the aphid Aphis gossypii (Glover) [20]. Stickiness causes significant losses to cotton producers and the textile industry since it can be an issue during mechanical harvest with stripper or spindle pickers [1] and cotton ginning [2, 29]. Physiological plant sugars, mostly sucrose, that are found in mature cotton fibre are equally distributed inside the lint and do not promote stickiness (Elsner, 1983).

According to the International Textile Manufacturers Federation (ITMF) cotton contamination surveys, the average degree of contamination with stickiness of world production was 18.83% from 2005 to 2016, with a minimum of 16% in 2009 and 2016 and a maximum of 23% in 2013 (ITMF, 2016). Stickiness remained practically constant in 2019 as compared to 2016 (i.e., 16% in 2016 vs. 15.7% in 2019) (ITMF 2020). This demonstrates that stickiness is still a global contamination issue.

Stickiness entails increased pest control expenditures and lower cotton marketability for producers. Stickiness might indicate unique handling and processing needs for ginners. Stickiness in the textile mill entails reduced processing efficiency, poorer yarn quality, and, in extreme circumstances, total closure [11].

Tesema [48] investigated the amount of stickiness and the kind of sugars discovered on commercially cultivated Ethiopian cotton types (*Gossypium hirsutum L.)* collected from the country's cotton-producing areas. The study was conducted both before and after saw ginning. The findings revealed that cottons from Bonta and

Tendaho farms were highly polluted with Trehalulose and Melezitose. Neither of these sugars is found in cotton plants [18]. As a result, their presence (in a dominating amount) on the investigated Bonta and Tendaho cotton lint indicates insect honeydew contamination. Trehalulose was shown to be the main sugar in the Tendaho cotton-growing region, whereas Melezitose was discovered to be the major sugar in the Bonta cotton-growing area.

In the study, the amount of insects' sugars in the rain feed cotton-growing areas, such as Metema, trehalulose (0.3%), melezitose (2.3%), Humera, trehalulose (0.2%), and melezitose (3.2%) showed least stickiness level as compared the major irrigated cotton-growing areas such as Tendaho, trehalulose (40.5%), melezitose (12.5%), Bonta, trehalulose (24.0%), and melezitose (22.5%). Trehalulose was shown to be the major sugar in the Tendaho cotton-producing area, demonstrating insect honeydew pollution in this growing location. Trehalulose was likewise shown to be responsible for the sticky residue on the machine parts in this study.

The working temperature of the spinning machinery parts (Carding machine: mean = 30 °C; Draw frame: mean = 43.5 °C; Roving frame: mean = 28.3 °C; Ring frame: mean = 26.5 °C; Rotor machine: mean = 34 °C) as well as their humidity (carding machine: RH = 60%; Draw frame: RH = 55%; Roving frame: RH = 55%; Ring frame: RH = 62%; Rotor machine: RH = 55%), may also induce favourable condition for the decomposition of *Trehalulose* as sticky residue on spinning machine parts.

Hequet et al. [20] investigated the influence of cotton stickiness on yarn quality. Twelve commercial cotton bales infected with insect honeydew were chosen for the investigation based on their stickiness level as determined by a high-speed stickiness detector (H2SD). In addition, five non-sticky bales from one module were chosen for mixing with the infected cottons in order to generate mixes with varying stickiness levels. The 12 tainted bales were broken and piled. Ten samples per bale were collected. The fibre characteristics of each sample were evaluated using a high-volume instrument HVI, H2SD, and high-performance liquid chromatography (HPLC). Tables 5 and 6 present the profile of the examined samples.

Following the preparation of the contaminated and non-contaminated mixture, which was found to have mild-to-moderate levels of stickiness as measured by H2SD, 30 Ne count and 22 Ne count cotton yarns were spun using ring spinning and rotor spinning, respectively. Figure 6 shows the mass coefficient of variation (as determined by USTER tester 3) vs the number of spinning doffing for ring spun yarn of count Ne 30. The results of the yarn quality tests were reported as an average of 17 different blends. The variations in the percentage of coefficient variation CV% of ring spun yarns with consecutive doffs are shown in Fig. 6. It demonstrates that the buildup of sticky deposits on the ring spinning frame had a little but considerable influence on the USTER tester 3 CV%. Researchers discovered identical findings for the USTER tester in three thin and thick places. They discovered no significant difference in yarn nep count or hairiness.

They discovered no stickiness influence on yarn tensile strength or count strength product (CSP). However, the number of ends down on Ne 30 ring spun yarn was

Table 5 HVI results on twelve contaminated bales and one uncontaminated module

Bale ID	UHML, (mm)	UI, (%)	Strength, (cN/tex	Micronaire, MIC	Reflectance (%)	Yellowness, +b
2	28.7	82.9	30.2	4.51	75.5	10.7
3	29.5	82.7	29.7	4.63	76.9	10.5
4	29.2	84.3	29.9	4.88	77.5	10.2
5	26.9	81.9	24.3	5.35	72.6	9.7
6	29.0	82.8	31.9	4.18	77.1	10.4
7	26.7	82.0	26.9	4.39	72.4	8.7
8	27.4	80.6	25.1	5.49	70.1	8.3
9	28.4	83.6	30.1	4.61	76.9	10.7
10	29.0	83.1	29.4	4.55	77.1	10.5
11	28.2	81.3	29.6	4.44	77.1	10.0
12	27.2	79.9	24.6	5.30	70.8	8.6
13	26.9	80.0	24.9	5.39	69.1	8.5
Average	28.1	82.1	28.0	4.81	74.4	9.7
Minimum	26.7	79.9	24.3	4.18	69.1	8.3
Maximum	29.5	84.3	31.9	5.49	77.5	10.7
1*	27.7	81.7	28.1	4.88	75.9	10.6

*Corresponds to the non-sticky module

found to be rather high, although this was unrelated to the stickiness level of the mix, as shown in Fig. 7.

There were no such time-related affects for the USTER tester UT3 CV%, number of thin places, or number of thick places for the rotor spinning. There was an extremely modest but statistically significant increase in the number of neps (280% sensitivity in the yarn). However, the examination of neps on USTER UT3 with the 200% sensitivity level did not show a significant difference. They claimed that these tiny amounts of stickiness have no effect on yarn hairiness, tenacity, or CSP.

The carded sliver H2SD values in the mixtures examined varied from 2 to 15.7. At these levels, a very minor but substantial short-term influence on spinning performance was seen for quality but not productivity (as assessed by ends down levels). Even at these very low levels of stickiness, the small loss in ring spun yarn, yarn quality after 72 h of spinning suggests a probable long-term influence on both quality and production. The authors determined that, in the short term and based only on H2SD measurements, less than 12 sticky patches (inside a mix, on raw material), with the stickiness caused mostly by aphid honeydew contamination, have no effect on production for either ring or rotor spun yarns. Even at the very low levels of stickiness evaluated, a minor but substantial long-term detrimental influence on ring spun yarn quality was discovered. Over lengthy durations of production, it is likely that the progressive accumulation of sugars on textile equipment will have a detrimental

Table 6 HPLC and H2SD results on twelve contaminated bales and one uncontaminated module

Bale ID	Glucose (a)	Fructose (a)	Trehalulose (a)	Sucrose (a)	Melezitse (a)	Total Sugars (b)	H2SD
2	45.3	32.9	1.7	6.8	12.8	0.234	2.0
3	47.6	32.5	1.7	4.3	13.9	0.231	1.9
4	42.9	15.2	2.7	8.9	29.5	0.112	2.3
5	18.4	19.0	30.1	0.6	31.9	0.163	12.6
6	25.6	36.5	15.5	2.3	19.7	0.386	18.9
7	44.4	33.1	3.0	0.8	18.8	0.133	14.5
8	18.8	15.3	39.8	0.0	26.1	0.176	22.4
9	19.2	27.0	18.1	7.2	28.7	0.792	38.2
10	18.4	27.9	18.3	8.1	27.1	0.705	42.8
11	42.8	29.3	6.5	6.9	14.5	0.276	41.5
12	13.6	22.8	34.7	0.2	28.9	0.536	64.7
13	17.6	23.3	34.8	0.0	24.2	0.541	69.9
Average	29.5	26.2	17.2	3.80	23.0	0.357	27.6
Minimum	13.6	15.2	1.7	0.00	12.8	0.112	1.9
Maximum	47.6	36.5	39.8	8.90	31.9	0.792	69.9
1*	44.4	23.6	3.2	10.60	18.1	0.216	2.5

*Corresponds to the non-sticky module; (a) Expressed in per cent of total sugars; (b) Expressed in per cent of the fibre weight

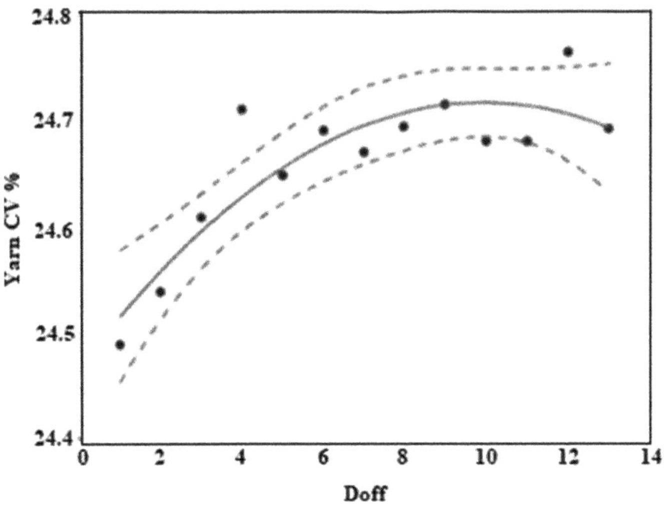

Fig. 6 UT3 CV%: variability between doffs for the ring spun yarn of 30 Ne: (UT3 CV% = 24.47 + 0.0496 (Doff number) – 0.00251 (Doff number)^2, Adjusted R^2 = 0.731 P (2, 10) = 17.27, $\rho <$ 0.001 Std error of estimate = 0.038)

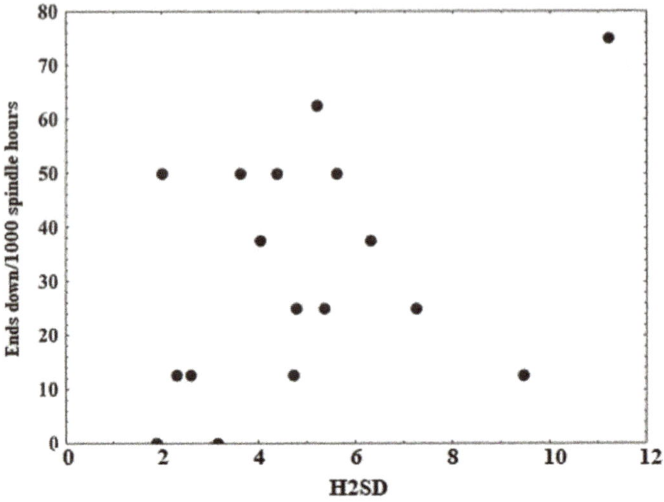

Fig. 7 H2SD versus number of ends down for 1000 spindle hours

Fig. 8 Zinser 351 with 1,680 spindles. *Source* I. Biermann, 29th International Cotton Conference, Bremen, 2008

influence on both productivity and yarn quality. The only method to mitigate such impacts would be thorough and time-consuming cleaning of the machinery.

5 Determination of the Technological Value of Cotton Fibre in Ethiopia

The current global textile industry demands cotton with strong and uniform fibres in order to create high-quality items at the high speeds required to repay capital expenses. The advent of High-Volume Instrument (HVI) assessment of cotton fibre quality has enhanced the relationship between cotton prices and characteristics on global markets [20]. Determining the technical value of cotton fibre is an intriguing area of textile research. The qualities of raw cotton have a significant effect (up to 80%) on the quality of finished yarn [24].

It is a well-known fact that many textile producers are increasingly interested in investing in textiles in underdeveloped countries. In recent years, the number of textile factories in Ethiopia alone has increased by more than 100% (the number of textile factories in Ethiopia in 2001 was 10 Government-owned; now, in 2022, there are 73, including 63 privately owned textile factories that are officially registered and producing).

The textile industry is labour-intensive and might need a large number of people. It is easy to see how important this sector is in resettling individuals in their own communities. As a result, understanding the technical value of cotton will be extremely beneficial to both cotton growers and processors. Farmers in Africa, particularly in the small and large-scale sectors, must fully grasp what sort of cotton fibres they will plant and how they will gain from the sale of their products. It is also necessary for local cotton growers to labour carefully to cultivate high-quality raw materials that can meet the world's current market. Hequet et al. [10] effectively describe the current issues faced by cotton breeders. They observed that contemporary-period cotton breeders are struggling to anticipate the processability of the raw material, which is the most significant responsibility.

It is easy to understand that even the cotton growers in developing countries like Ethiopia (where modern cotton breeding technologies are lacking) might face so many challenges in predicting the technological value of cotton. However, it is important to emphasize here that while cotton growers in developing countries face many challenges, there is a need to work consciously and come to a better prediction of the technological value of cotton. Therefore, the cultivation and production of high-quality cotton can offer better opportunities overall, especially for small and large farmers and all players in the cotton value chain.

Cotton grading is a system of standardized procedures used to determine the spinning value of cotton, which in turn determines its technological value. The properties of the cotton fibres of different genotypes would show a wide range of variation. Quantification methods providing single values combining multiple fibre quality characteristics would allow for a more accurate and reliable definition of the overall quality of cotton.

The main objective in constructing the index value is to provide a practical utility that can be used to make practically important distinctions when comparing different genotype categories [12, 22, 44, 52].

Cotton, according to Gulti [14], is a seed hair (cotton fibres or lint) that arises from the epidermal cells on the surface of the seeds. Cotton (Gossypium species) is a shrub endemic to tropical and subtropical climates worldwide. Seed hair, also known as cotton fibres or lint, is often off-white in colour, as seen in Ethiopian and Indian cottons. Although some existing varieties, such as Egyptian cotton, have naturally coloured fibres that are either white or creamy in colour. Furthermore, cotton species vary greatly in terms of other qualities such as strength, length, and fineness.

Kamal and Ragab [28] mentioned that due to the diversity of cotton fibre properties, the grading or classification of cotton according to its quality criteria will not be the same. Lord [32] also added that the quality of cotton fibres is defined by the physical properties related to their spinnability. Spinning consistency index (SCI) is one of the composite indices used as a quantitative tool to measure the technical value of cotton. The United States Department of Agriculture (USDA) emphasized that the SCI technique for classifying cotton utilizing the HVI system determines cotton fibre qualities (physical attributes) that impact quality. It defined SCI as a calculation for forecasting the spinnability of cotton fibre.

Lord [33] created an aggregated criteria of both cotton technical value and cotton yarn quality using a quantification approach known as the fibre quality index (FQI). The success of FQI can be ascribed to its ease of usage. The FQI model comes in various flavours. Samanta [39] developed a novel FQI formula that is based on HVI readings.

Tesema and Hussein [45] conducted research to define the best combination of quality traits that reliably indicate the technological value of each cotton variety grown in Ethiopia, India, and Egypt. The varieties used in their research have been widely grown in each of the above three countries, offering a wide variety of fibre quality characteristics. Twelve different varieties of the most commonly used cotton were selected from different locations in Ethiopia. These varieties were: DP-90, Acala SJ-2, Arba (commercial varieties), Cu-ok-ra, Lao Cara, Alber 637, BPA, Cucrova 1518, Bulk 202, Sille 1 (Stone Vile), Estamble, and R-36 (non-commercial varieties). All cotton varieties examined were measured with USTER 1000 HVI and their fibre properties and statistical parameters were evaluated using various quantification methods.

The different quantification methods used in their study were:

1. Spinning Consistency Index (SCI) (Uaster 1999)

$$SCI = -414.67 + 2.9FS + 49.17UHML + 4.74UI - 9.32FF + 0.65Rd + 0.36(+b) \quad (1)$$

2. Fibre Quality Index (FQI) [39]

$$FQI_{HVI} = \frac{FS * UHML * UI}{FF} \quad (2)$$

3. Multiplicative analytic hierarchy process (MI AHP) [12]

$$MI_{AHP} = \frac{FS^{0.394} * UHML^{0.380} * UI^{0.054}}{FE^{0.056} * FF^{0.059} * SFC^{0.054}} \quad (3)$$

4. Multiplicative analytic hierarchy process (MI AHP) [5]

$$MI_{AHP} = \frac{FS^{0.270} * UHML^{0.291} * UI^{0.145} * FE^{0.039}}{FF^{0.11} * SFC^{0.145}} \quad (4)$$

Fibre upper half mean length (UHML, mm), uniformity index (UI%), micronaire reading (MIC), fibre strength (FS-g/tex), fibre elongation (FE%), and fibre colour attributes, including reflectance (Rd%) and degree of yellowness (+b), were all determined using the USTER (HVI) in accordance with ASTM Designation (D-4605–86-1776–98). All fibre testing for the samples was performed in cotton laboratories under ASTM (D 1776–04)-regulated ambient conditions. SAS statistical software, version 9.1 SAS, (2004) was used to analyse the study's data.

Table 7 shows the values of Ethiopian cotton fibre qualities, as well as the computed values of the quantification techniques SCI, FQI, MIAHP-H, and MIAHP-M. The varieties were ranked in increasing order for each quantification technique. The highest technological value of each of SCI, FQI, MIAHP-H, and MIAHP-M was revealed by different varieties as follows 133.4-SCI (Bulk 202), 154.2-FQI (SJ-2), 12.03-MIAHP-H (Acala SJ-2), and 7.82-MIAHP-M (Acala SJ-2), whilst the lowest values were respectively 95.5-SCI (Alber 637), 110.7-FQI (Alber 637), 11.11- MIAHP-H (Lao Cara), and 7.32-MIAHP-M (Alber 637).

Table 7 The highest contributor representative positively and negatively of the quantification methods as a quality criterion determining the technological value of Ethiopian, Indian, and Egyptian cotton

Quantification method		Highest positive contributor	Highest negative contributor
Spinning Consistency Index (SCI)	$SCI = -414.67 + 2.9FS + 49.17UHML + 4.74UI - 9.32FF + 0.65Rd + 0.36(+b)$	+49.1 UHML	–9.32 FF
Fibre Quality Index (FQI)	$FQIHVI = \frac{FS*UHML*UI}{FF}$	FS, UHML, and UI	FF
Multiplicative analytic hierarchy process (MI AHP-H)	$MI_{AHP} = \frac{FS^{0.394}*UHML^{0.380}*UI^{0.054}}{FE^{0.056}*FF^{0.059}*SFC^{0.054}}$	$FS^{0.394}$ and $UHML^{0.380}$	$FF^{0.059}$
Multiplicative analytic hierarchy process (MI AHP -M)	$MI_{AHP} = \frac{FS^{0.270}*UHML^{0.291}*UI^{0.145}*FE^{0.039}}{FF^{0.11}*SFC^{0.145}}$	$FS^{0.270}$ and $UHML^{0.291}$	$SFC^{0.145}$

In the case of Ethiopian cottons, the FQI, MIAHP-H, and MIAHP-M displayed the same rank (R), with the corresponding values being as follows: 154.2-FQI (SJ-2), 12.03-MIAHP-H (Acala SJ-2), and 7.82-MIAHP-M (Acala SJ-2). It is acknowledged that, as shown in Table 8, fibre strength was the factor that contributed most positively to the technological values of FQI, MIAHP-H, and MIAHP-M. This was demonstrated by the cotton variety SJ-2, which among the examined Ethiopian cotton varieties displayed the highest fibre strength value (25.88).

The SCI highest value was for the UHML which was 49.1, while the lowest value was revealed by, i.e., the FF -9.32 which was the highest negative contributor (Table 8) [20].

There was no significant difference between the technological value of 133.4 (Bulk 202), which was the greatest quality, i.e., 12 in rank (R), and the technological value of 130.4 (Estamble), which was 11 in rank. Furthermore, the technological value of 128.3 (Sille1-S.V.) ranked 10 revealed no significant difference with both 11 and 12 ranks. That was accepted due to the occurrence of convergence in the values of the previously mentioned varieties.

6 Optimizing Cotton Blending Costs in Terms of Quality

In industrial practice, it is required to employ a blending of many cottons as a raw material for a variety of reasons. Most mills buy cotton throughout the year, not just once, and it is impossible to buy one sort of cotton for the whole season. Furthermore, several counts are spun in a mill, each requiring a particular quality of raw material.

Using multiple cottons in a blend has both advantages and disadvantages. Benefits include the ability to maintain a consistent blend over long periods of time, regardless of the availability of a particular cotton, and to keep the cost of the blend at a reasonably low level by blending more expensive cotton with cheaper cotton. It seems that these benefits are best achieved by using multiple types of cotton at the same time. However, including multiple types of cotton in a blend would also present some problems. The probability of a proportional representation of a mixture in the yarn cross-section decreases the more components are contained in the mixture. This leads to yarn segments with different fibre properties. Such a yarn would appear mottled during processing and dyeing if the variation in fibre properties used in the blend is significantly different. Mixing of two to three components is therefore preferable for practical reasons. In general, for optimal performance and yarn quality, the fibre properties of cotton used in a blend should not differ by more than the values given below.

As it is important to produce a very even mix of different bales, new machines have been developed to ensure a homogeneous blend. Developments in this area include automatic bale pluckers and automatic blending grabs. Broadly speaking, these machines take turns picking cotton from open bales that are lined up for ease of access. The smaller the quantities removed from each bale, the more effective the

Table 8 The technological values of Ethiopian cotton fibre properties determined by quantification methods, ranked (R) in ascending order, as well as the LSD at 5% for each quantification method

Ethiopian cotton fibre properties									The calculated Ethiopian technological values by quantification methods and their rank							
Variety	UHM	UI	SFI	FS	FE	MIC	+b	Rd%	SCI	R	FQI	R	MIAHP H	R	MIAHP M	R
DP-90	29.25	80.88	12.45	24.70	5.73	4.10	7.65	70.75	107.5	4	142.5	7	11.78	10	7.63	9
Cu-ok-ra	29.13	80.20	13.13	23.48	5.83	4.30	8.78	75.40	102.1	2	127.5	2	11.45	5	7.42	3
SJ-2	29.50	81.83	12.08	25.88	6.08	4.05	8.38	79.10	122.0	9	154.2	12	12.03	12	7.82	12
Arba	29.73	81.28	12.70	25.68	5.95	4.20	8.78	78.98	118.0	7	147.7	9	11.98	11	7.72	10
LaoCara	27.58	80.55	13.15	22.25	5.60	3.63	8.60	78.25	105.2	3	136.3	6	11.11	1	7.32	1
Alber 637	28.05	79.48	13.38	25.33	5.88	5.10	8.38	77.25	95.5	1	110.7	1	11.49	6	7.32	1
BPA	30.65	82.23	12.40	22.95	5.73	4.25	8.23	76.78	114.2	5	136.1	5	11.64	8	7.58	7
Cucroval518	29.61	84.45	12.55	21.70	6.03	4.13	7.75	70.50	116.0	6	131.5	3	11.23	2	7.44	4
Bulk 202	30.05	86.20	13.35	23.55	6.18	4.20	8.80	75.38	133.4	12	145.3	8	11.61	7	7.59	8
Sille1 (S.V.)	30.10	84.00	12.10	23.78	6.13	3.90	8.35	78.40	128.3	10	154.1	11	11.76	9	7.75	11
Estamble	26.58	85.05	12.35	24.50	5.93	3.73	8.78	78.53	130.4	11	148.7	10	11.40	4	7.56	6
R-36	27.91	84.15	13.15	24.00	6.13	4.23	8.00	77.35	121.6	8	133.4	4	11.37	3	7.45	5
Mean									116.2		139.0		11.57		7.55	
Least Significant Difference (LSD at 5%)									6.43		7.17		0.09		0.06	

The quantification methods: SCI = Spinning consistency index, FQI = Fibre quality index

MI AHP-H = Multiplicative analytic hierarchy process according to Hussein et al. [4]

MI AHP-M = Multiplicative analytic hierarchy process according to Majumdar et al. [12]

various processes are in producing a satisfactory blend. Blending different components of cotton has been an important manufacturing practice since the early days of the cotton textile industry. The effects of blending in minimizing varietal, geographic, growth, and production variables are well documented. Blending's role in controlling physical, aesthetic, and subjective properties has made it the single most important operation in cotton manufacture (Mogahzy 1992).

The economic impact of blending stems from its ability to reduce costs by allowing for the substitution of less expensive cotton in a blend without compromising quality or processing efficiency, by controlling the percentages of fibre components in the blend, and by allowing for the efficient use of available cotton qualities throughout the crop year without affecting processing performance or cost structure. Over the years, blending has been recognized as a combination of science and art, but in recent years, the tendency has been towards a completely scientific approach. This was made feasible by powerful tools such as high-volume instrument (HVI) systems and tailored fibre selection algorithms capable of supplying uniform fibre profiles on a daily basis (Mogahzy et al. 1988).

The recent shift to cotton marketing based on HVI measurements necessitates the development of a scientific, economic strategy that can link cost and quality while also allowing for cost optimization. Any strategy for determining a cost–quality relationship should, however, adhere to two key principles: it should be based on knowledge of the extent to which cotton is used in manufacturing and it should be fundamentally sound.

Mathematical linear programming is a well-known method for cost and profit optimization that is widely used in other industrial applications. The mathematical complexity of these analyses is now only a minor concern due to modern powerful computers and effective software. In order to assist spinning industries with the cost optimization of cotton blending, computer linier programming methodology has recently been introduced. The high-volume instrument (USTER HVI) measures the fibre quality parameters used in the computer-based linear programming method of cost optimization for blending.

7 Blending Cost Optimization by Computer Linear Programming Method

7.1 Objective Function of Cotton Cost

As discussed above, linear programming is a mathematical technique for maximizing or minimizing a linear objective function that specifies the utility or cost associated with each decision variable. When blending cotton, the decision variable is the proportion of a particular cotton to be used in the mix. The objective function is a linear function of the cost of cotton, which can be represented by the following expression:

$$\text{Minimize: } Z\ (\text{Cost}) \sum_{i=1}^{i=k} a_i c_i$$

where, Z is the total cost.

a_i = the proportion by weight of the ith component in the blend,

c_i = the cost of the ith component in the blend,

If the proportion of cotton A is 0.40 and proportion of cotton B is 0.60. Then the optimum cost Zmin = $0.62.

The total cost of raw materials generally consists of ordering and set-up costs (acceleration and testing), transportation costs (handling, insurance, warehousing, and data processing costs), and purchasing costs (price per kg, labour, and overhead). It is strongly recommended to consider these cost elements when defining the cost function.

7.2 Fibre Quality Constraints

A criterion suggests that understanding how fibre properties are translated in the blend should be the foundation for the use of fibre properties as quality constraints. In order to do this, it is necessary to look at the fibre characteristics that are most useful as well as whether or not the linearity assumption applies to fibre quality constraints.

Previous research indicates that the linearity assumption holds reasonably for various fibre qualities, provided the fibres in the mix are of the same type and structure, and the variability within each component is minimal. The validity of the linearity assumption is also dependent on the fibre quality under examination. Below is an explanation of the two key requirements in terms of the three basic fibre qualities fineness, length, and strength.

The constraint inequality for Fibre length may be written as:

$$\overline{FL_a} a_a + \overline{FL_b} b_b + \geq \overline{FL}_{desirable}$$

The constraint inequality for Fibre strength (tenacity) may be written in the following form:

$$\overline{FS_A} a_a + \overline{FS_B} b_b + \ldots \geq \overline{FS}_{desirable}$$

The constraint inequality for Fibre fineness (Micronaire) may be written in the following form:

Example: $\left(\frac{1}{MIC_A}\right) a_A + \left(\frac{1}{MIC_B}\right) a_B + \geq \ldots \left(\frac{1}{MIC}\right)_{desirable}$

Let two types of cottons, cotton from Middle Awash farm be 'A' and cotton from Metema farm be 'M', are used in the mix. Costs per USD of these two types of cottons are assumed to be \$0.5 and \$0.7, respectively. Thus, the objective function may be formulated by:

$$Minimize: Z(Cost) = \sum_{i=1}^{i=2} a_i c_i = 0.50a_A + 0.7a_M$$

Suppose this function is subject to the following constraints:
Fibre strength:
Let
$\overline{FL_A}$ = Fibre length component of cotton from Middle Awash farm = 29 mm.
$\overline{FL_M}$ = Fibre Length component of cotton from Metema farm = 27.5 mm.
where, a_A= Awash proportion.
a_M = Metema proportion

$$a_A + a_M = 1, a_A \geq 0, a_M \geq 0$$

29 a_A + 27.5 a_M + ...
The constraint inequality for Fibre strength may be written in the following form:

$$FS_A a_A + FS_M a_M + \ldots \geq FS_{desirable}$$

Let
$\overline{FS_A}$ = Fibre strength component of cotton from Middle Awash farm = 27 cN/tex.
$\overline{FS_M}$ = Fibre strength component of cotton from Metema farm = 24.5 cN/tex

$$27.0 + 24.5 \geq 26$$

The constraint inequality for Fibre fineness may be written in the following form:

$$\left(\frac{1}{Mic_A}\right)a_A + \left(\frac{1}{Mic_M}\right)a_M + \ldots \geq \left(\frac{1}{Mic}\right)_{desirable}$$

Mic_A = Fibre micronaire component of cotton from Middle Awash farm = 3.8
Mic_M = Fibre micronaire component of cotton from Metema farm = 4.3
Reciprocal of micronaire:
0.26 a_A+ 0.23 a_M $= \geq 0.25Mic^{-1}$
By using the simplex linear programme built-in Excel we can solve the above inequalities and find an optimum solution. The found optimum solution is:
Proportion of cotton from Middle Awash farm = 0.40.
Proportion of cotton in the blend from the Matema farm = 0.60.
The optimum cost of the blend is Z_{min} = 0.62 USD.

The solution simply indicates that if 40% of Middle Awash cotton and 60% of Metema cotton are blended together, the quality constraints above can be satisfied with an optimum cost of 0.62 USD per Kg.

8 Predicting of Ring and Rotor Spun Yarns Properties from Fibre Properties and Machine Parameters

Predicting the quality features of yarns, particularly tensile qualities, has been the focus of several studies throughout the previous century. Because of technical innovation, spinning, knitting, and weaving machines can now run at high speeds. This may result in excessive ends down for cottons with low tensile characteristics. Because breakages can result in work stoppages and reduced production, there has been increasing emphasis on forecasting spun yarn tensile qualities. One strategy for predicting the quality of spun yarns is to first instrumentally characterize the fibre qualities and then generate an adhoc curvilinear equation in which the cotton fibre attributes are blended into one integrated index [46].

Instrumental assessment of fibre characteristics should be viewed as dual functional, having benefits to both processing efficiency and product quality. Because of the dynamic nature of spinning (e.g., Ring, Rotor, Compact, Air-jet), whose efficiencies depend to some extent on the fibre properties and amount and properties of fibrous faults and foreign matters constituents to maintain specified levels of operation efficiency, consideration of the physical properties of fibres, fibrous faults, and amount and properties of foreign matters constituents should be mandatory. May and Taylor noted the need for breeding cottons with better tenacity to handle the increased strain on cotton yarns brought by increased knitting and weaving manufacturing speeds [30].

It has been documented that fibre quality attributes measured by High-Volume Instrument (HVI) and Advanced Fibre Information System (AFIS) have a strong relationship with the quality of spun yarns. El Mogahzy et al. for example, demonstrated that cottons with improved upper half mean length (UHML), tenacity, micronaire, and length uniformity can be used to produce yarns with a higher break skein factor. Using the advanced fibre information system (AFIS), [30] reported that yarn tenacity was affected by the amount of seed coat fragments in the raw material.

Predicting yarn tensile properties is important but offers an incomplete picture of how fibre quality impacts yarn quality. High-quality yarns should also have a low number of imperfections such as thin places, thick places, and neps. The number of thin and thick places in the yarn has been shown to be related to HVI micaronaire and uniformity. The relationships between these two yarn parameters with AFIS fineness and Scutter-Web array short fibre content (SFC) have also been documented.

The factors affecting yarn quality differ between ring, rotor, compact, air-jet and vortex spinning (Tables 9 and 10).

Table 9 Requirement during blending

Property	Difference
2.5% span length	5 mm
Micronaire	1.2
Trash	3%

Table 10 Factors influencing yarn quality in different spinning systems

	Ring spinning process	Rotor spinning process	Compact spinning	Air-jet spinning
Applied raw material	• Fibre length • Fibre tensile strength/elongation • Fibre fineness/maturity • Fibre purity/cleanliness	• Fibre tensile strength/elongation • Fibre fineness/maturity • Fibre purity/cleanliness • Fibre length	• Fibre length and length uniformity • Fibre strength • Fibre fineness	• Fibre fineness (fibre micronaire or denier) • Cleanliness (Neps, trash, and dust content) • Fibre strength • Length and length uniformity • Friction coefficients—'fibre-to-fibre' and 'fibre-to-nozzle'
Sliver/roving quality	Roving irregularity	Sliver irregularity	Roving irregularity	Sliver irregularity
Applied spinning components	• Ring and ring traveller • Drafting system • Spindle	• Rotor • Opening roller • Navel • Torque stop	• Ring and ring traveller • Drafting system • Spindle	• Drafting system • Spinning nozzles • Take up/winding rollers
Spinning settings	• Settings of drafting system • Spindle speeds	• Rotor speed • Opening roller speed	• Pneumatic system/aerodynamic device in the drafting system • Spindle speeds	• Drafting system • Nozzles • Delivery rollers

The yarn quality is determined by the raw material used, the quality of preparation, spinning parameters, and the spinning components used. Table 10 shows the ranking of influencing factors in different spinning systems.

9 Ring Spinning Process

When compared to the rotor spinning process, the ring spinning process involves more production steps. The later quality of the yarn is affected by each stage of the process. The primary focus of planning is CompACT3, along with traditional ring spinning and roving.

9.1 Roving Frame

In order to achieve an acceptable yarn quality during the ring spinning process, the drawn sliver count must be reduced in two steps. Because of the more pronounced effect of the drafting wave in roller drafting, the output irregularity increases as the draft increases (Eq. 5).

$$CV^2{}_o = CV^2{}_{in} + a(D - 1) \quad (5)$$

where $CV^2{}_o$, $CV^2{}_{in}$ = relative variances of irregularity of the output and input material, respectively, a slope of the line, D = applied draft.

Thus, the coefficient of variation attributed to the drafting wave is $\sqrt{a(D-1)}$.

Therefore, one advantage of a two-stage drafting operation (in roving and ring frame) is that the reversal of the material length at the second stage effectively provides a reversal in the drafting direction of the fibres. This tends to reduce the amount of bunching of short fibres to give a less pronounced drafting wave.

The first drafting stage is the final step in the process of preparing fibres for spinning. The intermediate product is known as roving (a continuous fibrous strand drawn from a sliver and given cohesion by inserting a tiny amount of twist) and is manufactured on twist insertion devices known as roving frame. The roving frame produces roving, which is a continuous fibrous strand drawn from a sliver and given cohesiveness by either adding a small amount of twist or compacting the fibres with an oscillating apron. It is drafted and twisted before being spun into yarn.

The machine parameters of the roving frame are dependent on the type of raw material to be manufactured (Table 11).

When processing cotton, the flyer speeds are typically kept lower due to its shorter length and weaker substance strength as compared to synthetic fibres like polyester, polyacrylic, or viscose. Additionally, the twist multiplier for cotton is usually higher. Unlike synthetic fibres, cotton has naturally lower fibre-to-fibre friction. However, the finishing agent can be used to manipulate this factor while processing synthetics. The fibre length plays a significant role in determining the break drafting zone and main drafting zone of the drafting system.

The break draft is highly dependent on fibre curling, particularly when processing synthetics. Because of the greater drafting forces in synthetics, the stress on the drafting system is increased. However, in general, this should be restricted to what

Table 11 Basic roving frame settings

	Carded Cotton Combed Cotton	100% Synthetics
Flyer speed (1/min)	1,100–1450	1,200–1,500
Lowering Flyer–nspi	100	50
Twist multiplier (α_m)	36–41	22–26
Delivery (m/min)	34	42
Breaking draft zone setting (mm)	65	65 fibres up to 45 mm 75 fibres up to 54 mm 85 fibres up to 63 mm
Main draft zone setting (mm)	46	49 fibres up to 45 mm 60 fibres up to 54 mm 76 fibres up to 63 mm
Break draft	1.19	1.25 to 1.31

is strictly necessary to reduce drafting system wear (for example, grinding intervals, bearing lubrication, and so on) to a minimum.

9.2 Ring Frame

The conventional ring spinning technique, which employs the ring and traveller spinning method, is now the most popular, accounting for an estimated 67% of the global market share, followed by rotor spinning (30%) and air-jet spinning (3%). Table 12 lists the yarn count ranges that can be generated using the ring, rotor, and air-jet spinning processes.

The quality of the yarn in ring spinning, rotor spinning, and air-jet spinning is, of course, heavily dependent on the quality of the raw material used. Significant research has been conducted to discover techniques for predicting yarn mechanical qualities like strength (tenacity) and elongation (strain) from raw material fibre parameters. Some are statistically based [11–20], while others are theoretically conceived [12–18].

Table 12 Yarn count ranges to be spun on ring and rotor spinning systems

Yarn counts	Ring spinning	Rotor spinning	Air-jet spinning
<Ne 10	Coarse yarn counts	Course yarn counts	Course yarn counts
Up to Ne 20	Coarse yarn counts	Medium yarn counts	Medium yarn counts
Up to Ne 40	Medium yarn counts	Fine yarn counts	Fine yarn counts
Up to Ne 100	Fine yarn counts	–	–
>Ne 100	Finest yarn counts	–	–

Table 13 clearly demonstrates the yarn count range that may be spun with which cotton, based on fibre fineness and staple length.

The fibre fineness determines the number of fibres in the cross-section at the respective process stage. If we look at cotton with 3.4 micronaire at Ne 80, we see that on average only 55 fibres remain in the yarn cross-section. At the same time, it becomes clear that yarn uniformity decreases in relation to the number of fibres contained in the yarn cross-section. The rule of thumb here is that the finer the spinning, the finer the fibres required. Considering that the processing interval increases with increasing yarn fineness, i.e., the roving bobbin change can extend by more than a month, the question arises whether the blend feed in spinning preparation is still the same after this period of time. In order to be able to guarantee consistent quality at all, the raw material must therefore be kept ready for several months. So there is a so-called marginal irregularity.

Table 14 illustrates the greatest potential unevenness based on the micronaire value. The sample presented here shows that a fineness of 4.8 MIC and a yarn count of Ne 30 cannot achieve a yarn unevenness of less than 9.80% under theoretically optimum conditions, regardless of whether this cotton variety is even capable of spinning.

Machine Parameters: The ring spinning machine features various machine components that have an active impact on yarn quality. The yarn path within the ring spinning machine and its components are shown below.

Table 13 Yarn and intermediate products counts to be predicted from fibre properties of ginned cotton

Fibre fineness Fibre length HVI	MIC 4.8 27.8 mm (1 1/32″)	MIC 4.0 28.6 mm (1 1/8″)	MIC 3.4 33 mm (1 3/8″)
Card output fineness	Ne 0.11	Ne 0.12	Ne 0.15
Draw frame output fineness	Ne 0.11	Ne 0.12	Ne 0.15
Roving frame output fineness	Ne 0.41	Ne 0.74	Ne 1.65
Spinning machine output fineness	Ne 12	Ne 30	Ne 80

Table 14 Number of fibres in the yarn cross-section and marginal irregularities

Fibre fineness (MIC) Staple length (″) HVI fibre length (mm)	4.8 1 1/32 27.8	4.0 1 1/8 28.6	3.4 1 3/8 33
Yarn count Ne 12 Fibres in yarn cross-section Marginal unevenness (%)	260 6.19	313 5.65	368 5.21
Yarn count Ne 30 Fibres in yarn cross-section Marginal unevenness (%)	104 9.80	125 8.94	147 8.25
Yarn count Ne 80 Fibres in yarn cross-section Marginal unevenness (%)	39 16.0	47 14.6	55 13.5

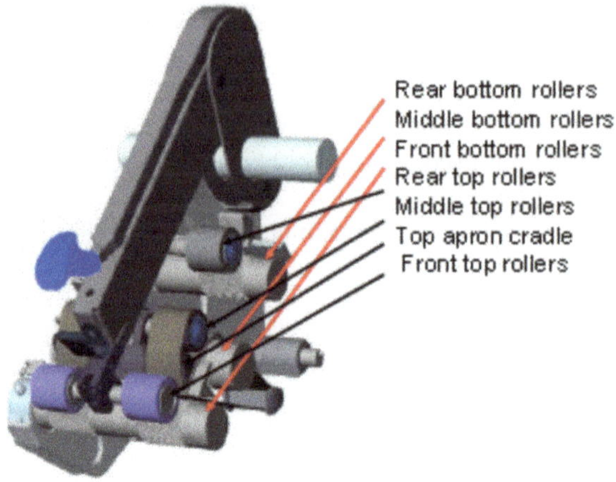

Fig. 9 Drafting system. *Source* I. Biermann, 29th International Cotton Conference, Bremen, 2008

Drafting System: A drafting system consists of steel and rubber rollers that guide roving at different speeds and draft it to the required yarn count (Fig. 9).

During drafting, irregularities caused by the properties of the fibre being processed and machine defects are superimposed on the input material, resulting in a more uneven drafted product. The following factors influence the unevenness of a single zone drafting wave: draft size, input material count, multiple inputs or doubling, roller drafting zone setting, and the degree of parallelism, length, and fineness of fibres in the input material.

The following are some essential drafting rollers parameters that are related to drafting irregularities:

Roller Coatings: with the proper choice of top roller coatings and their various coating hardness, it is possible to influence the different fibre properties and drafting results. In this case, the following rule of thumb applies (see Tables 5 and 15):

- The softer the coating, the better the IPI value, but the shorter the service life.
- Fine yarns run with coatings as soft as possible.
- Synthetics, especially polyester, prefer hard coatings due to higher draft forces.

Table 15 Influence of top roller hardness

Top roller coatings Hardness °Shore A	Yarn regularity IPI values	Survice life of coatings
85° Shore		
75° Shore		
63° Shore		

Table 16 Types of top apron cradle

Types of top apron cradles	Fibre length	Length classes
OH 62, OH 2022	Up to 40 mm	Cotton–short staple
OH 132, OH 2042	Up to 51 mm	Cotton–medium staple
OH 122	Up to 60 mm	Cotton–long staple
OH 6022	Up to 200 mm	Wool–worsted yarn

Top Apron Cradles: There are several kinds of top apron cradles, sometimes known as cages (Table 16). Various apron forms and sizes are required depending on the length of the fibre. The cradle, together with the deflection bar, impacts the main draft and hence the yarn quality. One of the most important influencing factors on yarn quality is the optimum setting.

10 Rotor Spinning Process

Rotor spinning has established itself as an efficient and important yarn manufacturing technology during the last 40 years. Rotor spinning provides advantages in terms of production rate, usage of floor space, and labour [31]. By separating twisting and winding in the yarn manufacturing process, open-end/rotor spinning eliminates most of the issues associated with ring spinning. The success of rotor spinning is attributable, on the one hand, to a significant increase in production, and, on the other hand, to the ability to fully automate the spinning process. This is feasible because rotor spinning integrates three manufacturing processes: speed frame, ring spinning, and winding.

The spinner is required to select appropriate raw materials. The following fibre categories are available for selection:

- Natural fibres
- Cotton secondary
- Synthetics
- Cellulosic
- Blends

The areas of textile application quoted in the recommendation are as follows:

- Outerwear
- Raised fabrics
- Classic denim
- Soft denim
- Fancy denim
- Knitwear
- Home textiles

- Technical textiles
- Workwear

The possibility of obtaining the qualitative and economic optimum is highest when a yarn is created with future usage in mind, such as an open voluminous yarn for raised fabrics or slender yarns with reduced hairiness for shirt fabrics.

Yarns may be actively designed using spinning components. Spinning components are primarily:

- Rotors
- Opening rollers
- Navels

Additional factors that impact yarn quality and running behaviour, but are not technically referred to as spinning components, are:

- Torque stop
- Channel plate or adapter

Depending on fibre material and textile use, the application recommendations indicate a recommendation for the.

- Choice of rotor
- Choice of adapter
- Choice of opening roller
- Choice of navel
- Choice of torque stop
- Yarn twist

11 Choice of Rotor

The fibres provided by the opening roller via the feed channel are collected again in the rotor, i.e., the fibres are redoubled in the rotor groove. The rotor also twists the previously untwisted fibres, resulting in rotor yarn (Figs. 10 and 11).

Depending on yarn application and production conditions, four parameters must be considered:

(1) Rotor type
(2) Rotor coating
(3) Rotor diameter
(4) Rotor speed

The spinner can select the first and second rotor types according to the textile application, fibre material, and yarn count (Table 17). The selection of rotor diameters is also shown in Table 18. For example, the yarn wanted for outerwear fabric production from 100% cotton and yarn counts of Ne 10 and coarser, the available first option rotor type is TT with diameters of 36 mm, 40 mm, 46 mm, or 56 mm.

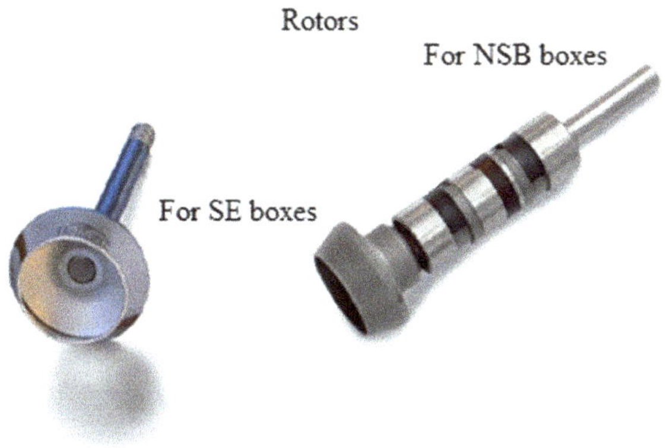

Fig. 10 Belcoro rotors

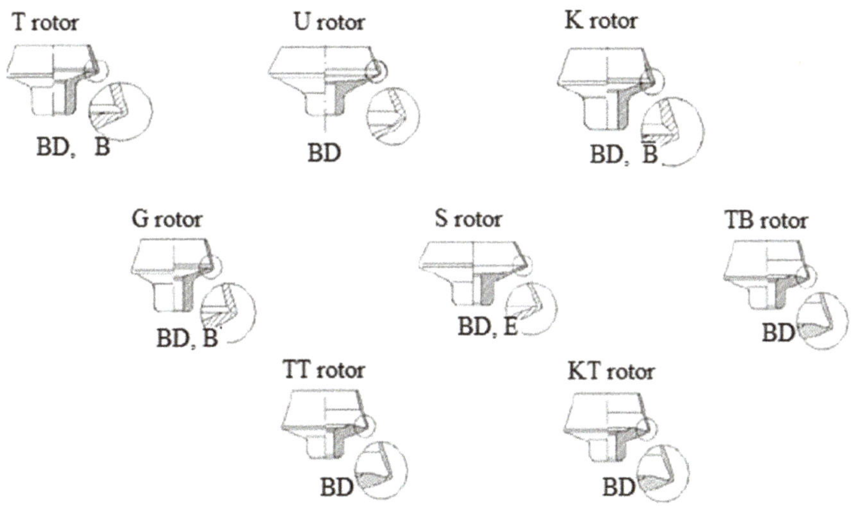

Fig. 11 Rotor types for Autocoro line

The diameter of the rotor used is determined by the length of the fibre being treated. In rotor spinning, it is generally accepted that the maximum fibre length (Lfibre) should not exceed the rotor diameter (Drotor), and thus their ratio (Lfibre/Drotor) should be between 0.5 and 1 to ensure proper processing and yarn formation and the production of acceptable yarn quality [4].

The rotor diameter must be proportional to the yarn count. Because of the higher fibre masses and lower absolute yarn twist, coarser yarns should always be produced on bigger-diameter rotors. Choosing the optimum rotor diameter may have a significant impact on critical yarn qualities and yarn factors such as volume, lustre, and handling. Furthermore, because it is directly related to yarn count and yarn twist multiplier, the rotor diameter has a significant impact on the selection of twist factor.

Table 17 Choice of rotor type for cotton outerwear

Rotor type	Real ∅, mm	Course count: Ne 10 and coarser	Medium count up to Ne 20	Fine count up to Ne 40
G - BD	28[1], 30, 31, 33, 36, 40, 46		○	●
G - B	36, 40, 46		○	
K - BD	31		○	○
S - BD	33, 36, 40, 46, 56	○		
T - BD	33, 34, 36, 40, 46	○	●	○
T - B	36[2], 40, 46	○	○	
KT - BD	26			○
TB - B	33, 34, 36, 40, 46			
TT - BD	36, 40, 46, 56	●	○	
U - BD	40, 46			
U - B	40, 46			

For SE spinning boxes. ● = 1st choice ○ = 2nd choice (Oerlikon Schlafhorst)

Table 18 Choice of rotor speed and rotor diameter from the application recommendation

Rotor diameter (mm)	Rotor speed (m/min)	Adapter type	Course count: Ne 10 and coarser	Medium count up to Ne 20	Fine count up to Ne 40
26[1]	170,000	A 26 SL			130,000–150,000
28[2]	150,000	A 28 SL		120,000–145,000	120,00–145,000
30	135,000	A 31		110,000–130,00	110,000–130,000
31	130,000	A 31		100,000–125,000	100,000–120,000
33	125,000	A 31		90,000–115,000	90,000–115,000
34	120,000	A 36/A 31		85,000–110,000	85,000–110,000
36	150,000	A 36/A 31	75,000–100,000	75,000–100,000	
40	100,000	‘A 40/A 36	70,000–90,000		
46	80,000	A 46/A 40	55,000–80,000		
56	70,000	A 56			

Table 19 Choice of opening roller type from the application recommendation

Spinning Navel	Course count: Ne 10 and coarser	Medium count up to Ne 20	Fine count up to Ne 40
KGG-A/KNN-A	○		
K 3-A/KN 3-A			
K 4-A/KN 4-A	●	●	○
KSS-A		○	●
KS-AB			○
KS K4-A	○	○	○
KS K6-A/KS K6-AB			○
KS 2R4-A			

Foreign fibres, dust, and trash particles have a stronger disruptive effect on smaller-diameter rotors than larger-diameter rotors. This results in more yarn breakage and less spinning stability. The 'recommendation of spinning settings' demonstrates the significance of merging all factors in connection to final product use.

The interaction of rotor speed and rotor diameter influences the centrifugal force with which the fibres are forced into the rotor groove, as well as the spinning tension, i.e., the yarn tensile force at the yarn detachment point. The centrifugal force increases in proportion to the speed, forcing the fibres to be forced deeper into the rotor groove and therefore increasing the torque required to create twist. This has an effect on yarn properties, particularly elongation and strength, as well as spinning stability. The spinning tension has a significant impact on the running behaviour and, in turn, the winding hardness of the packages. Spinning tension is affected by both rotor speed and rotor diameter. Because yarn tension cannot exceed a particular value due to the danger of increased yarn breakage, rotor diameter must be lowered if quicker speeds are necessary (Table 18). Fibre contact pressure in the rotor groove increases as the rotor size decreases and the rotor speed increases. This makes twisting more difficult. Furthermore, embedded trash particles may only be removed with increasing difficulty [50].

In terms of spinning process stability and yarn quality, rotor speed, rotor diameter, and yarn count cannot be considered separately. As a result, the 'recommendation of spinning settings' includes a variety of options. When choosing the rotor speed, performance limits owing to fibre strength and elongation must be considered. When this limit is exceeded, the yarn values decrease, particularly elongation, and the number of yarn breaks increases (Fig. 12).

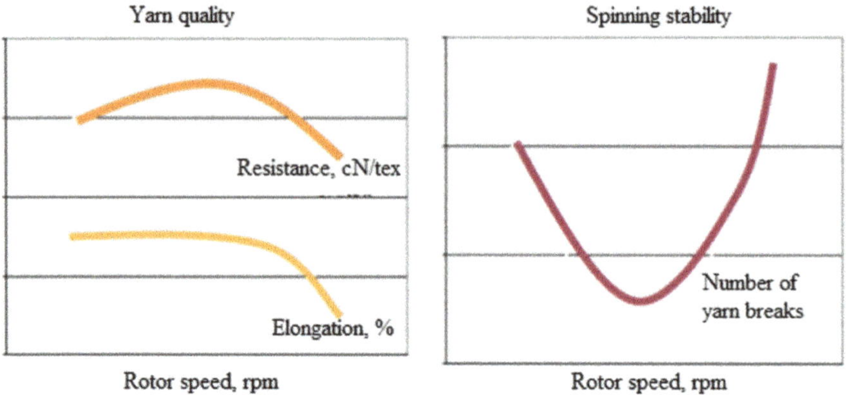

Fig. 12 Balance between rotor speed and yarn quality/spinning stability

Falling below the minimum speed also results in increased yarn breaks and, in general, lower yarn values. However, in addition to minimum speeds, there are also technical maximum rotor speeds. They depend on physical and metallic material properties. Most of the time, the adapter for spinning types SE 10 to SE 12, and the channel plate for spinbox types SE 7 to SE 9 were not considered spinning components in the usual sense. However, they ensure correct positioning of the navel in relation to the rotor groove, and targeted feed of fibres onto the rotor wall. As a result, selecting the proper adapter or channel plate is critical (Fig. 13).

In addition to the already described technical and technological explanations concerning the rotor, the configuration of the navel has also a decisive influence on spinning result and structure of textile products. As it was explained before, to insert twist into the fibre ribbon to produce the yarn, sufficient twist torque must be present at point ‘P’ in Fig. 14. This keeps the forming yarn from breaking at ‘P’ as a result of the high tension induced in AP by centrifugal forces. The rotor generates the twist torque as it carries the yarn tail AP through each revolution; QAP is, therefore, similar to a crank. The cranking action induces a twist in the length QA. The twist

Fig. 13 Choice of correct adapter type

torque builds up and propagates to P. In doing so, it has to overcome the barrier, at A, of the doffing tube and that is caused by the narrowness of the rotor groove. The spinning tension and the doffing tube geometry are therefore important factors. To overcome the twist barriers, a greater machine twist setting is required than in ring spinning. The middle section, known as the doffing tube navel, can, however, be adjusted to help reduce the twist level necessary to spin [31]. Figure 15 shows Schlafhorst-made ceramic navels.

Notches in the navel generate oscillations in the yarn. Since the already integrated piece of yarn briefly lifts off the rotor and groove, the friction between yarn and groove is reduced. Thus, the twist enters the fibre bungle with less effort. The navel with four notches K-4 or KN-4 has established itself as a universal navel. Navels with outer knurling increase the friction between the navel and yarn. The false twist is increased and displaced in the direction of the rotor groove as far as the binding-in zone. The spiral of the KS navel reduces the contact surface of the yarn, thus increasing the surface pressure between yarn and navel. Compared with a smooth navel, this results in increased false twist and improved spinning stability. Yarn

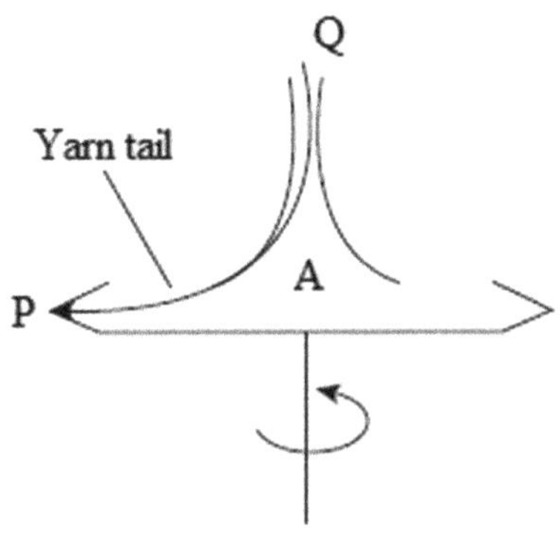

Fig. 14 Yarn tail inside the rotor and navel position

Fig. 15 Correct navel types

evenness, imperfections, and hairiness are not influenced by this. The KS navel has gained popularity above all for smooth weaving yarns in the medium and fine count range. It is not suitable for pure man-made yarns. Both notching and milling and the use of intermingling influence the visual appearance of the yarn. Compared with the smooth KG navel, the yarn appears coarser and has increased hairiness. However, spinning stability and consequently yarn break ratio are noticeably improved. By choosing the correct navel, the character of the yarn can thus be optimally matched to fit the corresponding application.

11.1 Choice of Opening Roller

The opening roller is responsible for cleaning the sliver and opening it all the way to the individual fibres, which are then transported into the rotor through the fibre guide channel. Such a degree of access to the individual fibre is required for trash separation, as well as for optimum yarn quality and running behaviour during rotor spinning. This process should be carried out with as minimum damage to the fibres as feasible. It is controlled by tooth shape, surface finish, assignment, material throughput, and opening roller speed (Fig. 16).

Belcoro opening rollers from Schlafhorst are designed for natural fibres such as cotton, cotton waste, and regenerated cotton fibres, as well as synthetic fibres with a staple length of 60 mm. In addition to the nickel-coated surface of the solid rings with the N designation, all clothing rings are formed of solid steel rather than wire clothing. Schlafhorst also manufactures opening rollers with 3d coating, a diamond coating with nickel plating (DN) for long service life and a high degree of fibre opening. Diamond-coated opening rollers can also be utilized entirely, depending on the application. Raw material parameters such as fibre length, distribution of

Fig. 16 Opening roller types

fibre length, short fibre content, surface structure of fibre, and friction value must all be considered during the opening roller selection process. Over the years, the most diverse systems for cotton processing have been developed to meet an extraordinarily broad variety of applications. These are the relevant types: B 174, NB 300, and B 20. With a few exceptions, type S 21 is utilized for the processing of synthetic fibres and their mixtures.

11.2 Opening Roller Speed

In addition to selecting the opening roller type, the opening roller speed is an important factor for fibre opening and yarn quality, depending on the raw material.

Table 20 shows the choice of opening roller types and their corresponding speeds in relation to raw material. Recommendations for choosing the speed of the opening rollers are given. In this regard, contrary to the often-quoted assumption of employing high-opening roller speeds with coarse count yarns and low-opening roller speeds with fine count yarns, the exact reverse is true. Finer yarns should always be spun at greater opening roller speeds than coarse yarns to produce good evenness, low degree of defects, and acceptable running behaviour. The supplied guidelines cover about 95% of all applications.

11.3 Choice of Torque Stop

While the navel generates a false twist, the torque stop causes blocking of the real twist. The ribs of the torque stop produce a twist blocking between the torque stop and the binding-in zone of the yarn in the rotor. The false twist is supported and briefly maintained. This results in an extension of the binding-in zone. The yarn strength in this area is increased. Thus, spinning stability is improved. As a result, it is possible to spin with a lower yarn twist at the same running behaviour or to reduce the yarn break rate when yarn twist is constant (Fig. 17).

Using the torque stop usually results in a slight increase in yarn hairiness. This is quite desirable for a voluminous yarn and the resulting soft handling. The application is particularly recommended for small rotor diameters.

11.4 Choice of Yarn Twist

The yarn twist has a significant impact not only on yarn strength but also on the basic yarn character. The yarn twist in the cotton industry varies according to the fundamental strength of the applied fibre material. As a result, recommending the appropriate yarn twist is quite challenging. However, if the twist is also considered

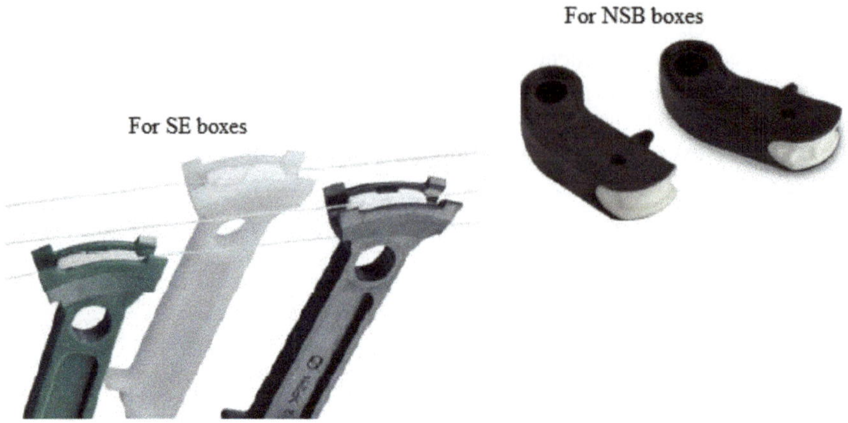

Fig. 17 Torque Stop types

Table 20 Choice of opening roller speed

Raw material type	Opening roller type	Opening roller speed
100% cotton Cotton noils Cotton waste Cotton regenerated fibers	B 174 N, DN B 20 N, DN B 300 N, DN	7,500-9,000 rpm (8,000 rpm)
Blends: CO/PES CO/CV CO/PAN PES/CV	S 21 DN	8,500-9,500 rpm (8,500 rpm)
Man made fibers: CV → PES → PAN →	B 20 DN S 21 N, DN S 21 DN, N	8, 000-9,000 rpm (8,500) rpm
PAN →	B 20 DN, N	7,500-8,500 rpm (8,000 rpm)

as the starting point of the relevant application, it might be of great assistance when producing a new fabric (Table 21).

12 Future Long Staple Cotton Requirements in Ethiopia

Fibre Length is usually understood to mean the Upper Half Mean Length (UHML). Uniformity Index (UI) and Short Fibre Index (SFI) are additional measures related to the fibre length distribution. Length is affected by genetics, environment during the growing season, and ginning. Instrumental measurements of UHML are usually similar to the results assigned by classers pulling staple. Classers assign staple lengths in 32nd of an inch, whereas instrument results are given in hundredths of an inch or millimetres and are more easily used in calculations of the mean or standard deviation over a number of samples.

Length is one of the most important parameters used in all segments of the cotton value chain. Length is the most important property in the production of ring spun yarn. Length affects the spinnability of cotton and influences the number of twists per inch of yarn required to achieve a given level of strength. Length is the most important property in setting drafting parameters within a textile mill. Length distribution strongly influences nearly all yarn quality parameters. UHML affects yarn strength. Length uniformity influences evenness, and SFI affects hairiness.

The influence of fibre length for short staple spinning is obvious. It has impact on spinning limit (yarn count), yarn strength and evenness, yarn hairiness, yarn breaks, as short fibres will cause a higher number of end breaks. In addition, characteristics of the final product will be affected, especially the handle and the lustre of the fabric [23].

Hence, fibre length is the driving factor for the yarn count range to be spun. It will also determine if the fibres will be processed into the carded or combed spinning line, and which end spinning system is the most suitable one to be used. Very short fibres will be utilized in Rotor open-end spinning, resulting in coarser yarn counts, whilst fibres with a longer staple will not only be used for finer yarn counts but will also pass through the combing process to further improve the already longer staple [53].

Fibre length is the most essential characteristic for predicting the highest attainable yarn count, along with fineness. The roller distances are calculated and set based on the fibre length in the process (Fig. 18). If the roller distance is too great, the fibres will 'float' and generate thick places or unevenness. If the roller distance is too short, fibres may break, resulting in an increase in short fibre content [41].

Fibre length has a substantial impact on yarn breaking in rotor and ring spinning. It effects not just the running behaviour but also the final yarn quality. As a result, a shorter fibre length with a higher proportion of short fibres lowers yarn evenness, defects (Thin, Thick places, and Neps), and increases yarn hairiness [53]. Because short fibres are not securely held in the drafting zone, they are floating and uncontrolled, resulting in increasing levels of faults in the succeeding processes—first in the sliver, and then subsequently turned into roving and yarn if no doubling method is utilized.

Ethiopia produces good raw cotton, with an upper half mean length of 26–31 mm. The country has the ability to develop long staple length cotton with enhanced seed and technology use. Table 22 compares the cotton quality metrics of Ethiopian cotton to one of the Indian cotton varieties (J-34).

According to the data in Table 22, Ethiopian cotton has the advantages of superior fibre maturity, length and evenness, and low contamination (trash content). The shortcomings include very low strength, a significant degree of yellowness in colour, and stickiness (not listed in Table 22). When compared to the Indian cotton J-34 type, which is utilized in spinning coarse yarn counts, Ethiopian cotton is suited for

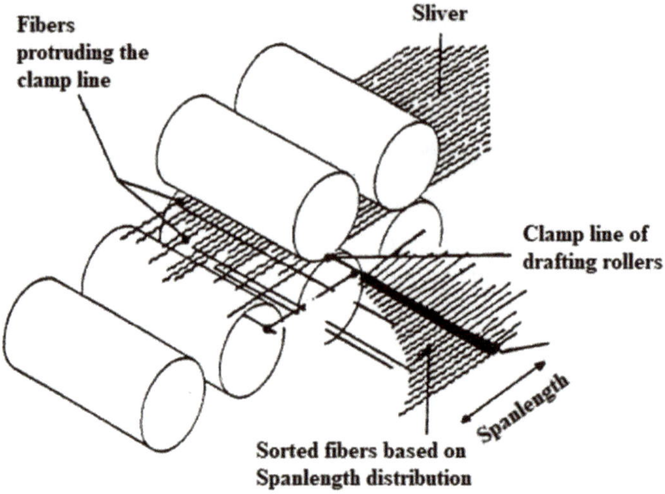

Fig. 18 Fibre length distribution in a drafting zone. *Source* ITMF–ICCTM guideline, 2020

Table 21 Choice of yarn twist from the application recommendations

Alpha$_{m}$	Alpha$_{\varnothing}$	Course count: Ne 10 and coarser	Medium count up to Ne 20	Fine count up to Ne 40
95	3.2			
100	3.4			
105	3.5			
110	3.6			
115	3.8			
120	4.0			
125	4.2	●		
130	4.4	●	●	
135	4.5	◉	●	●
140	4.6	●	◉	●
145	4.8		●	◉
150	5.0			●
155	5.2			
160	5.3			
165				

spinning coarse and medium yarn counts (Ne 20's–Ne 40's). Cotton cultivated in the Middle Awash region is a medium staple (28–30 mm) variety, but cotton farmed in the Humera and Metema regions is a short staple (25–27 mm) variety.

12.1 Future of Long Staple Cotton

According to the Ethiopian government plan provided by the Ethiopian Institute of Agriculture Research (EIAR, 2017), the long staple cotton need of spinning factories is predicted to increase over the next ten years. Table 23 shows the seed cotton requirements for long staple cotton from 2017 to 2030.

Table 22 Comparison of fibre characteristics of sample cotton varieties

Property	Ethiopian cotton	J-34 Indian cotton
Staple length (mm)	26–31	28.5
Micronaire (MIC)	3.5–5.3	4.5
Strength (cN/tex)	22–26	28.5
Uniformity index	79.2–87.1	83
Trash content (%)	3.0–6.0	4.5

Source Benchmarking of the Ethiopian Textile Industry, UNDO Draft report, 2010

Table 23 Strategic issues, research themes and planned strategic interventions to address critical issues facing national cotton research programme in the short (1–5), medium (6–10), and long (11–15) terms

Strategic issues	Strategic interventions from 2016 to 2030		
	In short term	In medium term	In long term
• Poor fibre quality characters (Length, strength, micronaire, etc.) below standard • Poor oil quality (high gossypol content and less fit for human consumption)	• Introduce the long staple cotton varieties from Egypt and conduct breeding programme • Develop varieties of cotton with less gossypol content and its oil fit for human consumption	• Improve the quality of oil content, long staple length, and others • Developing cotton varieties with combined traits of good quality, high oil content and desirable fibre quality • Continue developing cotton varieties with less gossypol content and more fit for human consumption	• Improve oil content, staple length and others • Use biotechnological tools for fibre quality breeding acceptable for netted fabrics • Use biotechnological tools for high oil content and oil quality breeding acceptable for human consumption

Source EIAR, 2017

13 Summary

Ring spinning technology has undergone several changes in recent years. Recent advancements in ring spinning are primarily focused on advances in drive systems, drafting arrangement, ring and traveller design elements, machine monitoring, and automation, which includes roving stop motion, roving bobbin transfer, doffing, and cop transfer to winding. Other advancements, such as compact, solo, double-rove, and core-sheath spinning, have increased the adaptability of ring spinning without altering the fundamental principle of ring spinning. The normal spindle speed of today's ring spinning machine is 16,000 rpm, compared to less than 7000 rpm fifty years ago. New spinning technologies such as open-end and air-jet are capable of producing yarn at rates much higher than that of ring spinning. In order to accommodate these mechanical capabilities, fibres must flow individually or in small bundles at very high speeds. They must also withstand the stresses caused by the opening and condensations included in the new spinning technologies. Because of the increased usage of synthetic fibres (for example, polyester) that may be adjusted to fit a certain spinning technique, determining the value of cotton (based on its contributing quality) is important for efficient blending and improved processing performance.

Cotton is produced primarily by five state-owned firms in Ethiopia. Tendaho, Middle Awashe, Upper Awashe, North Omo, and Abobo are among them. Cotton cultivation on a large scale takes place mostly in the Awash valley and three smaller areas, notably North Omo (Southern Regions), Ababo (Gambella Region), and Gode (Ogaden Region). Except for Ababa, cotton is grown with irrigation in other locations. In Matema and Humera, large private farms mostly produce cotton using rainwater. All these cotton-producing sectors (even small household farms) must be properly informed and act conscientiously to develop excellent quality cotton that meets the quality requirements of the global cotton market. Ethiopian Institute of Textile and Fashion Technology (EiTEX) must play a role in teaching cotton farmers in the country.

References

1. Brashears AD, Baker RV (2000) Comparison of finger strippers, Brush roll strippers and spindle Pickers on the Texas high plains. Proceedings of the Beltwide Cotton Conference 1:452–453
2. Carlson GA, Mohamed A (1986) Economic analyses of cotton insect control in the Sudan Gezira. Crop Prot 5(5):348–354
3. Chakraborty K, Ethridge D, Misra S (2000) Proc. Beltwide Cotton Research Conferences, 374
4. Chattopadhyay SK, Venugopal B (2020) Effects of Fibre Length Relative to Rotor Diameter On Yarn Tensions During Rotor Spinning and Its Correspondence with Yarn Quality. J. Inst. Eng. India Ser. E 101:45–52. https://doi.org/10.1007/s40034-019-00152-2
5. Chaudhry M Rafiq (1997) Harvesting and ginning of cotton in the world, Technical Information Section International Cotton Advisory Committee Washington, D.C. Proceedings of the Beltwide Cotton Conference Volume 2:1617–1619 National Cotton Council, Memphis TN
6. Chavan RB (2010) Ethiopian textile industry present status and future growth prospects. LAP Lambert Academic Publishing (2013–07–11), ISBN 13: 978–3–659–41735–1

7. Cocke JB, Kirk IW, Wesley RA (1977) Spinning Performance and Yarn Quality as Influenced by Harvesting, Ginning and Mill-processing Methods, USDA, MRR no. 1066 Cotton Outlook: February, (2015). http://www.usda.gov/oce/forum/
8. Deussen H, Neuhaus L (1988) Schlafhorst Dokumentation No 21
9. Dever JK, Gannaway JR, Baker RV (1988) Text Res J 58:433
10. Elsner O, Hani J, Lubenevskaya E (1983) The sugar content in lint of growing bolls. Cotton et Fibres Tropicales 38:223–227
11. Ellsworth PC, Tronstad R, Leser J, Goodell PB, Godfrey LD, Henneberry TJ, Hendrix D, Brushwood D, Naranjo SE, Castle S, Nichols RL (1999) Sticky cotton sources and solutions. The University of Arizona, Cooperative Extension. Publication, IPM series No. 13, AZ1156–12/1999
12. Fisher I (1967) The making of index numbers: A study of their varieties, tests, and reliability, 3rd edn. Sentry Press, New York, NY
13. Foreign Agricultural Service/USDA: Office of Global Analysis; December (2015) http://apps.fas.usda.gov/psdonline/circulars/cotton.pdf
14. Griffin AC, Columbus EP (1982) Dust in Cotton Gins: An Overview, Cotton Dust, American Chemical Society:27–36
15. Gulati AN (1957) The Physical Properties of Cotton (Asia Publishing House, Mumbai), 1957:201–205
16. Haimancioglu M, Ercarl MN (1981) Melliand Textilber
17. Harmancioglu M (1979) Textil-Praxis. 1466
18. Hendrix DL (1999) Sugar composition of cotton aphid and silverleaf whitefly honeydews. In P. Dugger and D. Richter (Eds.), Proceedings Beltwide Cotton Conferences, January 3–7, Orlando, FL:47–51. Memphis, TN: National Cotton Council of America
19. Hequet E, Abidi N (2002) High-Speed Stickiness Detector Measurement: Effect of Temperature Settings and Relative Humidity. The Journal of Cotton Science 6:68–76
20. Hequet EF, Abidi N, Ethridge D (2005) Processing sticky cotton: Effect of stickiness on yarn quality. Text Res J 75(5):402–410
21. Hequet, Kelly B (2012) Predicting yarn quality: An indispensable tool for cotton breeders, International Cotton Conference, Bremen, Germany
22. Hughs SE, Lalor WF, Wabha RT (1988) Text Res J 58:557
23. Hunter L (2006) Cotton Fibre properties: their impact on textile processing performance and costs
24. International Textile Manufacturers Federation (ITMF) (2016) Cotton contamination surveys (2005–2016). Wiedingstrasse 9, CH - 8055 Zürich. Web http://www.itmf.org
25. International Textile Manufacturers Federation (ITMF) (2020) Cotton contamination surveys (2007 2009 2011 2013 2016 2019). Wiedingstrasse 9, CH - 8055 Zürich. Web http://www.itmf.org
26. ITMF International Committee on Cotton Testing Methods (ICCTM) and by ICAC Task Force on Commercial Standardization of Instrument Testing of Cotton (ICCTC) (2020) A Guideline on Interpretation and use of Instrument Measured Cotton Characteristics. Version V 1.0 Issued on April 6
27. ITMF-ICCTM, Progress Report (2021) Proceedings of the virtual plenary meeting of the ITMF-International Committee on Cotton Testing Methods (ICCTM)
28. Kamal MM, Ragab MT (1995) Numerical quantification of lint grade in Egyptian cotton. Egy. J. Agric. Res. 73(3):723–736
29. Khalifa H, Gameel OL (1982) Control of cotton stickiness through breeding cultivars resistant to whitefly, Bemisia tabaci (Gene.) infestation. Improvement of oil seed industrial crops by induced mutations, IAEA:181–186
30. Krifa M, Gourlot JP, Drean JY (2001) Effect of seed coat fragments on yarn strength: dependence on fibre quality. Textile Res J. 71:981–986
31. Lawrence, CA (2003) Fundamentals of spun yarn technology. ISBN 1–56676–821–7
32. Lord E (1961) Manual of Cotton Spinning: The Characteristics of Raw Cotton. The Textile Institute, Manchester and London : 310–311

33. Lord E (1981) The Origin and Assessment of Cotton Fibre Maturity. International Institute of Cotton: 27
34. MacDonald, Stephen (2010) Markets, Institutions, and the Quality of Agricultural Products: Cotton Quality in India. Selected Paper prepared for presentation at the Agricultural & Applied Economics Association. AAEA, CAES, & WAEA Joint Annual Meeting, Denver, Colorado
35. May Lloyd O, Taylor RA (1998) Breeding cottons with higher yarn tenacity. Textile Res. J. 68(4):302–307
36. Militký J (2004) MATLAB Program for Complex Quality Evaluation. National Textile Centre Rept., Liberec
37. Mogahzy YEEl (1988) Selecting Cotton Fibre Properties for Fitting Reliable Equations to HVI Data, Textile Res. J. 58:392–397
38. Mogahzy, Yehia EEL (1990) Broughton, Roy JR., and Lynch, W. K., A Statistical Approach for Determining the Technological Value of Cotton Using HVI Fibre Properties., Textile Res J, 60(9):495–500
39. Murthy HV, Samanta SK (2000) A Fresh Look at Fibre Quality Index. The Indian Text. J. 111(3):29–37
40. Prakash Jai, Srivatsava SK, Bhargava GS (1991) Indian Text J., 101(5):126
41. Schenek A (2001) Naturfaserlexikon, Spinlab, Knoxville, Deutscher Fachverlag: 86. ISBN 3–87150–638–9
42. Shete DG, Sundaram V (1974) A Note on the Comparative Merits of Roller Ginning and Saw Ginning. Cotton Development 3:3–10
43. SICA News Letter, (1991) 13(18):9
44. Tan DW, Terrell J (1989) Business Statistics for Management and Economics, 5th edn. Houghton Mifflin Company, Boston, MA
45. Tesema GB, Hussein KMM (2015) Comparison of different quantification methods to define fibre quality of Ethiopian, Indian & Egyptian cottons. International Journal of Fibre and Textile Research 5(2):9–15
46. Tesema GB, Drieling A (2019) Statistical Analysis and Instrumental Characterization of commercial Ethiopian Cotton Varieties. Journal of Textile and Apparel, Technology and Management 11(1):1–16
47. Tesema GB, Drieling A (2020) Effect of Ginning on Tenacity, Elongation and other Fibre Quality Properties. The Journal of the Textile Institute 111(3):326–333
48. Tesema GB (2020) Cotton stickiness before and after saw ginning. The Journal of the Textile Institute 113(1):101–109
49. Townsend, Terry (2018) International Cotton Industry Consultant. Cotton in the World Economy. http://cottonanalytics.com/cotton-in-the-world-economy/
50. Trommer G (1995) Rotor Spinning, Deutscher Fachverlag GmbH, Frankfurt am Main. USTER News bulletin (1991), Measurement of the quality characteristics of cotton Fibre, 38(23):31
51. Weissenberger W, Legler F (1995) Text. Month, 41
52. Weisstein LF (2006) The Effect of Online Dynamic Pricing on Consumer Perceptions and Behaviors. Proceedings of the Fifth Workshop on e-Business
53. Werner Klein (2018) The Rieter Manual of Spinning (2018), Vol 1, Technology of short staple Spinning. 20–21, ISBN10 3–9523173–1–4

Technology of Fabric Manufacture (Traditional Cotton Woven Fabric Manufacturing in Ethiopia)

Million Ayele, Derseh Yilie Lemeneh, Mesfin Shiferaw Wubie, and Biruk Fantahun Adamu

Abstract Ethiopia is a country with diverse traditional and cultural identities. Traditional fabrics have been manufactured in Ethiopia for a long-time utilizing cotton fibre and hand weaving throughout the regions. It is a country with a long history and a strong artisan heritage that is yet mostly unknown to mass tourists. Hand weaving has a fascinating and somewhat hidden history in Ethiopia. Although there is no written record that indicates when and how traditional weaving first arrived in Ethiopia, evidence shows that it did so before the year 1000. People have been passing down the art of traditional cloth weaving from generation to generation by using traditional weaving machines. Ethiopia's economy has historically been heavily dependent on cotton and cotton weaving. Despite Ethiopia's strong agricultural sector, the country still has more than 200,000 handloom weavers. The most commonly used traditional clothing in Ethiopia varies by ethnicity and religion. Each ethnic group has created its own cultural clothing, which are unique and dynamic in their fabric patterns and designs and embody their socio-cultural values, especially during holidays when everyone dresses in their traditional best. Oromia, Sidama, Welayta, Gojjam, Gondar, Wollo, Siltie, Harari, and other regions of Ethiopia are well-known for their traditional clothing. The major objectives of this chapter are to examine traditional cotton weaving in Ethiopia, the history of weaving there, and several distinctive traditional handcrafted fabrics with various designs and patterns in various colour combinations.

Keywords Traditional fabric · Hand loom · Cotton · Weaving · Handmade cloth · Shema

M. Ayele (✉) · D. Yilie Lemeneh · M. Shiferaw Wubie · B. Fantahun Adamu
Ethiopian Institute of Textile and Fashion Technology, Bahir Dar University, Bahir Dar, Ethiopia
e-mail: milliti2000@gmail.com

K. Murugesh Babu et al. (eds.), *Cotton Sector Development in Ethiopia*, Textile Science and Clothing Technology, https://doi.org/10.1007/978-981-99-9149-5_11

1 Introduction

Weaving is a process of fabric production in which two sets of yarns or threads are interlaced at a right angle to each other. Ethiopia is a country with a rich traditional and cultural identity. Traditional fabrics in Ethiopia are manufactured from cotton fibre by hand weaving over a long period of time all over the regions, with distinct designs according to the culture and religion of the people living in that specific place. Cotton, which is cultivated across Ethiopia, is the principal raw material used to manufacture traditional fabrics [1, 2]. As seen in Fig. 1, raw cotton is purchased from a field or market, opened by hand, and spun (thread is manufactured) by hand using an Enzirt (spindle). Cotton yarn manufactured in factories is often used as the warp yarn, while hand-spun thread is utilized as the weft yarn.

In traditional weaving, the weft thread is wound on the pirn before being inserted in the wooden shuttle. While the shuttle carrying the pirn moves back and forth across the shed, the weft yarn is interlaced with the warp yarn. The inserted weft yarn is then pressed onto the fell of the cloth using the reed, and the process is repeated until the desired fabric length is attained.

Hand weaving and craft industries are important non-agricultural sources of income in Ethiopia. In rural and semi-urban areas of Ethiopia, hand weaving is still a common technique for making both specialty fabrics and everyday clothing. A new boutique designer market is also developing at the national and international level, offering great economic opportunities for select weavers. Despite the recognition of

Fig. 1 Hand spun cotton yarn production process

their talents and the support of many city projects, the majority of weavers still have a low status in society. Because of their unorganized work habits and frequent financial troubles, much of the younger generation is being persuaded to quit the industry in pursuit of more stable employment as a result of the industrial textile sector's recent expansion.

The development strategy for the sector aims to fully exploit the industry's potential for improving incomes and eliminating poverty. The Ethiopian government appears to be approaching development vertically, with emphasis on international and domestic markets at all scales, whereas this model typically risks focusing on mass industrialization and the expansion of large-scale farm operations as a result of foreign investment interests (https://www.crosspolynations.com/2019/09/14/ethiopian-cotton).

2 History of Traditional Weaving in Ethiopia

Ethiopia is a country with an ancient history and a rich handicraft tradition, but is still largely unknown to mainstream tourism. Ethiopia has a very strange and somewhat enigmatic heritage related to handweaving (https:http://store.hamlin.org.au/blogs/journal/ethiopian-textile-weaving-history-future (accessed 21 June 2022). Located in the Horn of Africa, Ethiopia has been a centre of cultural and commercial exchange for thousands of years. This history of influence and success has been firmly embedded in their famous traditional handwoven cotton cloth, both as a stimulus for its expansion and as the subject of Ethiopia's history.

One of the many cultural, religious, and historical values for which Ethiopia is known is the way its people dress in traditional costume. Weaving has a tremendous impact on traditional clothing. Although there is no written document that explains when and how weaving originated in Ethiopia, evidence suggests that it did so before the first century. (https://store.hamlin.org.au/blogs/journal/ethiopian-textile-weaving-history-future viewed on 21/06/2022; The History and Future of Ethiopian Textile Weaving, October 5 2021. https://store.hamlin.org.aujournalet, viewed on 22 02/2023).

Weavers spin cotton and produce traditional fabrics, making Ethiopia a proud country with its unique dress culture. They are also grateful to the nation, where people have passed on the expertise of traditional weaving fabric from generation to generation using traditional weaving machinery. Since ancient times, exquisite Ethiopian traditional clothing has been created by combining this skill with the indigenous techniques of weaving and embroidery. Cotton was cultivated on a small scale for many years, and all processing was done in the village's houses, including many individuals from the neighbourhood. The women would perform the majority of the auxiliary jobs, such as washing, finishing, fringing, and seam stressing, in addition to ginning and spinning cotton on a drop spindle. The actual weaving was a male-only occupation, and fathers would teach their sons how to do it. The pit loom is the most prevalent type of loom used for traditional weaving, and many of them are still in widespread use across the nation [2].

Another important cultural effect with origins in early trade routes may be found in the striking similarities between these looms and ancient Indian looms. A throw shuttle was used on these simple two-shaft looms, restricting the width of the cloth to the width of the weaver's arm span. If a wider fabric is needed, the strips can be connected along the selvedge. Many weavers choose machine-spun warp yarn from large, centralized spinning mills that have been constructed across the nation, whereas hand-spun cotton is usually selected for weft threads due to its distinctly soft and natural aesthetic character.

A common myth on the continent states that the advent of religion and the modesty norms it imposed encouraged people to wear in fabric, which began the development of handloom weaving in the region. The "Shamma" is the most well-known type of Ethiopian handwoven fabric. It is a collection of often gauzy white textiles worn as shawls or wraps and woven and worn throughout most of the country by both Christian and Muslim populations for everyday use and religious events. They were woven by nomadic Muslim and Falasha (Ethiopian Jewish) weavers who travelled between the residences of noblemen and set up their looms on their verandas. Originally, they were only worn by religious leaders and the upper crust of society. The clothing served as emblems of the society's intricate class and hierarchy structures as well as modesty aids. Later, the custom expanded rapidly throughout the country, making hand weaving a common home occupation and handloom cotton the national cloth for both daily wear and treasured ceremonial and religious purposes. Other examples include the "netela," a larger female version occasionally embellished with metallic thread "tibeb" patterns, the 'kuta', a male's thin gauzy shawl worn during warmer seasons, and the "gabi," a thicker wrap used by both women and men to ward off the chills at night and during the colder months. Yardage is commonly woven and stitched into clothing, such as the "Kemis," a traditional long women's dress.

Around the year 1898, a multicoloured border design called "tibeb" was added to the generally white fabrics. The once-simple, single-colour pattern that occasionally ornamented the edge was quickly replaced by the multicoloured tibeb, which gradually emerged as a distinguishing identifier and defining feature that is today associated with textiles from the region. The Tibeb is woven with an additional weft thread consisting of either naturally coloured locally grown cotton or imported silk or wool brought from the east by merchants. This demonstrates that Ethiopian weavers are conscious of the importance of design or pattern variances in expressing identity. They weave with warp, weft, and coloured yarn as basic materials.

In the nineteenth century, the colonial endeavour destroyed several African nations' handloom traditions. Local markets were overwhelmed with an excess of mass-produced, inexpensive clothing produced primarily in Manchester from Indian cotton and priced so low that traditional weavers couldn't compete. Despite the market continuing to be overrun with mass-produced imported cloth, this time from the east, Ethiopia has managed to maintain its symbolic independence through a persistent local demand for their own cloth, a symbol of cultural identity whose value to the community cannot be replaced.

The tenacity of the Ethiopian cotton handloom weaving has been demonstrated ever since it was founded. These fabrics were instrumental to documenting a number of changes in the ages, materially and symbolically. Unlike many other ancient methods of weaving, a constant demand on the market has preserved cultural significance and value; it appears to have positioned itself firmly for both durability as well as long-term potential (https://www.crosspolynations.com/2019/09/14/ethiopian-cotton/).

Weaving has seen an increase in demand since 1905. It is the period when scarves started to be used to indicate the social, economic, cultural, and political status of their wearers on various occasions and in various contexts. Ethiopia's economy has always been primarily reliant on cotton and cotton weaving. Despite Ethiopia's thriving agricultural industry, it is estimated that there are more than 200,000 hand loom weavers in the country.

The present state of hand weaving is that, in accordance with national policy, firms have been created in metropolitan regions to develop weaving talents and produce high career possibilities for young people in order to profit economically. It has also changed the lives of enterprises formed in Ethiopia's capital city of Addis Ababa, as well as other weavers working in the field. It is greatly assisting the community in the field of traditional weaving. Ethiopia's industrial textile sector is currently expanding. There has been a recent infusion of international investment, lured by an availability of cheap hydroelectric power, cheap labour, and "flourishing industrial parks." As a result, the number of factories being developed has expanded dramatically, creating positions that are particularly desirable for technical university graduates. It also provides knowledge, expertise, and technology to the industry, contributing significantly to economic growth. The expansion of the textile industry is also encouraging the growth of cotton farming, an industry on which the government is focusing on upgrading and aiding small-scale farmers, as well as farmers in the public and private sectors. In order to regulate this spread, attempts are presently ongoing to facilitate the establishment of organic cotton certification, which is not yet available.

3 Hand Weaving/Traditional Weaving

Traditional weaving is the method of creating cloth with diverse patterns on handlooms. Handweaving is completely done by hand in Ethiopia. It is more rigorous and time-consuming, requiring patience, talent, and physical power. The weavings are created with care, attention, and respect for traditional methods. A good quality warp and weft tread is required for the traditional weaving process.

3.1 The Preparation of Warp Thread

The process of warping begins after determining the types of thread required for weaving. Warping is the process of getting the warp thread ready for weaving. Warping is the process of preparing warp thread for weaving. The majority of the time and effort necessary for weaving is spent preparing the warp thread. The first stage in ensuring that the machine is precise and ready for operation is to understand its capacity. Before starting to prepare the warp yarn, it's crucial to determine the proper length, width, and number of rounds for the warp thread. After the warp thread preparation is complete, it must be loaded onto a traditional loom, and the tension must be properly set to prevent thread slackness. The preparation work is deemed finished once this is carried out correctly.

The first traditional warping process was performed outside the loom. The warping threads are wound onto a handheld warping reel, as seen in Fig. 2, and are then unrolled by the weaver around three to eight warping wooden posts that are buried in the ground in two parallel rows. A zigzag pattern is made between the wooden posts when the weaver unrolls the threads, according to the necessary number of ends and length.

The warping machinery used by producers of handwoven fabrics has recently been upgraded to produce longer lengths of warp yarn covering smaller regions. The required number of ends with the requisite length was produced in this modified warping machine, as shown in Fig. 3. As compared to the first traditional warping, this warping gives the following advantages:

- Easy maintenance of the machine as well as the ends
- Easy operations
- Minimum space utilization
- Preparation of more number of ends at the same time
- Saving in time

Fig. 2 Traditional warping (Photograph by author)

Fig. 3 Modified warping machine (Photograph by author)

3.2 Pirn Winding (Weft Thread Preparation Process)

Weft thread preparation is the process of unwinding weft thread from the cone or hank and winding it on hollow pieces of bamboo known as pirn. The shuttle should next be loaded with the pirn to prepare it for the weaving process (Figs. 4 and 5). The traditional pirn winding machine is made of wood or metal and is operated by hand. Weft winding must be done correctly to provide a trouble-free weaving operation.

Weavers utilize several types of yarn as raw materials depending on their fabric pattern. The most often used yarns were a factory produced warp known locally as "dir" or "mag," a weft spun largely by women at home, and "tilet" factory-manufactured coloured threads used for ornamental borders.

4 Developments in Traditional Hand Loom

Ethiopian weavers utilized pit looms that were either raised or hung. The first standard looms, represented in Fig. 6, had two harnesses that were operated by pushing pedals with the feet alternately up and down to interlace the threads. They range in width

Fig. 4 Traditional reeling machines (Photograph by author)

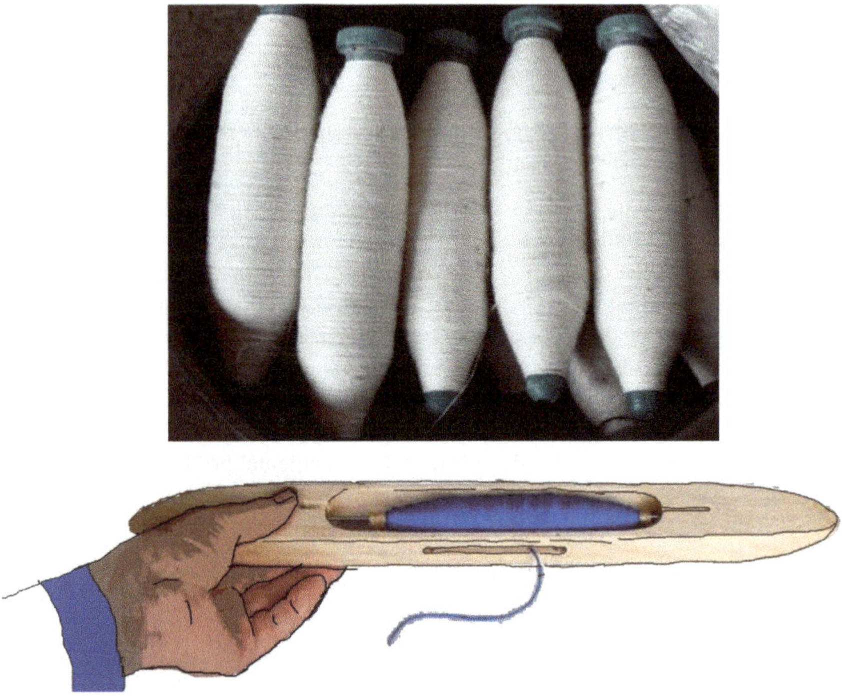

Fig. 5 Traditional pirns and shuttle (Photograph by author)

Fig. 6 Pit ground loom (Photograph by author)

from 70 to 90 cm and are found across the country. A ground loom digs a hole in the ground for a shed entrance that is controlled by foot. To weave the threads together, the weaver operates the shuttle by hand through the open shed. Every design and a drawing are carefully numbered and precisely tracked.

The majority of weavers in Dorze build their own looms out of locally available materials, notably bamboo and eucalyptus. The weavers set up their handloom adjacent to their traditional hut on the grounds of their farmhouse area for weaving. On a conventional, manual handloom, warp and weft yarn are used to make different types of fabrics. An Ethiopian handloom is made up of three parts: a hand loom frame, a reed, and a winding wheel.

The weaver often holds the weft yarn between his fingers while weaving different fabrics with attractive colours; this helps the weft strands to pass fast between the warp threads. This allows the weaver to change the colour and pattern of the thread more easily than if he had used a shuttle. The weaver may use a finger or a small stick to adjust the weft threads after they have been passed through the warp, then beat the weft yarn to the fabric's feel.

Weaving has been done on a variety of regional handlooms. The pit loom is the most commonly utilized by the Dorze (Fig. 7). The weaver in this loom sits on the edge of the pit above which the loom is situated, and he handles the treadles with his feet. The weaver throws the shuttle (mewerweria) from one side to the other with two hands and alternately raises and lowers the harnesses with two feet. This is followed by beating up the weft with the comb to the fell of the cloth and winding up the woven cloth on the wooden cloth roller.

Fig. 7 Dorze's pit loom (Photograph by author)

However, in order for hand-woven fabric manufacturers to succeed, new tools and processes must be developed. Many weavers now use metallic-frame handlooms, which are easy to disassemble and reassemble, have better output, are more design adaptive, produce wider fabrics, and can be transferred from one location to another. Several private groups were the first to introduce these handlooms. The action of the shuttle during weft insertion has been improved in looms with metal frames (Fig. 8). However, the handloom's basic technical features have not changed significantly throughout the years. (Ethiopian tradition of handloom weaving—Shimena https://www.shimena.net viewed on 22/02/2023).

5 Ethiopian Traditional Cotton Woven Fabrics

Ethiopia is home to more than 80 different ethnic groups, each of which has its own language, culture, traditional dress, and way of life. The country's ethnic and cultural diversity has given rise to a variety of distinctive and vibrant dress traditions. The specific patterns and motifs that set Ethiopian traditional textiles apart from other civilizations depend on the customs and way of life of the many ethnic groups. Ethiopian traditional clothing is produced with exquisite artisan work and absolutely unique artistry, mostly through weaving and embroidered skills. These traditional clothing symbolized religious as well as ethnic identity in various regions across the country. Unlike in many other regions of the world, the materials are primarily simple

Fig. 8 Handloom with metallic-frames (Photograph by author)

textiles made with modest equipment and worn everywhere. Traditional handwoven cloths have now become common. As a result, Ethiopia can become one of the world's leading fashion fabric producers by incorporating its own traditional bespoke fabrics into the production of garments with the 'Made in Ethiopia' label [3, 4].

6 Ethiopia's Ancient Cotton Woven Tradition

Traditional clothing was composed entirely of cotton. The fashion business is a product of the contemporary era. Prior to the mid-nineteenth century, most clothing was custom-made. It was produced for people, either from home or on request from dressmakers and tailors. Traditional clothes are essential in the tourist sector since they are distinctive and represent the identity and cultural wear of countries in every area of the country. These are solely indigenous patterns from the nation that demonstrate weaving technology (Fig. 9). There are no equivalent enhancements made to these trends to affect their values and distinguishing qualities. Their manufacturing technology has not yet been improved in order to expand the tourist business and enhance their distinctive qualities for global competitiveness [3].

7 Tibeb: The Art of Ethiopian Weavers

Traditional weavers in many sections of Ethiopia produce traditional clothing for their ethnic groups, with Dorze weavers being well-known and dominant in

Fig. 9 100% cotton woven robes worn in various forms in Orthodox religious churches (Photograph by the author)

the textile industry. (https://store.hamlin.org.au/blogs/journal/ethiopian-textile-weaving-history-future viewed on 21/06/2022). They can produce exceptionally beautiful weaving art by repeating geometric designs separated by different colours and lines, as well as organic shapes. With the use of basic looms, complex weaving patterns are employed to manufacture traditional clothing.

"Tibeb" is a term for the feature of a decorative border that describes patterns or decorations on a cloth made by hand loom with supplementary weft into the border of the shemma or made by hand embroidery in Ethiopia (Fig. 10). It is a pattern seen on handwoven 100% cotton fabrics. Ethiopian weavers are able to incorporate the "Tibeb" motif into larger designs. It is a traditional Ethiopian fabric that is almost commonly worn by Ethiopians across the country. Its elegance is well-known among Ethiopians and appreciated around the world. Tibeb is produced by weaving with a continuous extra weft thread of one or more different colours of either naturally dyed cotton yarn or silk, metallic, rayon (saba, which is twisted rayon), acrylic, and wool yarns. Cotton and pure silk were once used for Tibeb patterns, but these natural fibres have mostly been substituted by others, such as acrylic and rayon floss or twisted "art silk" painted with synthetic colours. The pattern augmentation was inspired by the many hues of the extra weft yarns. In general, Tibeb design characteristics in Ethiopia are a blend of the national palette of red, yellow, green, and black on a white fabric.

8 Traditional Woven Textile (Traditional Clothes) Common to the Country

Ethiopians use the bulko, gabi, netela, kuta, and qemis, which are the principal hand-woven textile articles of traditional clothing, regardless of ethnicity or religion. The motifs, designs, or "Tibeb" that have been applied to certain fabric elements to differentiate one ethnic group from the others are the major differences between these clothing and those of other ethnic groups. As a result, many Ethiopian weavers are familiar with the many motifs or patterns that define their identity.

Bullukko (Bulluko): a traditional garment that covers the full body and is used by ladies, men as overwear, and family members as nightwear.

Fig. 10 Weaving Tibeb Designs (Photograph by the author)

Netela: is a long, white gown composed of fine and delicate gauzy cotton worn by women (Fig. 11). It usually features appealing multicolour pattern bands at the margins, generally with silver or gold metallic threads, the so-called "Tibeb" design, which is the main source of its aesthetic value [5, 6].

Gabi: It is a thick cotton handwoven fabric created from coarse spun Ethiopian cotton yarn spun by hand by Ethiopian women. It contains four layers, comparable to a light-weight blanket worn solely by males. This incredibly soft blanket may be used

Fig. 11 Netela (Photograph by the author)

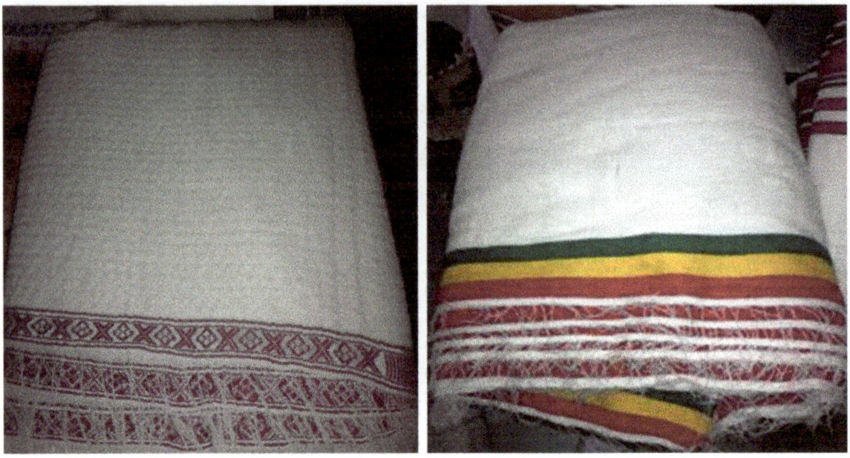

Fig. 12 Gabi (Photograph by the author)

as a blanket on a cold day or as a throw to cozy up around a TV on cold evenings. It is an extremely soft and high-quality hand-woven light-size gabi. It is warmer, rougher, and heavier than other traditional Ethiopian fabrics. Figure 12 depicts the many forms of Gabi with varying Tibeb at the borders [6, 2].

Habesha Qemis: It is the cloth worn by women regardless of their religion and is made from hand-spun cotton thread for both the warp and the weft without any size standard (length or width). It is a white robe decorated with tibet on the edges and waistband. The Habesha Kemis is a white, hand-woven cotton garment made from shemma. The shemma is hand-woven by traditional weavers, and handmade patterns known as tibeb are made using woven shiny threads and added to the waistband and edges. There are different designed patterns, either by using weaving or hand embroidery, in areas like around the collar, chest, and back. Besides, patterns may differ according to religion and ethnicity. The Ethiopian Orthodox people use a pattern of the cross in their kemis (The History and Future of Ethiopian Textile Weaving, October 5 2021. https://store.hamlin.org.au/journal/et viewed on 22 02/2023; Temesgen 2018). Women from Shewa, Gojjam, Wollo, and Gondor also wear white linen with embroidery on the cuffs, the middle, the bottom, or all three. There are typical patterns that anyone in Ethiopia might wear (Fig. 13d).

Figure 13a depicts a type of kemis that is worn by all Ethiopians and has the Ethiopian Flag colour at the edges (green, yellow, and red yarns of equal width). The dress contains green, yellow, and red colours at the edges and has a unique design and style. It is made by hand from 100% cotton yarn and can be worn on different holidays. Figure 13b, c shows types of kimes with a different tibeb design (in the design, cross symbols are shown) created by 100% cotton fabric. These types of clothes are worn by Orthodox Christians. Figure 13d shows a traditional kemis that can be worn by anyone, irrespective of religion.

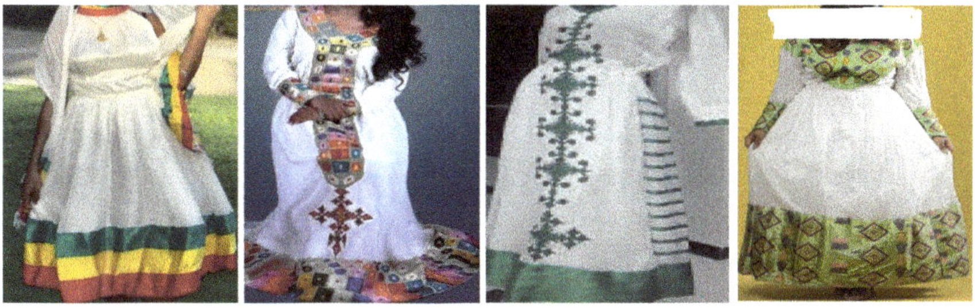

(a) (b) (c) (d)

Fig. 13 Ethiopian kemis (Photograph by the author)

Meqenet: is a cloth that is twisted around the woman's waist around Qemis and is about 3-m long and 70 cm wide, with a coloured pattern made by weaving at the end side or may be the entire part of the cloth, with thin stripes and simple geometric patterns (The History and Future of Ethiopian Textile Weaving, October 5 2021. https://store.hamlin.org.au/journal/et viewed on 22 02/2023; Temesgen 2018) (Fig. 14).

Fig. 14 Meqenet (Photograph by the author)

(a) (b) (c)

Fig. 15 **a** Jano cloth **b** Habesha kemis **c** wollo Kemis (Photograph by the author)

9 Traditional Fabric/Cloth Designs Based on Ethiopian Ethnicity

The most widely used traditional clothes in Ethiopia differ according to ethnicity and religion. Each ethnic group developed its own cultural clothing, which is distinct and dynamic in fabric designs and patterns and expresses its socio-cultural values, especially during holidays when everyone dresses up in their finest. Oromia, Sidama, Welayta, Gojjam, Gondar, Wollo, Siltie, Harari, and other regions of Ethiopia are well-known for their traditional clothing.

9.1 Traditional Woven Fabric in the Southern Part of Ethiopia

Jano is the name of a traditional cloth illustrated in Fig. 15. Jano is a brand on the woven cloth edges that are red (the red 'tilet' border) and are worn by Gonderians (Gondar is a city in Ethiopia's Amhara region). It refers to the tiled weave seen mostly on edges of kemis, gabi, kutta, and shirts.

9.2 Traditional Woven Fabric of Oromia People

Oromia traditional cloth is one of Ethiopia's traditional cloths, made from genuine 100% cotton by hand weaving with three colour combinations. The basic colours used are red, black, and white (Fig. 16). The clothes are handmade and woven by

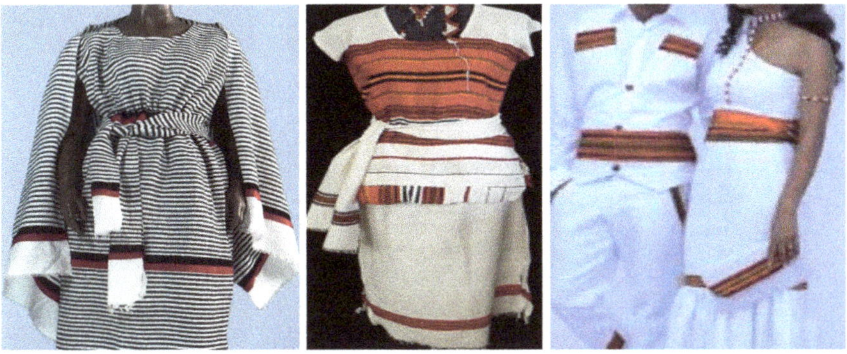

Fig. 16 Oromia people's fabric design and style (Photograph by the author)

traditional Ethiopian shemen. At a certain place in the cloth, mostly at the edges, there is a known colour combination of white, red, and black with definite patterns that identify the wearing styles of the Oromo nation.

9.3 Traditional Woven Fabric of Wolayta People

Wolayta people in northern Ethiopia create denguza patterns and motifs from known hues of red, yellow, and black (Fig. 17), with their unique cultural wearing customs. Figure 17a depicts the weave-repeat pattern.

(a) (b) (c)

Fig. 17 Wolayta people's fabric design and style (Photograph by the author)

Fig. 18 Women cloth fabric design and style of Silite people (Photograph by the author)

9.4 Traditional Woven Fabric of Silite People

Siltie cultural clothing is characterized by a fabric made of cotton or other yarns with combinations of three specific colours: red, broken white, and black design patterns as shown in Fig. 18.

9.5 Traditional Woven Fabric of Sidama People

The Sidama people's traditional garments are distinguished by a fabric made of cotton or other yarns with colour combinations of three distinct colours: red, black, yellow, and broken white (Fig. 19). Figure 19 depicts the design patterns cultivated in the Sidama people's materials and clothing styles.

According to the design description of Sidama cultural cloth, the warp yarn is a cotton white yarn with a different count than the weft yarn. The weave structure is essentially simple, and each pattern/design will be highlighted by a distinct colour combination. A weaving begins with a short distance of back-dyed colour yarn, followed by a short distance of brown colour dyed yarn. These dyed black and brown yarns are woven into a length of fabric. The next design pattern is a different colour combination.

10 Summary

In Ethiopia, a unique method of producing traditional cloth involves the interlacing of two yarns or threads at a right angle, resulting in distinct designs that are influenced by culture and religion. The primary raw material used for warp and weft yarns is cotton fibre, which is grown throughout the country. Typically, hand-spun thread is

Fig. 19 Fabric design pattern of Sidama people (Photograph by the author)

used for the weft yarn. Despite Ethiopia's ancient history and hand weaving tradition, it remains largely unexplored by mainstream tourism and is considered a secretive industry. Hand weaving is a significant part of the country's economy, with over 200,000 handloom weavers currently employed in the sector. The Dorze weavers, in particular, are known for their intricate weaving art using simple looms, such as pit looms and ground looms, and even constructing their own looms from bamboo and eucalyptus wood. To produce successful hand-woven fabric, modern advancements in warping, harnesses, weft insertion, and loom components are crucial.

Ethiopian traditional clothing displays unique patterns and motifs that are influenced by over 80 nationalities, showcasing religious and ethnic identity primarily through weaving and embroidery. Traditional cotton-based tourism fashions highlight cultural identity and national pride, promoting weaving technology and enhancing global competitiveness. Traditional weavers use simple looms to make elaborate weaving art such as Tibeb, a decorative border. Ethiopians wear hand-woven textile articles like bulko, Gabi, Netela, Kuta, and qemis to represent their ethnic identity, featuring unique motifs and designs. These traditional clothes reflect ethnicity, religion, and cultural values, showcasing Ethiopia's diverse heritage during holidays.

References

1. Riello G (2013) Cotton: the fabric that made the modern world. Cambridge University Press
2. Ethiopian tradition of handloom weaving – Shimena https://www.shimena.net/viewed on 22/02/2023
3. https://www.crosspolynations.com/2019/09/14/ethiopian-cotto

4. https://store.hamlin.org.au/blogs/journal/ethiopian-textile-weaving-history-future viewed on 21/06/2022
5. Adamu BF, Sitotaw DB (2021) Selected Indigenous Ethiopian Traditional Weaving Designs for Beauty and Diversity of Ethiopian People. TEXTILE 19(4):386–392
6. Mathiszig L (2014) The Dorze Weavers of Ethiopia. Textile 12(2):180–187
7. The History and Future of Ethiopian Textile Weaving, October 5 2021. https://store.hamlin.org.au/journal/et viewed on 22 02/2023

Chemical Processing of Cotton Textiles

Tamene Wagaw, Melkie Getnet, and Dehenenet Flatie

Abstract Cotton is the preferred option in the clothing industry because of its softness, absorbency, comfort, and resistance to chemicals. However, the production process is quite intensive, involving a series of preparatory, colouring, and finishing processes before the final product is ready for use. From cultivation to the final stages, cotton is the most water and chemical-intensive fibre. This chapter provides a concise overview of the three main categories of cotton processing. The preparation section covers the chemistry and structure of cotton fibres, as well as the sequential sub-preparatory processes that are used, including singeing, desizing, scouring, bleaching, and mercerization. The section also discusses process shortening through recipe formulation due to the high cost and negative environmental impact of the processes. Environmental issues related to the preparatory processes are also addressed. The dyeing, printing, and finishing sections follow a similar pattern and include information about the Ethiopian scenario.

1 Introduction

For over 4000 years, humans have been using cotton, which can be seen through archaeological evidence. It is a favoured choice in the apparel industry due to its softness, high absorption rate, and comfortable feel. Additionally, it has adequate strength and stability against chemicals. Cotton is a plentiful natural fibre made up of almost pure cellulose. Despite facing tough competition from synthetic fibres like polyester, cotton remains a significant and will remain a vital textile fibre. Its global

T. Wagaw (✉) · M. Getnet · D. Flatie (✉)
Department of Textile Chemical Process Engineering, Ethiopian Institute of Textile and Fashion Technology, Bahir Dar University, P.O. Box 79, Bahir Dar, Ethiopia
e-mail: Tamene.Wagaw@bdu.edu.et

D. Flatie
e-mail: Dehninet.Filatie@bdu.edu.et

M. Getnet
e-mail: tadesse@hs-albsig.de

K. Murugesh Babu et al. (eds.), *Cotton Sector Development in Ethiopia*, Textile Science and Clothing Technology, https://doi.org/10.1007/978-981-99-9149-5_12

consumption has risen in the past and is projected to grow by 3.1% from 2019 to 2024.

Coloured cotton products intended for informal wear undergo either dyeing or printing. The dyeing process typically employs anionic dyes such as direct, reactive, vat, sulphur, azoic, and others. Reactive dyes, which form primary bonds with fibre molecules, are the most commonly used due to their excellent fastness, wide colour range, brilliance, and relatively easy application. However, the low substantivity of reactive dyes and their hydrolysis results in inadequate fixation, necessitating excessive use of salt to enhance exhaustion and copious amounts of water to remove hydrolyzed dyes, leading to significant environmental damage.

Chemical processing of cotton can be broadly divided into preparatory [pretreatment] procedures that are aimed at eliminating natural and added impurities to enhance the ease of processing, appearance, and comfort; coloration [dyeing and printing] for aesthetic enhancement; and finishing [final finishing] to introduce non-inherent functional properties. This chapter is organized according to this classification, with Sects. 12.2–12.5 dedicated to preparatory procedures, Sects. 12.6–12.12 covering coloration, and finally Sects. 12.13 to 12.15 discussing finishing. The three categories of processes are explored in terms of their basic principles, sub-processes and their objectives, advancements made, environmental concerns, current conditions, and future prospects for Ethiopian mills.

This section will serve as a quick reference, particularly on the fundamental concepts of preparatory, dyeing, and finishing techniques, as well as on advancements made in easing effluent loads. Additionally, it can serve as an excellent resource on the current state of cotton processing industries in Ethiopia. The section is anticipated to be beneficial for scholars, educators, researchers, industry people, and other experts operating in the cotton processing sectors.

2 Chemistry and Structure of Cotton Fibre

Typical raw cotton fibre that has been ginned and mechanically cleaned is made up of approximately 95% cellulose molecules [71, 76, 83, 86]. Cellulose is a polymer composed of up to 20,000 glucose residues linked β-1 $\rightarrow$ 4 (according to [36, 52]). Once impurities and naturally occurring non-cellulosic materials are removed through preparatory processes, the cellulose content makes up over 99% of the fibre [71]. Considering the physical arrangement of cellulose in the fibre, cellulose content is primarily found in the secondary cell wall of cotton fibre. The major non-cellulosic components of cotton fibre, such as proteins, pectin substances, ash, wax, sugars, and organic acids, are mainly located in the cuticle and primary cell wall [29, 38].

When examining the structure of mature cotton fibres, it is apparent that they are composed of distinct components such as the cuticle, primary wall, secondary wall, and lumen. The cuticle is comprised of materials like waxes, pectins, and proteins that function as a protective coating against water and other elements. To prepare the fibre for chemical processing, this layer must be removed through scouring. The primary wall is a thin layer that acts as a membrane during plant cell development and is primarily made up of pectin substances with some cellulose present. The main part of the cotton fibre, the secondary wall, is composed of concentric layers of pure cellulose deposited at angles of 70 to 80 degrees to the fibre axis. These layers have fibrils that spiral in both S and Z directions. Finally, the lumen is a hollow canal that runs the length of the fibre, created when the protoplast dries up, and the lumen naturally collapses, leaving a central void or pore space in each fibre that contains protoplasmic remains [31, 33, 38, 54].

The fibre structure contains capillary spaces or pores of varying sizes between the fibrils of different sizes in each of the fibre components mentioned above. This is responsible for cotton's ability to absorb water and chemicals and its wickability [33, 71, 75]. Another significant aspect of cotton fibres is the orientation and density of packing of cellulose molecules. The crystalline region, which is inaccessible to water and chemicals, constitutes approximately 73% of the cellulose in the cotton fibre, with the remaining 27% being the accessible amorphous region [57, 113]. Cotton is regarded as the "king of fibres" or "white gold" and is the most essential natural textile fibre globally, owing to its morphology, microstructure, and associated properties of processability and comfort [10].

According to [71], natural impurities can make up to 12% of gray cotton, with the majority situated on the outer surface, cuticle, and primary cell wall, as previously mentioned. Since these impurities are hydrophobic and located on the surface, they can significantly impact chemical processing, aesthetics, and comfort. Therefore, pretreatment processes are crucial to remove these impurities and obtain consistent, high absorbency, and improved whiteness.

3 Pretreatment (Preparatory) Processes

The pretreatment process of cotton involves a series of procedures that aim to eliminate both the natural and added impurities that can hinder the intended chemical processes that enhance the performance and aesthetics of the fabric. Gray cotton material contains added impurities, including oils, size materials, and other residues from the plant growth and handling, which are intentionally added to facilitate spinning and weaving processes. Unfortunately, most of these impurities are hydrophobic, meaning they are not water-soluble and can compromise the comfort and further chemical processes. Therefore, the use of harsh chemicals that can dissolve or convert them into soluble form is necessary, which raises concerns about its environmental impact, presenting one of the challenges of cotton use [14]. To remove the impurities, the process involves breaking down the large insoluble starch in size materials into smaller, soluble molecules, converting insoluble waxes and fats into soap through saponification; emulsifying oils, and converting insoluble salts into soluble salts. All of these require appropriate chemicals, technology, and conditions. Consequently, the solubilized impurities can be dissolved in water and removed with it [65,

114]. The differences in the nature of impurities and the principles involved in their removal require the use of different chemicals, which lead to different and sequential processes.

3.1 Process Types and Their Sequences

The presence of impurities of various types, both inherent and added, in gray cotton fabric necessitated a variety of pretreatment operations, with the entire pretreatment processes for woven gray cotton fabric comprising singeing, desizing, scouring, bleaching, and mercerization [65]. The impurities to be removed, the chemicals to be utilized, and the concepts involved in impurity removal differ across different methods. In general, these activities are carried out independently and in prescribed sequences, unless changed procedures or technologies are employed, as will be detailed later.

Which of the pretreatment steps can be skipped depends on the type of impurities to be removed, the chemical processes to be used, the intended end use of cotton textiles, and so on. As a result, designing and selecting cotton pretreatment processes need a thorough understanding of cotton impurities, removal principles, cotton wet process requirements, and so on. The highlights of the procedures and the possibility of combining these different procedures are provided below.

3.2 Singeing and Desizing of Cotton

Singeing is the process of removing protruding fibres from fabric surfaces with the aim of enhancing the optical, end use, cleaning, printing, and other qualities of fabrics. Singeing technologies include passing the fabric over a gas flame or hot copper plates at optimal speeds and other conditions to burn away protruding fibres and then through a quenching liquid to avoid fire. The singeing process is typically coupled with desizing, with the two processes carried out in the same arrangement but at distinct phases, but using the same desizing and quenching medium [65, 114].

Desizing refers to the process of eliminating the starch substance incorporated into the warp yarns prior to weaving, to enhance their durability. The selection of chemicals to be employed in this procedure is dependent on the chemical constitution of the size materials. The typical size materials utilized for cotton yarns consist of natural and modified starch, as well as polyvinyl alcohol [1, 114]. In addition to the fundamental size material, the sizing solution may also contain various additives like lubricants, wetting agents, antifoaming agents, etc. that serve diverse objectives. The potential desizing techniques for cotton fabric include rot steeping, acid desizing, oxidative desizing, and enzymatic desizing, with the latter two being particularly significant.

As the name suggests, rot steeping involves the development of microorganisms through the immersion of the substrate in water to break down the size materials into a soluble form. On the other hand, the use of acid, oxidative, and enzymatic desizing methods relies on acids, oxidizing agents, and enzymes, respectively, to break down the size materials [114]. The following section provides more information on the primary desizing method used for removing starch.

Enzymatic Desizing: Enzymes are macromolecules that accelerate chemical reactions and are recognized for their specificity and gentler conditions. Their selective influence on only the desired chemical under mild conditions makes them exceptionally significant catalysts in reactions, particularly in intricate chemical reaction systems such as those found in cotton processing. In the desizing process, enzymes are extremely useful in eliminating starch, which is the common sizing material for cellulosic textiles. Various types of enzymes are available for different types of starch, such as amylase for breaking down amylose and amylopectin, and maltase for maltose. There are two different types of amylase enzymes that can act on different positions of starch: alpha amylase randomly breaks down the chain, causing a rapid reduction in molecular weight and facilitating complete removal. Beta amylase, on the other hand, starts at a chain end, eliminating one maltose unit at a time [7, 65].

Enzymatic desizing of cotton typically occurs under specific conditions, including a concentration of 3.37% alpha-amylase enzyme, a neutral pH, and a temperature of approximately 55 °C, for a duration of around 50 min [121]. Although enzymes generally work under mild conditions, with neutral pH and temperatures ranging from 40–75 °C, and without the use of harsh chemicals like reducing or oxidizing agents, certain enzymes may require the addition of salts to enhance their activity. For example, pancreatic amylase is ineffective without salt supplementation [21, 114]. If the optimum conditions are not met, enzymes may become deactivated or denatured. Compared to oxidative desizing methods, enzymatic desizing reduces fibre degradation, environmental impact, and associated effluent treatment costs.

3.3 Scouring of Cotton

Natural impurities such as oils, fats, waxes, minerals, leafy matter and motes present in cotton fibres can hinder dyeing and finishing processes if not eliminated. Fabrics may also get contaminated with additional oils, greases used for lubrication, mill dirt, temporary fabric markings and such during production. Scouring, which is a chemical process, is employed to remove these impurities and reduce their amount to achieve consistent and reproducible results during dyeing and finishing operations [65].

After desizing the warp yarns of gray woven cotton fabric to remove starch material, scouring follows to eliminate both inherent non-cellulosic materials and added impurities. Enzymatic desizing may not affect most of these impurities. Most of the impurities are expected to be removed during scouring, except for some residues and pigments. These may be removed, denatured or compensated for their light absorption at the blue region during the chemical and physical bleaching processes.

Both fats and waxes are esters of fatty acids, which are long-chain alkyl carboxylic acids. Fats are triesters of glycerine, while waxes are monoesters of fatty alcohols. These substances, along with impurities such as oils and greases, are hydrophobic and insoluble in water. To remove motes and leaf matter, they must be broken down and washed away. During scouring, chemicals are used to saponify waxes, emulsify oils, and convert insoluble salts into soluble ones through exchange reactions [114].

The textile industry uses various scouring methods, including alkaline, emulsion, and solvent-based methods. Scouring can be done in batch, semi-continuous, or continuous processes, and can use different technologies such as kier boiling or jigger type. Alkaline (NaOH) is the most commonly used scouring agent, and is typically used at temperatures between 90–100 °C [98].

A typical scouring recipe includes 2–4 g/l NaOH as a base scouring chemical for swelling and saponification, 1–3 g/l sequestering agent for softening hard water, 1–2 g/l wetting agent, and 1–3 g/l detergent. In extreme cases, solvents may also be used. NaOH converts fats and waxes into soap, while alkali solubilizes pectins and related substances and acts as a swelling agent. Sequestering agents remove minerals and heavy metals, while amino acids or proteins are solubilized by producing corresponding sodium salt. Oils are emulsified using an emulsification system [65].

3.4 Mercerization of Cotton

Mercerization, named after John Mercer who accidentally discovered the process, is the treatment of cellulosic textiles, specifically cotton, with concentrated solutions of NaOH ranging from 19 to 26% [114]. This treatment improves the strength of the fibre, its dimensional stability, lustre, dye affinity, and chemical reactivity [71]. Causticization is carried out with less concentration of NaOH, ranging from 10 to 16%, when the intention is only to enhance dye ability instead of mercerization [122]. When cotton fibres absorb NaOH solution, they swell, breaking hydrogen bonds and weakening van der Waal forces between cellulose chains. When the NaOH is removed, the expanded and freed chains rearrange and reorient, forming new bonds in the reorganized state. Tensionless mercerization causes the cotton fibre to swell, thickening the cross-section and shortening the length, resulting in denser, stronger, and more elastic fabric. However, if the process is carried out under tension, the coiled shape of the fibre is straightened, and the lumen almost disappears. The fibres become permanently round and rod-like in cross-section, and the fibre surface becomes smoother. The surface area decreases, reducing light scattering, and hence improving the fibre lustre [114].

Mercerization can be done at room temperature or hot, but hot mercerization allows better penetration of the alkali into the fibres. In cold mercerization, fast swelling increases the outer edge density of the fibre swiftly, restricting further penetration of NaOH solution, and the core of the fibre remains less mercerized. The viscosity of the cold caustic solution makes the penetration into the fabric more difficult. Therefore, mercerization is usually carried out at 60 °C to 70 °C for about 50 s to obtain optimum improvement in properties. The NaOH must be washed out after the fabric cools [65].

3.5 Bleaching of Cotton

Unprocessed cotton materials that have been desized and scoured have an off-white appearance and may even appear slightly yellowish due to the presence of natural pigments and residual impurities. However, the level of whiteness in cotton plays a crucial role in determining the colour of dyeings or printings. When dyeing light to medium shades, it is essential to have a consistent white base fabric. This is because it is much simpler to achieve shade matches on a uniform white background than on a fabric that varies in the amount of yellow. Additionally, cotton is often used in its white form, as seen in the traditional clothing of Ethiopia. Thus, whiteness is of great importance both technically and commercially.

To achieve white cotton, a chemical process called bleaching is employed. This process involves using bleaching agents like hydrogen peroxide or sodium hypochlorite to destroy the colouring matter and residual impurities while minimizing fibre degradation. Although the complete removal of colouring matter is not practical, physical bleaching can be used to incorporate optical brightening agents into the fabric, compensating for the blue energy absorbed by the residual impurities.

Hydrogen peroxide is currently the primary chemical bleaching agent for cotton. Its bleaching mechanism is based on the decomposition of hydrogen peroxide in water into hydrogen ions and per hydroxyl ions, which act as the active bleaching agent. However, the effectiveness of peroxide bleaching is influenced by various factors like pH, temperature, water softness, and time. Therefore, optimizing these parameters and using appropriate stabilizers is crucial for uniform bleaching and minimal fibre degradation.

4 Combined Pretreatment of Cotton

Performing desizing, scouring, and bleaching of cotton separately results in a significant consumption of energy and water, as well as the generation of a substantial amount of effluent. This places a great deal of pressure on the ecosystem and/or the cost of effluent treatment. Furthermore, it prolongs the processing time and increases fibre degradation. As a result, there is an urgent need to shorten the processing cycle by combining the desizing and scouring, scouring and bleaching, or desizing, scouring, and bleaching processes. Although various possibilities have been described [65] that involve enhancing the removal of impurities through the use of chemical recipes and different technologies, such as shaking, it appears that the combination of all three processes is not yet well-established. Other relevant works have also been published [81, 94, 123].

5 Environmental Issues Related to Cotton Pretreatment

Around 20% of the world's industrial water pollution is believed to stem from textile treatment and dyeing, which involves the use of roughly 8,000 synthetic chemicals. Although cotton is the most important fibre in the textile industry, it is also responsible for the largest amount of water consumption and pollution [24]. The preparatory, coloration, and final finishing processes are responsible for this, with preparatory processes such as desizing, scouring, bleaching, and mercerization having a critical impact on the ecology due to their use of large amounts of alkali, peroxides, emulsifiers, wetting agents, detergents, and significant quantities of rinsing water [106, 56]. In addition to the chemicals utilized in these processes, impurities from the fibre are removed into the water, resulting in high levels of BOD (Biological Oxygen Demand), COD (Chemical Oxygen Demand), DS (Dissolved Solids), SS (Suspended Solids), and other contaminants [91]. This environmental impact, in conjunction with the more stringent global environmental concerns, has prompted the sector to seek alternative processes to reduce the ecological burden.

6 Advanced Technology and Proccsscs of Cotton Pretreatment

Conventional preparatory processes for cotton have been facing challenges due to their significant water and energy usage, as well as their negative impact on the environment. Professionals in the field have been searching for alternatives that are more sustainable, with lower water and energy consumption and less ecological impact. As a result, various alternatives have been developed, such as water reuse, the replacement of harsh chemicals, shortening processes through combinations, and the use of high-tech to aid in preparation. The following is a summary of some of these alternatives.

All cotton products, whether woven or knitted, require pretreatment processes such as desizing, scouring, and bleaching to enable effective dyeing and finishing. These processes discharge a significant amount of pretreatment and wash liquor. The added chemicals are not fully reacted or consumed during the pretreatment operations. Reusing the liquor can save and enhance the effective use of both water and chemicals. Numerous research outputs are available on the possibilities of this.

The use of biotechnology, including enzymes from renewable resources, has enabled the preparation of cotton to be more environmentally friendly, with reduced energy consumption and the avoidance of harsh chemicals. Process parameters have also become milder. Enzymatic desizing of cotton has a long history, with the use of amylase enzymes for the removal of starch sizes being one of the oldest enzyme applications. Nowadays, scouring, bleaching, and combined processes are made possible with the help of enzymes, at least on a research level.

Another emerging technology in cotton pretreatment is the use of plasma, ultrasonic, microwave, ozone, and other energies to facilitate the removal of impurities. The principles involved in the application of these technologies include physical etching and chemical reactions at the surface, removing surface material, added and natural impurities, and forming volatile products. The cavitation and heating effects of ultrasonification, rapid, uniform, efficient, and easily penetrating microwave heating, and the bleaching effect of ozonation are successful because of the high oxidation potential of ozone. These technologies facilitate efficient removal of impurities with lower chemical, energy, and water consumption and with minimal impact on the environment.

7 Coloration of Cotton Fibres

7.1 Common Colourants Used for Colouring Cotton

There exist several categories of dyes that can be utilized for the purpose of colouring cotton. Nevertheless, sulphur (29%), direct (27.4%), vat (19.4%) and reactive (9.7%) dyes account for over 85% of the dyes that are employed for the dyeing of cotton [8, 18, 78, 79]. Direct cotton dyes have an inherent affinity for cotton and other cellulosic fibres, and they offer a relatively wide spectral range at an affordable price. Normally, they are employed in an aqueous solution in the presence of an electrolyte, such as NaCl (Sodium Chloride) or Na_2SO_4 (Sodium Sulphate) [70]. However, their fastness properties are not very good, especially with regards to wet fastness, and therefore, they require after-treatment to enhance their fastness performance [78, 79]. Sulphur dyes are widely used for the purpose of dyeing cotton fibres [53]. They are produced by melting particular nitro substituted phenols and arylamines with sulphur and sodium sulphide [78, 79]. The fastness properties of fabrics dyed with sulphur dyes are inferior to those of vat and reactive dyes, but superior to those of direct dyes. Due to the simplicity of their application and the wide range of products available, such as garments, furnishings, linings, and carpets, sulphur dyes are a popular and cost-effective category of dyes [53, 90].

Reactive dyes, as the name implies, react with cotton fibre in an alkaline medium [78, 79]. They are soluble anionic dyes which, in solution, are repelled by the negatively charged surface of the cotton fibre [66]. They can provide a wide range of colours with good brightness and good washing fastness. However, thorough washing of the dyed samples is required, and 50% of the cost of dyeing is related to washing off and effluent treatment [111]. Vat dye application involves a complex process of vatting, absorption, and oxidation for cotton [20]. However, cotton vat dyed textiles display exceptional fastness properties [99]. Unfortunately, these dyes are rather expensive and can only provide a limited colour range with very few bright colours [67].

7.2 Dyeing of Cotton Fibres

A dye is composed of two types of groups: chromophore and auxochrome. The chromophore is a colour bearing group and is responsible for the primary coloration while the auxochrome intensifies it [115]. When dyeing cotton, hydrogen bonds are formed due to the presence of oxygen and hydrogen groups in the dye's structure. Additionally, van der Waals forces are generated between the dye particles and fibres. These forces allow the dye molecule to be retained by the fibre. Cotton has many OH groups, which interact physically with the dye molecule during the diffusion process. When the van der Waals force attracts the dye molecule to the fibre, it gets anchored instead of forming an actual bond. This phenomenon is observed when cotton is dyed with vat, direct, and sulphur dye [67, 99, 117]. During the dyeing process, both fibre and dye are attracted to each other by chemical forces of action, leading to dye attachment on fibre through chemical bonding. The bond can be either ionic or covalent.

Reactive dyes are the most commonly used dyes due to their excellent fastness properties and ease of application [110]. The number and nature of the functional group present in the reactive dye molecule affect the dyeing behaviour. Dyes containing monochlorotriazine (MCT) and sulphatoethylsulphone (SES) moieties have higher fixation efficiency than dyes containing only one type of reactive group (i.e., a vinyl sulfone precursor). Cotton natural dyeing is problematic due to the low affinity and weak fastness properties of natural dyes on cotton fibre [49, 48]. Several studies have used metal ions, ultrasound energy, enzymes, mineral mordants, and biomordants of mordant to enhance the dyeability of cotton fibre with different natural colourants, making them insoluble in water [28, 119].

8 Progress in the Technology and Processes of Cotton Dyeing

Conventional dyeing of cotton with water necessitates vast amounts of water and chemicals, leading to a significant number of pollutants that have catastrophic environmental impacts. Recently, alternative methods for dyeing cotton have emerged that utilize non-aqueous reactive dyeing systems, such as solvent-assisted dyeing systems, reverse micellar dyeing systems, dye/solvent suspension systems, and non-nucleophilic solvent-assisted dyeing systems [112].

The textile coloration industry has recognized the use of solvents as a viable alternative to reduce effluent generation and avoid environmental costs. Over the past few decades, solvents such as tetrachloroethylene (PCE), trichloroethylene (TCE), and benzyl alcohol have been commonly used for solvent-assisted dyeing. The production and use of chlorinated solvents are linked to the growth and development of the synthetic and organic chemicals industry [32]. PCE offers numerous benefits in dyeing compared to water, such as lower energy requirements for heating and vaporization, faster fabric drying, quicker wetting, less swelling, lower risk of mechanical deformation, and lower cost [77, 129].

Reverse micelles are self-assembled spherical nanoscale aggregates formed by certain surfactants in non-polar media [72, 93]. They consist of a polar liquid phase encapsulated by a monolayer of surfactant molecules and evenly distributed within a non-polar oil phase (water-in-oil microemulsions) [116]. They can be used as dye carriers to facilitate the dyeing of textile fibres in non-aqueous media. For example, water-soluble dyes can be dissolved in non-aqueous media using a reverse micellar system, which can solubilize small amounts of water in the interior of a micelle to provide a stable and aqueous microenvironment, known as a "water pool," in non-aqueous media [103].

Most investigations into reverse micellar systems have been carried out using hydrocarbon solvents like heptane and hexane, which are hazardous to the environment and human health. Hence, the utilization of eco-friendly solvents in the dyeing process of cotton with reactive dyes has emerged as another research area. An effort was made to dye cotton in a suspension of reactive dye/decamethyl cyclopentasiloxane (D5) system. The findings revealed that the dye uptake in this system was over 99% in the absence of electrolytes, and the fixation rate was 6% higher compared to a conventional water-based dyeing system [73]. Further, an attempt was made to stabilize the use of the suspension system of dye/decamethyl cyclopentasiloxane (D5) for reactive dyeing of cotton fibre by using HPIS as a dispersant [41].

Another emerging solvent-assisted dyeing system involves the use of a non-nucleophilic solvent system for the reactive dyeing of cotton in a non-aqueous medium. It was reported that N,N-dimethylacetamide (DMAc) was used to preswell cotton at a high temperature (150 °C). The dyeing was carried out using a 40/60 mixture of DMAc and dimethyl carbonate (DMC), which is a co-solvent for dye exhaustion, and potassium carbonate was used for dye fixation. The results showed that excellent exhaustion and colour yield were achieved with the use of monochlorotriazine (MCT) reactive dye, and the application of a 10-cycle solvent-dyeing sequence can reduce organic waste and the use of electrolytes by over 99% [25]. In another experiment on the reactive dyeing of cotton in a non-nucleophilic solvent system, DMSO was used to swell cotton and solubilize the MCT dyes, with DMC utilized to promote dye exhaustion, and potassium carbonate used to promote dye fixation [26].

Presently, natural textile dyeing has garnered significant attention in research and textile industries due to the growing concerns about water pollution, sustainability of raw materials and processed products, biodegradability, and eco-friendly characteristics. Natural dyes can be dyed using conventional exhaustion methods [4] or non-traditional dyeing techniques such as ultrasonic, microwave, pad dyeing, High temperature High Pressure (HTHP) dyeing, contact dyeing, and other optimized approaches [108].

9 Current Scenario and Future Prospects of Cotton Dyeing by Ethiopian Mills

The beginning of Ethiopia's textile history dates back to 1939 in the city of Dire Dawa. The textile sector involves various processes like cotton production, ginning, spinning, weaving, knitting, dyeing, finishing, and garment manufacturing. Currently, there are 38 textile mills, most of which are government-owned and primarily produce workwear apparel for the local market [124]. As the textile industry is evolving, so is the pollution generated from textile processing. According to a recent report from the journal 'Eco Textile News,' the effluent discharged by Ethiopian Textile Mills is "highly contaminated" and exceeds the limits set by the National Environmental Protection Agency [84].

10 Environmental Issues in Cotton Dyeing

The textile sector is a major contributor to environmental pollution worldwide due to its high-water usage (Chen and Burns 2016). Currently, the process of dyeing cotton fabrics with direct and reactive dyes is both water- and energy-intensive and generates pollution. Cotton fibres carry negative surface charges in water, which repel anionic dyes and slow down exhaustion. To overcome this, high concentrations of electrolytes (such as sodium chloride and sodium sulphate) are added to the dye bath, along with extended dyeing times at elevated temperatures. These electrolytes serve to overcome the negative charges on the cotton fibre and reduce dye solubility. Multiple rinses and after-washes are required to remove unfixed dyes, and a polymeric "fixation" agent is often used to improve wet fastness properties. As a result, a significant volume of wastewater, containing dyes and chemicals, is discharged from cotton dye houses [51].

The use of reactive dyes is predicted to increase by 50% between 1992 and 2004. Manufacturers are now focusing on addressing environmental issues such as effluent colour and minimizing chemical usage associated with reactive dyes. Due to the negative effects of conventional water-based reactive dyeing on the environment, including the need for large amounts of water and chemicals and the generation of significant effluent, a non-aqueous solvent-assisted approach is an alternative way to reduce pressure on the environment [112]. Chavan has critically reviewed environmentally friendly dyeing of cotton with reactive dyes, highlighting innovations in dyes and dyeing processes for high dye bath exhaustion and fixation to reduce total effluent colour. Some approaches include using bifunctional reactive dyes with high exhaustion and fixation properties, low-salt reactive dyes, machinery developments for dyeing at a low liquor ratio, pad troughs with reduced volumes, replacing urea with dicyandiamide, and implementing the econtrol process [23].

Reducing agents such as sodium hydrosulphite and alkalis such as sodium hydroxide are commonly used to apply vat dyes. However, some byproducts of sodium hydrosulphite are acidic and require excessive amounts of sodium hydroxide to neutralize them. These byproducts, including sulphur compounds like Na_2S and NaHS, can pollute the air by forming H_2S. Additionally, salts of sulphur such as sulphate and sulphites (Na_2SO_4, $NaHSO_4$, $Na_2S_2O_3$) can contaminate sewage, lower its pH, and corrode concrete pipes. To address these issues, researchers have explored alternative eco-friendly reducing systems such as electrochemical reduction and the use of organic reducing agents like hydroxy acetone, iron pentacarbonyl compounds, and iron (II) complexes. Electrochemical reduction involves direct contact between the dye and electrode, but a reducing agent must be added to ensure stability of the reduced dye bath. However, this method requires a higher amount of dyestuff to produce a specific shade. Hydroxy acetone (CH_3–CO–CH_2OH) has been reported as an effective reducing agent for vat dyes like indigo. This system is biocompatible and results in 20% higher indigo uptake with lower consumption of assisting chemicals [12, 105].

11 Printing of Cotton Products

Cotton printing is the application of colour to cotton cloth in predefined patterns or motifs. The colour is bound with the fibre in correctly printed textiles to resist washing and abrasion. Printing is similar to dyeing, but unlike dying, when the entire cloth is evenly covered with one colour, printing applies one or more colours to the fabric in specific areas and in finely defined patterns. Water, dye or pigment, thickening agent, and other textile auxiliary agents are used in the printing of the paste in a wet process [34, 101].

12 Advancements in Printing Technology of Cotton Products

Presently, numerous printing technologies have undergone rapid developments. Among these, 3D printing is a rapidly emerging technology for additive manufacturing, offering cost-effective and flexible options for product development and production. This technology can also be employed in textile production to apply 3D structures on fabrics as an additional process [68]. The 3D-printing process involves depositing material layer by layer to build a product, eliminating the need for adhesives and providing the ability to decorate the fabric's surface [22, 45, 118].

Wu et al. have developed cotton-containing 3D-printing wires through 2D-braiding technology, which includes cotton, low-melting polyester (LMPET), and TPU. This fabric is produced by 3D printing and exhibits good flatness while containing a significant proportion of cotton powder. Furthermore, 3D-printed cotton-containing fabric displays greater softness, abrasion resistance, and tensile properties than fabrics printed with polylactic acid [128].

Ink-jet printing is a rapidly growing imaging technology that offers numerous benefits over conventional printing methods such as rotary and flat-bed screen printing. It provides unique advantages, including simplicity, reduced production costs, lower effluent waste, lower water and energy consumption, unlimited design combinations, and the ability to produce innovative personalized finished articles with creativity in design patterns and cleanliness in production for cotton products [40, 46]. Sodium alginate is a crucial thickener used for preparing printing paste for digital ink-jet printing due to its solubility and excellent stability even after high-temperature fixation treatments. Similarly, chitosan is a novel biomaterial widely used in textile printing due to its biocompatibility, biodegradability, and non-toxicity. Previous research has reported that chitosan can improve colour yield, colour fastness, and anti-bacterial function in textile printing [64]. Alkali treatment of cotton fibre increases its contact area with dyes, thereby enhancing its wettability and promoting ink wicking into the fibre. The cross-section of printed fabric indicates that the ink does not entirely penetrate the back of the fabric, enhancing the colouring ability of mercerized cotton fabric [130]

The utilization of sublimation thermal transfer printing on cotton is not commonly employed due to the absence of disperse dyes' attraction to cellulose fibres. However, there is a chance to employ disperse dyes on cotton commodities through sublimation printing technique. By altering the surface of cotton textile using POLAPPRET PU-S, it can enhance the longevity of prints and at the same time, retain the colour quality as anticipated. This can introduce a fresh outlook for digital transfer printing on cotton products [13].

13 Environmental Issues in Cotton Printing

The process of colouring textiles through printing involves a wet process that uses water, dye or pigment, thickening agents, and other auxiliary agents for preparing the printing paste. This leads to significant pollution of textile effluent due to unfixed colour and washed-off elements from the paste [2]. Although natural colourants were once popular for textile applications, the advent of synthetic dyes and pigments led to their decline. Synthetic dyes are preferred due to their ease of production in large quantities, affordability, and ability to produce a wide range of colours with very good colour fastness. However, the production of synthetic dyes has several environmental drawbacks.

Unlike synthetic dyes, natural dyes are mostly renewable and sustainable. Various natural materials can be used in printing cotton fabric through techniques such as direct printing with madder, ink-jet printing using reactive dye and discharge, and resist printing. The bio-mordant process increases the colour strength value for both printed formulations. Organic salts and alkali are more effective than traditional urea/alkali printing. The use of organic salts in reactive dye cotton printing is cost-effective and results in high-quality prints with sharp outlines, low penetration, substantial dyestuff savings, and low wastewater load. Digital textile ink-jet printing offers high-quality and scalable processing. Optimum recipe for discharge printing on cotton fabric involves 3% eco-friendly reactive dye, 25% gum tragacanth, and 20% Rongalite C. Resist printing on processed cotton fabrics is enhanced by adding chitosan to the resist-printing paste. The optimal chitosan concentration is 1.6%. Using tartaric acid printing agents yields better results than using each agent separately [34]. Water-based ink-jet inks for digital textile printing can be prepared using four natural dyes: annatto, cutch, pomegranate fruit rind, and golden dock. The inks produce prints that pass wash, light, and rub fastness tests and have colour consistency and fastness comparable to those of synthetic dyes. This opens up the possibility of producing environmentally friendly ink-jet inks using natural dyes for the digital printing of cotton through suitable printing techniques [101].

Georgios and colleagues conducted a study to explore the potential of using natural pigments as a replacement for synthetic pigments in the production of eco-friendly cotton prints. They aimed to develop a standard printing protocol using natural dyes/pigments that meet new environmental requirements. To achieve this, they used 11 natural pigments and employed the flat-bed screen printing technique to print cotton fabrics. Among these pigments, pomegranate fruit, indigo, cutch, lac dyes, and annatto showed excellent fastness properties. These findings could lead to the creation of a new range of natural and eco-friendly dyes/pigments that can be applied using traditional dyeing and printing methods on natural fibres [102].

14 Finishing of Cotton Fabrics

Cotton processing is completed with the crucial step of final finishing, which forms a vital component of the entire process. Despite its significance, cotton finishing is currently beset by several environmental challenges. The traditional process of finishing cotton fabric involves singeing, designing, scouring, bleaching, and mercerization (for specific applications), followed by drying. However, for further enhancement of its properties, cotton fabric must undergo final finding. Consequently, it is mandatory to subject the fabric to various physical and chemical treatments in order to impart the necessary properties. Among the essential properties to be imparted to cotton fabric are wash and wear finish [39], repellent finish [37], and proof finishes, among others.

The final finishes to be imparted to cotton fabric are vital, as they improve its appearance, comfort properties, repellent properties, and other aesthetic and functional properties [42]. This chapter focuses on the most important final finishing processes that have been applied to cotton fabrics.

14.1 Types of Finishing of Cotton Products

In the introductory section, it was mentioned that the final finishing of cotton fabric is introduced to alter its physical and chemical properties [9], resulting in the following changes:

1. The appearance of cotton fabric is significantly improved, leading to a change in its overall look.
2. The physical properties of the fabric are modified, thereby enhancing its comfort and feel.
3. The altered properties of cotton fabric allow for its diverse applications.
4. The finishing process improves almost all physical properties, including drapability and others.
5. Specific end uses or additional applications can be derived from the finished fabric.

To achieve the aforementioned properties, scholars have classified the functional finishing of cotton fabric into various groups [104] (Fig. 12.1).

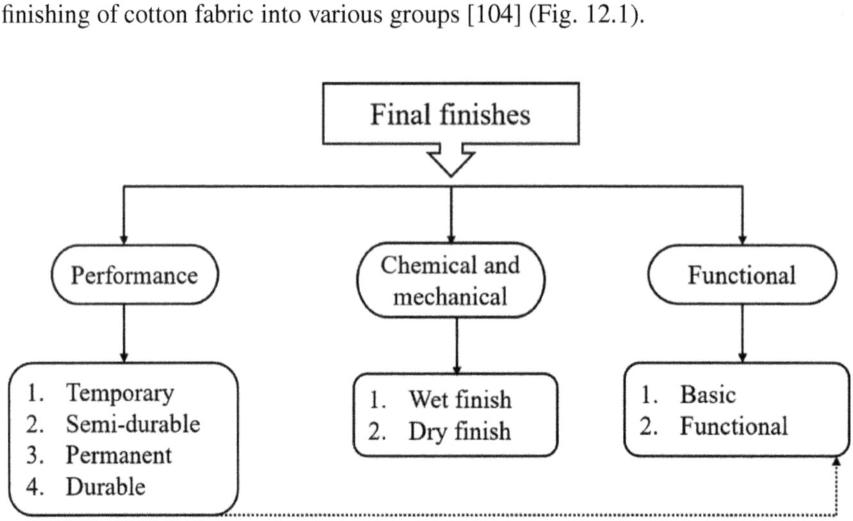

Fig. 1 Classification of finishing processes

14.2 *Common Finishing Processes*

The process of cotton finishing is utilized during the production of cotton fabric. Various final finishes are employed to modify the fundamental properties of the fabric and impart the necessary characteristics. This section delves into the latest and crucial final finishing methods that are applied to cotton fabric.

Final finishes can be categorized into two major groups based on their purpose, which are as follows:

(a) ***Basic Finishes***: These finishes are implemented to enhance the aesthetic appeal of cotton fabric. Examples of basic finishes include industrial ironing, steam ironing, and other decorative finishes that can be added to the fabric to improve its weight and appearance. It is recommended to frequently iron cotton fabric, especially after washing, to remove the hydrogen bond introduced by water, which leads to hydrogen bond cleavage and rearrangement, resulting in shrinkage [87].

(b) ***Functional Finishes***: Specific finishes are applied to improve the functional performance of cotton fabric. These finishes enable cotton fabric to serve various purposes beyond common dressing, such as UV-protection [3], fire protection [5], antimicrobial [35], chemical protection [17], and other areas that require specific finishes. The following paragraph outlines the most significant functional finishes for cotton fabric.

(i) **Crease-Resistance Finish**

The formation of wrinkles in cotton fabric is closely related to the presence of free hydroxyl (OH) ions in the amorphous regions of the cellulose structure [95]. When these hydroxyl ions come together, creases are formed. To prevent creasing in cotton fabric, the formation of hydrogen bonds between the hydroxyl groups must be either concealed or completely eliminated by creating a strong cross-link that removes water molecules. One effective cross-linking agent for this process is formaldehyde [15]. However, the use of formaldehyde has been discontinued due to its harmful effects on humans and the environment. Nowadays, there are formaldehyde-free agents available in the market for durable cross-linking, such as citric acid [95], alpha-Lipoic acid [58], itaconic acid [16], and fibroin and citric acid composites [74]. These crease-resistant finishing agents are either completely free of formaldehyde or contain low levels of formaldehyde release chemicals. Therefore, it is of utmost importance to treat cotton fabrics with crease-resistant finishes to ensure that they can be used without any creasing issues.

(ii) **Water Repellent Finish**

Cotton is a type of fabric made from cellulose. As a result of its composition, it has a natural affinity for water and is therefore considered hydrophilic. This quality makes it highly absorbent compared to synthetic fabrics. However, in situations where water resistance is important, cotton can be treated with a water repellent finish. There are several methods for achieving this effect. One of the most effective methods involves using silicone and urethane acrylates, which can be cured using UV light [37]. Ren and Zhao [96] have also explored the use of modified vegetable oils as a water repellent finish for cotton fabric. Another important finishing agent for repellent finish is fluorochemical chemical finishing [44]. Despite its excellent comfort properties and versatility, the lack of water repellency in cotton fabric can limit its technical applications. Therefore, adding a functional water repellent finish is crucial for expanding its potential uses.

(iii) **Other Functional Finishes**

Cotton is a natural fibre with inherent properties that provide high comfort levels (owing to its excellent moisture absorption), excellent conductivity (owing to its low static charge development and moisture absorption properties), and other similar attributes. However, favourable conditions for microbial production can easily damage cotton fabric. Therefore, antimicrobial finishing is crucial. Antimicrobial finishing of cotton fabric can be achieved using zinc oxide and its major composites [35], copper sulphate, polyvinylpyrrolidone, benzoic and salicylic acids [88], thyme essential oil [125], and various other methods [19, 60, 92]. Other essential properties, such as fire retardancy [78, 79], UV-protection [59], and antistatic and smart finishing, can also be applied to cotton fabric without compromising its natural properties, such as comfort.

On the basic of its degree of performance, finishes applied to cotton fabric can be classified into:

I. ***Temporary***: finishing which can be run off after the first wash can called temporary. When finishes are required for a short period of time, it can be applied temporarily on cotton fabric. An antimicrobial finish for the use and throw medical fabrics can be applied temporarily [47].
II. ***Durable Finish***: a finish imparted to cotton fabric that can be last up to the life of the fabric or that might be lost its effectiveness after several washes are called durable finish. Easy care finishes can be categories [30, 50].
III. ***Semi-durable Finish***: fabric finishes that might be stay for several washing considered as semi-durable finish. Some flame retardant finishes such as ammonium phosphates and inorganic non-phosphorus flame retardants [50]. They have poor resistance against washing.
IV. ***Permanent Finish***: Some finishes might be imparted chemically to the cotton fabric which also not only is a permanent finish but also alters its chemical structure. This category of cotton fabric finishing is under the classification of permanent finish. Chitosan based finishes which imparted antimicrobial [6, 109] can be permanent finishes. They long last until the life of the garment.

On the basis of processes involved in the finish application, there are dry and wet finishes.

(1) **Mechanical (Dry) Finishes**:

A finishing performed to improve a fabric's appearance, performance, and durability is referred to as mechanical fabric finishing. Mechanical finishing procedures change the fabric's qualities manually, in contrast to chemical finishing procedures. The various fabric names that are used, the types of dye, and the procedures utilized for fabric dyeing are only a few of the many variables that affect the type of finishing a cloth obtains. This is necessary since finishing must improve the fabric's overall appearance. Heat, pressure, and release of moisture have been involved in the mechanical finishing of cotton fabrics.

(a) **Calendering**

Although much attention has been given to chemical finishing, mechanical finishes for cotton fabric have still basic importance. Calendering is one of the most frequently used mechanical finishing for cotton fabric [43]. High pressure is used during the calendaring process to alter the characteristics of the fabric used for cotton textiles. As the technique it is more efficient with soft or open weave fabrics than with tightly woven or rigid fibres, these fabrics are typically calendared. Calendaring removes the wrinkles and crinkles from the fabric, making it feel smooth. Since cotton is highly affected by creasing, calendering is very important.

(b) **Heat Setting**

Thermal treatment Cotton fabrics and their composites undergo heat setting (thermal treatment) to stabilize their structure and release stress introduced during manufacturing [62]. This permanent finishing method is the most common in cotton fabric production and does not involve chemical agents.

(c) **Other Mechanical Finishes**

Various mechanical processes are employed during cotton fabric finishing, depending on the intended use. For instance, compacting is performed to reduce shrinkage by compressing the fabric's structure lengthwise [107]. Sueding, which involves mild raising and piling of the fabric, is also another method used [69]. The application of pressure, heat, and compaction affects cotton fabric characteristics such as stiffness, thickness, moisture content, tensile strength, resistance to abrasion, weight, and lustre, both positively and negatively [82]. Mechanical finishing processes can influence the final handling of the fabric, which is a concern for both customers and manufacturers. Although various physical finishes are still in use, development in this area is not as extensive as in chemical finishing.

(2) **Chemical Finishes (Wet Finishes)**

Chemical finishes, which involve the use of chemical agents to modify the chemical structure of cotton fabric, are a type of permanent finish. Several wet finishes have been discussed in the functional finishing section of this subchapter.

15 Advances in Cotton Finishing Technology

As previously discussed, finishing is performed on cotton fabric to modify its physical and chemical properties. Finishing improves serviceability and promotes dimensional stability, both of which help attract customers and increase profitability. However, traditional methods may not fully achieve these properties, making advancements in finishing technology necessary.

Cotton finishing, specifically chemical finishing, has seen significant advancements. One notable development in cotton fabric finishing is the use of nanomaterials [120]. According to Vigneshwaran et al. innovative and unconventional properties such as enhanced electrical conductivity, superhydrophobicity, antimicrobial properties, self-cleaning activity, UV-protection, and flame-resistance have been achieved in cotton using nanomaterials. Additionally, energy production using nanostructures on the surface of cotton fibres is being developed. Enzymes have also been used recently to finish cotton fabrics, contributing to environmentally friendly processing [100]. Enzyme processing of cotton fabric has several advantages, including no effect on strength or comfort. Plasma surface modification is an environmentally friendly alternative to conventional processing that promotes cotton fabric bulk properties [61]. Cationization is another development in cotton fabric finishing that promotes an environmentally friendly process by increasing the cotton fabric's zeta potential, allowing for salt-free reactive dyeing. Other advancements in cotton fabric finishing include microencapsulation, foam applications, wet finishing processes, and transfer methods [11].

In conclusion the share of the cotton fabric is still taking the lion share due many reasons. It can be seen that novel and more innovative technologies have been introduced in the development of finishing of cotton fabrics specially in chemical finishing streams.

16 Environmental Issues in Cotton Finishing

Conventional methods of finishing cotton fabric require a large amount of water and energy, but newer techniques have been developed that use minimal water and energy. Despite this, the majority of textile mills in Ethiopia continue to use conventional methods, which poses serious environmental concerns. The disposal of waste after cotton finishing contributes to pollution and can harm the environment, human health, and animal health. Cotton fabric finishing can leave a significant carbon footprint, leading to air pollution, water pollution, solid waste pollution, increased CO_2 emissions, greenhouse effects, maritime pollution, and fresh water pollution [89]. However, Parvin et al. suggest that this pollution can be reduced by using biodegradable chemicals, enzymes, effluent treatment, chemical chimneys, functional finishes, organic cotton production, and machinery modifications.

Durable press finishes are a common cause of environmental pollution and health problems associated with cotton fabric finishing [126]. This is due to the use of formaldehyde as a cross-linking agent, which can be harmful to health. However, low or formaldehyde-free cross-linking agents are now available [127]. Flame retardant finishes [97], durable water repellent finishes [55], and antimicrobial finishes [63] can also contribute to environmental pollution if the proper finishing agents are not used.

Overall, cotton fabric finishing can cause a range of environmental problems due to the chemicals and processes used. It is essential to prioritize health and safety measures in all industries. Without adequate monitoring, surveillance, and enforcement of laws, textile effluents can have a significant impact on the environment.

17 Summary

Cotton is the preferred choice in the apparel industry due to its softness, absorbency, and comfort for wear, along with its sufficient strength and stability towards chemicals. However, the process of preparing cotton for its end use is intensive and requires multiple steps, including preparatory, coloration, and finishing processes.

Preparatory processes serve to remove natural and added impurities from the cotton and modify its fibre structure chemically or physically to enhance processability, aesthetics, and comfort. Each step of the sequential multi-step process, such as singeing, desizing, scouring, bleaching, and mercerization, has a specific purpose. However, these preparatory processes require a significant amount of harsh chemicals, making the sustainability of cotton preparation questionable due to the environmental load and stringent environmental regulations. To address this, measures such as replacing harsh chemicals, reducing liquor ratio, using non-aqueous media, and adopting high-tech aided processes are being implemented.

Coloration processes, including dyeing and printing, are undertaken for aesthetic purposes. However, the sustainability challenge is just as critical due to the discharge of unused colour and auxiliary chemicals into the environment. Therefore, advancements are being made to alleviate environmental concerns from textile coloration houses. Finishing processes are no exception, and similar measures are being taken.

In the Ethiopian case, traditional methods are still being followed for cotton processing, which may lead to environmental devastation as the industry expands. Immediate action is necessary to address these concerns.

References

1. Ahmed T et al (2021) Evaluation of sizing parameters on cotton using the modified sizing agent. Clean Eng Technol 5:100320
2. Ahmed N, Nassar S, El-Shishtawy MR (2020) Novel green coloration of cotton fabric. Part II: Effect of different print paste formulations on the printability of bio-mordanted fabric with madder natural dye. Egypt J Chem 63(5):1669–1677
3. Alebeid OK, T. Zhao T, (2017) Review on: developing UV protection for cotton fabric. J Text Inst 108(12):2027–2039
4. Ali S, Hussain T, Nawaz R (2009) Optimization of alkaline extraction of natural dye from Henna leaves and its dyeing on cotton by exhaust method. J Clean Prod 17(1):61–66
5. Alongi J et al (2014) A comparative analysis of nanoparticle adsorption as fire-protection approach for fabrics. Polymers 7(1):47–68
6. Aly A et al (2007) Innovative dual antimicrobial & anticrease finishing of cotton fabric. Polym Plast Technol Eng 46(7):703–707
7. Aniş P, Davulcu A, Eren HA (2008) Enzymatic pre-treatment of cotton. Part 1: desizing and glucose generation in desizing liquor. Fibres Text East Eur 16(4):100–103
8. Aspland JR (1997) Direct dyes and their application. Text Dye Color AATCC 2:27
9. Bajaj P (2004) Finishing of textile materials. J Appl Polym Sci 3(3):631 659
10. Basra AS, Malik C (1984) Development of the cotton fibre. Int Rev Cytol Elsevier 89:65–113
11. Basyigit ZO (2021) Application technologies for functional finishing of textile materials. Text Funct Appl 101
12. Bechtold T et al (1997) Electrochemical reduction processes in indigo dyeing. J Soc Dye Color 113(4):135–144
13. Bemska J, Szkudlarek J (2013) Surface modification of cotton fabrics for sublimation printing. AUTEX Res J 13(3):67–70
14. Benli H, Bahtiyari MI (2015) Combination of ozone and ultrasound in pretreatment of cotton fabrics prior to natural dyeing. J Clean Prod 89:116–124
15. Blanchard EJ et al (1994) Dyeable cross-linked cellulose from low formaldehyde and non formaldehyde finishing systems. Ind Eng Chem Res 33(4):1030–1034
16. Boondaeng A et al (2021) Bio-synthesis of itaconic acid as an anti-crease finish for cellulosic fibre fabric. RSC Adv 11(42):25943–25950
17. Boufi S et al (2019) Cotton functionalized with nanostructured TiO2-Ag-AgBr layer for solar photocatalytic degradation of dyes and toxic organophosphates. Int J Biol Macromol 128:902–910
18. Broadbent AD (2001) Basic principles of textile coloration, vol 132. Society of Dyers and Colorists Bradford, UK
19. Budimir A, Bischof Vukusic S, Grgac Flincec S (2012) Study of antimicrobial properties of cotton medical textiles treated with citric acid and dried/cured by microwaves. Cellulose 19(1):289-296
20. Chakraborty JN, Chavan RB (2004) Dyeing of denim with indigo. Indian J Fibre Text Res 29(1):100–109
21. Chand N et al (2012) Enzymatic desizing of cotton fabric using a Ca2+-independent α-amylase with acidic pH profile. J Mol Catal B Enzym 83:46–50
22. Chatterjee K, Ghosh TK (2020) 3D printing of textiles: potential roadmap to printing with fibres. Adv Mater 32(4):1902086
23. Chavan RB (2001) Environment-friendly dyeing processes for cotton. Ind J Fibre Tex Res 26:93–100
24. Chavan RB (2018) Thirsty textile and fashion industry PART I: water distribution on earth and virtual water, water footprint concepts. Latest Trends Text Fash Des 2(5):249–269
25. Chen L et al (2015) Comprehensive study on cellulose swelling for completely recyclable nonaqueous reactive dyeing. Ind Eng Chem Res 54(9):2439–2446
26. Chen L et al (2015) Hydrolysis-free and fully recyclable reactive dyeing of cotton in green, non-nucleophilic solvents for a sustainable textile industry. J Clean Prod 107:550–556

27. Chen et al (2006) Environmental analysis of textile products. Cloth TextEs Res J 24(3):248–261 https://doi.org/10.1177/0887302X06293065
28. Daberao A et al (2016) Cotton dying with natural dye. Int J Res Sci Innov (IJRSI) 3(8):157–161
29. Dave H et al (2014) The removal of impurities from gray cotton fabric by atmospheric pressure plasma treatment and its characterization using ATR-FTIR spectroscopy. J Text Inst 105(6):586–596
30. Dehabadi VA, Buschmann H-J (2013) Gutmann JS (2013) Durable press finishing of cotton fabrics: an overview. Text Res J 83(18):1974–1995
31. Denham HJ (1923) VII-the structure of the cotton hair and its Botanical aspects. II-The morphology of the wall. J Text Inst Trans 14(4):T86-T113.
32. Doherty RE (2000) A history of the production and use of carbon tetrachloride, tetrachloroethylene, trichloroethylene and 1,1,1-trichloroethane in the United States: part 1-historical background. Carbon Tetrachloride Tetrachloroethylene Environ Forensics 1(2):69–81
33. Eichhorn S, et al (2009) Handbook of textile fibre structure: volume 2: natural, regenerated, inorganic and specialist fibres
34. El-Sayed E et al (2022) A short observation on the printing cotton fabric using some technique. J Text Color Polym Sci 19(1):17–24
35. El-Naggar A et al (2003) Antimicrobial protection of cotton and cotton/polyester fabrics by radiation and thermal treatments. I. Effect of ZnO formulation on the mechanical and dyeing properties. J Appl Polym Sci 88(5):1129–1137
36. Fang DD (2018) Cotton fibre: physics, chemistry and biology. Springer
37. Ferrero F et al (2008) Water-repellent finishing of cotton fabrics by ultraviolet curing. J Appl Polym Sci 107(2):810–818
38. Flint E (1950) The structure and development of the cotton fibre. Biol Rev 25(4):414–434
39. Foster S, Tashlick I (1962) Application and properties of a durable wash-and-wear finish for cotton. Text Res J 32(7):574–581
40. Fryberg M (2005) Dyes for ink–jet printing. Rev Prog Color Relat Top 35(1):1–30
41. Fu C et al (2015) A non-aqueous dyeing process of reactive dye on cotton. J Text Inst 106(2):152–161
42. Gargoubi S et al (2016) Enhancing the functionality of cotton fabric by physical and chemical pre treatments: a comparative study. Carbohyd Polym 147:28–36
43. Goldthwalt CF et al (1959) Study of improvement of smoothness of cotton fabrics. Text Res J 29(2):134–143
44. Grajeck EJ, Petersen WH (1962) Oil and water repellent fluorochemical finishes for cotton. Text Res J 32(4):320–331
45. Grimmelsmann N et al (2018) Adhesion of 3D printed material on textile substrates. Rapid Prototyp J 24(1):166–170
46. Gupta S (2001) Inkjet printing-a revolutionary ecofriendly technique for textile printing. Ind J Fibre Tex Res 26:156–161
47. Gupta D, Laha A (2007) Antimicrobial activity of cotton fabric treated with Quercus infectoria extract. Indian J Fibre Text Res 32(1):88–92
48. Haji A et al (2018) Cotton dyeing and antibacterial finishing using agricultural waste by an eco-friendly process optimized by response surface methodology. Fibres Polym 19:2359–2364
49. Haji A, Qavamnia SS, Bizhaem FK (2016) Salt free neutral dyeing of cotton with anionic dyes using plasma and chitosan treatments. Ind Text 67(2):109–113
50. Haule LV et al (2012) Surface and bulk chemical analysis of the durability of an easy care finish on cotton. Cellulose 19(3):1023–1030
51. Hauser PJ, Tabba AH (2001) Improving the environmental and economic aspects of cotton dyeing using a cationised cotton. Color Technol 117(5):282–288
52. Hessler LE, Merola GV, Berkley EE (1948) Degree of polymerization of cellulose in cotton fibres. Text Res J 18(10):628–634
53. Hickman WS (1995) Preparation. In: Shore J (ed) Cellulosics dyeing, vol 81

54. Hock CW, Ramsay RC, Harris M (1941) Microscopic structure of the cotton fibre. Text Res J 11(4):200–217
55. Holmquist H (2016) Environmental assessment of alternative durable water repellent chemicals. Chalmers University of Technology, Degree Thesis, Gothenburg, Sweden
56. Hoque MM et al (2022) Ecofriendly pretreatment of cotton knit fabrics by ultrasonic irradiation and reusing bath chemicals. Res Sq 1:1–13 https://orcid.org/0000-0001-7576-9269
57. Hu XP, Hsieh YL (1996) Crystalline structure of developing cotton fibres. J Polym Sci Part B: Polym Phys 34(8):1451–1459
58. Huang C et al (2021) Development of hydrophilic anti-crease finishing method for cotton fabric using alpha-Lipoic acid without causing strength loss and formaldehyde release problem. Prog Org Coat 151:106042
59. Ibrahim N et al (2009) UV-protecting and antibacterial finishing of cotton knits. J Appl Polym Sci 112(6):3589–3596
60. Ibrahim M et al (2013) Antimicrobial protection of pre-irradiated cotton and cotton/polyester fabrics printed with pigment colours. Polym Renewable Resour 4(4):185–202
61. Ibrahim NA, Eid BM (2020) Plasma treatment technology for surface modification and functionalization of cellulosic fabrics. In: Advances in functional finishing of textiles. Springer, pp 275–287
62. Islam S (2019) Attaining optimum strength of cotton-spandex woven fabric by apposite heat setting temperature. J Inst Eng (India): Ser C 100(4):601–606
63. Jain A et al (2014) Antimicrobial polymers. Advanced healthcare materials, 3(12):1969–1985Jelil RA (2015)A review of low-temperature plasma treatment of textile materials. J Mater Sci 50(18):5913–5943
64. Kan CW, Yuen CWM, Tsoi WY (2011) Using atmospheric pressure plasma for enhancing the deposition of printing paste on cotton fabric for digital ink-jet printing. Cellulose 18:827–839
65. Karmakar SR (1999) Chemical technology in the pre-treatment processes of textiles. Elsevier
66. Khatri A et al (2015) A review on developments in dyeing cotton fabrics with reactive dyes for reducing effluent pollution. J Clean Prod 87:50–57
67. Koh J (2011) Dyeing of cellulosic fibres, in Handbook of textile and industrial dyeing. Elsevier, pp 129–146
68. Korger M (2016) Possible applications of 3D printing technology on textile substrates. In IOP conference series: materials science and engineering. 141 012011. IOP Publishing
69. Kuo C-FJ, Lin W-T (2019) Optimum processing parameters of sueding fabric comfort by applying the Taguchi method and fuzzy theory. Text Res J 89(23–24):5165–5176
70. Lemin D et al (1946) The dyeing of direct dyes on cotton 62(5):132–150
71. Lewin M (2006) Handbook of fibre chemistry. CRC Press
72. Liu D et al (1998) Investigation on the conductivity and microstructure of AOT/non-ionic surfactants/water/n-heptane mixed reverse micelles. Colloids Surf A 135(1–3):157–164
73. Liu JQ, Miao HL, Li SZ (2012) Non-aqueous dyeing of reactive dyes in D5. Adv Mater Res 441:138–144
74. Lou J et al (2021) Preparation and application of polyaldehyde trehalose as a new hydrophilic anti crease finishing agent for cotton fabric. Text Res J 91(34):335–346
75. Mao Z et al (2014) States of water and pore size distribution of cotton fibres with different moisture ratios. Ind Eng Chem Res 53(21):8927–8934
76. McCall ER, Jurgens JF (1951) Chemical composition of cotton. Text Res J 21(1):19–21
77. Milicevic B (1970) Solvent dyeing: theory and practice. Text Chem Color 2(5):17–28
78. Mohsin M et al (2013) Performance enhancement of fire retardant finish with environment friendly bio cross-linker for cotton. J Clean Prod 51:191–195
79. Mohsin M et al (2013) Environment friendly finishing of sulphur, vat, direct and reactive dyed cotton fabric 53:341–347
80. Mojsov K (2012) Enzyme applications in textile preparatory process: a review. International J Manag IT and Eng (IJMIE) 2(9):272–295
81. Molla TA, Chavan RB, Nierstrasz V (2017) Combined pre-treatment and causticization of cotton fabric for improved dye uptake. Adv Res Text Eng 2(1):1016

82. Nabar G, Tawde G (1955) Some aspects of the mechanical finishing of cotton textiles. J Soc Dyers Colour 71(6):343–350
83. Nakanishi K et al (1999) Comprehensive Natural Products Chemistry: Carbohydrates and Their Derivates Including Tannins. Elsevier, Cellulose and Related Lignins
84. Nimkar U (2018) Sustainable chemistry: a solution to the textile industry in a developing world. Curr Opin Green Sustain Chem 9:13–17
85. Nionzima F et al (2023) Biotechnology of microbial enzymes microbial enzymes used in textile industry. Elsevier, pp 649–684
86. Nishiyama Y et al (2010) Looking at hydrogen bonds in cellulose. Acta Crystallogr D Biol Crystallogr 66(11):1172–1177
87. Okano T, Sarko A (1985) Mercerization of cellulose. II. Alkali–cellulose intermediates and a possible mercerization mechanism. J Appl Polym Sci 30(1):325–332
88. Ongabekova A, Burkitbay A, Kalzhan G (2020) Development of the composition for Antimicrobial protection of textile materials. In: ТЕХНИЧЕСКИЕ НАУКИ: ПРОБЛЕМЫ И РЕШЕНИЯ
89. Parvin et al (2020) A study on solutions of environment pollution and workers' health problems caused by textile manufacturing operations. Biomed J Sci Tech Res 28(4):21831–21844
90. Parvinzadeh M (2007) The effects of softeners on the properties of sulfur dyed cotton fibres. J Surfactants Deterg 10(4):219–223
91. Patel H, Vashi RT (2015) Characterization and treatment of textile wastewater. Elsevier
92. Pietrzak K et al (2016) Antimicrobial properties of silver nanoparticles misting on cotton fabrics. Text Res J 86(8):812–822
93. Pileni MP (1989) Structure and reactivity in reverse micelles. Elsevier
94. Prabaharan M, Rao JV (2003) Combined desizing, scouring and bleaching of cotton using ozone. Indian J Fibre Text Res 28:437–443
95. Ramachandran T et al (2009) Optimization of process parameters for crease resistant finishing of cotton fabric using citric acid. Indian J Fibre Text Res 34(4):359–367
96. Ren Q, Zhao T (2010) Synthesis and application of modified vegetable oils in water repellent finishing of cotton fabrics. Carbohyd Polym 80(2):381–386
97. Roberts D, Hall M, Horrocks A (1992) Environmental aspects of flame-retardant textiles-an overview. Rev Prog Color Relat Top 22(1):48–57
98. Rocky A (2012) Comparison of effectiveness between conventional scouring & bio-scouring on cotton fabrics. Int J Sci Eng Res 3(8):1–5
99. Santhi P, Moses JJ (2010) Study on different reducing agents for effective vat dyeing on cotton fabric. Indian J Fibre Text Res 35:349–352
100. Sarkar S et al (2020) Application of enzymes in textile functional finishing, in Advances in functional finishing of textiles. Springer, pp 115–127
101. Savvidis G et al (2014) Ink-jet printing of cotton with natural dyes. Color Technol 130(3):200–204
102. Savvidis G et al (2017) Screen-printing of cotton with natural pigments: Evaluation of color and fastness properties of the prints. J Nat Fibres 14(3):326–334
103. Sawada K et al (2006) Dyeing natural fibres in supercritical carbon dioxide using a nonionic surfactant reverse micellar system. Color Technol 118(5):233–237
104. Schindler WD, Hauser PJ (2004) Chemical finishing of textiles. Elsevier
105. Semet B, Grüninger GE (1995) Eisen(II)-Salz-Komplexe als Alternative zu Hydrosulfit in der Küpenfärberei. Melliand Text 76:161
106. El Shafie A, Fouda MM, Hashem M (2009) One-step process for bio-scouring and peracetic acid bleaching of cotton fabric. Carbohyd Polym 78(2):302–308
107. Shahid M et al (2010) Spirality in cotton knit fabrics before and after compacting using selected yarn count and stitch length. J Innov Dev Strat 4(2):11–17
108. Shahid M et al (2013) Recent advancements in natural dye applications: a review. J Clean Prod 53:310–331
109. Sharaf S et al (2011) Comparative investigations on the efficiency of different anchoring chemicals for the permanent finishing of cotton with chitosan. AUTEX Res J 11(2):71–77

110. Siddiqua UH et al (2107) Relationship between structure and dyeing properties of reactive dyes for cotton dyeing. J Mol Liq 241:839–844
111. Smith B (2003) Wastes from textile processing. In: Plastics and the environment. Eds. Anthony L. Andrady, Wiley, pp 243–309
112. Tang AYL, Kan CW (2020) Non-aqueous dyeing of cotton fibre with reactive dyes: a review. Color Technol 136(3):214–223
113. Timpa JD, Ramey HH Jr (1989) Molecular characterization of three cotton varieties. Text Res J 59(11):661–664
114. Tomasino C (1992) Chemistry & technology of fabric preparation & finishing. Citeseer
115. Trotman ER (1975) Dyeing and chemical technology of textile fibres. Griffin
116. Uskoković V (2007) Drofenik M (2007) Reverse micelles: inert nano-reactors or physico chemically active guides of the capped reactions. Adv Colloid Interface Sci 133(1):23–34
117. Valko EI (1957) The theory of dyeing cellulosic fibres. Text Res J 27(11):883–898
118. Vanderploeg A et al (2017) The application of 3D printing technology in the fashion industry. Int J Fashion Des Tech Educ 10(2):170–179
119. Vankar PS et al (2008) Ecofriendly sonicator dyeing of cotton with Rubia cordifolia Linn. using biomordant. Dye Pigment 76(1):207–212
120. Vigneshwaran N, Arputharaj Mohammad, A, Shahid Ravindra, Adivarekar (2020) Advances in functional finishing of textiles functional finishing of cotton textiles using nanomaterials. Springer, pp 43–56 https://doi.org/10.1007/978-981-15-3669-4
121. Vigneswaran C, Ananthasubramanian M, Kandhavadivu P (2014) Bioprocessing of textiles. Elsevier
122. Wagaw T, Chavan RB (2012) Optimization of caustic soda concentration for causticization of cotton. Open Access Sci Rep 1(9):6
123. Wagaw T, Chavan RB (2013) Causticization followed by combined desizing, scouring and bleaching of cotton fabric. Colourage 60(3):43–47
124. Wagaye BT, Walle GA (2018) Overview of Ethiopian textile industry. J Text Polym 6(2):117–120
125. Walentowska J, Foksowicz-Flaczyk J (2013) Thyme essential oil for antimicrobial protection of natural textiles. Int Biodeterior Biodegrad 84:407–411
126. Welch CM (1992) Formaldehyde-free durable-press finishes. Rev Prog Color Relat Top 22(1):32–41
127. Welch CM, Peters JG (1997) Mixed polycarboxylic acids and mixed catalyst in formaldehyde-free durable press finishing. Text Chem Color 29(3):22–27
128. Wu Mj et al (2022) Cotton-containing printing wires based on the two-dimensional braiding method for three-dimensional printing of clothing. Text Res J 92(9-10):1384–1393
129. Wuebbles DJ et al (2011) Three-dimensional model evaluation of the Ozone Depletion Potentials for n-propyl bromide, trichloroethylene and perchloroethylene. Atmos Chem Phys 11(5):2371–2380
130. Zhao H et al (2022) Insights into coloration enhancement of mercerized cotton fabric on reactive dye digital inkjet printing. RSC Adv 12(17):10386–10394

Role of Cotton in Ethiopia's Apparel Industry

Wassie Mengie, Sufiyan Derbew Tiku, Agmasie Bantie Kassa, and Solomon Tsegaye

Abstract Ethiopia has a long history of cotton agriculture and cotton garment production. Until the establishment of the Dire Dawa textile mill in 1939, the textile and garment business was limited to traditional textiles. Currently, both traditional and modern cotton industries supply a range of clothing options, creating ample job opportunities to meet the increasing demand for apparel in local and international markets. The Ethiopian textile and clothing sector primarily relies on cotton, with handmade cotton items and various cotton products from modern apparel industries being commonly produced. Cotton clothing is favoured for its aesthetic appeal, durability, versatility, comfort, and biodegradability, making it an ideal choice for various products like clothing, bedding, and textiles. Consequently, Ethiopian cotton apparel is popularly sold in the global market, with the United States, Germany, Italy, Turkey, the United Arab Emirates, Canada, Spain, and China being the main markets. However, the sector faces challenges in producing quality products that meet consumer needs, mainly due to the quality of the input cotton. This issue hinders the sector's competitiveness in the global market. Besides the economic benefits, cotton garments are also closely associated with social, cultural, and religious activities. Ethiopia aims to become an African hub for fabric manufacturing by 2025, investing in well-established industry parks and infrastructure across the country to attract international apparel buyers. Strategies such as value addition to indigenous cotton cultivation, sustainable cotton production, and availability of renewable green energy are being designed to enhance the sector's competitiveness in the global market.

W. Mengie (✉) · S. D. Tiku · A. Kassa · S. Tsegaye
Ethiopian Institute of Textile and Fashion Technology, Bahir Dar University, 1037, Bahir Dar, Ethiopia
e-mail: wassie215@gmail.com; Wassie.Mengie@bdu.edu.et

S. D. Tiku
e-mail: Sufiyan.Derbew@bdu.edu.et

A. Kassa
e-mail: Agmasie.Bantie@bdu.edu.et

S. Tsegaye
e-mail: Solomon.Tsegaye@bdu.edu.et

K. Murugesh Babu et al. (eds.), *Cotton Sector Development in Ethiopia*, Textile Science and Clothing Technology, https://doi.org/10.1007/978-981-99-9149-5_13

Keywords Cotton clothing · Textile and apparel sector · Cultural clothes · Sustainable cotton

1 Introduction

1.1 Overview of Ethiopian Clothing Industries

For over 3,500 years, Ethiopia has had a prosperous history of producing cotton-based fabric through traditional handloom techniques. Though the country's textile and clothing sector had a long history, cloth production was limited to handlooms until the early 20th century. In the early 1900s, Ethiopia's cotton and textile sector began to expand on a commercial level. The first modern, integrated textile mill, the Dire Dawa textile mill, was introduced in 1939 by the Italians. The first garment factory, Addis Garments, was established in the 1950s [8, 11, 13, 19]. Similar to traditional textiles and clothing, modern Ethiopian textiles also began with 100% cotton fibres. Due to the growth of cotton production, five additional large private integrated textile mills were established in the 1960s [19]. With significant support from the Ethiopian government, the sector has undergone rapid growth, with numerous new players, often foreign-owned textile and garment manufacturers, establishing production facilities to supply foreign markets [7]. Since 2000, the sector has experienced rapid expansion, and its contribution to the country's economy has improved. The Ethiopian textile and apparel sector is currently experiencing rapid growth because the Ethiopian government has given it high priority to transform an agricultural-based economy into an industry-led one [38]. The Ethiopian government has set various incentive packages to attract foreign direct investors. Some of the intensive schemes are:

- Loans available at interest rates starting from 3 to 4%.
- Exemption from custom duty on imported equipment.
- Exemption from custom duty payment on capital goods and spare parts.
- Corporate income tax (CIT) exemption for a period of 1–9 years.
- Income tax exemption ranging from 2 to a maximum of 7 years for manufacturing investments.
- Export incentives like duty draw-backs, vouchers, and export credit guarantee schemes.
- VAT of 15% is reimbursable on a monthly basis.
- No income tax for foreign staff engaged in knowledge transfer or expertise exchange for a minimum of 2–4 years.

Due to its extensive initiatives and advantageous factors such as abundant labour, land resources, cost-effective electricity, and water supply, Ethiopia is attracting foreign direct investment (FDI) in the textile and garment manufacturing sector. For instance, the inflow of FDI into Ethiopia has increased remarkably from $0.1 billion in 2008 to $3 billion in 2016 [14].

The textile and apparel sector in Ethiopia has witnessed remarkable growth in recent years (2010–2020), with a significant increase in apparel exports. The number of textile and apparel industries rose from 108 in 2014 to over 300 in 2021, indicating rapid progress despite the challenges faced by the sector (van der Pols 2015). Additionally, the export capacity of the sector has shown significant improvement, with Ethiopia exporting a total of 12, 112 and 353.5 million US$ worth of apparel in 2010, 2016, and 2020 respectively, as reported by the WTO [48]. These developments indicate that the Ethiopian apparel industry is well-positioned to become a global sourcing hub.

2 The History of Cotton in Ethiopian Clothing

Ethiopia, as the lone African country that was never colonized, had a highly developed civilization and culture dating back to the Axumite Kingdom in the third century B.C. Among the many advancements of this civilization was the well-established art of hand spinning and weaving, which made Ethiopia the centre of handmade textiles and clothing in East Africa [27, 32]. Research shows that cotton gradually replaced animal skins and plant-based textiles for clothing production, becoming one of the chief imported items for this purpose during the 1st century A.D. in the Axumite kingdom [10]. From ancient times to the present day, handwoven cotton textiles and garments have been integral to Ethiopian culture, religion, and tradition [32]. These handmade textiles and garments, produced through the art of weaving and embroidery, have played a significant role in the fluctuations of commerce, religious conflicts, wars, changes in fashion, production demands, and imports of foreign textiles [32].

The cottage industry is a prime example of Ethiopian indigenous knowledge, producing textiles and garments from cotton fibres. In ancient times, clothing was a symbol of class, with only high-class societies such as the nobility wearing decorative clothing until the 19th century. Prior to the 19th century, hand weaving was practised in southern Ethiopia, and the wearing of cotton clothing was limited to nobles and peasants. In the 19th century, hand weaving became common in northern Ethiopia, and the wearing of cotton clothing gradually spread to other societies, with only aristocrats wearing silk and brocade [10, 26, 30].

Up until the commencement of the 1900s, such bans were widespread across Ethiopia. Before 1957, the majority of clothing requirements for the people were met by small-scale industries [29]. At present, cotton attire is favoured by the majority of the populace, who adhere to customary outfits [10]. Broadly speaking, both vintage and current Ethiopian clothing relied heavily on cotton threads, and a variety of cotton-based garments were employed in the practice of cultural, religious, and other traditional customs.

3 The Significance of Cotton in the Growth of Ethiopia's Clothing Industry

Ethiopia has a rich history of producing its own clothing from cotton, dating back to the Axumite kingdom [25, 27]. Throughout the ages, cotton has remained the primary material used in Ethiopian clothing. In ancient times, Ethiopians would create 100% cotton fabrics through hand spinning and hand weaving, which they would then fashion into garments using hand stitching. This traditional method of clothing production remains a significant source of income for many people in both rural and urban areas, providing a means of meeting household needs and generating additional income [49]. The cotton-based clothing industry is one of the most important non-agricultural sources of income in Ethiopia, supplying various types of indigenous clothes and creating job opportunities for a large number of people [8]. Many individuals establish micro, small, and medium-sized cotton-based enterprises to generate income across the country [49].

However, with the rise in human population and living standards, the demand for clothing has increased, leading to the importation of fibres, yarns, and fabrics other than cotton, such as acrylic, polyester, nylon, and wool.

The first modern textile industry in Ethiopia was established in Dire Dawa in 1939, with a production capacity of 3.6 million yards of fabric and 3000 kg of yarns, all of which were 100% dyed and printed [13]. However, prior to this, Ethiopians produced their clothing using handlooms.

Today, cotton textiles and garments are produced in Ethiopia using both handlooms and modern weaving and knitting machines. Home spinners use approximately 1,684 t of lint cotton to produce 1,347 t of hand spun yarns, which they supply to the handloom sector for the production of Ethiopian cultural clothes (European Commission 2021). While hand-spun yarns are only used as weft yarns for handlooms, colourful decorative threads and warp yarns are imported or sourced from local textile industries. Although the capacity of the textile mills is not sufficient to meet the country's needs, cotton remains a vital material in the development of the Ethiopian clothing industry, particularly in the production of cultural clothes.

The cottage industry in Ethiopia is a sector of indigenous knowledge that specializes in creating traditional textiles and garments made from cotton fibres. Historically, cotton clothing was exclusively worn by nobility and peasants in southern Ethiopia, where hand weaving was practised. However, by the 19th century, hand weaving had become prevalent in northern Ethiopia and the use of cotton clothing gradually spread to other communities [30].

At present, the handloom industry produces a variety of cotton clothing items, including Netella, Gabi, Kemis, Buluko, Scarf, Mekenet, Kuta, shirts, skirts, and medium-sized blankets (as depicted in Fig. 1).

These traditional garments have distinct designs that differ based on factors such as region, faith, cultural customs, ethnic group, and the economic standing of the community.

Fig. 1 Common cotton based Ethiopian cultural clothes (Photograph by Author)

Figure 1 illustrates that Habesh Kemis is the prevailing traditional attire for women, crafted from white handwoven textiles and adorned with diverse Tibeb patterns and/or embroidery. Our research conducted in the Northern, Southern, and Central regions of Ethiopia revealed that cotton dresses with various Tibeb designs are created, as depicted in Figs. 2, 3, and 4.

Several research studies have concluded that the production of cultural garments made from cotton has a significant impact on the growth of the clothing sector in Ethiopia, providing a source of income for many people.

At present, there are numerous medium and large-scale textile and garment companies that manufacture clothing using domestically grown cotton for both local and global markets. Some of the major textile and garment companies in Ethiopia include Bahir Dar Textile S.C., Kombolcha Textile S.C., Ayka Addis Textile PLC, Else Addis Industrial Development PLC, Kanoria Africa Textile PLC, Maa Garment and Textile PLC, MNS Manufacturing PLC, Telaje Garment PLC, and many more [12]. As illustrated in Fig. 5, cotton-based clothing items such as T-shirts, denim trousers and jackets, bed sheets, shirts, dresses, skirts, towels, curtains, socks, and innerwear are among the most common cotton garments produced. The Ethiopian apparel industry primarily exports cotton-based clothing to the EU, US, and Chinese markets. This demonstrates that cotton-based clothing plays a significant role in the growth of

Fig. 2 Various types of Tibeb designs in Amhara region

Fig. 3 Tibeb designs collected in Axum and Mekelle (Photograph by Author)

the textile and garment industries. Due to their unique properties, the demand for cotton-based clothing continues to increase both locally and internationally.

4 Characteristics of Cotton-Based Clothing

Cotton textiles and apparel are widely utilized in our daily life, from household clothes such as t-shirts, shirts, towels, and bed sheets to commercial and industrial uses [21]. Cotton is a beautiful, comfortable, durable, adaptable, and biodegradable natural fabric that is ideal for a variety of end applications such as clothing, bedding, textiles, and a wide range of other items [6]. The most significant characteristics of

Fig. 4 Tibeb designs collected in Shiro Meda, Addis Ababa (Photograph by Author)

Fig. 5 Different cotton based apparel products produced in Ethiopia (Photograph by Author)

clothing are comfort, attractive appeal, durability, and performance in its intended purpose. Most of the time, these features may be obtained by carefully selecting raw materials and production parameters [15].

A human being demands pleasant surroundings: their daily activities are strongly tied to their level of comfort. As a result, the most essential means of human comfort is the wearing of environmentally suitable clothing. The subjective element of clothing comfort is connected to wear scenarios such as working, non-critical, and critical conditions [20]. Cloth comfort is the result of a combination of several textile features, including fibre chemical structure and morphology, mechanical capabilities, heat transport, and moisture management [34]. Wearer comfort of the cloth can be distinguished as psychological, thermo-physiological, and sensorial [33].

The thermo-physiological wearer comfort of clothing is defined by the heat and moisture management qualities of the textile material, as well as the ability of clothing to maintain heat balance under varying wearing conditions [1]. As a result, various aspects connected to the formed fibre type and qualities, yarn properties, fabric constructional parameters, and kinds of fabric finish applied to impact the transfer of heat, air, and moisture in liquid and vapour form [5, 35]. Cotton clothing has great thermo-physiological comfort features, including moisture control and heat balancing. Cotton clothing may be manufactured with a wide range of qualities that can be utilized in a variety of environmental circumstances. However, achieving this can be accomplished through meticulous engineering of cotton clothing parameters such as yarn specifications, fabric constructional features, and finishing techniques. Cotton garments are suitable for use in both cold and hot environments. In cold weather, thicker cotton fabrics such as sweaters and Gabbi serve as excellent thermal insulators and provide warmth to the wearer. Essentially, thicker cotton clothes prevent the dissipation of heat from the human body and block the entry of cold air to the wearer's skin, thereby providing comfort. Conversely, in hot environments, lightweight and porous cotton fabrics offer comfort to the wearer by absorbing moisture exuded by the human body and transferring heat and moisture in the form of water vapour. This also permits the flow of air to the wearer's skin, thereby ensuring comfort.

In Ethiopia, people wear a dense cotton blanket known as Gabbi (Fig. 6) to cover their upper body and shoulders during cold weather. This product is composed of four layers of fabric made from hand-spun weft yarns that are twisted at a low level, and it functions as an excellent thermal insulator. Gabbi is a widely used cloth in the highlands of Ethiopia, especially among the elderly.

Cotton attire is commonly employed for undergarments due to its soft texture, great moisture management, and breathability. Cotton is a natural fibre that is biocompatible with the human body and hypoallergenic. Therefore, cotton garments do not cause any irritation and are an ideal choice for individuals with allergies or respiratory conditions such as asthma. Cotton fabrics possess properties such as resistance to high-temperature washing, sterilization ability, and the capability to absorb odours released by the human body during daily activities, which make them perfect for use in kitchen clothes, towels, bedding, baby clothes, hygiene, and medical products. Cotton clothing is durable and can withstand high temperatures, making washing and maintenance easier.

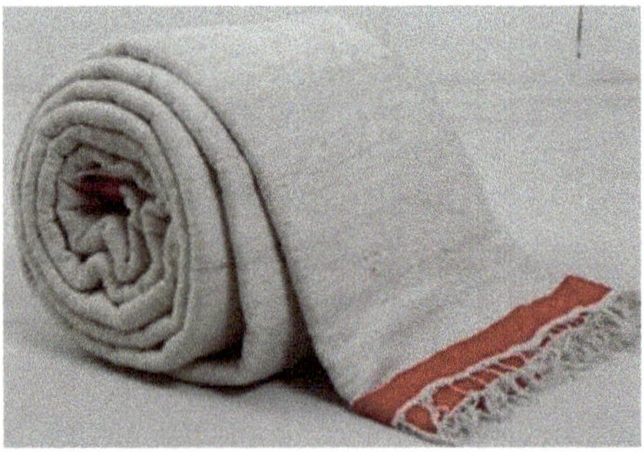

Fig. 6 Ethiopian cotton Gabbi Blanket (Photograph by Author)

Cotton's absorbent and durable properties make it an excellent material for industrial products such as tarpaulins, tents, hotel sheets, army uniforms, and high-tech products like astronaut clothing used in space shuttles. This demonstrates that cotton goods can be used in a wide range of applications, and their range of applications can be extended by blending them with other fibres such as wool and polyester and applying different finishing techniques [6]. For instance, cotton-polyester blends are commonly used for work clothes and military uniforms.

The Ethiopian textile and apparel industry produces a variety of cotton products, including bed linen, t-shirts, shirts, denim, towels, inner garments, and socks. Additionally, various cultural garments made from cotton have been created to cater to local and international markets. Due to the numerous benefits of cotton clothing, the Ethiopian export market is primarily dependent on cotton-based apparel. Cotton garments are made with a wide range of applications, and their properties vary according to their end use. As a result, cotton clothing and industrial products are in high demand in both local and global markets. To meet this demand, several factories produce 100% cotton products and cotton blend clothing.

5 Cotton Clothing Market in Ethiopia

The fashion industry is a crucial economic sector globally, serving as the largest and most valuable industry. It contributes significantly to the world economy, with China being the biggest exporter of textiles and clothing and the EU being the largest importer of such products [13]. Over the past nine years, the apparel market has generated an average revenue of more than 1.51 trillion USD, and this value is expected to increase to 1.94 trillion USD by 2027 [39].

Cotton is the most extensively used natural fibre in textile and apparel production, accounting for one-third of the total fibres manufactured worldwide. Its versatility, durability, and comfort make it a popular choice. The global market for textile and garment factories is heavily dependent on cotton, with a value of USD 748 billion and USD 786 billion, respectively, in 2016 [43]. The global cotton market was worth $38.54 billion in 2020, and it is expected to reach $46.56 billion by 2027 [36].

The textile and apparel industry is predominantly concentrated in Asian countries such as China, Pakistan, and India, which have long been known as the traditional powerhouses of textile and apparel manufacturing. These countries, along with the United States, are also the largest producers of cotton globally. However, buyers are exploring new clothing locations, such as East African countries like Ethiopia and Kenya, due to favourable trade agreements and their strategic geographic position [8, 36, 37].

Ethiopia has a rich history of cultivating cotton and utilizing it in the production of garments. For a long time, cotton consumption was confined to the traditional textile sector. However, the government is presently placing great emphasis on the textile and clothing industry, leading to a rapid rise in cotton consumption due to the expansion of textile industries [3].

Currently, the Ethiopian government is prioritizing the textile and clothing industry and is implementing several intensive programs to attract investors to the sector. More than 300 large textile and clothing industries operate in the country, producing fabrics and garments for both domestic and international markets using cotton, wool, and synthetic fibres such as polyester, acrylic, and nylon. As textile industries expanded, the demand for cotton fibres grew, and to meet the excess demand, textile industries began importing cotton. According to WTO figures, in 2020, Ethiopia imported cotton worth $25,600 [48]. It is challenging to estimate the amount of cotton clothing available in Ethiopia, as the textile industry can only meet 10% of the fabric demand for the garment industry, and the remaining 90% is met through imported fabrics.

There are approximately 100 foreign garment industries in Ethiopia that export products to the US and EU markets. For instance, in 2019–20, 60% of clothing products were exported to the United States, while the rest was exported to countries such as Germany, Italy, Turkey, the United Arab Emirates, Canada, Spain, and China. International buyers like H&M, PVH, Next, Walmart, Primark, Tesco, Marks & Spencer, Superior Uniform Group, and Decathlon source cotton clothing from Ethiopian apparel industries. Some of these companies, such as H&M and PVH, have their apparel industries in Ethiopia. These buyers purchase various types of clothing, including T-shirts, linens, shirts, tops, pants, shorts, jackets, household goods, sweatshirts, pyjamas, dresses, leggings, skirts, nightshirts, underwear, and more [7–9].

6 Socio-economic Benefits of Ethiopian Cotton-Based Apparel

The value chain of cotton-based clothing has social, economic, cultural, and environmental benefits.

6.1 Economic Benefits

The production of fabric and textiles using cotton is a crucial source of income for both urban and rural areas in the country [18]. The cotton garment value chain involves a large number of people, from cotton farmers to garment retailers, and provides employment opportunities to many individuals. [12] report highlights the presence of over 300 medium to large-scale industries in the ginning, spinning, weaving/knitting, finishing, and apparel sectors [12]. These industries primarily focus on processing cotton-based textiles. While cotton apparel is primarily consumed domestically, it is also exported to international markets. Ethiopian apparel exports are dominated by cotton-based products such as t-shirts, women's pants, shirts, and

Table 1 Export revenue of apparels (in million USD) [24]

No.	Product types	Budget year and export performances						
		2005/06	2006/07	2007/08	2008/09	2009/10	2010/11	2011/12
1.	Apparel	6.9	8	9.7	6.6	6.7	26.7	63
2.	Handloom products	0.1	0.2	0.4	0.4	1.7	3.4	4.4

bed sheets, which contributed to 35%, 18%, and 6% of total clothing exports in 2014, respectively [40–42]. As a result, the Ethiopian GDP benefits significantly from cotton garments, and these products are a significant source of foreign exchange. The Ethiopian apparel sector has an installed capacity of approximately 12 and 40 million pieces of woven and knitted garments, respectively, with only 60% of the installed capacity being utilized [14]. The potential output capacity of Ethiopia's garment industries is expected to reach around 62 million knitted garments in the near future (Table 1). Small and medium-sized enterprises also contribute to the production of cotton-based clothing.

6.2 *Social-Cultural Benefits*

Ethiopia is a diverse nation, with each ethnic community, religious group, and nationality having its unique attire. This implies that their customs and religious ceremonies are significantly influenced by their clothing, which is predominantly made of cotton. Clothing is intricately linked to the different religious, cultural, and societal traditions of the population. Since ancient times, cotton garments have been used in many of the cultural and religious traditions of the Ethiopian Orthodox Tewahedo Church (EOTC), as depicted in Fig. 7.

Fig. 7 EOTC devotees celebrate the holidays by wearing iconic cotton clothing [4]

Fig. 8 Leaders and elders deliberate on the date on which Fiche is celebrated by community members (Photograph by Author)

As depicted in Fig. 7, various religious groups such as Sunday school participants, clergy, and women observe and honour their faith by donning traditional Ethiopian attire. Specifically, during the Ethiopian Epiphany, individuals adorn white cotton garments embellished with various Tibetan patterns and embroidery, adding vibrancy to their religious practices and customs. This cotton clothing is also utilized in other native cultural and societal ceremonies, including nuptial festivities, New Year revelries, customary music and cinema, and distinguished titles. These garments symbolize diverse cultural, social, and religious traditions, constituting an essential aspect of one's identity.

The Sidama people have been commemorating the Fichee-Chambalaalla cultural event for centuries, marking the key New Year celebration among the community. During this festivity, participants wear a range of cotton-based attire, such as the Seemma costume adorned by adult men during song and dance performances, as portrayed in Fig. 8 [40].

7 Challenges and Opportunities for Expanding Ethiopian Cotton Clothing's Global Market

Cotton is the primary raw material for fabric manufacture, which is expanding in tandem with global population and economic expansion. As seen in the graph, global consumption of cotton fibres has progressively increased since 1960, whereas chemical fibre usage has increased significantly [28]. Cotton will account for 24% of total global fibre consumption in 2020 (Fig. 9).

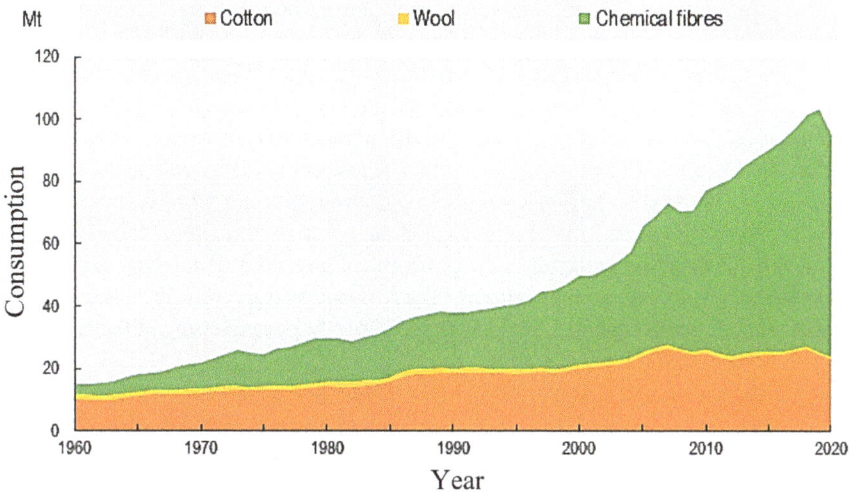

Fig. 9 Historical trends in global consumption of textile fibres

The Ethiopian textile industry relies solely on locally produced cotton as its input material, which is crucial for the production of textiles and garments for various purposes [44]. Therefore, any fluctuations in the cotton sector, whether positive or negative, have a direct impact on the textile and apparel industry. Despite significant economic growth in Ethiopia's textile and clothing industry over the past two centuries, the main challenge remains the sustainable supply of input materials, particularly cotton fibres. Ethiopia possesses around 2–3 million hectares of arable land with adequate resources (such as agroecological features, land and water resources, available labour, a long-standing tradition in cotton cultivation and processing, and regular and dependable air transport to potential market destinations of cotton products) that are potentially suitable for cotton production [6]. The availability of labour contributes to the cultivation of sustainable cotton, which enables international buyers/retailers to source cotton-based apparel. The opportunities presented by cotton cultivation also encourage international buyers to produce sustainable cotton apparel products by cultivating their own cotton in Ethiopia and supporting farming for their sustainable cotton and cotton garment supply.

Presently, the cotton production in Ethiopia falls short of meeting the escalating demand of textile industries [3, 17]. Additionally, the quality of Ethiopian cotton does not cater to the production of competitive international cotton products, posing a major challenge for the textile and apparel sector, hindering its global market share. However, this challenge can be overcome by enlarging the crop and enhancing the current yield in both quantity and quality. The sector confronts several hindrances in utilizing diverse unilateral and bilateral agreements. Some of these challenges that impede competitiveness are limited access to global information and technology, societal instability, exorbitant input costs, and inadequate local textile and accessories suppliers.

The Ethiopian textile and garment markets attract international buyers, offering enormous business prospects. Some of the benefits include the availability of trainable labour, sustainable and cost-effective energy supply, various incentive packages, strategic location near Asia and Europe, and duty-free access to the European Union, in addition to over 16 bilateral trade agreements, involving China, India, Turkey, and Russia. The Ethiopian government also provides incentives to investors involved in the Textile and apparel sector, such as preferential trade deals, land policies, up to nine years of tax holidays, and duty-free imports of machinery, equipment, and construction materials. The government has initiated reforms to diminish trade barriers in this trade environment. These business opportunities allure foreign investment, and consequently, FDI has played an indispensable role in the country's economic development [8, 42].

Ethiopia is striving to establish itself as a prominent centre for light manufacturing in Africa by 2025. To achieve this goal, various industrial parks have been developed in different regions of the country, with a focus on the textile and apparel industries. This move is essential in expanding the global market share of Ethiopian apparel products, as noted by Vallejo and Mekonnen [42].

In today's market, there is a growing demand for sustainable cotton products, which support socially responsible and eco-friendly cotton production practices, as highlighted by Voora et al. [44]. Ethiopia has responded by engaging in sustainable cotton cultivation, with the assistance of African-made cotton (CmiA) (Cotton made in Africa) standards. The CmiA standard ensures that cotton is grown with a focus on quality and quantity, while also promoting social justice for cotton farmers and workers in the ginning factories. This approach also prioritizes healthy living conditions and environmental protection. As a result, clothing made from sustainable cotton can be easily marketed internationally, thereby contributing to the growth of Ethiopia's cotton clothing brands, as noted by Preuss [31]. This presents another opportunity for Ethiopia's textile and clothing sector to increase its global market share.

8 Strategies for Developing Globally Branded Ethiopian Cotton Products

The fashion industry places significant emphasis on sustainable and eco-friendly procedures and products to tackle various environmental issues like pollution, greenhouse gases, and global warming [23]. The modern generation prioritizes sustainable fashion, natural fibres, and eco-friendly fabrics. Studies indicate that consumers have a favourable attitude towards buying organic products, and retailers influence customers to buy organic clothing items, thereby increasing their purchase intention [2]. Consumers are motivated by the belief that purchasing organic clothing is advantageous. This highlights the role of organic cotton clothing in supporting organic farming and promoting the environmental benefits of purchasing such clothing [16].

Therefore, cotton is the ideal solution for creating sustainable fashion items for both fashion retailers and customers. Enhancing traditional practices in the cultivation of native cotton and traditional textiles plays a critical role in creating Ethiopian cotton-based clothes that are globally recognized. Sustainable cotton is produced in Ethiopia with the support of African Made Cotton (CmiA) and the Competitive African Cotton Initiative (COMPACI). These organizations are working towards addressing the value-addition challenges of cotton in Ethiopia and laying the groundwork for establishing a sustainable textiles and apparel sector [45].

Due to the distinctive and exclusive designs, ability to manufacture small batches, and use of eco-friendly materials, there is a high demand for handmade indigenous products both domestically and internationally. Discerning retailers seek a reliable source for a consistent supply of authentic handloom products. Therefore, enhancing the value and delivery time of handmade cotton garments can be beneficial.

Consumer awareness regarding global environmental issues has increased significantly. Consequently, customers are conscious of the environmental impact and human welfare involved in purchasing sustainable, eco-friendly, and ethical products, and are willing to pay extra for them [22, 46, 47]. In this context, Ethiopian handmade cotton products and sustainable cotton clothing are ideal for environmentally conscious customers. Ethiopian textile and apparel industries are also exploring the possibility of supplying organic cotton clothing to the international market, which can help establish a good reputation. In general, value addition to indigenous cotton cultivation and handmade cotton garments, sustainable cotton production, and the availability of renewable green energy can facilitate the production of environmentally friendly cotton-based products that meet customers' requirements.

9 Summary

Cotton has long been the sole input material in Ethiopia's textile and garment industry. It is critical to the growth of the textile and garment industries as well as the country's economy. Cotton clothing is in high demand in both domestic and international markets due to its versatility, durability, and comfort. The quality of Ethiopian cotton, on the other hand, is insufficient for the production of competent international cotton products. Other challenges to competitiveness include a lack of access to international information and technology, societal instability, high input costs, and a lack of local textile and accessory providers.

Presently, the government is confronting the obstacles and executing various comprehensive schemes to allure investment in the textile and garment industry. Despite the fact that cotton-based clothing constitutes the majority of total clothing exports, the exports to the world market are insignificant. Nonetheless, the mounting worldwide demand for cotton garments, the presence of skilled labour, and a sustainable and cost-effective energy supply are prospects that aid in enhancing the global market share of this domain. Generally, cotton clothing assumes a crucial part in the growth of the nation's economy, specifically in the textile and garment sector, and also holds significant importance in cultural, social, and religious aspects of the society.

References

1. Abel PJ, Unmar R, Rosunee S (2022) Engineering the comfort parameters of 100% cotton plain woven fabrics. J Text Inst 113(9):1806–1815
2. Abrar M, Baig SA, Bashir M, Shabbir R, Ayub M (2018) Consumer attitude and purchase intention towards organic textile products. Amazon Investiga 7(17):472–485
3. Addis T, Kachi A, Wang J (2021) A review of current state and future directions of cotton production in Ethiopia. Cogent Food Agric 7(1):1880533
4. ARCCH (2018) Ethiopian Ephiphany
5. Atalie D, Gideon R, Melesse G, Ferede E, Getnet F, Nibret A (2022) Thermo physiological comfort of half bleached woven fabrics made from different cotton yarns parameters. J Nat Fibres 19(13):5034–5049
6. Brazilian Cooperation Agency (2022) An Overview of the Cotton Sector in Africa and Brazil; CA (2022). The features and benefits of cotton
7. de Haan E, Theuws M (2017) Quick scan of the linkages between the Ethiopian garment industry and the Dutch market. Centre for Research on Multinational Corporations (SOMO), vol 1
8. EIC (2020) Realizing new productive capacity in Ethiopia's textiles and apparel sector: strategy and policy recommendations. Ethiopian Investment Commission
9. Eropean Commisssion (2021) Cotton value chain analysis in Ethiopia. Brussels
10. ETC (2005) Ethiopian Costumes. Addis Ababa, Ethiopia, Ethiopian Tourism Commission
11. Ethiopian Investement Commission (2020) Realizing new productive capacity in Ethiopia's textiles and apparel sector: strategy and policy recommendations. Ethiopian Investement Commission
12. ETIDI (2021) Ethiopian textile and apparel industry profile. In: E T. I. D. Institute (ed.)
13. Fashion United (2022) Global Fashion Industry Statistics. The Global Apparel Market
14. Gera R, Taye H, Bekele A (2017) Implementation of statistical quality control for spinning processes of dire dawa textile factory, pp 1–14
15. Gokarneshan N (2019) Comfort properties of textiles: a review of some breakthroughs in recent research. Juniper Online J Mater Sci 5(3):1–5
16. Hustvedt G, Dickson MA (2009) Consumer likelihood of purchasing organic cotton apparel: influence of attitudes and self-identity. J Fash Mark Manag 13(1):49–65
17. ITA (2022) Ethiopia—Country Commercial Guide
18. ITAB CONSULT PLC (2015) Value chain analysis for weaving products. Addis Ababa
19. ITC (2016) Textile & clothing value chain roadmap of Ethiopia. International Trade centre, Geneva, Switzerland
20. Khanzada H, Khan MQ, Kayani S (2020) Cotton based clothing. In: Wang H, Memon H (eds) Cotton science and processing technology. Springer, pp 377–391
21. Latif W, Basit A, Ali Z, Baig SA (2018) The mechanical and comfort properties of cotton and regenerated fibres blended woven fabrics. Int J Cloth Sci Technol 30(1):112–121
22. Lin S (2010) A case study in Hawaii: who will pay more for organic cotton? Int J Consum Stud 34(4):481–489
23. Mahmoud TO (2018) Impact of green marketing mix on purchase intention. Int J Adv Appl Sci 5(2):127–135
24. Markos G (2020) The rising flow of FDI in Ethiopia: a critical perspective on the textile and apparel industry. J Glob Econ 8(4):1–5
25. MoI (2013) Ethiopian industrial development strategic plan (2013–2025): ministry of industry Addis Ababa, Ethiopia

26. Munro-Hay S (1993) Aksum, An African civilisation of late antiquity. JSTOR, France
27. Nelson HD, Kaplan I (1981) Ethiopia, a country study, vol 28. US Government Printing Office
28. Nicholson GE (1960) The production, history, uses and relationships of cotton (Gossypium spp.) in Ethiopia. Econ Botany 14(1):3–36
29. OECD (2022) OECD-FAO Agricultural Outlook 2021–2030. from he Organisation for Economic Co-operation and Development (OECD)
30. Ofcansky TP, Berry LB (1993) Ethiopia: a country study
31. Pankhurst A (2000) Indigenous knowledge systems in craftwork: the cases of iron smelting, tanning and weaving in Ethiopia. Paper presented at the Indigenous Knowledge Systems in Ethiopia: Report of Ethiopia National Workshop, Addis Ababa university, Addis Ababa, Ethiopia
32. Preuss S (2015) Ethiopia joins cotton made in Africa initiative
33. Purdy JM (2015) Made in Ethiopia. In: 17th Annual Africana studies stduent research conference, "Emerging Perspectives in Africana Studies". Bowling Green State University, Bowling Green, OH, p 43403
34. Raccuglia M, Sales B, Heyde C, Havenith G, Hodder S (2018) Clothing comfort during physical exercise–determining the critical factors. Appl Ergon 73:33–41
35. Ravandi SH, Valizadeh M (2011) Properties of fibres and fabrics that contribute to human comfort improving comfort in clothing. Elsevier, pp 61–78
36. Raybagi A, Kadole P (2021) Thermophysiological comfort: an essential requirement of clothing fabrics. J Text Cloth Sci 4(1):27–32
37. Research and Markets (2021) Global cotton market & volume by consumption, production, export, import countries, companies, forecast. Dublin Ireland
38. Rundassa MW, Azene DK, Berhan E (2019) Comparative advantage of Ethiopian textile and apparel industry. Res J Text Apparel
39. Smith P (2022) Global revenue of the apparel market 2014–2027. Statista
40. SRS (2014) Ethiopia Patent No.: M. o. C. a. T. The Federal Democratic Republic of Ethiopia, Authority for Research and Conservation of Cultural Heritage
41. Staritz C, Plank L, Morris M (2016) Global value chains, industrial policy, and sustainable development—Ethiopia's apparel export sector. Geneva, Switzerland
42. Staritz CW, Lindsay (2017) Made in Ethiopia: the emergence and evolution of the ethiopian apparel export sector. Roskilde Universitet
43. Vallejo B, Mekonnen TG (2021) Foreign direct investment (FDI) and learning in Ethiopia's textile and garment sector. Working Paper 25: FDI and learning in Ethiopia's Textile and Garment Sector. Open Air African Innovation Research
44. Voora V (2015) Cotton made in Africa and competitive African cotton initiative 2015 Stakeholder conference. Paper presented at the Cotton Made in Africa and Competitive African Cotton Initiative 2015 Stakeholder Conference, Addis Ababa, Ethiopia
45. Voora V, Larrea C, Bermudez S (2020) Global market report: cotton. JSTOR
46. Wang CC, Tsai S (2010) Consumer attitudes towards sustainable and environmental strategies in fashion clothing: University of Manchester
47. Weiner HE (2017) Wearing your ethics: investigating consumer purchase intention of ethically produced fashion products
48. WTO (2022) WTO trade connectivity heatmap. From World Trade Organization
49. Zhang X, Moorman L, Ayele G (2011) Infrastructure and cluster development: a case study of handloom weavers in rural Ethiopia. J Dev Stud 47(12):1869–1886

Cross Cutting Issues in Cotton Sector

Cottonseed Production Technology

Michael Kebede and Merdasa Balcha

Abstract In cottonseed production, the cottonseed quality could be both affected at pre-harvest and post-harvest operations playing major roles; respectively in the field and laboratory cottonseed quality standards. The pre-harvest operations all related to the production practices and weather conditions. While, the post-harvest operations that affect cottonseed quality include harvesting, processing, and storage conditions. Performance of different methods of cottonseed processing with accuracy and efficiency will enhance the performance of cotton seed crop in the field. Before planting, the process of making high-quality cottonseeds begins in the field. Field selection is important from several standpoints and majorly could be categorized as a specific and general field requirement. Cotton yield and fibre quality are to a great extent determined by air temperature, soil moisture, and soil fertility conditions during the growing season. Cotton harvesting begins about six months after planting and is the most expensive operation of cotton cultivation. Cotton could either be handpicked or harvested mechanically by using either spindle cotton pickers or stripper cotton harvesters. After harvesting the cotton product should be carefully handled, for maintaining the produced cottonseed quality through major post-harvest handling activities of seed cotton ginning, cottonseed delinting, cottonseed treatments, and proper cottonseed storage. It is also important for growers to plant high quality cotton seed crop varieties adapted to their farm situations, management styles, and intended market uses. Cotton seed crop growers or producers must be aware of that the classes of cotton seed crop grown or produced fulfill the following national minimum quality field and laboratory standards for conventional and hybrid cotton seed crop varieties being produced, as to legally marketing or distribute the cottonseed to the consumers.

M. Kebede (✉)
Ethiopian Institute of Agricultural Research (EIAR), Melkasa Agricultural Research Center, P. O. Box 436, Adama, Ethiopia
e-mail: kebedeyomichael@gmail.com

M. Balcha
Werer Agricultural Research Center, Ethiopian Institute of Agricultural Research (EIAR), P.O. Box 2003, Addis Ababa, Ethiopia

K. Murugesh Babu et al. (eds.), *Cotton Sector Development in Ethiopia*, Textile Science and Clothing Technology, https://doi.org/10.1007/978-981-99-9149-5_14

1 Introduction

Cotton (*Gossypium* spp.) is the predominant natural fibre and oil seed crop grown worldwide. It is a crucial commercial crop, playing a pivotal role in the economic, political, and social spheres of the world. The cotton and cotton textile industries are vital to the economic growth of both developed and developing countries. For thousands of years, cotton cultivation has been a significant economic activity across all continents. Despite occupying only 3% of the world's crop acreage, cotton is one of the most substantial crops in terms of land use after food grains and soybeans [10].

Cotton is a vital source of cash income for millions of smallholder farmers and their families in over 20 countries across all regions of Sub-Saharan Africa. From 2004 to 2014, the African continent contributed 6% to the global production of seed cotton (world production was approximately 1.4 million metric tons) [10]. However, cotton is also one of the more labour-intensive and expensive crops to produce, and the cotton industries are subject to several risks, such as input price fluctuations and severe pest attacks [30]. In addition to these challenges, farmers also face other issues such as low seed germination, insufficient seed storage facilities, low soil fertility, low literacy rates, and lack of training. To maximize profits from cotton, African countries must increase seed availability, support agricultural research, and adopt a capacity-building strategy [2]. By doing so, cotton production in Africa can be increased, and smallholder farmers can reap the benefits.

In general, the quality of cottonseed can be influenced by pre-harvest and post-harvest activities. The pre-harvest activities involve farming practices and weather patterns. The impact of weather and physiological factors on seed development is noteworthy, resulting in varying seed quality depending on the time of picking [5]. The last picking produces smaller and less mature seeds than earlier pickings. On the other hand, post-harvest operations that affect cottonseed quality include harvesting, processing, and storage. The methods used in seed processing have a significant impact on cottonseed quality. The efficient and accurate implementation of these methods can enhance cottonseed performance in the field [14].

Consequently, it is crucial to provide farmers with adapted and high-quality cottonseeds to ensure a productive and profitable cotton industry for all stakeholders. Cottonseeds should have good germination rates, high yields, and adaptability to climatic and agronomic conditions. Additionally, cottonseeds must produce high-quality fibres that meet international market standards to ensure the best possible returns for farmers. These factors are fundamental in determining whether farmers can make profits and improve their livelihoods [29].

2 Cotton Variety and Seed Selection

Selecting high-quality cottonseed varieties is a crucial aspect of cotton farming that can lead to improved yields and better cotton quality.

2.1 Cotton Variety Selection

The selection of a suitable cotton cultivar (varieties) and high-quality seeds has a long-lasting impact on the crop's vigor in the early stages and the overall health and uniformity of the plant during the initial forty days. The potential yield and fibre quality at harvest are determined by the cultivar selection and seed quality [27]. The choice of cultivar is a crucial step in cotton cultivation and achieving optimal seed yields and fibre quality. Currently, cultivars are selected based not only on genetic performance or seed yield potential but also on other technology packages. Cotton Growers are advised to minimize risks by planting multiple cultivars or a cultivar suitable for their agro-ecological conditions. Considerations for cultivar selection should include planting dates, seedling vigor, water regimes (irrigated vs. rainfed and degree or efficiency of irrigation), maturity classes, and plant growth characteristics. It is equally important to note that some cultivars may perform better in certain situations than others [11]. Choosing the right type of cultivar to plant is one of the most significant decisions that cotton growers need to often make. There are many differences among cultivars, and it is therefore essential to evaluate which characteristics are most important for their production area.

The average lifespan of cotton varieties is becoming significantly shorter; thus, growers have little time to gain experience with existing underproduced cotton cultivars. Growers must adapt quickly to new cotton cultivars and gain as much experience with them as possible within a short time frame. Cotton cultivar selection at the grower level should be based on research data and local varieties well in a specific location for a year, but their average performance in other environments with identical conditions may be significantly lower, indicating inconsistent behavior or poor stability. Cotton varieties that perform well across a wide range of environments indicate a high degree of stability, which suggests that these varieties may perform well across planting dates, soil types, rainfall patterns and irrigation practices, grower management practices, and other factors.

2.2 Cottonseed Selection

The initial growth of the cotton harvest, together with the overall health and consistency of the cotton plant within the first forty days, depend heavily on the choice of cultivar and the quality of the seeds. The selection of the right variety and seed quality sets the stage for the final yield of seeds and the quality of the fibres at harvest. The success of the cotton crop in the first forty days and in the long-term hinges on the use of high-quality seeds. It is crucial to ensure that the seeds germinate and emerge quickly, as this minimizes the risk of cotton crop seedling diseases and pests. To enhance the vigor of the cotton crop seedlings, it is essential to prepare the land optimally, control soil moisture and temperature, choose the right planting date, manage seedling diseases, control pests, and maintain fertility [27].

3 Cotton Field Selection and Inspections of the Cottonseed Crop

Cotton field selection, like cotton variety and cottonseed selection, is critical to cotton yield and quality. As a result, the following sections have addressed some of cotton crop field selection requirements and inspections standards.

3.1 *Land Requirements for Cottonseed Crops*

The guidelines for certifying seeds of various crops comprise of specific conditions pertaining to the land requirements. Certified crops seeds could not be cultivated on fields that do not satisfy the stipulated land requirements. The selection of the field is crucial from various perspectives and can be classified as specific and general requirements [19].

3.1.1 Specific Requirements

The area designated for growing cottonseed crops must adhere to certain specifications, which comprise being free of uncontrollable weed growth that produces seeds, being free of volunteer crops, off-type or ratoon cotton crop plants, and maintaining a minimum permissible isolation distance from neighboring cotton crop plants. Additionally, a field map of the produced or grown cotton crop plant seeds and registration of the cottonseed crop plants varieties are mandatory. The ensuing sections provide a brief overview of the specific requirements.

(i) ***Free from Inseparable Seed Producing Weed***

When selecting a location for cotton planting, it is important to consider the weed situation (history). Weeds can cause various issues, such as significantly reducing cottonseed yield and interfering with harvest. Often, weeds will remain green even after the cotton crop has been defoliated and is ready for harvest. When seed cotton is module and kept for a few weeks before ginning, the green material can have a significant impact on the moisture content of the seeds. This transfer of moisture from green plant matter to seed cotton can lead to heating, which can decrease the quality of the cottonseed.

Certain stubborn weed seeds are particularly challenging to remove from cottonseeds. The use of a gravity table and acid delinting can aid in their removal. In general, it is best to avoid having weed seeds in cottonseed fields. If weed seed plants are present in small numbers, manually pulling them out is a good practice. While this may be costly, it can prevent weed seed contamination of cottonseeds [23].

(ii) ***Free from Volunteer, Off-type or Ratoon Cotton Crop Plants***

Volunteers, Off-types or Ratoon harvest plants are plants that differ from the under produced variety in morphological or other characteristics, which may arise due to segregation, mutation, cross-pollination or mechanical mixing. Such cotton plants must be either eliminated (rogued) or nurtured (as possible improvement of the variety). They are generally easily recognizable as taller or smaller, earlier or later blooming, or plants with traits that are distinctly different from the norm.

Fields that previously grew similar cotton crops should be avoided to reduce the risk of contaminating the current specific cotton variety with the previously grown varieties that emerge from volunteer, off-type or ratoon cotton crop plants. Crop rotations also help prevent disease buildup. Ratoon and volunteer cotton plants need to be controlled as they may host next season's crops and could also transmit cotton crop specifically related diseases. Different crop plants may have varying field and laboratory standards in different countries. For example, for various cottonseed classes, the currently applicable Ethiopian cottonseed field minimum field standard of is listed on Table 1 [9].

(iii) ***Preserve the minimum allowed distance of isolation from other cotton plants***

Organizations responsible for certifying seeds have established specific distances for isolation standards, which are contingent upon the species, type, and size of the seed field. Regardless of whether the varieties are proprietary or public, certified or noncertified, distance or time isolation is critical for the production of cottonseed.

When different variations of the same species are grown in close proximity, plants that rely on cross-pollination, such as cotton crops, can be cross-pollinated, resulting in contamination of the variety. To maintain the purity of frequently pollinated plants, including cotton crops, a variety of isolation techniques must be employed. Commonly utilized methods include distance and time isolation, both of which are used to improve the genetic quality of cotton seed during multiplication [20].

The distance between the seed crop and any potential contaminant for varietal purity is greatly affected by the type and mode of pollination. For cotton crops, which are often-pollinated plants, the minimum standard of isolation distances for cottonseed production are listed for different cottonseed classes on Table 1 [9]. Seed crops can also be protected against potential sources of foreign pollen by growing them before or after the same species crop. The period of separation depends on the crop and projected weather conditions, as well as the time it takes for the crop

Table 1 Field standard for conventional and hybrid cottonseed varieties

Category	Pre-basic seed	Basic seed	Certified seed
Rotation (min, year)	3	2	2
Isolation (min, meters)	300	200	100
Off-types and Other Cultivar (max %)	0.01	0.02	0.03

Source ESA [9]

to mature or its variety. For most crops, waiting at least 21 days between plantings will ensure adequate isolation distances [20]. However, unlike other crop plants, time isolation should not be practiced with cotton plants due to their indeterminate fruiting habit.

(iv) ***Field Map of the Produced or Grown Cotton Crop Plants Seed***

During the inspection of a land site, it is crucial to accurately chart all cottonseed crops based on their planting times and areas, including neighboring units. This is especially important in cottonseed fields and isolated cotton crops, where knowledge of the neighboring crop's genetic profile can be vital. Another useful tool for growers is a fertility map of the field, which allows the cottonseed producers or growers for better decision-making regarding irrigation, nitrogen fertilization, and defoliation of the cottonseed plants. Additionally, farmers can use aerial maps to monitor their farms and receive real-time information on their crops.

Overall, field mapping is essential for identifying the agro-ecological domains of various cotton crop varieties, determining field size and isolation distances, and maintaining accurate records. This information helps local seed supply systems estimate the potential market size for each cotton crop variety they wish to produce. Having this knowledge at hand, cottonseed growers or producers could make informed decisions about investing in a selected and certified cottonseed variety.

(v) ***Registration of the Cottonseed Crop Plants***

Making sure that the enrolment of the seed harvest was completed just before planning to expand any cottonseed fields, as the seed regulation authorities always require the seed fields to be registered shortly after establishment.

According to different types of national laws on seed production and distribution, varieties for business seed multiplication must undergo prescribed testing and registration for a particular variety's right of ownership. Testing involves assessment for value for cultivation and use (VCU) and distinctness, uniformity and stability (DUS). As a DUS test is necessary to ensure that the variety is new. DUS also verifies the following: first, that the variety is different from any existing one on the market; second, that it is uniform for concise description and identification; and third, that it is genetically stable. This guarantees that the variety maintains its originality.

Crop variety registration is the authorized registration of a new crop variety as sufficiently distinct from other similar crop varieties to be eligible for plant variety protection in its own right. However, this does not necessarily imply that a registered variety would be released for general multiplication [20].

3.1.2 General Requirements

The overall prerequisites for growing cotton crops involve the availability of the cottonseed fields, size of the cottonseed fields, cultivation and past usage of the cottonseed fields, and consistency of the cottonseed fields. Every aspect of these general requirements has been elaborated in the subsequent subsections.

(i) ***Accessibility of the Cottonseed Fields***

The cottonseed plantation should be conveniently situated and in close proximity to the roadside for effortless monitoring, administration, and consistent examination during the entire season. The location of the cottonseed field should also be easily reachable by markets, laborers, and transportation for the smooth delivery of essential supplies such as cottonseeds, pesticides, and fertilizers, as well as for the transportation of the harvested cottonseed to customers.

(ii) ***Size of the Cottonseed Fields***

The size of the fields for each stage of multiplication and increase of a cotton crop variety may differ considerably, based on factors such as the current or expected market share of each variation, the extent of testing during the increase phase, and the projected yield of the cottonseeds. Additionally, the field sizes will be influenced by the availability of labor, the requirements for quality planting materials, and the ability of the country, organization, or agent to meet the demand.

(iii) ***Cropping and Field History of the Cottonseed Fields***

It is important to take into account the previous crop rotation when considering the field's cropping history, as this can reduce the risk of unwanted volunteer plants from the same or related species contaminating the seed crop [21]. To minimize the potential for re-growth of a different cotton variety from the same plants in the cottonseed field during the specified period outlined in the cotton crop standards, it is recommended that the previous crop for the produced cottonseed not be cotton. However, if cottonseed growers or producers are able to produce the exact same cotton variety of equal or lower class, they may plant the same cotton variety on the same field. Successive cotton crops of the same variety and certification class may be grown on the same area without any time interval, provided that the cotton crop varietal purity is maintained.

If crop rotation is being implemented, the minimum standard for cottonseed production is to rotate 2–3 times for different cottonseed classes [9]. Ideally, the land used for cottonseed production should be rotated in a way that minimizes or avoids the build-up of weeds and insect pests. The primary goal of cottonseed crop husbandry practices is to provide an optimal growing environment for the cottonseed crop that ensures high-quality cottonseed production.

(iv) ***Uniformity of the Cottonseed Fields***

The selection of fields for cultivating cottonseed is crucial for various reasons. One of the key considerations is the uniformity of the seed fields. The evenness of the cottonseed fields is reflected in the uniformity of the maturity of the cotton crop. Variations in soil types, low-lying areas, and other factors can cause uneven maturity of the cotton crop. This may result in some areas opening earlier or later than others, creating "green spots" at harvest time, which can be problematic from the perspective of green plant material. Achieving uniformity in cotton crop maturity and boll opening during harvest can significantly improve the quality of harvested cottonseed.

3.2 Climatic Requirements for Cottonseed Crop

The factors to be taken into account for the growth of cottonseed crops are the climatic conditions such as appropriate temperature, moisture, and soil conditions.

(i) *Temperature Requirements*

The production of cotton and the quality of its fibres are significantly impacted by the air temperature throughout the cultivation period. Cotton thrives best when the weather conditions are favorable, and the summer temperatures remain above 25 °C. Any temperature below 20 °C can affect the growth of the cotton plant, particularly during the flowering and boll stages. In reality, low temperatures right after planting, when germination takes place, as well as cold night temperatures during any growth phase, can pose a potential threat to the cotton plant. The emergence of cotton plants can be adversely affected by cold weather. For optimal cottonseed germination, the temperature in the cotton seedbed should be at least 18 °C. It is essential for growers to have a comprehensive understanding of weather patterns during sowing to ensure the most advantageous growing season for the cotton crop.

(ii) *Moisture Requirements*

Globally, the amount of moisture required to ensure an average harvest varies from approximately 500–1,250 mm. Cotton is a plant that can tolerate drought and can still provide relatively good yields under dry conditions, even in areas where the annual rainfall is less than 500 mm. However, for profitable yields that meet quality standards, a higher amount of rainfall that is evenly distributed is necessary. Excessive air moisture during the ripening phase and the period leading up to harvest can cause boll decay. All seeds require moisture to sprout and grow. Insufficient moisture during planting stages can harm the cotton crop as well. The period from the flowering stage to boll maturity is the most crucial for the cotton crop in terms of moisture requirements. Inadequate moisture during this period can result in excessive shedding of flowers and bolls. Excessive moisture can also lead to unwanted, excessive plant growth. Both insufficient and excessive moisture can result in lower yields.

(iii) *Soil Requirements*

The most suitable conditions for the cotton plant are found in deep, rich, sandy loam soils that have good drainage. The growth of cotton is hindered in sandy or heavy clay soils, as the latter can cause issues with seedling germination. Ideally, the soil should be at least one meter deep and free from any impenetrable layers, such as plough soles, stone reefs or a high-water table. If the soil is poorly drained, it should be avoided, as cotton is vulnerable to waterlogging. While cotton can tolerate varying pH levels, the best results are achieved when the soil pH is between 5.5 and 7.5 [7].

3.3 *Inspections of the Cottonseed Field and Seed Crop*

An official from the "Certification Agency" or their assigned representative, as well as the seed producer or grower, are obligated to carry out a field examination of seed crops to ensure they satisfy the necessary criteria. Seed crop inspection is a routine and standard practice for seed authentication. Seed regulators will make 3 to 6 visits to the seed field during the growing season to verify that the cottonseed crop grown or produced meets the designated field standards for certification requirements of a given country. Seed inspectors must be granted unrestricted access to the cottonseed field and all documentation. Certification will be expedited by compliance and adherence to guidelines concerning the cottonseed crop grown or produced.

3.3.1 Major Criteria for the Inspections

During inspections to confirm the quality of cottonseed fields, several crucial factors need to be evaluated. These include ensuring that the cotton seed crop is cultivated from an approved source (retaining labels is necessary to establish this), verifying that the field meets the required land standards for the previous crop, ensuring that the prescribed isolation standards meet the minimum requirements, verifying that the grown or produced cotton seed crop adheres to national standards for rogueing, and confirming that the grown or produced cotton crop accurately reflects the varietal specifically identified characteristics.

3.3.2 Stages of the Inspections

The inspection of a farmland and seed crop is carried out at diverse periods of crop development to conduct different confirmations and assessments of different pollutants, deviations, diseases, and weeds. The following are the objectives that need to be achieved during a particular phase of cotton field and seed crop inspection:

(i) ***Sowing Time***

The primary goal of scrutinizing the sowing time is to ascertain the source and genetic authenticity of the cultivated cotton plants. Additionally, it aims to confirm that the field is free of any accidental plants and that the prescribed isolation gaps are upheld. Furthermore, it involves validating the quality of seed bags, planting densities, and adherence to the requisite sowing schedule.

(ii) ***Pre-flowering***

The pre-flowering or vegetative phase evaluation aims to verify the presence of any aberrant plants and the effective removal of such plants from the cultivated cotton fields. Additionally, it is conducted to ensure that the vicinity surrounding the cotton crop is maintained in a tidy manner.

(iii) ***Flowering Stage***

These stage checks whether off-type or volunteer other cotton plants are still present and that any grown or produced cotton plants that are diseased are effectively removed from the fields.

(iv) ***Post-flowering (Pre-harvest) Stage***

When conducting inspections after flowering or before harvesting, it is important to perform detailed counts to ascertain the level of different impurities in the seed field. Additionally, it is crucial to identify and eliminate any off-type or volunteer plants, as well as other cotton plants that were previously indistinguishable, before they begin to shed pollen.

(v) ***Harvest Time***

This is the final inspection stage that must be carried out on a cottonseed crop field. In this examination, detailed counts are made for different aspects, and isolation requirements are ultimately verified to ensure that the seeds have fulfilled all the necessary standards for cottonseed fields. If the cottonseed field fulfils the criteria for cottonseed crop certifications, the cottonseed grower or producer ought to be provided with essential guidance concerning the safety measures to be implemented while harvesting, ginning and delinting.

3.3.3 Conditions for Cottonseed Field and Cottonseed Crop Re-Inspection and Rejection

The cottonseed fields and cottonseed crops that fail to meet the required standards will be rejected if:

i. The cottonseed field size exceeds the registered size;
ii. There is no crop cultivation in the registered cottonseed field;
iii. The cottonseed crop is affected by floods or poor crop management and differs from the sowing report;
iv. The cottonseed field and crop fail to meet any of the required standards; and
v. The cottonseed grower or producer fails to submit applications for inspection of the cottonseed field and crop within the specified time before sowing or planting.

If the cottonseed fields and crops do not conform to the prescribed standards for certification at any stage of inspection, the certifying agency may allow for re-inspection upon the request of the grower or producer. The sources of contamination in the cottonseed field and crop must be removed within the prescribed isolation distances and/or contaminated plants in the cottonseed field (if directed by the certifying agency) before re-inspection can be performed. The certifying agency may also perform additional re-inspections if deemed necessary.

4 Cultivation Practices for the Cottonseed Crop

The cotton crop is grown in a diverse range of soils, although it thrives best in deep, fertile soils that have excellent drainage, ample organic matter, and a high capacity to retain moisture. However, some countries cultivate cotton in soils with cracking clays. *G. barbadense* is more tolerant to salt than *G. hirsutum*, as reported by Ashour and Abd-El'Hamid [3]. Salinity stress can negatively affect cottonseed germination and emergence, with the most common outcome being stunted cotton plant growth, according to Ashraf [4]. Irrigation is a viable option for growing cotton in poor-quality soils, since it provides the necessary moisture and nutrients in a controlled manner [6]. The flooding of furrows at ground level is the primary irrigation method, which necessitates proper field levelling, as observed by the OECD [22].

4.1 Soil and Seedbed Preparation

The aim of primary tillage is to enhance soil aeration, promote irrigation saturation, and integrate ample amounts of plant residue into the soil. Effective and economical tillage requires low soil water content. Improper moisture levels during tillage can lead to soil structure breakdown. Primary tillage is indispensable for virgin soils or any soil type to eliminate debris and break up the plough layer. All plant remains must be thoroughly incorporated into the soil. A uniform seedbed with appropriate planting depth and spacing, optimal germination, weed control, and sufficient moisture retention is crucial for high-quality cotton yields and cottonseed qualities.

Proper seedbed preparation is crucial for cotton crop emergence and population. Research indicates that half of the variable costs and all of the annual fixed costs are spent during or prior to the first 40 days of cotton crop production. Cotton crop growth commences well when seeds are sown in favorable seed to soil contact, warm soil temperatures of at least 25 °C, sufficient soil moisture, and fertility. Early prevention and management of weed seed and host plants are also essential for good crop growth [7].

4.2 Planting Dates

Planting cotton requires careful preparation of the soil to ensure sufficient moisture for favorable germination and root development. It is recommended to use pre-prepared ridges that provide proper drainage and optimal pre-determined temperature. The ideal planting time varies depending on the temperature and specific conditions of each season. Temperature is the primary factor that affects the plant's growth and yield [1, 24]. Planting can begin when the soil temperature at a depth of 10 cm is above 14 °C for at least three consecutive days [22].

The appropriate planting time for cotton varies greatly between countries and even within regions of the same country, depending on the specific agro-ecological zones. In Ethiopia, for example, the optimal planting time for "Irrigated Upper Awash" areas is between the first week of April and the first week of May, while "Irrigated Middle Awash" areas are best planted between the third week of April and the last week of May. "Irrigated Lower Awash" areas should be planted between the first week of May and the last week of June, "Irrigated South Omo" areas between the third week of June and the last week of July, and "Irrigated Arba Minch" areas between the first week of May and the last week of June. In "Rain-fed" areas, planting dates depend on the arrival of rainfall, but most commonly occur between June and July [18]. Early planting while there is still moisture in the soil increases the chances of successful planting in rainfed cotton crop producing areas. However, early planting also comes with risks, such as seedling vigor and disease problems associated with cool and/or wet periods, premature cutout due to early fruiting and drought, and late-season boll rot due to expected rains in late seasons. Late planting exposes the cotton plant to insect pest attacks. Boll rot is common in areas where boll opening coincides with rainfall, high humidity, and overcast conditions. Seed sprouting from the exposed seedcotton can also be a problem during the fall of some years if similar conditions prevail. Additionally, significant yield loss and quality degradation can occur when lint is exposed to rainfall and wind [11].

4.3 *Plant Population/Seeding Rate*

Likewise, the cotton plant density and seeding rate vary across countries based on the distinct agro-ecological zones of specific regions. In the United States, the final stand of cotton plants is planned to have 8–12 plants per meter of row. The planters are calibrated to sow 8–16 seeds/m, depending on the soil type. If the seed quality is poor or if fields are expected to have seedling diseases, soil crusting, or poor emergence, the planting rate may be increased. The planters can be calibrated for each variety to be planted, as different varieties have seed sizes ranging from 8,000 to 13,000 seed/Kg, which significantly affects the number of seeds planted. Therefore, the final rates may range from less than 15 to more than 20 kg/ha. With proper management of nitrogen, moisture, and insects, satisfactory yields can be achieved, although dense stands tend to increase the node number at which plants begin fruiting [11]. On the other hand, in Ethiopia, the recommended planting cotton plant density and seeding rate for agro-ecologically rain-fed and irrigated areas of the country are spaced 80 cm between rows and 25 cm spacing between plants, and 90 cm between rows and 20 cm spacing between plants, respectively, providing approximately between 50,000 and 55,555 plant populations per hectare [18].

4.4 Water Management and Irrigation

While diverse cotton growing areas must take into account various factors such as environment and soil types, a key requirement for a successful cotton crop is ample moisture throughout the growing season. This moisture can be obtained from rainfall, supplemental irrigation, or full irrigation. Proper irrigation management has the greatest impact on cotton yields compared to other inputs, and it also maximizes the benefits of other crop inputs used for pest control. As a perennial plant grown annually, cotton has a complex relationship with applied irrigation. It is a luxury consumer, meaning that excess water will result in strong vegetative growth at the expense of yield. To keep the cotton crop's root zone below field capacity and above allowable depletion, irrigation scheduling should determine the duration and frequency of irrigation. This ensures that the crop's roots are exposed to an ample supply of easily available water with sufficient oxygen to promote healthy root growth.

As cotton accumulates biomass, its water requirements increase, with taller plants demanding more water. Avoiding water stress at the beginning of the first square is critical in establishing adequate plant structure to facilitate yield goals. The growth rate is represented by the length of the internodes, which should be 3.5–5 cm from the first square to peak bloom. If the length exceeds 5 cm, the plant is receiving excess water or fertilizer. If the length is less than 3 cm, the plant is under stress (source: www.netafimusa.com).

4.5 Crop Rotation

By making wise management decisions, a cultivator can significantly decrease the negative effects caused by various insect pests, plant diseases, and undesirable plants on their harvest. Even though it may prove challenging for some growers, a well-planned crop rotation strategy that involves non-host crops for the primary cotton crop is one of the most effective ways to disrupt the life cycles of numerous insect pests, plant diseases, and weeds that lead to decreased yield and quality losses in cottonseed production. Crop rotations are also utilized to reduce the growth of volunteer cotton plants, which in turn decreases the need for manual weeding. Additionally, crop rotations are employed to optimize soil nutrition and fertility [11].

4.6 Rouging

Achieving varietal purity during seed production is a challenging task that demands meticulous efforts. It involves the timely and painstaking removal of off-type plants, volunteers, weeds, diseased plants, and contaminants that do not conform to the varietal characteristics from the cotton field's seed production. This process, known as rouging, must be carried out at various stages by closely monitoring the seed crop's development. The unwanted plants must be identified and removed based on visual inspection in the field. Any plant that deviates from the cotton crop's produced variety in terms of morphological features is considered an off-type and included in the rogue category. The rouging process should be completed before the flowering stage to prevent pollinating insects from causing genetic contamination.

5 Cottonseed Crop Harvesting

The process of harvesting cotton commences approximately half a year after sowing and is the most expensive aspect of cotton farming. It is crucial to harvest the cotton as soon as the boll splits open and maintain its cleanliness to prevent weather-related harm that may diminish its quality and output. If the cotton is left in the field for an extended period, it may fall out or be destroyed by rainfall. The optimal time to harvest cotton is when the seed-cotton mass contains 12% moisture or less and is free of leaves or other plant fragments that could pollute the cottonseed. Cotton can be harvested either by hand or by using mechanical means [7].

5.1 Harvesting by Hand (Manual Picking)

Around 70% of the global cotton production, which exceeds 100 million bales, is manually harvested [13]. While a few countries have adopted machine harvesting, it is only in developed nations like the US and Australia where the cotton crop is almost exclusively (>99%) harvested mechanically due to economic considerations. In contrast, approximately 75% of the world's cotton is still hand-picked, boll by boll, especially in less developed countries and those with lower labor costs [26].

Hand-picking is carried out over a period of at least two months since not all bolls ripen at the same time. It is preferable to collect dry cotton that is free from debris. Although manual harvesting is slow and requires considerable labor and time, it helps maintain the fibre characteristics of cotton. The level of foreign matter in harvested seed cotton typically ranges from 1 to 5% for hand-picked cotton before it is processed in a gin. Overall, hand-picking results in much cleaner cotton compared to mechanical harvesting.

5.2 Machine Harvesting

Cotton varieties with a stunted and condensed growth pattern that exhibit premature and simultaneous opening of bolls are best suited for mechanical picking. There are two fundamental kinds of automated cotton harvesters: spindle cotton pickers and stripper cotton harvesters. The process of machine harvesting is akin to manual harvesting, except that instead of human hands, machines are used to extract seed cotton from the plant. All harvesting methods use air to transport and lift the seed cotton into a storage compartment known as a basket. Once the basket is filled, the stored seed cotton is emptied into a boll buggy, trailer, or module builder. In comparison to manual harvesting, machine harvesting, whether spindle or stripper, results in the collection of debris, leaves, bits of stems, and burrs, which have to be removed at the gin.

To ensure effective machine harvesting of cotton, the cotton machines must be in excellent condition before being deployed in the field. Replace any spindles that are excessively worn or damaged. The alignment and adjustment of spindles to moisture pads and doffers have a significant impact on the efficiency of a cotton machine harvester. Poorly adjusted spindles will leave some cotton on the spindle, resulting in spindle twist and reduced quality and harvesting efficiency. A well-adjusted machine harvester and operational speed will collect cotton with minimal debris, especially bark. The picking units and basket grates must be cleaned each time the basket is emptied. The accumulated debris and low-quality fibre should be discarded and not mixed with the under grown or produced specified cottonseed variety.

Machine harvesting should begin when the dew has dried and end after the dew has formed. Using a meter to assess the moisture of the seed cotton helps; if one is not accessible, the seed must be bitten, and if they crack, the moisture is probably low enough for machine harvesting.

5.2.1 Spindle Cotton Pickers

The majority of the cotton that is harvested mechanically is harvested using spindle cotton pickers. These machines are designed to harvest the cotton from bolls that are open, while leaving partially opened bolls and empty burs on the plant to be harvested later. The spindles on the picker rotate and have barbs that catch and release the cotton fibres once they have separated from the boll. This type of picker is selective and uses spindles that have tapered, barbed spindles to remove the cotton from the bolls. The spindle machines only remove the lint and seed from the plant, leaving the burs, unopened bolls and the plant itself intact. The barbed spindles wrap the cotton fibres around them, which are then removed by brushes and deposited into a basket. The spindle cotton picker can result in field losses ranging from 5 to 15%, as the cotton burr typically remains attached to the stalk during harvesting. Ideally, the height of the plant should not exceed 1.2 m for efficient spindle picking. Since the spindles gently pluck the lint from the bolls, rather than stripping them, they are particularly well-suited for long staple varieties [17].

5.2.2 Stripper Cotton Harvesters

Mechanical cotton harvesters referred to as stripper harvesters are utilized to collect the whole plant, including both mature and immature bolls, and are discarded after the crop is fully grown. These machines are equipped with boll separators or bur extractors to distinguish between mature and immature bolls. Stripper harvesters have rollers or mechanical brushes that detach the entire boll from the plant. The cotton stripper is indiscriminate, collecting not only well-opened bolls but also cracked and unopened bolls, burrs, and other plant fragments. Stripper harvesters are most productive when collecting cotton varieties that mature uniformly. During harvesting, cotton burrs are removed along with the seed cotton, resulting in field losses ranging from 2 to 5%. Mechanical harvesting with a spindle cotton harvester can be performed multiple times during the season, resulting in more uniformly mature seeds than with a stripper and a superior product. On the other hand, strippers can harvest more area per hour than spindle pickers with comparable operating widths. The height of the plant should not exceed about 0.9 m for stripper-harvested cotton because too much foreign matter will be collected [17].

6 Post-harvest Handling of the Cottonseed

Following the harvest, it is imperative to handle the cottonseed with great care to ensure the preservation of its quality.

6.1 Seed Cotton Ginning

The value and usefulness of cotton are only realized after the separation of the fibre from the cottonseed and impurities through the process of ginning. Cotton ginning involves the extraction of cottonseed from the fibre using a specialized machine called a cotton gin. The fibre is the primary agricultural asset of cotton, with the highest quality cotton producing long, robust, and fine fibres. Half of the cottonseed is fibre, while the other half has significant economic value. Following the ginning process, the seed is composed of 50% kernel, 11% linters, and 38% hull [17].

6.1.1 Saw-Type Gin

Gin stands of the saw type usually come with saws that have a diameter of 12–18 inches (30.5–45.7 cm) and are placed at intervals of 1/2–1 inch. A single mandrel can hold as many as 198 saws. These saws extend through the ginning ribs, grip the fibres, and remove them from the seeds, which are too big to pass through the opening in the ginning ribs. The productivity of a single gin stand has increased from less than 1 bale per hour to over 15 [13].

6.1.2 Roller-Type Gin

The first mechanical device used for separating extra-long staple cotton lint from seed was the roller-type gin. This technology consisted of saws projecting through a leather ginning roller, a stationary knife held tightly against the roller, and a reciprocating knife that pulled the seed from the lint while the lint was held by the roller and stationary or rotary knife [13]. Roller-type gins are classified as double roller, single roller, and rotobar rotary knife roller ginning.

(i) ***Double roller gin*** is the most preferred technology due to its ability to gin most varieties at low costs. In the past decade, the use of double roller ginning has increased significantly, and currently, around 35% of the world's cotton is being ginned on roller gins, with double roller ginning alone having a 30% share [25].
(ii) ***Single roller gin***, on the other hand, is becoming outdated due to its higher ginning costs.
(iii) ***Rotobar rotary knife roller gin*** has a ginning rate that is about 20% of the saw-ginning rate per unit of length. This technology is suitable mainly for black-seeded cotton, where the attachment is lowest. However, efforts are being made to make it suitable for fuzzy-seeded cotton as well.

The gin stands, whether saw-type gin or roller-type gin, pull the fibre from the seed. They are the heart of the ginning system, and the capacity of the system, as well as the quality and potential spinning performance of the lint, depend on the operating condition and adjustment of the gin stand. Gin stands must be properly adjusted, kept in good condition, and operated at or below design capacity. Overloading gin stands can reduce the quality of the cotton. Increasing the ginning rate above the manufacturer's recommendation can increase yarn imperfections, and seed damage can result from increasing the ginning rate, especially when the seeds are dry. High ginning rate and low seed moisture cause seed damage ranging from 2 to 8% of the seed in gin stands. Therefore, it is essential to keep the gin stand in good mechanical condition, gin at recommended moisture levels, and not exceed the capacity of the gin stand or other components of the system [12].

It is important for ginners to be knowledgeable about the selection criteria for ginning technologies when purchasing ginning machinery. This should be based on information available in their area and other guiding factors, rather than solely on the suitability of the ginning technology for their cotton. Ginners should also understand the role of cotton fibre parameters, such as length, micronaire, strength, and trash content, in selecting suitable ginning technologies. It is recommended that ginners purchase ginning machines based on the suitability of the technology to the fibre in their area [25].

Table 2 Comparisons between Saw-type Gin and Roller-type Gin Machines

Advantages of roller-gins	Situation in saw-gins
– *Much lower investment costs* – *Less damage to cotton fibre* – *Simple and robust design, easy to operate* – *No major repair and maintenance problems* – *No major need for specialized manpower*	– *Considerably higher investment costs* – *Damage to cotton fibre (fibre shortening)* – *More complex design, not so easy to operate* – *Major repair and maintenance problems* – *Need for highly specialized manpower*
Disadvantages of roller-gins	*Situation in saw-gins*
– *Low ginning output* – *Not very efficient in cleaning thrash* – *Higher cost of ginning per bale*	– *Comparatively higher output* – *Efficient in cleaning thrash* – *Lower ginning cost per bale*

Evaluation of ginneries engaged in cottonseed focuses on their performance and seed quality. Saw gin is considered more effective than roller gins for ginning seed cotton. Results show that saw gin causes about 1–1.5% seed damage and 7–8% lint remains with the seed, while roller gin causes 3–4.5% seed damage and 12 to 14% lint remains with the seed. Ginning at a rate of 4.5–5.5 kg of lint per 100 cm saw length per hour guarantees maximum seed quality [14]. Table 2 illustrates a comparison of saw-type and roller-type gins.

6.2 Cottonseed Delinting

The process of removing the residual seed coat fibres after ginning is referred to as cottonseed delinting. This can be achieved through three different methods: chemical, mechanical, and flame. Among these, chemical delinting is the most widely used technique, particularly when the cottonseeds are intended for planting. When conducting chemical delinting, it is crucial to exercise caution. This involves combining the cottonseeds with concentrated sulfuric acid (H_2SO_4) or anhydrous hydrochloric acid (HCI), both of which are commercial grade chemicals. In chemical delinting, there are two primary methods: dry gas delinting and wet acid delinting [17].

6.2.1 Flame Delinting

The technique of flame delinting (also known as flame "zipping") involves applying kerosene to the fuzzy cottonseed, heating it, and then exposing it to a direct flame for short periods of time [29]. This method burns off more of the remaining seed fuzz than mechanical delinting, but it can also lower the quality of the seed by exposing it to high temperatures [5, 17]. Although flame delinting can be highly effective, it is a challenging process. The flamed seed must be cooled rapidly, and improper cooling can cause damage that leads to poor germination and an inadequate stand. While flame delinting significantly improves the flowability of cottonseed, the results do not meet the standards required for modern precision planters. As a result, this method is no longer used due to the inability to control the quality of treated cottonseed. However, because the cottonseed are heated as they pass through the flame and burning linters, it is critical to quickly and effectively remove sparks and cool down the seed. Failure to do so could result in severe heat damage to the cottonseed [8]. Modifications have been made to eliminate this problem.

6.2.2 Dry Gas Delinting

During the process of dry gas delinting, a rotating drum filled with fuzzy seeds is injected with dry anhydrous hydrochloric acid (HCI) gas. The burners heat up the drum to achieve a seed temperature of 49 °C, leading to hydrolysis. The hydrolyzed lint is then broken down in a scalper. Any remaining traces of acid are neutralized by introducing ammonia gas. This dry acid technique is estimated to be around 50% less expensive than alternative methods [14].

6.2.3 Wet Acid Delinting

In Ethiopia, the most common and straightforward method for delinting cottonseeds is by using concentrated sulfuric acid (H_2SO_4). The process involves treating the ginned seed with concentrated sulfuric acid (100 ml Kg^{-1} of fuzzy seed for 2–3 min), followed by thorough washing with water and an alkaline solution to remove any acid residues. The seed is then washed again with water to remove any alkaline residues, dried to 8–9% moisture content, graded, and packed [14]. However, the wet acid delinting process produces a large amount of acidic wastewater that can cause environmental pollution and health hazards. Therefore, an alternative method called dry gas delinting has been developed and gained popularity in recent years.

Procedure to Be Followed for Wet Acid Delinting

- Choose plastic laminated containers or baskets for chemical delinting, as earthenware and metal vessels can corrode and pose a handling risk.
- Add commercial concentrated H_2SO_4 at a ratio of 1 L acid per 10 kg of fuzzy seed (100 ml Kg^{-1} of fuzzy seed) to the container with the required amount of seeds.
- Stir vigorously and continuously with a wooden stick for 2–3 min until the fuzzy material sticking to the seeds is completely digested, and the seed coat turns a dark brown color.
- Add water to fill the container, drain the acid water, and repeat the washing process 4–5 times to thoroughly rinse the delinted cottonseeds with clean water until all acid residues are removed to prevent seed damage.
- Soak the entire washed seed in water at a ratio of 1:10 to remove any ill-filled and floating seeds while the healthy and good seeds remain at the bottom (sinkers).
- Drain the water completely and dry the delinted sinkers seeds in the shade.

Advantages of wet acid delinting

- Eliminates some externally seed-borne pathogenic organisms.
- Kills insect pests' eggs, larvae, and pupae.
- Helps to remove immature, ill-filled, and damaged seeds.
- Improves plantability in plate planters and allows for greater ease of mechanical cleaning.
- Improves germination percentage by removing inhibitors in the pericarp.
- Makes seed dressing more effective and easy.
- Reduces the volume of seed storage.

Disadvantages of Wet Acid Delinting

- Extreme corrosiveness of chemicals to machinery and dangerous to handle.
- Consume large amount of energy to dry the seed.
- Cause both soil and water pollutions.

6.3 Cottonseed Treatments

Microbes that cause disease, known as pathogens, are frequently present in or on cotton seeds and in the soil. These microbes can lead to diseases that devastate cottonseeds or seedlings. It can be easy to underestimate the impact of diseases and nematodes on cotton production, as the cotton plant tends to be less susceptible to disease than other crops, and nematode symptoms can be mistaken for other issues. To offset potential losses from poor stands caused by seedling disease, cottonseed growers or producers should acknowledge the cost of combating diseases and nematodes, rather than increasing seeding rates [11]. Protective and systemic fungicides, as well as insecticides and occasionally antibiotics, can be applied to cottonseeds to prevent or reduce damage caused by many pathogens. In many developed nations, it is common knowledge that treating cottonseeds with both protective and systemic fungicides maximizes the emergence of cotton seedlings. The chemistry used to treat cottonseeds has significantly improved in recent years [15]. However, many cotton farmers do not use chemical treatments on their cottonseeds before planting due to the cost of protective and systemic fungicides, resulting in untreated cottonseeds being almost exclusively utilized in Ethiopia.

When possible, it is advisable for cottonseed growers or producers to always treat their cotton seeds with fungicides. Fungicides are typically used to combat seed rots, surface molds, and seedling blights as well as other pathogenic fungi caused by organisms found in the seed and soil. Some treatments, like carbendazim, thiram, and captan at 2 g/kg of cottonseeds, are protectant fungicides that shield the cottonseeds from fungi present on the seeds or in the soil surrounding the cottonseeds. Other treatments, such as Vitavax (carboxin), baytan, metalaxyl (Allegiance), and mefenoxam (Ridomil Gold), have systemic activity and offer some protection to the cotton seedling immediately following germination, as they are absorbed by the seedling. Cottonseed growers or producers can significantly reduce the impact of cotton seedling diseases by avoiding conditions that increase the risk of fungal pathogen damage to cottonseeds/seedlings [11].

Guideline for utilizing chemical treatments on cottonseeds

i. An ideal pesticide ought to be highly efficient in managing a particular disease-causing organism.
ii. The chemicals employed must be non-toxic to cottonseeds, possess long-term stability, user-friendly, and low in toxicity to both humans and animals.
iii. Every cottonseed that undergoes chemical treatment must have a tag indicating the chemical utilized. Depending on the level of danger of the cottonseed treatment to humans, the label must state either "Caution," "Poison," or "Poison Treated."
iv. According to the law, chemically treated cottonseeds are strictly required to be colored with a bright color.

6.4 Cottonseed Storage

Most cropseeds are typically stored for a certain period of time, during which they may undergo considerable deterioration. This is because seed deterioration is an inevitable and irreversible process that occurs due to spontaneous metabolic activities within the seeds, which function as living organisms. While good storage conditions can slow down the rate of deterioration, they cannot improve seed germination and vigor, regardless of the quality of the storage facilities used. Therefore, seeds that are intended for planting are managed in a way that minimizes their deterioration and maintains their viability and vigor as high as possible before their quality declines to unacceptable levels [20].

Expensive storage problems often arise during field exposure, harvesting, and conditioning of the seed. Excessive harvesting delays, mechanical injuries, and improper drying techniques, followed by poor storage conditions, can lead to rapid deterioration of seed germination and vigor. Studies have also shown that seeds do not store well for long periods, especially for those crop species that possess high seed oil content and require seed scarifications (delinting of cottonseeds), such as cottonseed crop.

For cottonseeds, the best storage conditions involve maintaining moisture contents between 6 and 8% [17]. When planning for safe seed storage, temperature and relative humidity should also be taken into consideration, as seed deterioration can be greatly slowed down when seeds are stored under cool, dry conditions. A general rule of thumb for safe seed storage is that the sum of the temperature (in degrees Celsius) and relative humidity should not exceed 100 units. For instance, if the temperature is 40 °C, the relative humidity should not exceed 60%. In general, seeds should not be stored for extended periods under conditions of high temperatures and relative humidity.

Steps for Proper Seed Cotton Handling Before Storage

i. Reduce insect infestation in the field by harvesting and collecting seed cotton from the field on time.
ii. Remove insect-infested seed cotton before storing; this effectively removes future sources of infestation or contamination.
iii. Dry the seed cotton sufficiently to prevent unforeseen ignition of fire threats, microorganism growth, and cottonseed respiration rate.
iv. Select an appropriate storage method and environment for the seed cotton, as well as a time frame for seed cotton storage. Sanitation of the seed cotton storage containers and of the warehouse storage, so that insects or larvae and weed seeds are removed prior to storage of newly-harvested seed cotton.

Requirements for an Ideal Cottonseed Storage Facility

i. The storage should provide maximum possible protection from ground moisture, rains, insect pests, fire and other calamities.
ii. The storage should provide the necessary facility for inspection, disinfection, loading, unloading, cleaning and reconditioning.
iii. The storage should protect cottonseed from excessive moisture and temperature favorable to both insect and mould development.
iv. The storage should be economical and suitable for a particular situation.

Steps for Correct Handling of Stored Cottonseed

i. Place bags containing cottonseed on pallets or tree branches arranged in a lattice formation to prevent moisture from the floor affecting the seed's moisture content, deterioration rate, and germination.
ii. Check both the interior and exterior of the building for drainage or erosion issues, signs of rodent pathways and holes, and the presence of household or other waste or weeds. Remove them from around the building to block the entry of rodents and insects.
iii. Whenever possible, store cottonseed for the shortest possible time, particularly in hot or humid conditions. If it must be stored for an extended period, move it to more favorable storage conditions.
iv. Properly test, label, and tag the produced cottonseed before distribution, if possible.

6.5 Cottonseed Marketing

The process of seed distribution, from its production on the farm to its use by the consumer, is known as seed marketing. Depending on the type of seed and its proximity to its intended use, the marketing process can be a simple exchange between farmers or a complex transaction involving multiple intermediaries and a highly organized seed industry. Therefore, in order for successful cottonseed marketing, there must be a demand for cottonseed, a mechanism for timely supply, and an organizational structure to ensure the production of high-quality cottonseed [16].

Previously, seed production and marketing in Ethiopia were centralized, with seed producers given quotas and distribution handled by public seed enterprises, cooperative unions, and primary cooperatives. However, the seed marketing system is gradually shifting towards being governed by market forces, with the seed-marketing directive endorsed by the Ethiopia Ministry of Agriculture in March 2019. This will enable the transformation of the seed sector through the development of new seed laws, regulations, and directives. Currently, Direct Seed Marketing (DSM) is being piloted, allowing seed companies to take full responsibility for selling seed directly to farmers through agents or their own shops in designated districts.

Despite the seed-marketing directive in Ethiopia, the marketing of cottonseed is largely conducted through local/informal seed exchange and trade, which is more extensive than in commercial crops such as maize, wheat, sorghum, and *teff*. Obtaining certified cotton planting seed has always been a challenge, with farmers obtaining uncertified seed from various sources or saving their own. There is currently no legally certified cottonseed producer enterprise in the country, with basic and pre-basic seed production being handled by the Ethiopia Institute of Agricultural Research (EIAR), Werer Agricultural Research Center (WARC) at their own cotton research farms. They remain the only major suppliers of cotton planting seed.

7 Cottonseed Quality Testing (Laboratory Standard)

Seeds are the fundamental basis of agriculture. Farming's day-to-day operations have been modernized by technology, but the production of high-quality seeds remains crucial for the maintenance of yields and crop quality. The quality of cottonseeds is particularly important for cottonseed crop production, with high-quality cottonseeds being those with high germination and vigor potential, which are essential for establishing excellent plant stands. While most growers or producers are only familiar with standard germination, which measures a seed's ability to produce a healthy seedling under ideal conditions [15]. Other characteristics such as purity, trueness to variety, moisture content, number of weed seeds, and appearance are essential for farmers planting cottonseed crops.

To achieve profitable cottonseed production, it is essential to have a uniform plant stand of healthy and vigorous cotton crop seedlings. Therefore, every professional certified cottonseed grower or producer should aim to achieve and maintain high cottonseed quality. Understanding the basics of seed, seed quality, and seed standards can help seed producers, farmers, and growers or producers recognize the crucial role of superior cottonseed production. Additionally, it is important for growers to plant high-quality cottonseeds that are suitable for their farm situations, management styles, and intended market uses.

Table 3 Laboratory standards for conventional and hybrid cottonseed varieties

Category	Pre-basic seed	Basic seed	Certified seed
Pure seed (min %)	99	98	97
Other crop seed (max %)	0.5	0.1	0.3
Weed seed (max %)	N.S*	0.2	0.3
Infected/infested seeds (max %)	N.S*	N.S	N.S
Inert matter (max %)	1	2	2
Germination (min %)	85	80	80
Verification of species cultivar	–	–	–
Moisture content (max %)	8	8	8

N.S = Not Specified*
Source ESA [9]

Cottonseed growers or producers should be aware that classes of cottonseeds must meet national minimum quality standards (as outlined in Table 3) to legally market or distribute the produced cottonseeds to consumers.

8 Summary

Cotton crop is one of the more labor-intensive and expensive crop to produce. The cottonseed supply of adapted and quality cottonseed to farmers is essential to ensuring productive and remunerative cotton sectors for farmers, traders and states. Cottonseeds are needed that will germinate well, yield well and are adapted to climate and agronomic conditions. Cottonseed quality plays an important role in the production of cottonseed crops. High quality cottonseeds are those seed lots with high germination and vigor potential which are critical for establishing good stands. The most familiar seed quality laboratory standard testing is the standard seed germination test, which is a measure of the seed's ability to produce a normal, healthy seedling when conditions are ideal. Yet, characteristics such as trueness to variety, purity, number of weed seeds, moisture content and appearance are important to farmers planting cotton seed crops.

High-quality cotton crop seed is critical to success in the first forty days and the crop's ultimate performance. Choosing which variety to plant, which seed to sow, where to sow are one of the most critical first steps to be taken in to account in producing a cotton crop and achieving optimal seed yields and fibre qualities. Considerations for cotton variety and seed selection should also be catered to a range of general requirements like: planting dates, seedling vigor, water regimes (irrigated versus rainfed and degree or efficiency of irrigation), maturity classes, temperatures, soil moisture conditions, cotton seed crop's climatic requirements and plant growth characteristics. While, the land to be used for cotton seed crops cultivation also need to fulfill the specific requirements including: free from inseparable seed producing weed; free from volunteer crop, off-type or ratoon cotton crop plants; maintaining minimum permissible isolation distances from another neighboring cotton crop plants; field map of the produced or grown cotton crop plants seed; and registration of the cottonseed crop plants must be considered.

Inspections of the cottonseed field and seed crop inspection of the cotton seed crops have to be conducted at least at five stages namely: sowing; pre-flowering; flowering; post-flowering (ore-harvest); harvesting stages are the most common detrimental stages to ensure that the grown or produced cotton seed crop meets the prescribed seed crop's field standards certification. Cottonseed fields and cottonseed crop not conforming to prescribed standards for certification at any stages of inspection, the certifying agency either shall or could allow for the re-inspection, or may reject incase if the field standards are not maintained. The cottonseed producers or growers need to maintain prescribed minimum standards in order to meet minimum cottonseed qualities standards to be used for sowing.

Cottonseed qualities majorly further influenced by the pre- and post-harvest handling processes to be conducted. Cotton can either be handpicked (manual picking) or harvested mechanically (machine harvesting). Cotton is best harvested when the seed-cotton mass is at 12% moisture or less and with no leaves or other plant parts to contaminate the cottonseed. The cotton crop must be harvested as soon as the boll opens and kept as clean as possible, before weather can damage or completely ruin its quality and reduce yield. While the major post-harvest handling of the cottonseed crop includes: ginning, delinting, cottonseed treatments and cottonseed storage. Ginning is required as soon as possible after picking, is a separation process of the cottonseed from the cotton fibre using a cotton gin machine. Cottonseed delinting is the removal of the fine seed coat fibres that remain after ginning. Chemicals applied to the cottonseed can prevent or reduce the harmful attacks of many pathogens. These treatments include fungicides, insecticides, and, occasionally, antibiotics. Cottonseed treatment chemistry has improved dramatically in recent years. Nevertheless, many cotton growing farmers do not chemically treat cottonseed before planting due to the cost involved with protective and systemic fungicide chemicals. Cottonseeds store best when maintained at moisture contents between 6 and 8%. In planning for safe seed storage, temperature and relative humidity should be also be considered, because seed deterioration could greatly slowed when seeds are stored under cool, dry conditions.

References

1. ACCRC (2001) NUTRIpak—A practical guide to cotton nutrition. In: Rochester I (ed) Australian Cotton Cooperative Research Centre, Narrabri, New South Wales, p 60
2. Amanet K, Chiamaka EO, Quansah GW, Mubeen M, Farid HU, Akram R, Jatoi WN (2019) Cotton production in Africa (Chapter 17) In: Cotton production. https://doi.org/10.1002/978 1119385523
3. Ashour NI, Abd-El'Hamid AM (1970) Relative salt tolerance of Egyptian cotton varieties during germination and early seedlings development. Plant Soil 33:493–495
4. Ashraf M (2002) Salt tolerance of cotton: some new advances. Crit Rev Plant Sci 21:1–30
5. Chaudhry MR, Guitchounts A (2003) Cotton facts. Technical Paper No. 25 of the Common Fund for Commodities (CFC), International Cotton Advisory Committee (ICAC). Library of Congress Control No. 2003107604. ISBN: 09704918-3-2. http://www.icac.org/
6. Cothren JT (1999) Physiology of the cotton plant. In: Smith WC, Cothren JT (eds) Cotton: origin, history, technology and production. Wiley, New York, pp 207–268
7. Cotton Production Guideline (2016) Cotton production guideline. Department: Agriculture, Forestry and Fishery, Pretoria, Republic of South Africa, pp 1–20
8. Delouche JC (1981) Harvest and post-harvest factors affecting the quality of cotton planting seed and seed quality evaluation. In: Proceedings of the beltwide cotton conference. Mississippi Agricultural and Forestry Experiment Station, vol 39762. Mississippi State University, Mississippi State, Mississippi, pp 1–17
9. Ethiopian Standards Agency (ESA) (2012) Ethiopian standard, ES 441:2012. Cotton-specification: ICS:65.020.20, 2nd edn., pp 1–6
10. FAOSTAT (2017). http://www.fao.org/faostat/en/#data
11. Georgia Cotton Production Guide (2016) Georgia cotton production guide. In: Cooperative extension/The University of Georgia College of Agricultural and Environmental Sciences. UGA, Cotton Web Page, pp 1–148
12. Gérald E (2008) Comparative analysis of organization and performance of African cotton sectors. In: Quality and marketing of cotton lint in Africa. Africa Region Working Paper Series No. 121
13. International Trade Centre (ITC) (2007) Cotton exporter's guide. In: Product and market development, International Trade Centre, UNCTAD/WTO Geneva
14. Karivaradaraaju TV (2007) Processing of cotton seed at SIMA CD & RA. model training course on "Cultivation of Long Staple Cotton (ELS)". Central Institute for Cotton Research, Regional Station, Coimbatore, pp 15–22
15. Keith E, Guy C (2020) Cotton seed quality and planting decisions (Chapter: 6). In: 2020 Cotton information: NC: North Carolina Cooperative Extension, pp 51–53
16. Larry OC, Miller BM (1999) Seed marketing (Chapter 15). In: Seed science and technology, 3rd edn. Springer Science Business Media, New York, pp 352–361
17. McDonald MB, Copeland LO (1997) Seed production principles and practices of seed production. In: Chapter 14: Oil seeds. Springer Science+Business, Media Dordrecht, New York, pp 253–288. https://doi.org/10.1007/978-1-4615-4074-8
18. Merdasa B, Nurhussien S, Yonas B, Michael K, Donis G, Zemedkun A, Samuel D, Arkebe G, Seleshi G, Fikremariam T, Mekashaw A, Tamiru D (2022) Cotton production guideline in Ethiopia. Werer Agricultural Research Center (WARC), Ethiopian Institute of Agricultural Research (EIAR), Addis Ababa, Ethiopia. ISBN: 978-99944-3-509-8. http://www.eiar.gov.et/
19. Minnesota Crop Improvement Association (MCIA) (2013) Seed certification handbook. Agronomic Seed Certification Standards Supplement, pp 1–38
20. Neuendorf O (ed) (2004) Seed manual—Rural household seed security. In: A comprehensive approach to improve quality seed availability for small scale farmers. Harare. ISBN: 0-7974-2958-1
21. OECD (2012) OECD seed schemes, guidelines for control plot tests and field inspection of seed crops. Paris

22. Organisation for Economic Co-operation and Development (OECD) (2010) "Section 1—Cotton (Gossypium spp.)". In: Safety assessment of transgenic organisms, vol 4. OECD Consensus Documents, OECD Publishing, Paris. https://doi.org/10.1787/9789264096158-5-en
23. Randy B (2016) Evaluating cotton seed quality. Soil Crop Sci (SCS) Commun SCS-2005-16(6-05):1–2
24. Robertson B, Bednarz C, Burmester C (2007) Growth and development—First 60 days. Cotton Physiol Today 13(2):1–5
25. Sharma MK (2014) New trends in cotton ginning & cotton seed processing. In: Presented at 6th Meeting of the Asian cotton research and development network, at Dhaka, Bangladesh
26. Supak JR, Snipes CE (eds) (2001) Cotton harvest management: use and influence of harvest aids, The Cotton Foundation Reference Book Series, No. 5, The Cotton Foundation, Memphis, TN
27. The Cotton Foundation (2007) The most critical period in cotton production. In: Expert recommendations of best management practices for an efficient, Cost-effective cotton production system through fruiting to finish. A publication of the National Cotton Council of America and the Cotton Foundation, pp 1–19
28. TRAIDCRAFT (2011) Cottonseed supply for planting in Africa: a study into the functioning of current structures for research, breeding, multiplication and distribution and their impacts on cotton farmers. Prepared for Traidcraft Policy Unit, pp 1–100. http://www.traidcraft.co.uk/
29. United States Department of Agriculture—Agricultural Marketing Service (USDA-AMS) (2014) Limited scope-hydrogen chloride use petition. https://www.ams.usda.gov/sites/default/files/media/Hydrogen%20Chloride%20TR%202014.pdf
30. Vitale J, Greenplate J (2014) The role of biotechnology in sustainable agriculture of the twenty first century: the commercial introduction of Bollgard II in Burkina Faso. In: Songstad DD, Hatfield JL, Tomes DT (eds) Convergence of food security, energy security and sustainable agriculture, biotechnology in agriculture and forestry, vol 67. Springer, Berlin, pp 239–293

Food and Nutrition (Cotton as a Feed and Food Crop)

Michael Kebede

Abstract Cotton is primarily grown for the fibre, used in the textile industry and the oil from the cottonseed. However, cottonseed is considered a major source of vegetable oil worldwide, both as a food crop for humans and a source of protein for ruminant and, to some extent, non-ruminant animals. However, its use is limited due to the presence of a significant hindering polyphenolic anti-nutritional and toxic enzyme called gossypol, which is a terpenoid aldehyde, as well as cyclopropenoid fatty acids (CPFAs) and tannin compounds that make it toxic to humans and monogastric animals. Commercially, cottonseed is widely used for its oil, as well as other by-products and as planting material "seed" by farmers. The most well-known by-products of cottonseed are cotton linters, hulls or husks, cakes and meals, oils, and stalks and gin wastes. To maximize the use of these by-products, which are obtained after the challenging process of cottonseed oil extraction, they must undergo additional processes such as caustic or alkali refining, bleaching, winterization, fractionation, hydrogenation or hydration, interesterification and deodourization before they can be used as edible oil or finished consumer goods. The composition of cottonseed oil is mainly made up of fatty acids, and the physicochemical functional and nutritional characteristics of oils; while the fats are largely depend on their fatty acid composition and triglyceride structure. These fatty acids can be classified as saturated if they do not contain double bonds, monounsaturated if they contain one double bond, and polyunsaturated if they have two or more double bonds in the molecule. Cottonseed oil typically contains 47–81% unsaturated fatty acids, including 13–21% monounsaturated (oleic) and 34–60% polyunsaturated (linoleic), as well as 18–35% saturated (palmitic and stearic) fatty acids.

M. Kebede (✉)
Ethiopian Institute of Agricultural Research (EIAR), Melkasa Agricultural Research Center, P.O. Box 436, Adama, Ethiopia
e-mail: kebedeyomichael@gmail.com

K. Murugesh Babu et al. (eds.), *Cotton Sector Development in Ethiopia*, Textile Science and Clothing Technology, https://doi.org/10.1007/978-981-99-9149-5_15

1 Introduction

Cotton (*Gossypium spp.*) is an arborous plant from the Malvaceae family. It is one of the earliest plants that were farmed by humans and has been utilized for over 4,000 years. It is mainly grown for its fibre, which is used in the textile industry, and the oil extracted from the cotton seed [54]. Over 20 million farmers in developing Asia and Africa rely on cotton as a source of income. The primary objective of its cultivation is to meet the essential need for cotton textiles. For every kilogram (Kg) of cotton fibre, 1.65 kg of cottonseed is produced, making cottonseed a significant by-product of the cotton plant [94].

Cottonseed is mainly utilized for its oil extraction and feed, but it also contains several essential nutrients, including high-quality protein, hull, and linters, which constitute 25–35% of the seed and 15–25% of the oil. Gossypol, a phenolic compound that accounts for about 1% of the seed weight in the kernel, is harmful to humans and animals with monogastric digestion [30, 34]. Soybean, sunflower, rapeseed, cotton and peanut oil are the primary sources of edible seed oils globally. Edible Oilseeds add important nutritional value to the diet due to high-quality protein and vegetable oil, along with oil-soluble vitamins like vitamin A. Oilseeds are also used in animal feed because of their high protein content. Their seeds contain energy for the sprouting embryo mainly as oil, compared to cereals, which contain the energy in the form of starch [72].

Oil derived from the pressing of soybeans and other oilseeds, commonly known as vegetable oil, accounts for approximately 55% of global vegetable oil production. Palm oil (35%), palm kernel oil, coconut oil, and cottonseed oil also contribute to this aggregate [81]. Cottonseed oil, which is extracted from the entire cottonseed, is considered the second-largest source of vegetable oil worldwide and is a by-product of cotton [8]. This oil is highly regarded as a heart-healthy oil due to its absence of cholesterol. Cottonseed oil has a distinct flavor and cooking quality due to its specific combination of saturated and unsaturated fatty acids [3].

Human consumption and technical or industrial uses are the two primary respective uses of vegetable oils and fats. The majority of vegetable oils utilized in food are either mono- or polyunsaturated. Vegetable oils are the primary dietary sources of vitamin E. Nowadays, the majority of vegetable oils are traded as commodities for the production of margarines, cooking oils, and processed foods intended for human consumption. However, there has been a recent surge of interest in the potential use of oil crops for various non-edible items, such as high-value medicines, biodegradable plastics, and cosmetics [28].

For instance, in Ethiopia, the oilseeds sector is experiencing robust growth due to increasing demands from both domestic and global markets. This sector plays a significant role in the country's overall growth and development and provides employment and income generation opportunities for many small-scale farmers and businesses involved in trading, oil crushing and transportation [124]. Oilseeds are cultivated on more than 3 million farms (30% of all farms) on 800,000 ha (8% of total acreage) and contribute 5% to the nation's total production of grain products [120].

According to the Central Statistical Agency (CSA) [21], the country produced 0.65 million tons of oilseeds, providing employment and income for 3.3 million people. The primary oilseed crops are sesame, linseed, and noug seed, which together account for 86% of the national oilseeds production. Ethiopia's diverse agro-climate provides a natural comparative advantage and ample potential to cultivate other oilseeds, such as groundnut, cotton, safflower, sunflower, rapeseed, and soybean [124]. In the past, cottonseed oil was more crucial as a vegetable oil than it is now, probably as a results of the by-products obtained from cottonseed processing.

Similarly, the global market initially replaced cottonseed oil with soybean oil shortly after the Second World War, and since the 1980s, palm oil has become the predominant product, particularly as a primary ingredient in various unhealthy foods. Nevertheless, Ethiopia has not made any significant efforts to utilize the valuable ingredient found in cotton, which can be used not only as a fibre crop but also as an oilseed crop, as it has been done in the United States for the past two centuries [79]. When selecting healthy sources of oil, several factors need to be taken into account, including the fatty acid composition, healthy minor components, and the extraction process of the oil. Availability, price, food application, and stability are other essential factors that must also be considered.

2 Cottonseed By-Products

While cotton is primarily cultivated for its lint, there are various by-products that can be obtained to enhance the value of the sector and benefit stakeholders such as farmers, ginners, oil millers, and others downstream users. Cottonseed is a crucial commodity in the industry due to its oil and other derivatives. The seeds are also sold to farmers for planting, although they are often owned by the ginners. Some ginners keep some seeds to sell for planting, while those with milling facilities process cottonseed for its oil. The ginners without mills sell their cottonseed to oil millers. On the farm level, cotton stalks are typically discarded after the seed cotton is harvested. At subsequent processing nodes, ginners produce cottonseed alongside lint, which millers process into linters, hulls, cottonseed oil, and cake. The primary cottonseed by-products are as follows:

2.1 Cotton Linters

The fibres that remain on the cottonseed after the ginning process, known as fuzz, are typically short and covered in short hairs. Once the lint has been removed from mature seeds, which can be done manually or mechanically, the remaining cottonseed may or may not have fuzz. If it does not, it is referred to as "naked." The fuzz that is removed from the seed is called linters and can make up as much as 10% of the seed's weight. However, for naked-seed varieties, linters only contribute about 1%

of the total seed weight. Linters are composed of approximately 75–80% cellulose [52].

After the ginning process, linters, which are short, fine, or curly fibres that are typically less than 3 mm in length, can be recovered from the fuzzy seeds using a de-linting machine at the cottonseed oil mill or at the ginnery before processing the cottonseed for oil or black seeds for planting. Different types of linters can be produced depending on the number of times the cottonseed passes through the de-linting machines. These linters have various uses, including the manufacturing of cellulose products like cellulose acetate, viscose rayon, carboxymethyl cellulose, microcrystalline cellulose, and cellulose nitrate, as well as the production of specialty-grade paper. Additionally, linters can be processed into various by-products such as pulp, bandages, ear buds, cotton balls, pads, and cushion material [116].

2.2 *Cottonseed Hulls (Husks)*

The outer covering of the cottonseed, known as the seed coat (hull), is initially soft and permeable before the boll opens. However, before and during boll opening, numerous layers of the seed coat are joined together to create a hard and impermeable layer that provides extensive protection to the embryo compared to other agricultural crops. The hull is typically black and consists of six layers, comprising around 20–50% of the seed weight, which can vary depending on the seed size and thickness of the seed coat. The seed hull is composed of approximately 35–60% alpha cellulose, 19–27% pentosans, 15–20% lignin, and 5% ash, proteins, fat, among other substances [52].

The husks (hulls) are removed from the cotton kernels before oil extraction, and they are a direct by-product of the dehulling process that exposes the kernel. The recovery of hulls from cottonseed can range from 20 to 30% of the seed weight. With advancements in technology, the hulls can be utilized in various economic applications, such as in the production of synthetic rubber and petroleum refining, as well as incorporated into the mud utilized in oil well drilling. De-hulling is a crucial process that removes the linters from the fuzzy cottonseed, and the hulls are then separated from the kernel through screening [115].

2.3 *Cottonseed Cake and Meal*

Although "cottonseed cake and cottonseed meal" are often used interchangeably, they actually refer to distinct products that are co-products of the oil extraction process. Cottonseed cakes, which are not ground and have a high residual oil content, are produced by crushing whole seeds and have a nutrient content that is approximately 40% protein and 5–7% fat. In contrast, cottonseed meal is produced by crushing kernels and is ground, dried, and pulverized. It has a lower residual oil content and a higher protein content of approximately 50% and a fat content of 1–2% [53]. Cotton oil cakes and oil meals are by-products of seed oil extraction. These can be classified as either edible or non-edible. Edible cotton oil cakes are highly nutritious, with protein contents ranging from 15 to 50% [92].

Cottonseed cake is a suitable animal feed for ruminants but not for non-ruminants due to the presence of the enzyme gossypol, which inhibits nutrient absorption. This limits the market for cottonseed cake because non-ruminants such as poultry, fish, and pigs cannot consume it [116]. The demand for animal feed, particularly cottonseed cake, is increasing in Africa. To meet this demand and take advantage of relatively good prices (compared to other oilcakes), oil mills and cotton companies must increase their crushing rates of cottonseed and improve its quality. As a co-product of the seed, the oil will also need to be marketed [53].

Cottonseed cake is also used to improve soil fertility as an organic source of soil nutrients, which can reduce the cost of chemical fertilizers for crop and cotton production. However, detailed analysis of the macro- and micronutrient composition of cottonseed cake is lacking, and further studies are needed to make appropriate recommendations for integrated nutrient management to maintain soil health and productivity [79]. Proper storage is essential for seed cakes. Cottonseed cakes should have to be well aerated to prevent heating and excessive dryness, which can shorten the shelf life of cottonseed cakes. It should also be protected from moisture during transport and storage. According to traders, cottonseed cakes have a shelf life of approximately 2–3 months. Piled cottonseed cakes do not easily overheat [15].

Cottonseed meal is produced by grinding the flakes after most of the cottonseed oil has been extracted. It is a residue that remains after the oils have been removed from cottonseeds. This ingredient is sometimes used in the production of cookies and is commonly found in livestock feeds. Cottonseed meal is a versatile ingredient that complements a variety of plant and animal proteins, containing 40% bypass protein. It is a sought-after ingredient in feed for ruminants, such as dairy and beef cattle, goats and sheep. Different methods are used to extract cottonseed oil, resulting in various types of cottonseed meal. Unlike other major oilseeds like soybean and sunflower, there is no dominant process for cottonseed meal production, leading to a wide range of products varying in protein, fibre and oil content [50].

2.4 Cottonseed Oil

Cottonseed oil has been widely used since at least the middle of the nineteenth century [61]. This highly valuable by-product is obtained through the crushing of the cottonseed kernel, which contains the inner flesh of the seed after the tough outer hull has been removed. The cottonseed processing industry exclusively employs local seeds to produce oil for various purposes, including both edible and industrial applications like margarine and soap [116].

Cottonseed kernel oil, also known as heart oil, is one of the most unsaturated edible oils available. It does not require full hydrogenation for many cooking applications, unlike some of the more polyunsaturated oils. The oil has a polyunsaturated to saturated fatty acid ratio of 2:1 and comprises approximately 65–70% unsaturated fatty acids, including 18–24% monounsaturated (oleic) and 42–52% polyunsaturated (linoleic), as well as 26–35% saturated (palmitic and stearic) fatty acids. Scientists describe cottonseed oil as naturally hydrogenated due to its levels of oleic, palmitic, and stearic acids [74].

2.5 Cotton Stalks and Cotton Gin Waste (Gin Trash)

Cotton stems are residual outcomes of the collection of raw cotton seeds. It comprise the remaining parts of the cotton plants that are not utilized after the seed cotton has been picked. In many developing countries where cotton is grown, the locals lack the necessary expertise and technology to enhance the value of cotton stems. To prevent cotton pests and diseases from invading the next season's cotton plants, farmers in these countries burn the stems during the preparation of their farms. As a result, the amount of cotton stems available for alternative uses such as animal feed or fuel is relatively low [116]. Conversely, in countries where cotton farming is mechanized, farmers dispose of the stems at the end of each season by cutting them at ground level, shredding them, and incorporating them into the soil [53]. In terms of volume, cotton stems are the primary by-product of cotton farming, and they represent an additional source of income for farmers after the sale of seed cotton. Apart from converting cotton stems into briquettes for domestic heating, they can also be processed into particle and hard boards that are useful in construction. Research has shown that cotton stems are comparable to most species of hardwood. Therefore, cotton stems have enormous potential as a substitute raw material for the production of valuable products such as particle boards, pulp and paper, hard boards, pellets and briquettes for heating, mushrooms, compost, manure, corrugated boards, and boxes [116].

Aside from cotton stems, which are an established by-product of the ginning process, another cotton by-product is created, commonly referred to as cotton gin waste, gin trash, or gin motes. This by-product consists of undersized, broken or immature seeds with attached fibres. The term gin motes encompasses any gin waste that can be used for its fibre content. It is primarily composed of lint cleaner waste, but may also include motes from the gin stand (where the gin separates the motes from the mature, whole seeds). The amount of such residue is significant, particularly with machine-harvested cotton. The need for seed cotton and lint cleaning is increased by mechanical harvesting. For example, a typical gin in the United States produces about 180 kg of by-products for each bale (217.7 kg). Most of the by-products (or trash) generated by the gins were traditionally discarded back into the fields at a cost, serving as a soil amendment. However, motes (semi-processed or eventually re-ginned) can be utilized to produce some of the same nonwoven items that are made using linters [53].

Moreover, ruminants can be fed with cotton gin waste, which contains 90% of the nutritional value of the hulls [35]. It can also be utilized as a bulking agent to enhance the effectiveness of animal manure composting [18]. In the United States, cotton trash has been examined for its potential as a source of fuel. The cotton stalks possess a comparable specific energy (17.1–18.1 mJ/kg) to wood [25], indicating that the waste could be utilized as an industrial fuel for a power plant [44] or mixed with pecan shells to produce BBQ briquettes [25]. Additionally, there has been some curiosity in fermenting cotton waste to produce ethanol [55].

3 Challenges and Opportunities in Developing Cotton By-Products

In contemporary times, cottonseed was initially utilized as a by-product of the cotton plant in 1665, when residents of the British West Indies created oil from the cotton flower for medicinal purposes. The production of cotton and, consequently, cottonseed was relatively small due to the challenge of manually separating seed and fibre. The predicament of disposing of the copious amounts of seed produced became evident. Although methods of crushing seeds to extract oil existed, cottonseed posed a particular issue. The kernel, which holds most of the oil, is encompassed by a sturdy hull, making it arduous to grind. Furthermore, this hull is coated with short fibres that absorb the expressed oil, ultimately reducing yields [40].

Numerous food processing technologies are available and well-known to ensure food safety for a chosen food process, thereby effectively controlling any potential food safety hazards. The advent of new refining technology has created opportunities in developing cotton by-products, making it possible to produce oil products tailored to almost any commercial need. From the 1960s and 1970s, the introduction of expander technology to the oilseed industry led to a 50% reduction in free gossypol levels. Surveys conducted by the National Cottonseed Products Association (NCPA) in the early 1990s and again in 2000 demonstrated that the levels of free gossypol in meal manufactured with an "expander-solvent technology" remained low (<0.18%) [38, 117].

3.1 Challenges and Constraints in Developing Cotton By-Products

Compounds found in cotton plants can have negative impacts on the health of humans and animals [80]. The seeds, in particular, contain anti-nutritional and toxic compounds that can be harmful if consumed in large amounts [2, 107, 113]. Gossypol, a terpenoid aldehyde, cyclopropenoid fatty acids (CPFAs), and tannins are the most significant compounds in terms of biosafety and human health. Gossypol is a polyphenolic compound that occurs naturally in cotton and is typically found in pigment glands throughout the plant, with a higher concentration in the seed. All cottonseed contains gossypol, with levels ranging from 0.40 to 2.0% [63]. The amount of gossypol present depends on various factors, including species, variety, fertilization, growing conditions, and insect pressure.

Different species and varieties of cotton have varying concentrations of gossypol in their seeds [82]. For example, cottonseed meal can contain between 1.16–1.52% total gossypol and 0.024–0.82% free gossypol [77, 99, 109]. Likewise, the free gossypol content in whole cotton seeds varies among cotton varieties, with gossypol concentrations ranging from 0.02 to 6.64% [91]. The concentration of gossypol in cottonseed can exceed 14,000 mg/kg of total gossypol and 7,000 mg/kg of free gossypol [4], making it toxic to both humans and non-ruminant animals. Although there are methods to remove gossypol, such as refining the oil, gossypol toxicity limits the use of cottonseed in animal feed. The equipment needed to reduce gossypol content in refined cottonseed oil is expensive and not readily available in Africa [53].

The primary concern in cottonseed is the harmful effects of gossypol toxicity on mono-gastric animals, resulting in decreased growth rate, fertility depression, internal organ abnormalities, feed conversion, and low protein digestibility. However, gossypol can be extracted from cottonseed and used for medicinal purposes such as anti-cancer, antiseptic, anti-fertility agents, and antiviral activity [106]. Acute toxicity due to free gossypol causes moderate harm to animals, leading to constipation, dyspnea, anorexia, and weight loss. Repeated exposure to lower doses of gossypol can affect the testis in males (reduced sperm motility, inhibited spermatogenesis, and depressed sperm counts) and reproductive organs and embryo development in females [37].

Gossypol toxicity can manifest in various ways, and there is no specific diagnostic test for it. Clinical signs such as dyspnea, decreased growth rate, anorexia, weakness, and gastroenteritis, along with a history of consuming cottonseed products above the recommended levels, are major indicators. In mature cattle, gossypol toxicity can cause decreased dry matter intake, decreased milk production, panting, elevated heart rate, ruminal stasis, severe abomasitis, hemoglobinuria, and sudden death [98]. Gossypol ingestion has also been linked to decreased hematocrit and hemoglobin concentrations and increased erythrocyte fragility.

The primary impact of gossypol on beef cattle is its effect on the reproductive function of male cattle, which is the most commonly reported aspect. There is no evidence of long-term or permanent reproductive issues in female cattle. Research has shown that pubescent and growing bulls experience abnormal or reduced sperm motility due to gossypol [63]. Although gossypol does not significantly affect the milk or growth production of adult animals, high levels of cotton seeds can lead to negative impacts on gonads and reproductive health in both male and female ruminants [93].

3.1.1 Opportunities in Developing Cotton By-Products

The agribusiness of cotton by-products have the potential to enhance the quality of livestock in sub-Saharan Africa, particularly in the arid and semi-arid agroecological areas, in response to the growing demand due to population growth and urbanization rates. In African countries, there is a surge in consumer demand for oilseed products, including animal feeds for the dairy, livestock, and poultry sectors and edible oils for human consumption. Currently, cottonseed production in sub-Saharan Africa is around 2.2 million tons, which yields about 400,000 tons of oil (18% optimization rate) and 500,000 tons of protein (23%) [53].

As long as sub-Saharan African nations are not self-sufficient in edible oil, it is unlikely that cotton oil will be used to produce biofuel [53]. Nonetheless; Vegetable oils, such as cotton oil, have been proposed as a potential source of income for commodity producers in Sub-Saharan Africa due to the growth of global demand for biofuels, which now accounts for roughly 9% of global vegetable oil production [57]. Cottonseed oil can also be used as biofuel, either alone or mixed with fuel in certain diesel engines ranging from 5 to 100 kWs. However, direct fuel-injection engines cannot use natural vegetable oils, thus, cottonseed oil must undergo etherification to be transformed into biodiesel [53].

The utilization of cottonseed as a protein supplement in animal feed is limited by the presence of gossypol and cyclopropenoid fatty acids (CPFAs). However, ruminants are less affected as these compounds are detoxified during digestion in the rumen [62]. To maximize the use of cottonseed, the free gossypol content can be significantly reduced (63–85%) through various processes such as roasting, cracking, or extrusion. The toxicity of gossypol can be mitigated by adding iron salts, which can bind to gossypol. The addition of iron sulphate (up to 500 mg/g DM of the diet) to steer diets has been shown to decrease gossypol in plasma [100]. To improve growth performance, different ratios of iron to free gossypol have been used, including 1:1 for cattle, pigs, and Nile tilapias, 2:1 for broilers, and 4:1 for layers [36].

Additionally, crude cottonseed oil can be refined to produce edible oil and fatty matter, which can be used to make soap stock for soap production, bakery fat for bread production, and margarine [115].

4 Cottonseed Oil Processing

In general, cottonseed oil processing is deemed to be a more challenging substance to handle compared to other prevalent plant-based oils, owing to its lengthy processes as well as its relatively elevated concentrations of free fatty acids and a more vivid colour due to the presence of gossypol and related pigments that are carried forth during the pressing process.

4.1 Cleaning

Typically, oilseeds are blended with a variety of foreign substances such as sand, stones, stalks, weed seeds, foliage, and more during the processes of harvesting, handling, and transportation. It is best to clean the seeds before storing them. When seeds are mixed with stone, iron, and wood pieces, they can damage mechanical equipment during processing. Also, foreign matter can reduce the protein content and increase the fibre content of meal residue after oil extraction. Furthermore, foreign matter mixed with oilseeds may contain high moisture content, which can cause overheating during storage. Proper cleaning of dried seeds is also necessary to eliminate sand, dirt, dust, leaves, stems, weed seeds, stones, metal pieces, and other extraneous matter before storing [11].

If there is high moisture content while storing, or if it is not detected and corrected by aeration or rotation, local hot spots in the oilseed can damage the quality and pose a fire hazard. Additionally, cleaning oilseeds before storage eliminates the need for further cleaning for processing and saves the trouble of double handling of seeds. In summary, adequate cleaning of oilseeds can increase the crushing capacity of oil expelling units, decrease in-plant maintenance, and enhance the quality of oil and cake [11].

4.2 Delinting

Delinting is a process of removing the fine fibres of the seed coat left after ginning cotton. This process yields two products—the seeds and the linters—while also enhancing the quality of the seeds [116]. Delinting can be done using three methods—chemical, mechanical, and flame. Chemical delinting involves immersing the seed in a mild sulfuric acid solution, followed by rinsing it with water. In areas with low relative humidity, some chemical delinting plants also use an anhydrous HCI scarification technique. Mechanical delinting uses equipment similar to that used by cotton oil mills to remove the seed fuzz before crushing the seed. Flame delinting, on the other hand, exposes the seeds to brief periods of direct flame to burn off the remaining seed fuzz. Among these methods, chemical delinting is the most commonly used [73].

4.3 Dehulling or Decortication

After the removal of lint, a machine equipped with a set of knives is used to dehull the seed. The knives gradually score the hulls, separating the tough outer layer that encases the cotton kernel. The fibrous hulls of oilseeds contain a low percentage of oil, which varies depending on the type of oilseed. Extracting oilseeds by dehulling is beneficial because the hulls can reduce the overall oil yield and the capacity of extraction equipment [11]. A series of shaker screens is used to facilitate the separation of the hulls from the kernel. Once this process is complete, the hulls can be sold in bulk or pellet form as a singular ingredient, or mixed with roughly 35% cottonseed meal to create a product that has several advantages in terms of protein content, ease of handling, and transportation [63].

4.4 Cottonseed Oil Extraction Methods

The extraction and processing of oils from seeds is a large-scale business with high production capacity. The wide range of residual oil in cottonseed meal can be attributed to the different techniques employed for oil extraction [50]. The cottonseed oil extraction rate, which can be 10–16% of the cottonseed's weight, depends on the crushing method used [16]. The cottonseeds may undergo dehulling, cracking, drying, or heating before being fed to the press. The resultant cake is dried, ground, and processed into large pellets [7]. The mechanical extraction of cottonseed is not very effective, and up to 20% of the seed oil may remain in the pressed cake, depending on the technology employed [53, 84]. Another method of extraction is mechanical pressing, which is not efficient in oil extraction. The cakes produced from these methods are considered to be of higher quality. Seed cakes must contain some oil to prevent them from becoming too tough for animals to consume.

Based on the transport infrastructure availabilities, equipment, solvent method, and skilled workforces applied in different parts of the world; cottonseed could be well extracted using all three methods *viz.* screw-press (expeller-pressing) method, hydraulic press (expander-press) method, and solvent extraction method. The preferred method of cottonseed processing varies from country to country and even from region to region within the same country [39, 83].

4.4.1 Hydraulic Press (Expander-Press) Method

The hydraulic press technique, also known as the expander-press method, has been in use for a long time and continues to be utilized globally for cottonseed oil extraction. This traditional method employs a circular mortar to extract oil from cottonseed, which has a lower oil content compared to most other oilseeds. However, the hydraulic press method is not very effective in terms of both the quantity and quality of cottonseed oil production. With this method, only 10 L of oil can be obtained from 100 kg of cottonseed, which is considerably less than what can be achieved with modern technologies [53]. Over time, cottonseed oil extraction has evolved from edgestone to wedge press to hydraulic press. Hydraulic pressing was the primary method of separating oil from cottonseed during the nineteenth century.

As cotton spinning, weaving, and ginning processes improved during the eighteenth century, more cottonseed became available for crushing. Consequently, the labor-intensive hydraulic press method was quickly replaced by the continuous screw press in the early 1900s. Since edible oil commanded a decent market price and both hydraulic and screw presses left almost 20% of the available oil in the press cake, research was conducted to find a suitable solvent to extract the remaining oil from the cake [39, 83].

Rolling is a more satisfactory way of obtaining flake particles of 0.13–0.25 mm for hydraulic pressing of cottonseed, as opposed to the irregular shape obtained by grinding. In the preparation of oilseed for expression in the screw-press method (expeller-pressing), the production of thin particles or flakes is not necessary because heat is generated, and seed particles are broken by shearing stress developed in the barrel of the screw-press method during oil expression. Small oilseeds such as sesame, rapeseed/mustard, and linseed, as well as medium-sized oilseeds such as cottonseed, are usually rolled before processing with the screw-press method in large-scale commercial plants [11].

4.4.2 Screw-Press (Expeller-Pressing) Method

The expeller-pressing technique involves the use of a screw press, which comprises a stainless-steel cylinder screen that encloses a large-bore screw. The screw and cylinder have a narrow gap between them, and an adjustable backpressure is present at the end of the chamber. Breaker bars are also incorporated to disrupt the compressing mesh. The screw presses (expellers) are designed to have a rotating screw shaft within a cylindrical barrel, allowing oil to flow between the flat steel bars of the barrel while retaining the solid material or press cake. The barrel is closed, except for a single hole through which the extracted oil drains.

In this method, the oilseeds undergo pressure exertion in increasing order through screw presses (expellers) that have rotating screws or worms. The generated pressure and heat result in oil drainage from oilseeds, and the cake is ejected out of the barrel. The efficiency of oil extraction depends on seed preparation. In an efficient screw press (expeller), a single pressing produces a cake containing 5–7% oil. Double pressing is rarely performed, as the cake obtained contains approximately 4% oil. The choice of double pressing depends on the processes' economics, oilseed type, and the cake's end use [11].

Mechanical extraction of oil involves the use of manual ram pressing or engine-driven screw pressing. Engine-driven screws have an oil content efficiency of 68–80%, while manual pressing can achieve an efficiency of 60–65%. The challenge of using mechanical pressing is that it is only applicable to specific seeds [95]. However, pre-treatment procedures can improve oil recovery for screw presses, providing 89% efficiency in a single pass and 91% for a double pass [110]. The screw-press method is relatively simple and not capital-intensive. Screw presses (expellers) are available in a wide range of processing capacities. The screw-press method preserves most of the natural qualities (colours, flavours, and nutritional qualities) of cottonseed oil, which is free of solvent or chemical residues. However, even the most potent presses cannot remove more than 70–80% of the oil from cottonseed, and the residual oil level in cakes cannot be reduced below 3–5% [53].

The expeller-press method, also known as the screw-press method, utilizes heat to enhance the connection of the gossypol protein, which transforms free (harmful) gossypol to bound (non-toxic) gossypol. As a result, the produced meals have the least amounts of free gossypol (200–500 mg/kg) [75, 84]. Although heating may lead to a decrease in protein quality due to the binding of gossypol to lysine at high temperatures, the screw press's shearing effect deactivates gossypol at temperatures that do not affect protein quality [111]. Generally, for oil extraction, the screw-press (expeller-pressing) method is preferred over hydraulic presses since it provides a continuous process, higher capacity, requires less labor, and removes more oil.

4.4.3 Solvent Extraction Method

The procedure of solvent extraction involves the utilization of a solvent to extract oil from cottonseed oilcake that has been crushed, resulting in the production of crude extracted cottonseed oil and deoiled cottonseed meal. Once the separation has taken place, the solvent is evaporated, leaving behind the oil. The extraction process employed can also impact the quality of the oil. Large and modern edible oil mills use the organic solvent extraction method, which extracts most of the oil, but the seed cakes produced are of inferior quality. For optimal efficiency, it is recommended that the decorticating method be used when extracting cottonseed oil using solvents. This method involves removing the hull from the cottonseed, using only the kernel/meat for solvent extraction. The cottonseed is dehulled, cracked, dried, heated, or flaked, then screw-pressed or expanded, and the pressed flakes or pellets are finally subjected to solvent extraction [50].

The use of solvent extraction technology has the potential to increase cottonseed oil production, as it can extract oil from the seeds more fully, resulting in a higher recovery rate of up to 97–99% [52, 116]. Compared to other seed oil extraction methods, the solvent extraction method is more efficient and can achieve a more complete recovery of oil. However, this technology is only suitable for large processing capacities due to its complexity and high capital requirements. The elimination of residual chemical solvents can affect the natural qualities of the oil [53]. The pre-press solvent extraction method combines a mechanical extraction step, which reduces the oil by half to two-thirds of its original level, with a solvent extraction, resulting in a 97% oil extraction rate [50].

Out of all the techniques available, the most commercially feasible method to remove gossypol from cottonseed meal is the solvent-based approach, as it dissolves the solute (gossypol) in acidified solvents [27, 68]. Several commonly used solvents include butanol, ethanol, hexane, isopropanol, methanol, acetone, chloroform, and pentane [102]. The combination of solvents has been found to significantly decrease the total gossypol concentration [86]. Gerasimidis et al. [43] found aqueous acetone to be very effective in lowering the free gossypol levels via a two-stage solvent extraction process, which results in a 72% protein concentrate [43]. Cottonseeds contain roughly 0.7–1.5% (7,000 mg/kg) free gossypol and 2–4% (14,000 mg/kg) bound gossypol. Crude oil contains about 0.6% gossypol through solvent extraction, 0.14% through the expander-solvent method, and approximately 0.06% when the extraction process involves mechanical pressure and heat treatment [53, 78].

In the direct solvent-extraction technique, the oil is extracted using a solvent, typically hexane, which is the most commonly used solvent but is criticized due to concerns about the environment, health, and safety. Ethanol is a new solvent that can inactivate a higher proportion of gossypol, resulting in safer cottonseed meals [103]. The resulting cake is heated to eliminate the solvent and then ground into meal [53].

The solvent extraction method, which does not require high temperatures, limits the gossypol-protein binding, and the residual free gossypol content may be up to ten times higher than in screw-pressed meals (1,000–5,000 mg/kg) [75]. In the pre-press solvent extraction method, the amount of free gossypol is in the 200–700 mg/kg range, comparable to that of screw-pressed meals [84].

4.5 Refining Cottonseed Oil with Caustic or Alkali

Cottonseed oil, in its natural state, has a dark red color and an unpleasant odour compared to most other edible oils due to the presence of residual gossypol and related substances. However, refining and bleaching techniques can be employed to achieve a lighter color before it can be sold to the food industry in bulk or as a final consumer product [34, 53, 115]. Refining refers to any treatment aimed at purifying cottonseed oil by removing free fatty acids, phospholipids, gossypol, and other impurities. The process also aims to eliminate most of the free fatty acids present in the crude oil. The process involves treating crude cottonseed oil with an alkali solution, which converts free fatty acids to soap [40]. The refining process has a significant impact on the quality and economic performance of vegetable oil compared to other processes used in converting crude oil to a finished product. Poorly refined oils adversely affect the operation of subsequent processes and the quality of the final product. Poorly refined oils require additional processing and handling, increasing the cost of finished products beyond those produced from properly refined good quality oil [39].

Currently, there are two distinct methods employed for refining vegetable oils: chemical and physical. However, physical refining has limitations and most vegetable oils are refined using this method. Nevertheless, cottonseed oil cannot be effectively purified by physical refining due to the presence of non-glyceride materials. On the other hand, alkali refining is highly effective in removing gossypol and related pigments from cottonseed oil by combining them with alkali/caustic soda (NaOH) [39, 47]. In the process of alkali refining, NaOH solution is added to the oil in sufficient quantity to neutralize the Free Fatty Acid (FFA) content. Research has shown that alkali refining and bleaching can reduce the gossypol content in cottonseed oil to less than 1 ppm from 0.05 to 0.42% in solvent-extracted oil and from 0.25 to 0.47% in screw-pressed oil [60]. However, caustic refining may also result in the loss of 10–20% of tocopherols, and during deodourization, 30–60% of the remaining tocopherols may be lost [39].

4.6 Bleaching

The process of bleaching is commonly known as the removal of color, either partially or completely, and is considered to be accurate. Nevertheless, it is also used as a means of purification to prepare oil for further processing. The typical method of bleaching involves the adsorption of pigments and other impurities, aside from glycerides, found in most vegetable oils on bleaching earth [39, 40].

During a standard process, bleaching materials are introduced into the oil in a stirred vessel either under a vacuum or at atmospheric pressure. The oil is then heated to a temperature of 160–230 °F (70–110 °C) and held for a certain period to allow contact time with the bleaching earth. Once the adsorbent has effectively captured the impurities, it is taken out from the oil using a filtration system [39].

An independent bleaching process is necessary for oils that have undergone hydrogenation, fractionation, or interesterification for three reasons. Firstly, it eliminates all traces of any catalyst that was used during the previous process. Secondly, it eliminates any undesirable colors that were generated during the previous process. Finally, it removes any peroxide and secondary oxidation products. Post-bleach systems are often batch systems to accommodate a wide range of products [39]. Bleached cottonseed oils are almost colorless and have a peroxide value close to zero [20].

4.7 *Winterization*

It is possible for oils to contain small quantities of waxes and fully saturated triglycerides, which can cause precipitation or haziness when stored at lower temperatures. Although these compounds have no adverse health or flavor effects, consumers view the haziness unfavourably [34]. In order to produce salad oil that remains clear at lower temperatures, such as those found in refrigeration, cottonseed oils must undergo winterization. This involves removing a significant portion of the more saturated triacylglycerols to prevent the oils from becoming cloudy and solidifying. If the more saturated glycerol esters are not removed from cottonseed oil, it will solidify and become cloudy at temperatures commonly found in refrigeration, which is approximately 7.2 °C (45 °F) [20, 39].

The stearin from fats in warm, dry, refined, and bleached cottonseed oil is normally removed by quick freezing and filtration. The separation of stearin requires slow cooling to form crystals that are large enough to be removed through filtration or centrifugation. Locally, cottonseed oil is winterized within a tank, with stearin separated from the liquid oil through filtering with plate and frame presses. The oil is fed following gravity to prevent the crystals from breaking up during filtration. However, most processors have made improvements to their equipment and processes to enhance efficiency, such as using jacketed and enclosed tanks equipped with programmable cooling and agitation, better filtration, and improved pumping methods [39].

4.8 Fractionation

The melting points of cottonseed oil range from 13.3 to 35 °C (8–95 °F) due to its composition of triacylglycerols. This limited range of melting points restricts the potential applications of cottonseed oil. Fractionation, a process that separates liquid and solid fractions of vegetable oils, is based on the reduced solubility of more unsaturated triacylglycerols in comparison to more saturated ones as the temperature decreases. Therefore, cooling the oil results in gradual crystallization of triacylglycerols. As crystallizing triacylglycerols exhibit diverse polymorphic forms, it is crucial to regulate crystallization conditions to minimize the inclusion of liquid (uncrystallized) material in the crystallized phase. This necessitates slow cooling of the oil that needs fractionation. Filtration is typically used to separate the crystallized material from the liquid phase [14, 39].

Three commercial processes for fractionating edible fats and oils are dry fractionation, solvent fractionation, and aqueous detergent fractionation. Among these, solvent or aqueous detergent fractionation processes provide better separation of specific fractions for the more refined fats and oils products, such as salad oil and some highly specialized edible oil products made from cottonseed oil. Solvent fractionation technology can produce high stability liquid oils, with active oxygen method (AOM) stability of 350 h minimum without the benefit of added antioxidants, and cocoa butter equivalents [39].

4.9 Hydrogenation (Hydration)

The process of hydrogenation is carried out on cottonseed oil to transform its liquid state into a semi-solid or plastic fat that is more appropriate for making margarine or shortenings. The primary objective of hydrogenating cottonseed oil is to meet the requirements of manufacturers who desire specific mouth feel, stability, melting point, and lubricating qualities that are well-suited to their products. Furthermore, hydrogenation enhances the oxidative stability by converting unsaturated fatty acids into saturated fatty acids [20, 39]. Hydrogenation involves treating the oil with hydrogen in the presence of a catalyst (nickel) under controlled pressure (100–300 kPa) and temperature conditions (150–200 °C), resulting in a semi-solid or plastic fat that is suitable for various food applications [34, 42]. The hardening process comprises two stages. In the first stage, hydrogen is added to the double bonds, enhancing the oil's stability to oxidation during frying and increasing its melting point. In the second stage, some of the double bonds are converted from the naturally occurring *cis* form to the *trans* form. *Trans* fats have higher melting points than *cis* fats. When liquid oils are treated to reduce the number of double bonds, the hydrogenation process is made selective to ensure that the hydrogen reacts first with the highly unsaturated linolenic acid [34].

Cottonseed oil has an advantageous fatty acid composition compared to other linoleic oils, as it requires less hydrogenation to achieve the same level of hardness. In its unhardened form, cottonseed oil mainly consists of 2.5% stearic and 22–23% palmitic as the principal saturated fatty acids. The unsaturated fatty acids in cottonseed oil consist of roughly 18% oleic, 54% linoleic, and less than 1% of both linolenic and palmitoleic. During hydrogenation, the original stearic and palmitic fatty acids remain unchanged. However, if the hardening reaction were entirely selective, the linoleic fatty acid would need to be entirely converted to oleic or other 18:1 isomers before any hydrogen reacts with oleic acid. Once the linoleic acid disappears, the oleic acid and its isomers absorb hydrogen to become fully saturated stearic fatty acid. Although this complete selectivity is never reached in practice, how closely it is approximated depends on factors such as catalyst type and dosage, temperature, and pressure [39].

4.10 Interesterification

The interesterification process involves the reorganization and redistribution of fatty acids within triacylglycerol molecules. This process shuffles the fatty acids and randomly replaces them on the glycerol esters, resulting in changes to the oil or fat's physical, structural, and functional properties. The resulting products can be used to create shortenings, margarines, high-stability liquid oil products, and other specialty products that differ from the original oil or fat properties. In cottonseed oil, the rearrangement and redistribution of fatty acids are not random, as saturated fatty acids are typically found in the *sn*-1 or *sn*-3 positions, while unsaturated fatty acids are found in the *sn*-2 position. This is significant because the fatty acid distribution in triglycerides and the nature of the fatty acids present both impact the physical characteristics of fats.

Random reorganization alters the distribution of fatty acids to achieve a typical rise in melting point at high temperatures (<200 °C) for extended periods, but it is more effective to utilize catalysts that hasten the process (30 min) and reduce the required temperature (50 °C). Sodium methoxide is the most favoured catalyst for this process [42]. Optimal desired functional properties of a product can be attained by choosing appropriate reaction conditions. By conducting the reaction at a relatively low temperature with directed interesterification, a solid fat can be produced by using natural cottonseed oil as the raw material. The remaining unsaturated portion can be used as a salad oil, as most of the saturated fatty acids have been eliminated [60].

4.10.1 Deodourization

The final step in the refining of edible oils is deodourization, which eliminates unpleasant and volatile odourous compounds from vegetable oils using vacuum-steam distillation at high temperatures. The aim is to obtain bland oil that has a low level of free fatty acids and a peroxide value of zero. In the process, a small amount of natural antioxidants, such as tocopherols and sterols, is also partially removed. Cottonseed oil can be deodourized at lower temperatures, resulting in more tocopherols being retained from the deodourizer distillate in a separate process [20, 87]. Deodourization can be carried out continuously or semi-continuously in larger facilities or in batches in smaller plants [87]. Most commercial deodourizers operate at a temperature range of 245–275 °C (475–525 °F) under a negative pressure of 2–10 mm Hg [41, 60]. During high-temperature deodourization, polyunsaturated acids may be transformed into *trans* isomers. However, this is less of an issue with cottonseed oil, as it is less likely to occur with linoleic acid than with linolenic acid [39].

5 The Composition and Physicochemical Characteristics of Cottonseed Oil

Cottonseed oil is usually made up of fatty acid, which is most often analyzed using gas chromatography (GC) device, following *trans*-methylation (conversion of fatty acids into their respective fatty acid methyl esters). The predominantly identified eight (8) fatty acids present in cottonseed oil are broadly described with their characteristics and shapes in Table 1.

The physical and chemical characteristics of a specific cottonseed oil depend on the type of cotton grown and the environmental conditions in which it is cultivated, such as temperature, soil quality, fertilizers, and rainfall. Additionally, the way the oil is handled and stored after harvesting can also affect its properties. Various factors, including geographic regions, climate, fertilizers, seed handling practices, and storage conditions, contribute to the variations in the properties of cottonseed oil before it is crushed or extracted from the seed [23, 60]. Breeding and agronomic conditions can also be utilized to beneficially modify the physical and chemical properties of cottonseeds in order to enhance their suitability for oil, feed, and food products without compromising the quality of the fibre [58]. However, the specific functional and nutritional characteristics of oils and fats are primarily determined by their fatty acid composition and the structure of their triglycerides, i.e., the arrangement of the fatty acids on the glycerol backbone [45, 96]. Like all oils and fats, cottonseed oil is composed of glycerol esters with a smaller proportion of non-glyceridic substances [39]. It is the chemical composition that ultimately determines the physical and chemical properties of all fats and oils, which, in turn, dictate the suitability of the oil for various processes and applications. The fatty acid composition, peroxide value, iodine value, saponification value, and unsaponifiable matter are among the major physicochemical properties that determine the processability and applicability of the oil, amongst others.

Table 1 Fatty acid composition of main Cottonseed oil (vegetable oil)

No	Common (Trivial) names	Systematic name	Symbols (Lipid no)	Formula	Melting point (F°)	Percentage (%)	Classification
1	*Myristic*	Tetradecanoic	C14:0	$CH_3(CH_2)_{12}COOH$	130°F (58 °C)	0.5–1.3	Saturated
2	*Palmitic*	Hexadecanoic	C16:0	$CH_3(CH_2)_{14}COOH$	145°F (65 °C)	17–31	Saturated
3	*Stearic*	Octadecanoic	C18:0	$CH_3(CH_2)_{16}COOH$	157°F (70 °C)	1–3	Saturated
4	*Palmitoleic*	*cis*-9-Hexadecenoic	C16:1 (n–7)	$CH_3(CH_2)_5CH{:}CH(CH_2)_7COOH$		<1	Monounsaturated
5	*Oleic*	*cis*-9-Octadecenoic	C18:1 (n–9)	$CH_3(CH_2)_7CH{:}CH(CH_2)_7COOH$	61°F (13 °C)	13–21	Monounsaturated
6	*Linoleic*	*cis*, *cis*-9,12-Octadecatrienoic	C18:2 (n – 6)	$CH_3(CH_2)_4CH{:}CHCH_2CH{:}CH(CH_2)_7COOH$	20°F (–5 °C)	34–60	Polyunsaturated
7	*Linolenic* –Alpha-linolenic –Gamma-linolenic	All-*cis*-9,12,15-Octadecatrienoic	*C18:3* C18:3 (n – 3)	$CH_3CH_2CH{:}CHCH_2CH{:}CHCH_2CH{:}CH(CH_2)_7COOH$ $CH_3(CH_2)_4CH{:}CHCH_2CH{:}CHCH_2CH{:}CH(CH_2)_4COOH$	9°F (–11 °C)	<1	Polyunsaturated
8	*Arachidic*	Eicosanoic	C20:0	$CH_3(CH_2)_4CH{:}CHCH_2CH{:}CHCH_2CH{:}CHCH_2CH{:}CH(CH_2)_3COOH$	168°F	<0.5	Saturated

Source CICR [20] and Encyclopedia of Food Grains [34]

5.1 Free Fatty Acid

The primary way to classify the composition of fatty acids (FAs) in oils is based on the number of double bonds present in the molecule. Fatty acids can be categorized as saturated if double bonds are absent, monounsaturated if they contain a single double bond, and polyunsaturated if two or more double bonds are present in the molecule. Fatty acids are denoted using an abbreviated nomenclature system that specifies carbon chain length, the number and location of double bonds, and the degree of unsaturation [67]. The nutritional, industrial, and organoleptic properties of an oil are determined by its FA composition. FAs are typically expressed as a percentage of all fatty acids in the oil [69]. Most vegetable oils consist of straight-chain homologous series of saturated and unsaturated carboxylic acids with even-numbered chain lengths ranging from C_8 to C_{24}. The unsaturated series generally contains one to three methylene-interrupted, *cis*-double bonds. Although some vegetable oils lack typical FAs, their presence in nutritious edible oils is generally considered undesirable [14].

The major part of consumable oils consists of 95–98% triacylglyceride (TAG), while the minor part comprises complex combinations of minor compounds such as free fatty acids (FFA), *mono*-, *di*-, and *tri*-acylglycerides, tocopherols, sterols, phospholipids, pigmented compounds, etc. On the other hand, refined vegetable oils are almost completely devoid of FFA. In healthy plant cells, FFA is not present in significant amounts, as it is either esterified to lipid or associated with protein or other cell constituents [14]. Oil chemists have known since the 1880s that FFA levels are a reliable indicator of crude cottonseed oil quality. When triacylglycerols undergo hydrolysis, the molecule splits at the ester linkage to produce FFA, *di*- and *mono*-acylglycerols, and eventually free glycerol. This reaction is typically triggered by the presence of moisture and accelerated by heat, and can also be promoted by certain enzymes (lipases). The liberated FFA have a distinct flavor and odour, which become more unpleasant when the fatty acid chain length is less than 14 carbons. Cottonseed oil, which mostly contains C_{16} and C_{18} fatty acids, does not become unappetizing until the FFA level exceeds 1.0% [39]. This poses a particular challenge for the edible oil industry, as FFA must be removed by refining, since the refining process neutralizes or reduces FFA.

5.2 Peroxide Value

The deterioration of oils is largely caused by oxidation. The primary products of this reaction are hydroperoxides, which are formed when unsaturated fatty acids react with oxygen. Although hydroperoxides do not have any flavor or odour, they rapidly break down into aldehydes that often have a strong and unpleasant taste and smell. The level of oxidative rancidity can be determined by measuring the concentration of peroxides, which is usually expressed as peroxide value (PV). PV measures the millimoles of peroxide (or milliequivalents of oxygen) absorbed by a kg of oil/fat sample (mEq/kg) that oxidizes potassium iodide to iodine [39]. This is determined through a redox titrimetric analysis, under the assumption that the compounds reacting in the test are peroxides or similar products of lipid oxidation. Fats and oils of high quality and recently deodourized will have a peroxide value of zero. A peroxide value greater than 20 corresponds to fats and oils of very poor quality that typically have significant off flavors [104]. The peroxide value is determined according to the method and procedure outlined by the Association of Official Agricultural Chemists [9], and is calculated as follows:

$$\text{Peroxide value} = \frac{\mathrm{V} \times \mathrm{N} \times 100}{\mathrm{W}} \times 1000$$

where

V Volume in ml of sodium thiosulphate needed for titration
N Normality of sodium thiosulphate solution
W Weight of oil sample in gram.

5.3 *Iodine Value*

The measure of unsaturation in a fat or vegetable oil is expressed as the iodine value, which is the number of grams of iodine absorbed by 100 g of oils/fats when determined using Wijs solution. A higher iodine value indicates greater unsaturation or more double bonds in the fatty acids, as well as the oil's stability towards oxidation [76]. Cottonseed oil's iodine value results can vary depending on the year, region, and growing season. A cooler season leads to oil with a higher linoleic fatty acid (18:2) content and a lower oleic fatty acid (18:1) content, while a warmer season reverses this trend, affecting the number of double bonds and thus the iodine value [39]. The AOCS method and procedure [9] are used to determine the iodine value, which is then calculated as follows:

$$\text{Iodine Value} = \frac{(B - S) \times N \times 126.9}{W}$$

where

Iodine Value	Iodine in g to be absorbed per 100g of sample
B	Volume of titrant (ml) for blank
S	Normality of $Na_2S_2O_3$ (Sodium thiosulfate)
126.9	MW (molecular weight) of iodine (g/mol)
W	Weight of oil sample in gram.

5.4 Saponification Value

The saponification value is the amount of potassium hydroxide (mg) required to saponify 1 g of oil/fat. This number is indicative of the average molecular weight of the fatty acids present in the oil/fat and is closely linked to the molecular mass of the oil/fat, which neutralizes both the fatty acids and the fatty acids present as acyl glycerol [76]. Measuring the alkali-reactive groups, the saponification number is also a valuable tool in predicting the type of triacylglycerols present in an oil/fat. Glycerol esters containing short-chain fatty acids have higher saponification numbers than those containing longer-chain fatty acids. The saponification numbers of cottonseed oil range from 189 to 198, with an average of 195. When no other analytical measurements are available, saponification numbers have too much overlap to distinguish individual oils/fats. Most oleic and linoleic classification oils have saponification numbers in the range of 180–200 [39]. Typically, the saponification number is determined using the method and procedure outlined by the Association of Official Agricultural Chemists [9]. The saponification number is calculated as follows:

$$\text{Saponification Value} = \frac{(S - B) \times N \times 56.1}{W}$$

where

Saponification Value	KOH in mg required to saponify
B	Volume of titrant (ml) for blank
S	Volume of titrant (ml) for sample
N	Normality of HCl
56.1	MW (molecular weight) of KOH (mg/mmol)
W	Weight of sample in gram.

5.5 Unsaponifiable Matter

Unsaponifiable matter refers to the substances found in oils/fats that can dissolve in nonpolar solvents but not in water, and are not glycerides. These substances are resistant to saponification when exposed to strong alkalis. They include lipids of natural origin like sterols, hydrocarbons, tocopherols, pigments, and other high molecular weight materials that are insoluble in water [39]. The amount of unsaponifiable material varies depending on the type of oil/fat. For instance, in cottonseed oil, it ranges from 0.5 to 0.7% and is mainly composed of phytosterols, particularly-sitosterol and campesterol [23, 24, 64]. In deodourized oils, the content may be slightly lower due to the reduction of sterols during alkali refining and high-temperature deodourization [39]. The unsaponifiable material is determined using the method and procedure outlined by the Association of Official Agricultural Chemists [9], and is calculated as follows:

$$\text{Unsaponifiable Matter} = \frac{100(A - B)}{W}$$

where

A Weight in g of the residue
B Weight in g of the free fatty acids in the extract
W Weight in g of the sample (Weight in g of the free fatty acids in the extract as oleic acid—0.282 VN: Where, V—Volume in ml of standard sodium hydroxide solution; and N—Normality of standard sodium hydroxide solution).

6 Nutritional and Health Attributes of Cottonseed

Today, it is widely acknowledged that to sufficiently provide for the rapidly growing global population, a greater amount of plant-based protein should be incorporated into human diets. An achievable solution is to utilize protein derived from the cotton crop. Prioritizing cotton production is essential as it not only yields fibre, a renewable resource for clothing manufacturing, but also offers consumable oil and protein for human and animal consumption [19]. Cottonseed presents a practical substitute for animal feed as it boasts significant levels of protein (12–32%), oil (17–27%), and energy (approximately 90% of total digestible nutrients) [32, 85, 105, 126]. Cottonseed protein is highly degradable and soluble, with albumin and globulin making up 75% of total proteins. Rumen protein degradability values usually exceed 70% [6].

6.1 Cottonseed Attributes for Human Use

Cotton is mainly cultivated as a fibre crop. The harvest is obtained in the form of seed cotton, which is then processed through ginning to separate the seeds and lint. The ginned seeds are covered with short, soft fibres called linters. Before the seeds are crushed for oil, these linters must be removed. They are utilized in various products including food items. The linters are produced as either first-cut or second-cut linters. The first-cut linters are primarily employed for different purposes, while the second-cut linters have a shorter fibre length and are used as a significant source of cellulose for both food and chemical applications. They function as a cellulose base in high-fibre dietary products and as a viscosity enhancer (thickener) in salad dressings and ice cream.

6.1.1 Cottonseed Oil

The seeds of cotton are crushed to produce three distinct products: oil, meal, and hulls. The oil is primarily utilized for cooking and salad dressing, and it ensures stable frying without additional processing, thanks to the balanced amounts of stearic, oleic, and palmitic acids present within it [70]. The significance of cottonseed oil in human nutrition is on the rise, and the demand for this valuable commodity is increasing due, in part, to a general shortage in the supply of edible oils [20]. Cottonseed oil has one of the healthiest nutritional profiles among vegetable oils. Like other vegetable oils, it does not contain cholesterol in its natural, unhydrogenated state. Cottonseed oil is highly nutritious and healthier than palm oil, consisting of 70% unsaturated fatty acids. In contrast, palm oil is highly saturated with fatty acids [53]. Cottonseed oil produces relatively stable frying oil with potential health benefits due to the distribution of saturated and unsaturated fatty acids within the oil component [84]. Nevertheless, it does contain over 50% *Omega-6* fatty acids and only trace amounts of *Omega-3* fatty acids [53].

Cottonseed is a potential source of nutritional supplements for human consumption [65, 66]. However, the presence of the toxic compound gossypol (2,20-Bis(formyl-1,6,7-trihydroxy-5-isopropyl-3-methylnaphthalene) in cottonseed has hindered the development of such products. The Food and Drug Administration in the US (FDA) has set a limit of 450 ppm for free gossypol in human food products and ingredients, while the Food and Agriculture Organization (FAO) and World Health Organization (WHO) recommend maximum guidelines of 600 ppm for free gossypol and 12,000 ppm for total gossypol [34]. Despite its harmful effects, gossypol and its derivatives have potential therapeutic applications. In vitro studies have shown that these compounds have antiviral properties against some viruses, such as human immunodeficiency virus [88, 123] and H5N1 influenza virus [122, 123], as well as antibacterial and antifungal effects [5, 71]. Gossypol has also been found to be a promising treatment for various types of cancer, including leukemia [10], lymphoma [59], colon carcinoma [118], breast cancer [90, 125], myoma [48], prostate cancer [56], and other malignancies [114, 121]. In the future, biotechnological advancements in cottonseed oil production aim to increase oil content and quality, improve the health benefits of cottonseed oil, and develop gossypol-free cotton [74].

6.1.2 Cottonseed Meal or Flour

Flour made from cottonseed, which is free from gossypol or has had gossypol removed or reduced to low levels, is sometimes consumed by humans [26]. Glandless cottonseed flour has the potential to serve as a raw ingredient for producing protein products with a textured structure. This can be accomplished through genetic modification that eliminates the harmful substance gossypol from the cottonseed [112].

6.2 *Cottonseed Attributes for Animal Use*

Cottonseed is employed as a raw material for the extraction of oil, while the residue in the form of cake or meal is employed for the production of animal feed [46]. Cottonseed meal is a secondary product that competes with other types of meals and is primarily used for animal feed. The husks or hulls, which are the outer portion of the seed, are either blended with meal for animal feed or used to produce energy for the processing facility [57]. The cottonseed meal that is obtained through scientific methods has an insignificant oil content and a high protein content of 40–42%, which is of the bypass type, and more energy than soybean meal or peanut meal [20, 53]. However, the presence of gossypol, a toxic substance, poses a significant challenge to the use of cottonseed as animal feed. The effects of gossypol on animal health and performance are dependent on the type of animal. Therefore, it is necessary to process cottonseed meal appropriately to reduce gossypol toxicity before using it as animal feed. Additionally, the feed rations given during the initial period of feeding must be appetizing to encourage consumption and fortified with higher levels of protein, energy, minerals, and vitamins.

6.2.1 Ruminant Animals

Cottonseed meal is a viable option for feeding ruminant livestock, especially dairy and beef cattle, sheep, and goats. Adult ruminants, which are more tolerant to gossypol than young ones with an immature rumen and non-ruminants, are mainly fed with cottonseed meal [53]. The mature microbial population in their rumen aids in detoxifying gossypol by binding it to amino acids, leading to its deactivation, binding, and degradation [16, 89, 101]. However, this detoxification mechanism may fail when excessive gossypol content is combined with low protein concentration in the rumen [97]. Therefore, cottonseed meal should only be used in limited amounts in the diet and introduced gradually to avoid toxic effects [16].

Whole cotton seeds, with their white and fuzzy appearance, are rich in protein (about 22% DM) and oil (about 20% DM), making them a good source of gross energy. They also contain high crude fibre content (about 28% DM), which makes them a suitable feed for ruminant animals [51]. The percentage of unprocessed cottonseeds used as cattle feed varies among countries. The oil content of the whole cotton seed ranges from 5 to 22% of fuzzy seed in most cultivated varieties. The kernel has an oil content of 28–35%, and there is a negative correlation between the levels of oil and protein [53]. Whole cotton seeds are increasingly used in the dairy industry since adding them to the early-lactation cow's diet increases energy intake, which often results in larger milk yields [53].

6.2.2 Non-ruminant Animals

Glandless cottonseed meal, which is almost devoid of gossypol [53], is a suitable feed for non-ruminant animals such as pigs, poultry, and fish. Unlike ruminants, non-ruminants are more susceptible to gossypol toxicity, as they are unable to detoxify gossypol in their rumen [22]. However, recent studies have shown that effective processing techniques can reduce and detoxify the gossypol content of cottonseed meal [49, 108]. Feeding glandless cottonseed meal at different levels to broilers has been shown to decrease the level of gossypol toxicity, improve feed intake, and promote a better feed conversion ratio (FCR). A comparable weight gain was achieved at a diet level of 30%, compared to a diet containing low levels of free gossypol cottonseed meal (0, 10, and 20%) [1, 119]. However, Diaw et al. [31] reported that including cottonseed kernels in broiler diets, even at levels as low as 6.25%, significantly decreased feed intake and growth performance. Nevertheless, adding iron or lysine to the diet could alleviate the effects of gossypol. Further studies have also demonstrated that adding ferrous sulfate ($FeSO_4$) to dietary cottonseed meal reduced the adverse effects of gossypol on chickens [17].

Whole cotton seeds contain high levels of fibre, making dehulled cotton seeds (kernels) a preferable option for poultry feed due to their high energy content, which is mainly attributed to their oil content and low levels of gossypol, a toxic substance for monogastric animals (such as pigs, poultry, and fish) found in cottonseed meal. Raw cottonseed kernels may contain between 0.6 and 2.0% free gossypol. To reduce the cost of broiler feeding, the poultry industry is seeking alternatives to major feed ingredients, as seen in studies by Batonon-Alavo et al. [12, 13]. In the US, the feed industry has established a maximum limit of 100 ppm for broilers and 40 ppm for laying hens for free gossypol levels in poultry diets [34]. Dehulled cotton seeds and cottonseed meal are commonly used separately or in combination as livestock, poultry, and fish feed. However, whole cotton seeds are not as widely used in pig and poultry diets as cottonseed meal [51]. Various methods have been developed to make cottonseed meal a wholesome protein supplement for poultry rations, including physical, chemical, and biological treatments, solvent extraction, fortification with amino acids, and biotechnological approaches [29].

7 Summary

Cotton is primarily cultivated for the fibre used in the textile industry, as well as for oil and as a feed for ruminant animals and to some extent non-ruminant animals. Cottonseed is known to have several important components, including approximately 15–25% oil, 25–35% high quality protein, 25–35% hull, and 10% linters. Cottonseed oil, a major by-product of cotton, is extracted from the whole or kernel cottonseed and is a significant source of vegetable oil worldwide. Among vegetable oils obtained from crushing soybeans and other oilseeds, including cottonseed oil accounts for approximately 55% of global vegetable oil production, while kernel oils from palm, coconut, and cottonseed together make up about 35% of global vegetable oil production. Vegetable oils and fats have two main uses: human consumption and technical or industrial applications. Most vegetable oils used in the food industry are unsaturated, either polyunsaturated or monounsaturated.

Although cotton is primarily grown for its lint, various by-products can be derived from cotton to increase value in the sector and benefit stakeholders such as farmers, ginners, oil millers, and other downstream actors involved in processing. Along with lint, ginners produce cottonseed, which is processed by millers into linters, hulls, cottonseed oil, and cake. The main cottonseed by-products include cotton linters, cotton hull, cottonseed cake and meal, cottonseed oil, and cotton stalks and gin. Various methods are used to extract cottonseed oil, resulting in different types of cottonseed meal. Cottonseed cake and meal are obtained through a process involving dehulling and solvent extraction. While cottonseed cake and meal are suitable animal feed for ruminants, they are not suitable for non-ruminants due to the presence of an enzyme called gossypol, which inhibits nutrient absorption. The presence of this enzyme in cottonseed cake and meal reduces its marketability for non-ruminant animals such as poultry, fish, and pigs.

Cottonseed oil is the most valuable by-product, obtained by crushing the kernel of the cottonseeds inside the meat of the seed that remains after removing the hard outer hull. However, crushing cottonseeds to extract cottonseed oil can be challenging. The tough hull surrounding the kernel makes it difficult to grind, and the hull is covered in short fibres that absorb the oils, resulting in lower yields. Additionally, cottonseed oil is considered more challenging to process compared to other vegetable oils because it has higher levels of free fatty acids and is more intensely colored due to the presence of gossypol and related pigments. After the seed cotton goes through the fibre removal processes of cleaning, delinting, and dehulling, the cottonseed is ready for oil extraction. Globally the existing cottonseed oils extraction methods, such as screw-press, hydraulic press, and solvent extraction, vary depending on factors like transportation infrastructure, hardware, solvent availability, and skilled labor. Once the cottonseed oil is extracted, it naturally has undesirable physical characteristics such as dark color, unpleasant smell from residual gossypol, cloudiness, solidness, and waxiness at temperatures below the melting point of cottonseed oil. To control potential food hazards, cottonseed oil undergoes extensive refining processes including caustic refining, bleaching, winterization, fractionation, hydrogenation, interesterification, and deodourization.

In general, the physicochemical properties of a specific cottonseed oil depend on the type of cotton grown and the environmental conditions during growth, such as temperature, soil quality, fertilizers, and rainfall. These properties are also influenced by the handling and storage methods after harvesting. Breeding and agricultural practices could also be used to improve the physicochemical properties of cottonseeds without affecting the quality of the fibre. However, it is the chemical composition that ultimately determines the physicochemical properties of fats and oils, which in turn affects their suitability for different processes and applications. The fatty acid composition, peroxide value, iodine value, saponification value, and unsaponifiable matter are some of the major physicochemical properties that determine the quality of oils and fats.

The composition of free fatty acids (FFA) in most vegetable oils consists of a series of straight-chain carboxylic acids with even-numbered chain lengths ranging from C_8 to C_{24}. The majority of edible oils are made up of 95–98% triacylglycerides (TAG), with a small fraction (2–5%) containing complex mixtures of minor compounds such as FFA. Cottonseed oil, which mainly consists of C_{16} and C_{18} fatty acids, remains palatable as long as the FFA level does not exceed 1.0%. The iodine value is a measure of the unsaturation of fatty acids in a fat or vegetable oil, indicating the number of double bonds present in the fatty acids. It is expressed as the amount of iodine absorbed by 100g of oil/fat when tested with Wijs solution. On the other hand, the peroxide value measures the concentration of peroxides, which are formed by the reaction between oxygen and unsaturated fatty acids. It is expressed as the amount of milli-moles of peroxide (or milli-equivalents of oxygen) absorbed by 1 kg of oil/fat sample (mEq/Kg). The saponification value, measured in milligrams of potassium hydroxide (mg), indicates the amount of potassium hydroxide required to saponify 1 g of oil/fat. This value is used to predict the types of triacylglycerols present in an oil/fat by measuring the alkali-reactive groups.

The significance of cottonseed oil in human nourishment is growing and the demand for this valuable commodity is increasing due in part to a general shortage in supply of edible oils. Cottonseed oil has one of the healthiest nutritional profiles among the vegetable oils. Like other vegetable oils, it does not contain cholesterol in its natural unhydrogenated form. However, the utilization of cotton as a feed and food crop, from the products and by-products of cottonseed has been limited worldwide, primarily due to the existence of the toxic compound gossypol. Nevertheless, cottonseed is believed to have the potential to develop as a nutritional feed and food crop supplements, both for animal feed and human foodstuffs being particularly valuable as an edible oil for humans; but also as a cottonseed meal or flour for both animals and humans only when it is derived from gossypol-free varieties, or if the gossypol has been adequately extracted through extensive and thorough processing or unless otherwise, that it is present in the food at low or permissible levels.

References

1. Abdulrashid M, Josephl ZO, Mohammed A, Adamu HY et al (2013) Response of broiler chickens fed cottonseed meal based diets. Int J Adv Agric Res IJAAR 1(2013):62–65
2. Abou-Donia MB (1976) Physiological effects and metabolism of gossypol. In: Gunther FA, Gunther JD (eds) Residue reviews – residues of pesticides and other contaminants in the total environment, vol 61. Springer Verlag, New York, pp 126–160
3. Agarwal D, Singh P, Chakrabarty M, Shaikh A, Gayal S et al (2003) Cottonseed oil quality, utilization and processing. Cicr Tech Bull 25:5
4. Alexander J, Benford D, Cockburn A et al (2008) Gossypol as undesirable substance in animal feed. EFSA J 908(2008):1–55
5. Anna Y, Medentsev AG, Krupyanko VI et al (2012) Gossypol inhibits electron transport and stimulates ROS generation in yarrowia lipolytica mitochondria. Open Biochem J 6:11–15
6. Arieli A (1998) Whole cottonseed in dairy cattle feed: a review. Anim Feed Sci Technol 72(1–2):97–110
7. Ash MS (1992) Animal feeds compendium. AER-656, Commodity Economics, Division, Economic Research Service, USDA
8. Ashokkumar K, Ravikesavan R (2013) Genetic variation and heterotic effects for seed oil, seed protein and yield attributing traits in upland cotton (*Gossypium hirsutum* L.). Afr J Biotech 12:5183–5191
9. Association of Official Analytical Chemists (AOAC) (2000) Official Methods of Analysis. Association of Official Agricultural Chemists (AOAC), 17th edn. Washington, D.C., USA. http://www.aoac.org
10. Balakrishnan K, Wierda WG, Keating MJ, Gandhi V et al (2008) Gossypol, a BH3 mimetic, induces apoptosis in chronic lymphocytic leukemia cells. Blood 112(5):1971–1980
11. Banshi DS, Prabhat KS, Ram KG et al (1992) Oilseeds processing technology. Central institute of agricultural engineering, Nabi Bagh, Berasia Road, BHOPAL—462 018, INDIA
12. Batonon-Alavo DI, Bastianelli D, Lescoat P, Weber GM, Umar FM et al (2016) Simultaneous inclusion of sorghum and cottonseed meal or millet in broiler diets: effects on performance and nutrient digestibility. Animal 10:1–11. https://doi.org/10.1017/S1751731116000033
13. Batonon-Alavo DI, Umar Faruk M, Lescoat P, Weber GM, Bastianelli D et al (2015) Inclusion of sorghum, millet and cottonseed meal in broiler diets: a meta-analysis of effects on performance. Animal 9:1120–1130. https://doi.org/10.1017/S1751731115000282
14. Benjamin C (ed) (2003) Encyclopedia of food sciences and nutrition. Academic Press, Maryland, USA
15. Berhanu G, Adane H, Kahsay B et al (2009) Feed marketing in Ethiopia: results of rapid market appraisal. Improving Productivity and Market Success (IPMS) of Ethiopian farmers project Working Paper, 15. ILRI (International Livestock Research Institute), Nairobi, Kenya, p 64
16. Blasi DA, Drouillard J (2002) Composition and feeding value of cottonseed feed products for beef cattle. Kansas State University Agricultural Experiment Station and Cooperative Extension Service, MF, 2538, p 22
17. Boling SD, Edwards HM, Emmert JL, Biehl RR, Baker DH et al (1998) Bioavailability of iron in cottonseed meal, ferric sulfate, andtwo ferrous sulfate by-products of the galvanizing industry. Poultry Sci 77:1388–1392
18. Brampton M (2001) Research turns trash to treasure. USQ News. University of Southern Queensland (2001)
19. Cater CM, Mattil KF, Meinke WW, Taranto MV, Lawhon JT et al (1977) Cottonseed protein food products. J Am Oil Chem Soc 54(2):A90–A93
20. Central Institute for Cotton Research (CICR) (2003) Cottonseed oil quality utilization and processing. CICR Technical Bulletin, No: 25. http://www.cicr.org.in
21. Central Statistical Authority (CSA) (2010) Agricultural sample surveys (2005–2009). http://www.csa.gov.et/

22. Chenoweth PJ, Risco CA, Larsen RE, Velez J, Tran T, Chase CC Jr et al (1994) Effects of dietary gossypol on aspects of semen quality, sperm morphology and sperm production in young Brahman Bulls. Theriogenology 42:1–13
23. Cherry JP (1983) Cottonseed oil. J Am Oil Chem Soc 60:360
24. Cherry JP, Leffler HR (1984) Chapter 13: Seed. In: Kohel RJ, Lewis CF (eds) Cotton No. 24 in Agronomy series; American Society of Agronomy, Inc., Crop Science Society of America, Inc., Soil Science Society of America. Madison, WI, pp 511–569
25. Coates W (2000) Using cotton plant residue to produce briquettes. Biomass Bioenerg 18:201–208
26. Coutinho EM (2002) Gossypol: a contraceptive for men. Contraception 65:259–263
27. Dechary JM, Kupperman RP, Thurber FH, Altschul AM (1952) Removal of gossypol from cottonseed by solvent extraction procedures. J Am Oil Chem Soc 29:339–341
28. Denis J (2004) Genetically modified oil crops. In: Robert MG (ed) Encyclopedia of plant and crop science. 1st edn. Marcel Dekker, Inc., pp 509–512. https://doi.org/10.1081/E-EPCS120010439
29. Devanaboyina N, Savaram VR, Arun KP, Vadali RB et al (2007) Cottonseed meal in poultry diets. A review. J Poult Sci 44:119–134
30. Devesh KS, Sharma SK. Sambi SS et al (2011) Comparative extraction of cottonseed oil by n–Hexane and Ethanol. ARPN J Eng Appl Sci 6(1):84–89
31. Diaw MT, Dieng A, Mergeai G, Sy M, Hornick JL et al (2010) Effects of the substitution of groundnut cake by conventional cottonseed kernel on broilers production in Senegal. Tropicultura 28(3):139–147
32. Dowd MK, Boykin DL, Meredith WR, Campbell BT, Bourland FM, Gannaway JR et al (2010) Fatty acid profiles of cottonseed genotypes from the national cotton variety trials. J Cotton Sci 14:64–73. http://www.cotton.org/journal/2010-14/2/upload/JCS14-64.pdf
33. El-Saidy DM, Gaber MM (2004) Use of cottonseed meal supplemented with iron for detoxification of gossypol as a total replacement of fish meal in Nile tilapia, Oreochromis niloticus (L.) diets. Aquacul Res 35(9):859–865
34. Encyclopedia of Food Grains (2016) The world of food grains. In: Encyclopedia of food grains, 2nd edn., vol 1. ISBN 978-0-12-803537-5. http://store.elsevier.com
35. Ensminger ME, Oldfield JE, Heinemann WW et al (1990) By-product feeds/crop residues. In: Feeds and nutrition, 2nd edn. Ensminger Publishing Company, Clovis, California, USA, pp 433–490
36. European Commission (2003) Opinion of the scientific committee on animal nutrition on undesirable substances in feed. European Commission, Health and Consumer Protection Directorate-General, Directorate C. Scientific Opinions. C2 Management of Scientific committees; scientific co-operation and networks, SCF/CS/CNTM/MYC/25 Rev 6 Final
37. European Food Safety Authority (EFSA) (2008) Gossypol as undesirable substance in animal feed. European Food Safety Authority (EFSA) Panel on Contaminants in the Food Chain (CONTAM). EFSA J 908:1–55
38. Forster LA, Calhoun MC (1995) Nutrient values for cottonseed products deserve a new look. Feedstuffs 67(44):16
39. Frank DG (2002) Vegetable oils in food technology composition, properties and uses. Blackwell Publishing Ltd. ISBN 1-84127-331-7
40. Frederick JF (1999) Wiley encyclopedia of food science and technology, V4. 2nd edn
41. Galvin AM (1991) Association of official agricultural chemists (AOCS) press. In: Wan PJ (ed) Introduction to fats and oils technology, Champagne, IL, pp 137–164
42. Geoffrey C (2009) Food science and technology. Blackwell Publishing Ltd.
43. Gerasimidis K, Fillou DT, Babatzimcpoulou M, Tassou K, Katsikas H et al (2007) Preparation of an edible cottonseed protein concentrate and evaluation of its functional properties. Int J Food Sci Nutr 58:486–490
44. Gomes RS, Wilson N, Coates WE, Fox RW et al (1997) Cotton (*Gossypium*) plant residue for industrial fuel: an economic assessment. Ind Crops Prod 7:1–8

45. Gunstone DF (2011) Vegetable oils in food technology: composition, properties and uses, 2nd edn. Wiley
46. Hamilton KA, Pyla PD, Breeze M, Oslon T, Li M, Robinson E et al (2004) Bollgard II cotton: compositional analysis and feeding studies of cottonseed from insect-protected cotton (*Gossypium hirsutum* L.) producing the *Cry* 1Ac and *Cry* 2Ab2 proteins. J Agric Food Chem 52:6969–6976
47. Hamm W (2001) Regional differences in edible oil processing procedures. 1. Seed crushing and extraction, oil movements, and degumming. 2. Refining, oil modification, and formulation. Lipid Tech., vol 13, pp 81–84, 105–109
48. Han ML, Wang YF, Tang MY et al (1987) Gossypol in the treatment of endometriosis and uterine myoma. Contrib Gynecol Obstet 16:268–270
49. Heidarinia A, Malakian M (2011) Nutritional evaluation of cottonseed meal with and without ferrous sulfate for broiler chickens. Res J Poult Sci 4:14–17
50. Heuzé V, Tran G, Hassoun P, Bastianelli D, Lebas F et al (2016) Cottonseed meal. Feedipedia, a program by INRA, CIRAD, AFZ and FAO. http://www.feedipedia.org
51. Heuzé V, Tran G, Hassoun P, Brossard L, Bastianelli D, Lebas F et al (2015) Cotton seeds. Feedipedia, a programme by INRA, CIRAD, AFZ and FAO. http://www.feedipedia.org
52. International Cotton Advisory Committee (ICAC) (2003) Common fund for commodities (CFC): cotton facts, 1st edn. ISBN 0-9704918-3-2
53. International Trade Centre (ITC) (2020) Beyond the fibre: capturing cotton's full value in Africa. ITC, Geneva
54. Ivana C, Nayanna B, Silvia C, Marília M, Benito S et al (2014) Gossypol toxicity from cottonseed products. Hindawi Publishing Corporation. Sci World J 2014:11, Article ID 231635. https://doi.org/10.1155/2014/231635
55. Jeoh T, Agblevor FA (2001) Characterization and fermentation of steam exploded cotton gin waste. Biomass Bioenerg 21:109–120
56. Jiang J, Slivova V, Jedinak A, Sliva D et al (2012) Gossypol inhibits growth, invasiveness, and angiogenesis in human prostate cancer cells by modulating NF-κB/AP-1 dependentand independent-signaling. Clin Exp Metas 29(2):165–178
57. John B (2010) Markets for cotton by-products, global trends and implications for African cotton producers. The World Bank Development Prospects Group & Africa Region. Policy Research Working Paper-5355
58. John P, Russell J, Lynn A, William et al (1986) Food and Feeding Quality of Cottonseed. Chapter 37. In: Jack R, James M (eds) Cotton physiology. The cotton foundation reference book series, No. 1. Tennessee, U.S.A, pp 557–595
59. Johnson PW (2008) New targets for lymphoma treatment. Ann Oncol 19(4):iv56–iv59
60. Jones LA, King CC (1996) Cottonseed. In: Hui YH (ed) Bailey's industrial oil and fat products, 5th edn, vol 2. Wiley, New York, NY, pp 166–171, 178, 182–186, 198, 192–209
61. Jones LA, King CC (eds) (1993) Cottonseed oil. National Cottonseed Products Association and the Cotton Foundation, Memphis, Tennessee, USA, p 60
62. Kandylis K, Nikokyris P, Deligiannis K et al (1998) Performance of growing-fattening lambs fed whole cotton seed. J Sci Food Agric 78:281–289
63. Kansas State University (2002) Cottonseed Feed Products for Beef Cattle, Composition and Feeding Value, Kansas State University, Contribution No. 02-426-E from the Kansas Agricultural Experiment Station
64. Kritchevsky D (1998) Phytosterols. In: Kritchevsky D, Bonfield C (eds) Dietary fibre in health and disease. Plenum Press, New York, pp 235–243
65. Kumar M, Paruthi P (2019) A nutrient-rich healthy drink from cottonseed: an Indian delicacy. J Ethn Foods 6:1–6
66. Kumar M, Tomar M, Punia S, Grasso S, Arrutia F, Choudhary J, Singh S, Verma P, Mahapatra A, Patil S et al (2021) Cottonseed: A sustainable contributor to global protein requirements. Trends Food Sci Technol 111:100–113
67. Leonardo V, Begoña P, José M et al (2004) Grain quality in oil crops grain quality in oil crops. Chapter 12. In: Roberto L, Rodolfo A (eds) Handbook of seed physiology applications to

agriculture. Food products press and the Haworth reference press, imprints of The Haworth Press, Inc., 10 Alice Street, Binghamton, NY, pp 13904–1580
68. Li Z, Smith KH, Stevens GW et al (2016) The use of environmentally sustainable bio-derived solvents in solvent extraction applications-A review. Chin J Chem Eng 24:215–220
69. Luis A, Pierre M, Gustavo P, Natalia I, Vincent A et al (2009) Management and breeding strategies for the improvement of grain and oil quality. In: Victor O, Daniel F (eds) Chapter 16: Crop physiology-applications for genetic improvement and agronomy. Elsevier Inc. ISBN: 978-0-12-374431-9
70. Malik TH, Ahsan MZ (2016) Review of the cotton market in Pakistan and its future prospects. Oilseed Fat Crops Lipids 23:D606
71. Margalith P (1967) Inhibitory effect of gossypol on microorganisms. Appl Microbiol 15(4):952–953
72. McKevith B (2005) Nutritional aspects of oilseeds. Nutr Bull 30(1):3–26
73. Miller BM, Lawrence OC (1997) Seed production principles and practices. Springer Science+Business Media Dordrecht. ISBN 978-1-4615-4074-8 (eBook). https://doi.org/10.1007/978-1-4615-4074-8
74. Mohamad H, Wenli S, Qi C (2020) Considering white gold, cotton, for its fibre, seed oil, traditional and modern health benefits. J Biol Environ Sci 14(40):25–39
75. Morgan S (1989) Gossypol toxicity in livestock. Oklahoma Cooperative Extension Service, VTMD-9116
76. Nadia I, Shahzad M, Khalil U et al (2002) Evaluation of vigour and oil quality in cottonseed during accelerated ageing. Int J Agric Biol 4(3):1560–8530
77. National Cottonseed Products Association (NCPA) (2002) Cottonseed and its products. 1255 Lynnfield Road, Suite 143, P.O. Box 172267, Memphis, TN. http://www.cottonseed.com/publications/cottonseedanditsproducts.asp
78. Nicholson SS (2012) Cottonseed toxicity. In: Gupta RC (ed) Veterinary toxicology: basic and clinical principles, 2nd edn. Academic Press, London, pp 1161–1165
79. Nicolay G, Estur G, Walsh C, Desalegn P et al (2020) Cotton value chain analysis in Ethiopia. Report for the European Union, DG–DEVCO. Value chain analysis for development Project (VCA4D CTR 2016/375-804), 128 p + annexes
80. Office of the Gene Technology Regulator (OGTR) (2008) The Biology of *Gossypium hirsutum* L. and *Gossypium barbadense* L. (Cotton), Ver. 2. Office of the Gene Technology Regulator, Australian Government Department of Health and Ageing, Canberra, p 91
81. Organisation for Economic Co-operation and Development (OECD)/Food and Agriculture Organization of the United Nations (FAO) (2019) OECD–FAO Agricultural Outlook 2019–2028. OECD Publishing, Paris/Food and Agriculture Organization of the United Nations, Rome. https://doi.org/10.1787/agr_outlook-2019-en
82. Ostil NP, Pandey SB (2006) Use of whole cotton seed and cotton seed meal as a protein source in the diet of ruminant animals: prevailing situation and opportunity. In: Proceedings of the 6th national workshop on livestock and fisheries research
83. O'Brien RD, Wan PJ (2001) Cottonseed oil: processing and utilization. In: Wilson R (ed) Proceedings world conference and exhibition on oilseed processing and utilization. AOCS Press, Champaign, IL, p 16
84. O'Brien RD, Jones LA, King CC, Wakelyn P J, Wan PJ et al (2005) Cottonseed oil: Chapter 5. In: Bailey's industrial oil and fat products, 6th edn. Wiley
85. Paim TP, Viana P, Tilburg MFV, Moura AA, Souza JR, McManus C et al (2019) Feeding effects of cottonseed and its co-products on the meat proteome from ram lambs. Sci Agric 76:463–472. https://doi.org/10.1590/1678-992x-2018-0072
86. Pelitire SM, Dowd MK, Cheng H et al (2014) Acidic solvent extraction of gossypol from cottonseed meal. Anim Feed Sci Technol 195:120–128
87. Phillip J (2001) Food industry—Solvents for extracting vegetable oils. In: George W (ed) Handbook of solvents. ChemTec Publishing, 38 Earswick Drive, Toronto, Ontario M1E 1C6, Canada. Chem Tec Publishing. ISBN 1-895198-24-0

88. Polsky B, Segal SJ, Baron PA, Gold JW, Ueno H, Armstrong D et al (1989) Inactivation of human immunodeficiency virus in vitro by gossypol. Contraception 39(6):579–587
89. Poore M, Glenn M, Rogers DVM et al (1998) Potential for gossypol toxicity when feeding whole cottonseed. North Carolina State University, North Carolina Cooperative Extension Service, Department of Animal Science
90. Van Poznak C, Seidman AD, Reidenberg MM et al (2001) Oral gossypol in the treatment of patients with refractory metastatic breast cancer: a phase I/II clinical trial. Breast Cancer Res Treat 66(3):239–248
91. Price WD, Lovell RA, McChesney DG (1993) Naturally occurring toxins in feedstuffs: center for veterinary medicine perspective. J Anim Sci 71(9):2556–2562
92. Ramachandran S, Singh SK, Larroche C, Soccol CR, Pandey A (2007) Oil cakes and their biotechnological applications—A review. Bioresour Technol 98(10):2000–2009
93. Randel RD, Chase CC Jr, Wyse SJ (1992) Effects of gossypol and cottonseed products on reproduction of mammals. J Anim Sci 70(5):1628–1638
94. Rathore KS (2007) Reducing Gossypol in cottonseed may improve human nutrition, Department of Social and Crop Science, Texas A and M University College Station TX
95. Rattanaphra D, Srinophakun P (2010) Biodiesel production from crude sunflower oil and crude jatropha oil using immobilized lipase. J Chem Eng Jpn 43:104–108
96. Reske J, Siebrecht J, Hazebroek J et al (1997) Triacylglycerol composition and structure in genetically modified sunflower and soybean oils. J Am Oil Chem Soc 74:989–998
97. Risco CA, Seebohm S, Adams AL, Staples CR, Van Horn HH, Thatcher WW (1998) Effects of gossypol from whole cottonseed on lactation & reproductive performance of dairy cows. In: 35th Florida dairy production conference, pp 23–30
98. Rogers GM, Poore MH, Paschal JC (2002) Feeding cotton products to cattle. Vet Clin North Am: Food Anim Pract 18(2):267–294
99. Salas C, Ekmay R, England J, Cerrate S, Coon CN et al (2013) The TMEn, proximate analysis, amino acid content and amino acid digestibility of glandless and commercial cottonseed meal for broilers. Int J Poult Sci 12(4):212–216. https://doi.org/10.3923/ijps.2013.212.216
100. Santos JEP, Mena H, Huber JT, Tarazon M (2005) Effects of source of gossypol and supplemental iron on plasma gossypol in Holstein steers. J Dairy Sci 88(10):3563–3574
101. Santos JEP, Villasenor M, DePeters EJ, Robinson PH, Baldwin BC et al (2002) Type of cottonseed and gossypol in diets of lactating dairy cows: lactation performance and plasma gossypol. J Dairy Sci 85(6):1491–1501
102. Satankar V, Singh M, Mageshwaran V, Jhodkar D, Changan S, Kumar M, Mekhemar M et al (2021) Cottonseed Kernel powder as a natural health supplement: an approach to reduce the gossypol content and maximize the nutritional benefits. Appl Sci 11:3901. https://doi.org/10.3390/app11093901
103. Saxena DK, Sharma SK, Sambi SS et al (2011) Comparative extraction of cottonseed oil by n Hexane and ethanol. ARPN J Eng Appl Sci 6(1):84–89
104. Sean F, Oscar A (2010) Fat characterization. Chapter 14. In: Suzanne N (ed) Food analysis, 4th edn. Springer Science+Business Media, LLC. https://doi.org/10.1007/9781-4419-1478-1
105. Silva RVMM, Carvalho GGP, Pires AJV, Pereira MLA, Pereira L, Campos FS et al (2016) Cottonseed cake in substitution of soybean meal in diets for finishing lambs. Small Rumin Res 137:183–188. https://doi.org/10.1016/j.smallrumres.2016.03.014
106. Singh S, Sharma SK, Kansal SK et al (2018) Batch extraction of gossypol from cottonseed meal using mixed solvent system and its kinetic modeling. Chem Eng Commun 206:1608–1617
107. Smith CW (1995) Cotton (*Gossypium hirsutum* L.). In: Smith CW (ed) Crop production: evolution, history, and technology. Wiley, New York, pp 287–349
108. Sterling KG, Costa EF, Henry MH, Pesti GM, Bakalli RI et al (2002) Responses of broiler chickens to cottonseed and soyabean meal-based diets at several protein levels. Poult Sci 81:217–226

109. Tang JW, Sun H, Yao XH, Wu YF, Wang X, Feng J et al (2012) Effects of replacement of soybean meal by fermented cottonseed meal on growth performance, serum biochemical parameters and immune function of yellow feathered broilers. Asian Aus J Anim Sci 25(3):393–400. https://doi.org/10.5713/ajas.2011.11381
110. Tang DYY, Yew GY, Koyande AK, Chew KW, Vo DVN, Show PL et al (2020) Green technology for the industrial production of biofuels and bioproducts from microalgae: a review. Environ Chem Lett 18:1967–1985
111. Tanksley TD Jr (1990) Cottonseed meal. In: Thacker PA, Kirkwood RN (eds) Nontraditional feed sources for use in swine production. Butterworth, Boston, pp 139–151
112. Taranto MV, Meinke WW, Carter CM, Mattil KF et al (1975) Parameters affecting production and character of extrusion texturized defatted glandless cottonseed meal. J Food Sci 40(6):1264–1269. https://doi.org/10.1111/j.13652621.1975.tb01068.x
113. Tumbelaka LI, Slayden O, Stormshak F (1994) Action of a cyclopropenoid fatty acid on the corpus luteum of pregnant and nonpregnant ewes. Biol Reprod 50:253–257
114. Tuszynski GP, Cossu G (1984) Differential cytotoxic effect of gossypol on human melanoma, colon carcinoma, and other tissue culture cell lines. Can Res 44(2):768–771
115. UNCTAD (2019) Cotton and its by-products in Eastern and Southern Africa, analysis of cotton by-products survey in Uganda. Project: 1617K—Funded by the United Nations Development Account 2016–2019
116. United Nations Conference on Trade and Development (UNCTAD) (2017) Cotton and its by products in the United Republic of Tanzania: Analysis of cotton by-products survey. UNCTAD, Special Unit on Commodities, Geneva, Switzerland
117. Waldroup P (2000) National cottonseed products association (NCPA) oil mill industry survey 1999 data collection. National Cottonseed Products Association. Summary, Industry Data. NCPA Technical Files
118. Wang X, Wang J, Wong SCH et al (2000) Cytotoxic effect of gossypol on colon carcinoma cells. Life Sci 67(22):2663–2671
119. Watkins SE, Skinner JT, Adams MH, Waldroup PW et al (1993) An evaluation of low gossypol cottonseed meal diets for broiler chickens. 1. Effect of cottonseed meal level and lysine supplementation. J Appl Poult Res 2:221–226. https://doi.org/10.1093/japr/2.3.221
120. Wijnands J, Biersteker J, van Loo E (2009) Oilseeds business opportunities in Ethiopia. Ministry of Agriculture, Nature and Food Quality, The Hague, the Netherlands
121. Wong FY, Liem N, Xie C et al (2012) Combination therapy with gossypol reveals synergism against gemcitabine resistance in cancer cells with high BCL-2 expression. Plos One 7(12), Article ID e50786
122. Yang J, Chen G, Li LL et al (2013) Synthesis and anti-H5N1 activity of chiral gossypol derivatives and its analogs implicated by a viral entry blocking mechanism. Bioorg Med Chem Lett 23(9):2619–2623
123. Yang J, Zhang F, Li J et al (2012) Synthesis and antiviral activities of novel gossypol derivatives. Bioorg Med Chem Lett 22(3):1415–1420
124. Yared S, Marieke R, Youri D, Myrtille D (2011) Small-scale edible oil milling operations: alternative business models for Ethiopia. LEI Memorandum 11-005, Wageningen UR (University & Research centre)
125. Ye W, Chang HL, Wang LS et al (2007) Modulation of multidrug resistance gene expression in human breast cancer cells by (-)-gossypol-enriched cottonseed oil. Anticancer Res 27(1):107–116
126. Yu J, Yu S, Fan S, Song M, Zhai H, Li X et al (2012) Mapping quantitative trait loci for cottonseed oil, protein and gossypol content in a *Gossypium hirsutum* × *Gossypium barbadense* backcross inbred line population. Euphytica 187:191–201

Cotton Research, Extension, and Promotion

Tewodros Duressa

Abstract The Ethiopian government currently finances and supports cotton research, which is carried out through a network that includes Assosa, a rain-fed research center in Benshangul Gumuz Regional State, and Werer Research Center (WARC), a primary federally funded irrigated research center, in Afar Regional State. There are no additional centers or sub-centers that cover all of the different agro-climatic zones used to cultivate cotton in Ethiopia. The Ministry of Agriculture and Natural Resources (MoANR) stopped providing extension services for cotton production in 2014 when the cotton sector was transferred to the Ministry of Industry (MoI). Because the major commercial farmers who could offer extension services do not provide sufficient funding, such out grower programs are often nonexistent in the cotton sector. Ethiopia's significant potential for cotton production remains untapped, as only 3% of the country's suitable cotton production area is currently being cultivated. However, compared to other African countries, Ethiopia's agricultural research on cotton is relatively new. Since the establishment of the National Agricultural Research Center (NARC) to coordinate research and resources across the country's entire research system, a more favorable research environment is being developed as the apex organization. The Ethiopian government is currently pursuing a participatory demonstration and training extension system. The goal of this approach is to disseminate the complete package of agricultural technology and practices (such as cultivars, pesticides, fertilizers, and improved cultural practices) for a specific crop. Agricultural extension programs are implemented at the Woreda (district) and Keble (village) levels, which is why cotton research has been underperforming for a long time. The public sector plays the most significant role, especially in the local distribution of inputs and agricultural extension services to smallholder cotton farmers.

Keywords Cotton research · Extension · Promotion · Production · Agriculture

T. Duressa (✉)
Cotton Extension Researcher, Textile and Garment Industry Development and Research Center, Addis Ababa, Ethiopia
e-mail: teddye2005@gmail.com

K. Murugesh Babu et al. (eds.), *Cotton Sector Development in Ethiopia*, Textile Science and Clothing Technology, https://doi.org/10.1007/978-981-99-9149-5_16

1 Introduction

Since 1964, the Werer Agriculture Research Centre (WARC) has been involved in cotton research in Ethiopia. Currently, the Ethiopian government finances cotton research, which is carried out through a network of two primary federal research centers: Assosa, a rain-fed facility in Benshangul Gumuz Regional State, and WARC, an irrigated facility located in the Afar Regional State. There are no additional centers or sub-centers that cover all the different agro-climatic zones suitable for cotton cultivation. However, there are six main agro-climatic zones in the country where cotton can be grown. For a long time, cotton research was neglected, leading to frustrating effects on cotton yield and quality.

Ethiopian cotton production is still primarily centered in WARC, although it has expanded to several regions of the country, such as Benishangul-Gumuz, Gambela, South Nation and Nationalities Peoples Reginald State (SNNPRS), and others. While the large-scale irrigated regions have received more attention, cotton grown by small-scale farmers using rainwater has traditionally been given less priority.

The "National Cotton Research Plan (2016–2030)" was developed by the Ethiopian Agricultural Research Institute (EARI) at the end of 2016. The main objective of the strategy was to increase cotton yield and productivity, as well as improve the quality of cotton lint and its by-products (oil and seed cake) through the implementation of a multidisciplinary and participatory research approach, along with the necessary tools [1]. However, in the meantime, the government established a cotton development sector within the Ethiopian Textile Industry Development Institute (ETIDI), and the institute shifted its focus to research by hiring cotton researchers and collaborating with various universities to address the research gaps in cotton. This approach has proven effective in bridging the knowledge gaps in cotton production and quality enhancement.

2 Cotton Production Extension

Particularly when it comes to the local delivery of supplies and agricultural extension services to smallholder cotton farmers, the public sector is currently the most dominant participant. Because of this, the Ministry of Agriculture and Regional Bureaus of Agriculture are the primary institutions in the country that provide agricultural advisory service. Increasing the human capital's capacity to provide farmers with agricultural guidance was the initial step in the Ethiopian government's goal of creating the largest agricultural extension network in Sub-Saharan Africa. Additionally, certain private agro-industries, like the sugar industry, offer technical assistance and extension services to farmers under contractual market agreements. Such out grower initiatives are mostly nonexistent in the cotton sector due to the lack of funding from larger commercial farmers who could provide extension assistance. Farmers' cooperatives in Ethiopia primarily distribute agricultural inputs and farm loans rather than providing their members with direct access to agricultural advisory services.

After the cotton sector was transferred to the Ministry of Industry (MoI) in 2014, the previous Ministry of Agriculture and Natural Resources (MoANR) ceased to provide extension services for the cotton crop. While large commercial cotton farms often fail to provide satisfactory services to smallholders, the expansion of private extension services remains limited. There are few connections to research, and farmers generally have little knowledge of what researchers are doing. Surprisingly, Ethiopia has one of the world's densest extension networks but does not utilize it for the cotton industry. However, the ETIDI's cotton development sector provides some assistance and technical support to cotton producers in cooperation with the Federal Ministry of Agriculture and the sub-district (Zone to Woreda) institutions.

2.1 Cotton Production Trends in Ethiopia

With 2,697,640 million hectares of land suitable for cultivating cotton, Ethiopia has a land area similar to Pakistan's, the fourth-largest producer globally [2]. A total of 65% of the suitable cotton farming land is located in 38 cotton-producing regions with high potential, while the remaining 35% is spread across 75 districts with medium potential. In contrast, according to a SOFERCO scoping study (2016) that has not yet been published, 30% of Ethiopia's cotton crop is currently irrigated, and 70% is grown on rain-fed land. Only 3% of the total suitable land for cotton production is currently being cultivated in Ethiopia [1, 2]. The yearly production of seed cotton was approximately 120,000 tons from 2011 to 2013 with an overall yield of 1.42 ton/ha [3]. Presently, the majority of cotton farming takes place in the Awash Valley, although there is also some cultivation in Gambella, Humera, and Metema. Only 35,000 of the 84,000 ha utilized for cotton cultivation in Ethiopia are irrigated, despite the country's ability to produce irrigated cotton.

Some of the most promising areas for cotton cultivation in Ethiopia are the Omo, Ghibe, Wabi-Shebelle, Awash, Baro-Akobo, Blue Nile, and Tekeze river basins [4]. As depicted in Fig. 1, Ethiopia has an average cotton production of 33,842.10 metric tons from 2000 to 2018. Ethiopia had the lowest cotton production during the specified time period in Fig. 1 (14,000 metric tons in 2001) and the highest amounts (66,000 metric tons in 2015).

2.2 Limitations on Cotton Production

Ethiopian cotton farmers have been predominantly using one primary seed variety for the past few decades, but it is increasingly vulnerable to pests and diseases. The California and Delta varieties of cottonseed, which are currently being used, were imported from the United States over 20 years ago, according to a report by

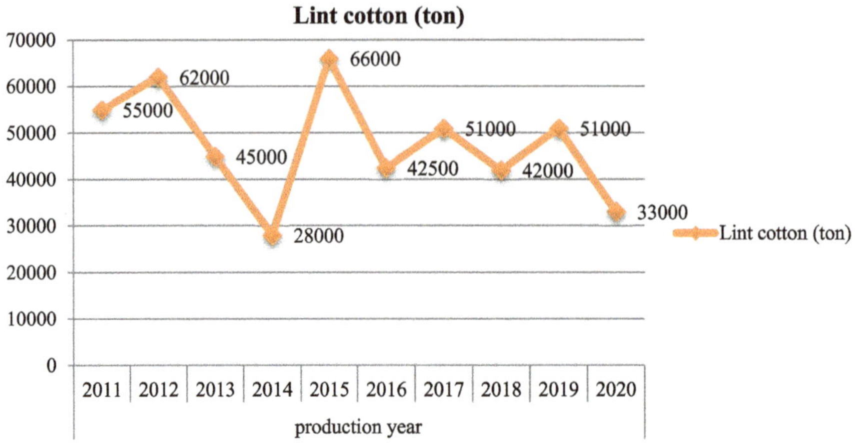

Fig. 1 The quantity of lint cotton production in Ethiopia from 2000 to 2018 years

ETIDI in 2015. Farmers experienced significant crop losses due to Mealy bug, Flea Beetle, Pink Bollworm, and Bacterial Blight during the most recent crop season. In some instances, these losses occurred after unfavorable weather conditions persisted despite multiple rounds of pesticide application. Meanwhile, there are environmental factors, challenges in accessing inputs, financial issues, rising production costs, and ineffective marketing strategies. Additionally, the Ethiopian government (GoE) has prioritized increasing sugar production to become one of the top 10 global producers of the commodity. Consequently, the Tendaho cotton farm, once a major cotton producer in the country, transitioned to sugarcane cultivation several years ago. Furthermore, many other smaller cotton enterprises have also shifted to sugar production, following the example of Tendaho.

In addition to the aforementioned challenges in cotton production, the lack of high-quality inputs, such as seeds and fertilizer, as well as pests, have hindered the potential growth of cotton production. Natural disasters like floods, especially those along the Awash River, and land ownership rights have also had a negative impact on the expansion of cotton farming. The previous ban on exporting excess cotton from 2010 to 2012, along with the fact that other crops, like sesame, were more profitable to grow, are two other commonly cited reasons for the decline in cotton output in the country. Due to the export restrictions, farmers reduced the amount of land used for cotton and instead chose to cultivate sesame or other cash crops. The production of cotton is generally limited by factors such as a lack of improved seed varieties, a lack of technical inputs, including labor, a lack of extension services, and limited irrigation techniques.

3 Cotton Technology Promotion and Development

"The Green Revolution is largely a result of technological innovation in the global public sphere where Western and developing country governments, publicly-funded non-profit national and international agricultural research organizations, universities, international aid agencies, and Western charitable groups collaborated to enhance agricultural productivity," asserts Parayil in 2003. Consequently, a variety of agricultural literature documents the historical agricultural achievements accomplished over the past 200 years. One noteworthy example is the transition from traditional or shifting cultivation to rotational fallow and eventually permanent cultivation; Green Revolution, Gene Technology. Shifting cultivation played a crucial role in maintaining soil quality and fertility through rotational fallow, while also preserving the environment and aiding in weed and disease control. While various studies on the impact of agricultural technology on poverty suggest that physical infrastructure and human capital complement each other effectively [5] (Canning and Bennathan 2000; Datt and Ravallion 1997, 1998; Omilola 2009), this is not always the case.

By discussing this quote from [6] (Mellor 1976), Babatunde Omilola argues that "the predominant literature on the poverty linkages of agricultural growth during the 1970s tends to demonstrate that agricultural technological change leads to increased production, which in turn leads to higher incomes for land-owning households and reduces poverty." According to Omilola (2009), individuals with jobs are believed to spend a significant portion of their income from agricultural products on labor-intensive goods and services, thereby increasing employment opportunities for the unemployed and ensuring food security for the poor.

Throughout the globe, there have been numerous alterations to cotton production. These modifications encompass the widespread cultivation of genetically modified cotton variety (Bt cotton) genetically modified through the insertion of one or more genes from a common soil bacterium, (*Bacillus thuringiensis*). The rise of pest resistance, consumer preference for environmentally friendly cotton, quality-conscious consumers, and demands for sustainable agriculture. Insecticide resistance has become a significant issue for global crop protection programs, posing a risk to pest control techniques specifically for cotton crops. The introduction of *Bt* cotton has significantly transformed Integrated Pest Management (IPM) perspectives. To ensure the sustainability of this powerful transgenic technology, it is crucial to properly understand the phenomenon of insect resistance to *Bt* toxins. Breaking yield barriers is vital for enhancing productivity per unit area and reducing production costs. Newer concepts of agronomy, such as high-density planting systems, have emerged. Additionally, advancements in science and technology have facilitated improved water and nutrient management.

Except for the two types of *Bt* cotton, the Ethiopian government has not authorized the commercial production of any genetically modified organisms (GMOs). The two cotton varieties imported from India have been successful and approved. However, unlike other countries, Ethiopia does not allow other *Bt* crops. Ethiopia has traditionally opposed GMOs despite persistent lobbying from international corporations.

However, in 2014, it did permit the field testing of genetically modified (GM) maize and the commercial production of *Bt* cotton. After two years of limited field testing, *Bt* cotton received "environmental release" certification from the Ethiopia Ministry of Environment, Forest, and Climate Change in June 2018 [1].

4 Cotton, Research, Extension, Status and Achievement

Approximately 85% of the population in Ethiopia relies directly or indirectly on agricultural products, such as livestock, for their means of living. This sector contributes to about 45% of the country's GDP [7]. The importance of agricultural research and its impact on Ethiopia's growth cannot be emphasized enough. The initiation of agricultural research in Ethiopia can be traced back to the establishment of the Ambo and Jimma Colleges of Agriculture in 1947, as well as the Imperial College of Agriculture and Mechanical arts (now known as Haramaya University) in 1953 [7]. Nonetheless, compared to other African nations, Ethiopian agricultural research is relatively new. The Institute of Agricultural Research was established in 1966, marking the commencement of organized agricultural research operations and the actual correlation between agricultural research and development [7]. Several other agricultural research facilities have been established subsequently, including:

i. The Plant Protection Research Center at Ambo—established in 1972 and merged with the Institute of Agricultural Research in 1995.
ii. Plant Genetic Resources Center—founded in 1974 and later became the Biodiversity Institute.
iii. Forestry Research Center—established in 1975.
iv. Wood Utilization Research Center—established in 1979.
v. National Soils Laboratory—established in 1989.
vi. Institute of Animal Health Research (IAHR)—established in, 1992.

4.1 The Present System of National Agricultural Research

The Institute of Agriculture Research (IAR) distributed some of its centers in 1993 to establish autonomous research organizations under the supervision of the relevant regional governments. Subsequently, these institutions were renamed as Regional Agricultural Research Centers (RARCs), and the corresponding regional agriculture departments were entrusted with their management. The Ethiopian Institute of Agricultural Research (EIAR), which succeeded the Ethiopian Agricultural Research Organization (EARO), was established in June 1997 by merging all the aforementioned agricultural research organizations.

i. The original IAR research centers at Holetta, Nazreth, Jima, Bako, Melka Werer, Ambo, Kulumsa and Pawe;

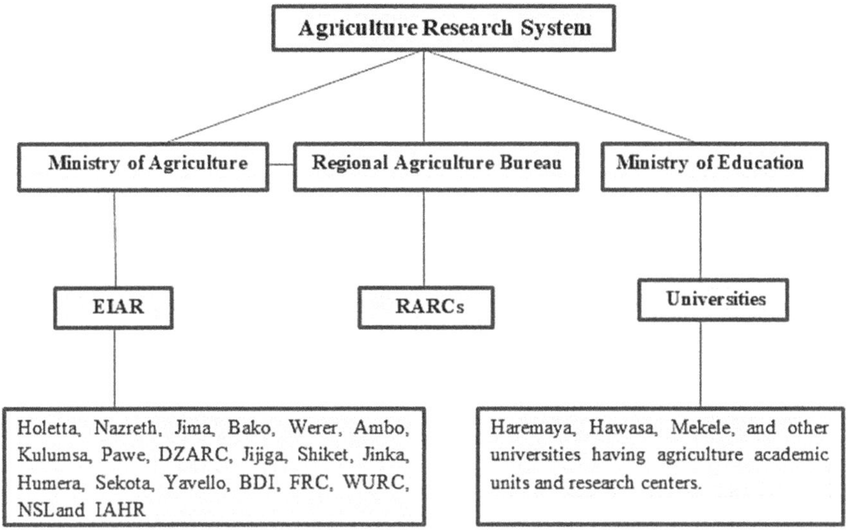

Fig. 2 Organizational Structure (1997/98) of Agricultural Research in Ethiopia

ii. Debre Zeit Agricultural Research Center, of which previously under Haramaya University, the Biodiversity Institute, the Forestry Research Center, the Wood Utilization Research Center, the Institute of Animal Health Research, and the National Soils Laboratory, all were previously under the Ministry of Agriculture;
iii. The recently established research centers with the financial backing of World Bank Agricultural Research and Training Program (ARTP). These agricultural research centers have been set up in Jijiga, Shiket, Jinka, Humera, Sekota, and Yavello, and depicted in Fig. 2. Ethiopian Institute of Agricultural Research (EIAR) is supervised by a committee which is led by MoA. It is managed by a director general appointed by the committee and has two assistant directors for research and administrative matters.

The RARCs continued to be governed by their own regional departments of agriculture. According to the Federal Negarit Gazeta, published in 1997, the EIAR's mission is to develop, advance, and adapt technologies as well as to coordinate, support, and promote research efforts in order to meet the nation's immediate and long-term agricultural requirements. The higher education institutions conduct agricultural research on livestock or national cash crops that is relevant to their local conditions under contractual agreements with EIAR. In Ethiopia, the private sector conducts either very little or no agricultural research. Despite the fact that the National Agricultural Research Services (NARS) do not include higher education institutions.

Some foreign agricultural research organizations either collaborate with the NARS through their various networks or are represented through their branch offices in Ethiopia. These include International Center for Agricultural Research in the Tropics (CIAT), International Center for Wheat Maize Improvement (ICWMI), International

Potato Improvement Center (IPIC), International Center for Agricultural Research in the Dry Area (ICARDA), International Center for Research in Agro Forestry (ICRAF), International Center for Research in Semi-Arid Tropics (ICRSAT), and International Livestock Research Institute (ILRI).

The EARI and the RARCs account for 86% of the total agricultural research activities in the country and 97% of the total financial resources allocated to agricultural research by the federal government. With only 3% of the federal agricultural funding allocated, the contribution of higher education institutions to agricultural research is only 13%. Since the NARC was established to coordinate research and resource management across the entire research system of the country, a more favorable environment is currently being created for research as a top organization. By comparing the experiences of Brazil, South Africa, and India, the NARC plans to revolutionize Ethiopian agricultural research by involving countries with more developed agricultural research, etc. A greater collaboration as well as personnel and capital capacity are created by the NARC's membership of the numerous research institutes and higher education institutions, putting them in a better position to produce improved agricultural research outputs and technology.

4.2 Research and Extension

The participatory demonstration and training extension system is the current extension approach being pursued by the Ethiopian government. The objective of this approach is to transfer the complete package of agricultural technology and practices (varieties, insecticides, fertilizers, improved cultural practices, etc.) for a specific crop. Regional agricultural bureaus are responsible for extension activities and have full responsibility for the design, implementation, supervision, and evaluation of extension programs.

The extension Office of the Federal Ministry of Agriculture is responsible for coordinating inter-regional extension activities and providing policy recommendations on national agricultural extension issues. Additionally, they provide advice to regional agricultural bureaus on extension management and administration, develop extension training materials, and plan agricultural extension training programs for regional extension staff. Extension has undergone the same decentralized process as agricultural research. The Research and Extension Liaison Committee (RELC) has been established in all regions. The RELC consists of representatives from the regional offices of the MoA, the research centers in the region, and farmers. The RELC members discuss and find solutions for issues such as production constraints, research programs and findings, and effective ways to disseminate these findings. Recently, the "Input Coordination Units" have replaced RELC as the link between research and extension at all levels in West and North Africa, the National Agricultural Research System [8].

Within the Ministry of Agriculture, there are limited operational connections between the different research institutes and between the research, extension, and training departments. There are multiple research institutions conducting research on various crops, topics, and issues without a formal coordination system, leading to duplication and overlap in the technology generation process and inefficient use of the available human and financial resources. Even though the research budgets of departments, institutes, and businesses appear to cover most of the current expenses, they still represent a relatively small portion of the institutions' overall budgets. These allocations would be a significant constraint if research volume and quality were to significantly increase from their current low levels.

The number of highly qualified research personnel with advanced degrees in the field of cotton is still limited, although it has changed significantly since the Agriculture Sector Review (ASR) was conducted in 2021. However, individuals with such degrees typically hold management positions and do not carry out research initiatives.

4.3 Impacts of Cotton Research Extension

Werer Agriculture Research Center has accumulated over 50 years of knowledge in cotton research since its establishment in 1964. Throughout this period, numerous studies have been carried out on different aspects of cotton production and administration. As part of the "national cotton research project," cotton research currently plays a vital role in the Ethiopian Institute of Agricultural Research/Crops Research Directorate/Pulses, Oilseeds, and Fiber Crops process. There are six fields of study:

i. Cotton Breeding and Genetics,
ii. Agronomy Physiology,
iii. Cotton Protection (Entomology, Weed, and Pathology),
iv. Extension,
v. Agricultural Economics and Gender Sensitization, and
vi. Mechanization.

All of the different climatic regions where cotton is grown are currently represented by only one main federal research facility for irrigation (WARC) and one research center for rain-fed areas (Assosa). The government currently provides funding and support for cotton research through a network. In reality, cotton research has been neglected for a long time, and the following reasons are given for its poor performance:

i. Research findings are often not shared with end users and are only relevant to the Middle Awash irrigated areas.
ii. There is a lack of human resources (very few researchers) with limited capacity and experience in research.
iii. Cotton research is still primarily based in Werer, Afar Region, while cotton production in Ethiopia has largely shifted to other regions.

iv. There is a lack of collaboration between the various research departments (breeding, agronomy, socioeconomics, irrigation, etc.) and the Ministry of Agriculture extension department, MoI, and ETIDI.
v. The WARC, which was originally established for cotton research, has recently expanded its programs to include over 20 other crops and animal research. This has reduced the focus and output of cotton research in terms of competition for qualified personnel, field trials, and facilities.
vi. While small-scale and rain-fed cotton cultivation have been abandoned for a long time, research has mainly focused on large-scale irrigated regions.

There is a lack of support and involvement from stakeholders (cotton growers, textile mills, ETIDI, Ministry of Agriculture, regional research institutes, NGOs, universities), outdated and sometimes non-functional laboratory equipment, and a structured high-level coordination mechanism. During the terms of different Ethiopian administrations, agricultural extension has always been a central aspect of policies related to the sector. Currently, the public sector dominates the agricultural extension service and operates in a decentralized manner, with agricultural extension being implemented at the district and village levels. When it comes to providing inputs and agricultural extension services to small-scale farmers at the local level, the public sector plays the most significant role [1].

The establishment of human capital to offer agricultural guidance services to farmers was the initial phase in the Ethiopian government's dedication to developing the largest agricultural expansion system in Sub-Saharan Africa. According to the presently accessible information, 11,000 Farmers' Training Centers (FTCs) had been built by the end of 2014, and at the "Kebele (village)" level, there were 45,712 Development Agents (DAs) working. Once all FTCs are constructed and operating at maximum capacity, it is expected that the number of frontline extension employees will increase to approximately 60,000.

The Ministry of Agriculture (MoA) and Regional Bureaus of Agriculture are major providers of agricultural advisory services in the nation. The MoA is accountable for developing and enhancing the general national agricultural and rural development strategies and policies for the nation, with input from the regions and other stakeholders. Within this strategy, the MoA establishes the overall national extension policy, offers financial backing for the extension system, and supports the regions with training and other capacity-building activities.

Each region has an Agriculture Bureau responsible for executing agricultural extension programs in their respective region. The bureaus are responsible for implementing, coordinating, and evaluating agricultural and rural development policies. Each bureau has a leader and a number of technical and administrative personnel, including department heads. These personnel offer technical and administrative assistance, as well as supervision and monitoring for the Woreda and Kebele level offices. The primary agro-ecological zones are utilized to internally divide each region's agricultural advisory support, providing more specific technical and administrative assistance as required, especially for the larger regions. Some regions, such as the South Nationalities and Nations Peoples Regional State (SNNPRS), which have numerous languages and ethnic groups, utilize zonal administration to a greater extent than others [1].

5 Future Trends

The initial step to improve the quality of advisory services provided to cotton farmers is to delegate responsibility to the MoA extension service. The paradox lies in the fact that Ethiopia possesses the foremost extension network in Africa and one of the most densely populated in the world, yet it is not utilized for cotton. The Ethiopian Cotton Development Authority (ECDA), which will establish a research-extension division, should supervise the cotton advisory service. Similar to the production department of ETIDI, this division will be responsible for creating customized extension packages for farmers based on their specific needs and agricultural practices. Capacity-building programs should be developed in collaboration with researchers and distributed to extension agents at the regional, zonal, and district levels.

The Ethiopian Cotton Producers, Ginners, and Exporters Association (ECGPEA), which could establish a "technical advisory division," may offer advisory services to large-scale commercial farmers. It is worth considering seeking specialized advisory services from ECPGEA because some of their issues are quite specific. The possibility of establishing a dedicated cotton Agricultural Development Partners Linkage Advisory Council (ADPLAC), managed at the MoA level and overseen by the NDCA, should be taken into account in terms of the Research-Extension-Farmers linkage. ADPLAC and cotton research centers and sub-centers should have close connections, and researchers should participate in MoA staff capacity-building programs. Universities should review their curriculum, improve their facilities, and persuade policymakers to support a robust agricultural mechanization program. Additionally, Agricultural Machinery dealers should consider factors beyond profit, such as the sustainability of the system. In general, cotton research should encompass the following areas:

i. Transitional changes: Medium-scale motorized level of mechanization.
ii. Reverse engineering.
iii. Business incubation.
iv. Strengthening of public-private partnerships (scale-out technologies).

Investments made by farmers in sustainable technology and their adoption of such technology were found to be positively influenced by enabling policies and programs, market connections, access to institutional support, and finance (Bayisag 2014). All stakeholders in the agricultural mechanization system must take an integrated approach. To maintain a sustainable agricultural production system, farmers, educational institutions, the research-extension system, policymakers, dealers, and manufacturers must all utilize their full potential.

6 Summary

The cotton industry is severely underutilized as the extension system is primarily focused on promoting food crops. The majority of small-scale cotton producers use inferior varieties, inadequate inputs, and ineffective management techniques, resulting in low yield and substandard lint. Extension services for cotton development have been significantly lacking compared to other important crops. Since cotton is considered a valuable commodity and a crop of government interest, funders have no desire to finance financial and technological research in the field. Therefore, there is minimal public funding available for research in this area. Additionally, cotton is cultivated and studied in a challenging climate, and the research system experiences a high turnover of qualified personnel. While the government has acknowledged the gaps in extension, research, and implementation of cotton cultivation and has transferred responsibility for organizing cotton cultivation to the Ministry of Agriculture, further research and study will be necessary from all parties involved for improvement.

References

1. NCDS (National Cotton Development Strategy (2018–2032) and Road map (2017)
2. EIA (Ethiopian Investment Agency) (2012) Cotton Production and Ginning in Ethiopia
3. Bayrau A, Bekele F, Assefa B, Mihiretu M (2014) An institutional assessment of the cotton and sugarcane commodities in Ethiopia: the climate change perspective. Addis Ababa, Ethiopia
4. ICAC (International Cotton Advisory Committee) (2014) World Cotton Research Conference
5. Biswanger et al (1993) Adoption of technologies for sustainable farming system: wageningen workshop proceedings, France, pp 1–149
6. Mellor JW, Paul D (2010) Agriculture and the Economic Transformation of Ethiopia, ESSP2 Working Paper 010, International Food Policy Research Institute, Washington, USA, pp 1–45
7. Tsedeke A, Deressa A, Molla A (2004) The Ethiopian NARS: evolution, challenges and opportunities. ASARECA Strategic planning paper No. 1, Addis Ababa, Ethiopia
8. WANA NARS Study (2010) WANA NARS Study. The National Agricultural Research System of Ethiopia. World Bank Program, Cereals, Pulses and Oil seed Balance Sheet Analysis for Ethiopia
9. ETIDI report in 2015 (2015) Ethiopian textile industry development institute cotton development directorate annual report
10. Ademola AA, Morris EJ, Parayil G (2013) Status of development, regulation and adoption of GM agriculture in Africa: views and positions of stakeholder groups. Food Policy 43:159–166 (Elsevier)
11. ECPGEA (2017) Ethiopia's cotton production, productivity and ginneries sector's key challenges and interventions
12. FDRE-Federal Negarit Gazeta (1997) Proclamation No. 79/1997 to Provide for the Establishment of the Ethiopian Agricultural Research Organisation
13. Babatunde O (2009) "Estimating the impact of agricultural technology on poverty reduction in rural Africa:" IFPRI discussion papers 00901. International Food Policy Research Institute, Washington, USA, pp 1–29

Challenges in Cotton Production, Quality, and Future Aspects

Mesele Mekuria

Abstract Ethiopia is fortunate to have a cotton-growing nation, sufficient irrigation water, and laborers to work in the cotton-producing sector. Traditional weaving, the establishment of spinning mills, integrated industrial parks, the widespread availability of affordable electricity, and the improvement of market access for cotton marketing and textile items are all important elements for cotton development in this country. Despite efforts to ensure the supply of cotton, which serves as the main resource for the strategic crop and the textile sector, it is becoming increasingly challenging to meet the growing demand for cotton in terms of quantity and quality. Another major issue is that the facilities for cotton research that may enhance the sector's growth and competitiveness are understaffed, underfunded, and unable to develop in line with the regional environment. In general, problems with cotton output are harming the nation's cotton-textile industry's ability to survive. The potential for the cotton sector to contribute to the nation's economy is limited by a number of difficulties in meeting local demand and competing on the global market. According to the review conducted, there is a lack of sufficient research and development activities in the cotton sector, poor actor integration within the value chain, limited loan access, inadequate government support, weak strategic implementation, and absence of appropriate strategic planning for cotton escort development are the main ones. To support the cotton industry going forward, the strategies should consider the components of the support package for cotton producers, the arrangement of policy framework to effectively connect the value actors along the value chain, the enhancement of cotton's research and development (R&D) activity, and the creation of favorable conditions to attract substantial foreign direct investment (FDI) to the cotton production along with the value chain industries.

Keywords Challenges · Research · Cotton production · Value chain · Government

M. Mekuria (✉)
Ethiopian Textile Industry Development Institute, P.O. Box No. 5637, Addis Ababa, Ethiopia
e-mail: meselemekuria8@gmail.com

K. Murugesh Babu et al. (eds.), *Cotton Sector Development in Ethiopia*, Textile Science and Clothing Technology, https://doi.org/10.1007/978-981-99-9149-5_17

Abbreviations

DBE	Development Bank of Ethiopia
ETIDI	Ethiopian Textile Industry Development Institute
EIAR	Ethiopian Institute of Agriculture Research
GDP	Gross Domestic Product
MOA	Ministry of Agriculture
SHF	Small Holder Farmers
WARC	Werer Agricultural Research Center

1 Introduction

Although cotton is typically cultivated as an annual crop, it has the potential to last for an extended period, similar to other perennial crops. It can be grown on both small and large farms, utilizing irrigation and rainfall. Cotton plays a crucial role as a crop for developing countries like Ethiopia. Due to its preference for warm climates, it thrives at elevations between 1000 and 1400 m above sea level, with temperatures ranging from 27 °C to 32 °C, and requiring at least 180 frost-free days during its growth cycle. While cotton can grow in various soil types, it thrives best in fertile soil with a pH range of 6–7. Ethiopia stands as an ideal country for cotton cultivation due to its abundant and suitable land resources, ample water supply for irrigation-based farming, and a sufficient workforce engaged in cotton production. Traditional weaving, which involves over 10,000 households, primarily utilizes cotton as the main material for creating cultural artifacts. Moreover, cotton serves as a vital input for the country's 19 spinning plants and 14 integrated industrial parks.

Consequently, there is an increasing local demand for cotton nationwide. This growing demand presents significant opportunities for cotton farming, enabling the country to enter the global market as a competitive cotton exporter. By meeting the local demand, expertise in cotton production is gained, further supporting the nation's textile industry. In order to fully benefit from cotton production and meet the rising demand in the textile industry, it is necessary to identify and address the challenges faced by the cotton sector. This will allow Ethiopia to export high-quality cotton to the global market.

2 Ethiopia's Cotton Production History

Since ancient times, cotton has been grown and used in Ethiopia, and manual spinning and weaving is still a well-known and popular craft in the country today. Several domesticated crops, including cotton (*Gossypium hearbaceum*), are said to have originated in Ethiopia [1]. Moreover, one of the cotton species from the Old World is still cultivated in Ethiopia. Ethiopia is one of the centers for the domestication and diversification of many cultivated plants, in addition to cotton. Cotton (*Gossypium spp.)* is the most important source of natural textile fiber and oilseed in the world. It is considered one of the most valuable gifts from nature to humanity. The fabric made from cotton is the gentlest on the skin compared to any other natural material on Earth.

The Italians and Germans conducted a historical cotton research project in Ethiopia from 1901 to 1910 in Upper Awash using Egyptian cotton, *Gossypium barbadense L.* However, after seven years of research, the project was terminated without success. During the second Italian occupation of Ethiopia, cotton research was resumed. After a few years of research, the Italians proved the feasibility of large-scale cotton production in Ethiopia. Furthermore, cotton research was resumed at what is now known as the Werer Agricultural Research Center (WARC), formerly the Melka Werer Agricultural Research Station, in 1964.

Commercial cotton farming began to expand widely in the 1960s when the British company Mitchel cot expanded cotton plantations in the Awash Basin area. The cotton land cover increased from a previously low level to 140 thousand hectares in 1972. When Ethiopia transitioned to socialism in 1974, the cotton cover remained for a while, along with the transfer of private farms to the government, particularly in the southern Blate area. Later, efforts were made to expand government cotton plantations in the northwest of the country around the Humera area, but they were not successful.

When the country's government changed in 1991, large cotton plantations, especially those in the Awash basin, which were previously owned by the government, were transferred to private investors. Since then, the coverage has changed significantly instead of continuing to expand. Despite efforts to strengthen the sector, the average coverage has currently decreased to 57 thousand hectares for various reasons.

3 Structure and Types of Stakeholders in the Cotton Sector

Given that cotton is a crop by nature, similar to other crops, it has historically been under the jurisdiction of the Ministry of Agriculture. For a significant period of time, cotton in Ethiopia was primarily used as a raw material for traditional weaving, which was predominantly carried out by small-scale industries. With the introduction of agriculture-led industrialization by the government in 2002, as part of the first development and transformation plan, the Ministry of Agriculture has made efforts to increase both the quantity and quality of cotton production. Since the 2012 planting season, the Ministry of Industry (MoI) has taken over the management of the cotton industry, while the Ministry of Agriculture continues to oversee cotton research, agricultural extension services, seed and chemical supply, and crop regulation activities. However, this approach has not yet resulted in the expected increase in cotton production in terms of quantity and quality. As a result, the cotton sector has recently been transferred back to the Ministry of Agriculture.

The cotton sector in Ethiopia is known for its diversity, incorporating both irrigation-based and rain-fed farming, large commercial farms as well as smallholder farms, saw and roller ginning, a well-developed textile industry, and traditional hand-loom weaving. The value chains for cotton-textile, textile, and apparel are complex and intricate. Various stakeholders are involved in the cotton industry, including smallholder farmers, commercial cotton farmers, gins, spinning mills, cotton dealers, and input suppliers.

3.1 Research and Extension

Cotton research is currently funded by the Government and conducted through a network of one main irrigated federal research Centre, WARC, and one rain-fed research Centre in Assosa. However, there are no other centers or sub-centers representing all of the cotton-growing agro-climatic zones in Ethiopia. This indicates that national cotton research is weak and has been neglected for a long time, leading to poor performance in the sector.

While cotton production has mainly been focused on other regions of the country, such as Benishangul-Gumuzi, Gambela, South Nations Nationalities Peoples, and Amhara National Regional States, cotton research is still primarily concentrated at WARC. The existing research facilities have mostly disregarded smallholders and rain-fed cotton cultivation in favor of large-scale irrigated areas.

The public sector is currently the most important player, particularly in terms of providing inputs and agricultural extension services to smallholders at the local level. The Ministry of Agriculture and Regional Bureaus of Agriculture are the main providers of agricultural advisory services in the country. The first step in the Ethiopian government's goal of creating the largest agricultural extension system in Sub-Saharan Africa was to build the necessary human capital to provide agricultural advice services to farmers. However, the cotton industry has been lacking in this regard, as it was previously under the MoI of the Industry and continues to lack out grower programs. This is because major commercial farmers who could provide extension services do not have the financial means to do so.

3.2 Access to Inputs

Currently, there are no existing supply mechanisms in place, but both large-scale and small-scale farmers rely on informal channels for input distribution. Some licensed large-scale farmers are able to produce cotton planting seed and supply it to other large-scale and small-scale farmers. However, there is no organized system for the supply of inputs, especially seed. Seed production requires specific criteria, requirements, and licensing, but until recently, there were no certified seed producers. The Werer Agricultural Research Centre (WARC) directly supplies a small amount of basic seed to selected farms, which then multiply and distribute it for the next season. In Ethiopia, seed from annual crops is often used. As cotton is a crucial raw material for the textile and garment industries, which produce goods for export, it is important to prioritize the import of agricultural inputs for cotton cultivation to contribute to the country's foreign currency balance.

Large commercial farms and agribusinesses have the option to obtain financial loans through the commercial banking sector, while smaller farmers have limited choices. Some smaller farmers can access short-term trade credit through their relationships with agribusiness or co-operative buyers. There is a possibility of obtaining loans from the Development Bank of Ethiopia through monitoring teams that evaluate project progress. However, these teams lack sufficient manpower to effectively carry out their tasks, and a significant amount of the allocated funds for agricultural projects have been redirected elsewhere, particularly for land development loans, where costs are difficult to control. It is hoped that there will be increased supervision of investments, projects, and loans.

4 Overview of the Cotton Sector in Ethiopia

In this section, the present condition of the cotton industry in Ethiopia is assessed by considering cotton cultivation, the value chain, and the pattern in production and consumption.

4.1 Area Cultivated and Yields

Cotton is still partially the traditional smallholder cash crop cultivated from the beginning of the rainy season (mid-June–end-September) until harvest (September–December) in Ethiopia. More than 70% of the country's lint is produced by large-scale "commercial" and medium-scale cotton farms, much of which produce under irrigation (Awash and Omo valleys), while Western Ethiopia (Humera, Metema, Benishangul-Gumuz, Gambela) relies on rainfall. In contrast to large-scale irrigated or mechanized farming, smallholder production, which accounts for 30% of the total production, is primarily hindered by poor technical productivity, the effects of climate change, and competition from more profitable cash crops.

According to the most recent data on cotton output collected by the Ethiopian Textile Industry Development Institute (ETIDI), an average of 57,000 hectares were cultivated with cotton, with 44% of the area located in Amhara, 15% in Gambella, 13% in Afar, 1% in SNNP, 8% in Tigray and Benishangul, and 1% in Oromia regional states. The average national production of 37,000 tons per year consists of yields that range from 2 to 3 t/ha of raw cotton in irrigated cultivation to 1.2–1.7 t/ha in rain-fed farms.

4.2 Farming and Production System

There can be multiple ways of classifying farms, based on their respective goals. It would be intriguing to consider a categorization that can be beneficial for the growth of the cotton industry, in terms of strategies to support each identified farm category. In fact, one could argue that there are three classifications of cotton farms in Ethiopia: Large, medium, and small farms, with the primary distinguishing factor being their size. It is generally accepted that there are two types of cotton farms in Ethiopia: Smallholders and large-scale "commercial" farms. Both large-scale and small-scale farms have both irrigated (25%) and rain-fed (75%) cotton production systems [2].

Mono-cropping has remained the dominant production system in most major cotton production areas in Ethiopia, with minimal or no addition of external input. This is often known to result in the depletion of essential plant nutrients, as well as the degradation of soil physical, chemical, and biological properties. Cultivation practices vary significantly between large mechanized farms and smallholder farms. In the case of the latter, irrigated cropping in the Afar Region and South Omo area follows mono-cropping, while the cropping system in the rain-fed areas of Northern and Western Ethiopia is mixed (i.e., sesame, sorghum, and maize).

Manual cotton picking is the prevailing harvesting method in Ethiopia. Large-scale farmers on irrigated fields typically pick twice, with the first harvest taking place after 70% of the bolls are open, and the remaining cotton being harvested three weeks later. Unlike the irrigated area, large-scale and small-scale rain-fed farmers usually harvest their cotton once all of the bolls are open.

4.3 Cotton Value Chain Integration

The Ethiopian cotton industry is highly diverse in comparison to other countries in Africa and the world. It encompasses various cultivation methods, including both irrigation and rain-fed systems, as well as a mix of large commercial farms and smallholder farms. The sector also includes different ginning methods, such as saw and roller ginning, and has both a modern textile industry and traditional handloom weaving. Ethiopia has a unique advantage in being able to grow a wide range of cotton varieties in different agro-climatic zones, either through rain-fed or irrigated methods. However, despite this potential, the majority of cotton cultivation in the country is currently dominated by a single variety, DP 90, which is grown in over 90% of the cotton areas. This creates a paradox as it limits the diversity of cotton varieties being grown. Furthermore, there is a mismatch between the characteristics of the dominant variety and the requirements of most spinning mills in the country. This creates challenges in terms of producing high-quality yarn and meeting the needs of the textile industry. The contractual arrangements between smallholder farmers and ginners are also weak, primarily due to a lack of transparency on the part of the ginners and inefficient farmers' organizations. This weak relationship is further complicated by the presence of middlemen or traders.

Overall, the Ethiopian cotton sector lacks coordination between different segments, leading to inefficiencies and gaps in quality. There is also a lack of transparency along the entire value chain, particularly in terms of price and volume marketing, grading, and pricing systems.

4.3.1 Cotton Value Chain Progression

Seed is the initial point for the cotton-to-clothing value chain, and high-quality seeds enhance yields and enhance quality. Major lint quality parameters, such as fiber length, thinness, strength, and ginning outturn are determined by the seed variety. Enhancing the quality of seed cotton produced in Ethiopia is of the utmost importance to provide lint matching the quality requirements of the textile industry both domestically and internationally. Ethiopia is characterized by strong value addition along the entire cotton and textile value chain, with advanced local processing of cotton lint into textile and apparel products. The cotton-textile and textile and garment value chains are very intricate. Ethiopia is increasing mill use starting from 2005/6 and utilizing on average 50 tons lint in 2015/16 became the second in value addition from Africa, next to Egypt which consumed 125 tons of lint cotton (NCDS 2017).

The demand for traditional hand woven products is increasing in domestic and international markets. However, the lack of clarity and weak connections between ginners, spinners, and weavers results in inefficiencies and missed opportunities. Traditional weavers express concerns about the availability, quality, and price of cotton yarn.

4.3.2 Ginneries as a Center in Cotton Value Chain

Ginning, in its precise sense, refers to the process of separating cotton fibers from the seeds. The cotton gin has its primary function as the transformation of a field crop, seed cotton, into a marketable commodity. Therefore, it serves as the link between cotton production and cotton processing. There are 26 Ginneries, of which 8 are roller and the remaining 18 are saw gins. However, most of the technologies used in ginning factories are outdated and obsolete. Some of the ginneries are currently not operational due to a lack of spare parts, and most of the ginneries are operating at reduced capacity and are unable to gin even the smallest amount of cotton produced in a season in a timely manner.

Inadequate infrastructure and outdated technology have a negative impact on production and maintenance costs, as well as the quality of the lint and seed. Seed-cotton is manually picked with a high amount of trash and contaminants. Other factors that have a negative impact on ginning include a lack of storage infrastructure, missing equipment along the processing line (such as feed control, lint cleaning, moisture restoration, poor quality bale covers, and limited fire protection), and a lack of quality control.

Ginneries face numerous challenges in their operations due to inadequate geographical capacity distribution. Frequent long-distance transportation of raw cotton leads to high costs and insufficient moisture content at ginning, which affects fiber quality. There is also a problem of raw cotton contamination with stones and polypropylene. Additionally, there is a shortage of labor due to low wages, frequent power cuts, difficulty accessing spare parts (and foreign currency), long payment terms, poor support from banks, and a lack of curriculum and resources for staff training.

4.4 Trend in Cotton Production and Consumption

Currently, there are 14 cotton spinning factories operating in Ethiopia, and if these factories operate at their maximum capacity, they require more than 120 thousand tons of lint cotton per year. However, due to various factors such as lack of funding and resources, their current utilization only amounts to 50 thousand metric tons. This creates a gap compared to the average annual production of 37 thousand metric tons of cotton in the country. Specifically, the factories require certain types of cotton, such as organic cotton and long staple length cotton, which are not widely produced in Ethiopia and need to be imported at a significant cost. Additionally, the quality of domestically produced cotton is low, which not only disrupts the production process of the ginning and spinning mills but also negatively affects the quality of the final products.

The handloom weaving sector in Ethiopia is the most dynamic in Africa. Ginning factories manually produce lint cotton, which is then spun into yarn by women spinners either individually or in spinning cooperatives. Some raw cotton is also hand-ginned and spun by women to create yarn for weaving. There is a growing demand for traditional handwoven products both domestically and internationally. However, the lack of transparency and weak connections among ginners, spinners, and weavers lead to inefficiencies and missed opportunities. Traditional weavers express concerns about yarn supply, pricing, and quality.

5 Cotton Production Potential and the Future Aspects

Cotton has been cultivated for centuries in dry to semi-arid areas of 300–1800 m above sea level in lowland Ethiopian ecosystems in Afar, Amhara, Benshangul, Gumz, Gambella, Oromia, Somali, Southern, and Tigray regions. Ethiopia is one of the countries in the world with a significant untapped potential for cotton production. As per the research conducted by ETIDI in collaboration with the Ethiopian Geological Service in 2019, the identified areas suitable for cotton production are known. Considering that the land capacity appropriate for cotton and the land allocated to cotton investors will be utilized, the long-term average productivity of cotton production capacity in Ethiopia will be 2.5 metric tons. By considering one ton of ginned cotton per hectare, it is possible to achieve a level where a total of 5 million tons of raw cotton or 2 million tons of ginned cotton can be produced on 2 million hectares of land after 15 years.

The land is also suitable for other crops that may be more desirable, therefore it is uncertain whether the entire area highly suitable for cotton will actually be used for cotton cultivation. A more realistic estimate of the potential for cotton output is three million metric tons of seed cotton or 1.2 million metric tons of lint cotton. No other country in Africa has the same capacity for cultivating a wide range of cultivars, both on a large scale and by smallholder farmers, in different agro-climatic zones, whether rain-fed or irrigated. This will enable Ethiopia to effectively supply the textile industry with locally grown cotton, create employment opportunities along the entire value chain, and increase the economic contribution of the cotton industry to GDP (Fig. 1).

Ethiopia is fortunate to have a cotton-growing region, capacity for irrigating water, and a sufficient workforce for cotton production. Traditional weaving, the growth of spinning factories, integrated industrial parks, the widespread availability of affordable energy, and increased market access for cotton and textile products are all factors contributing to the development of the cotton industry in the country. These

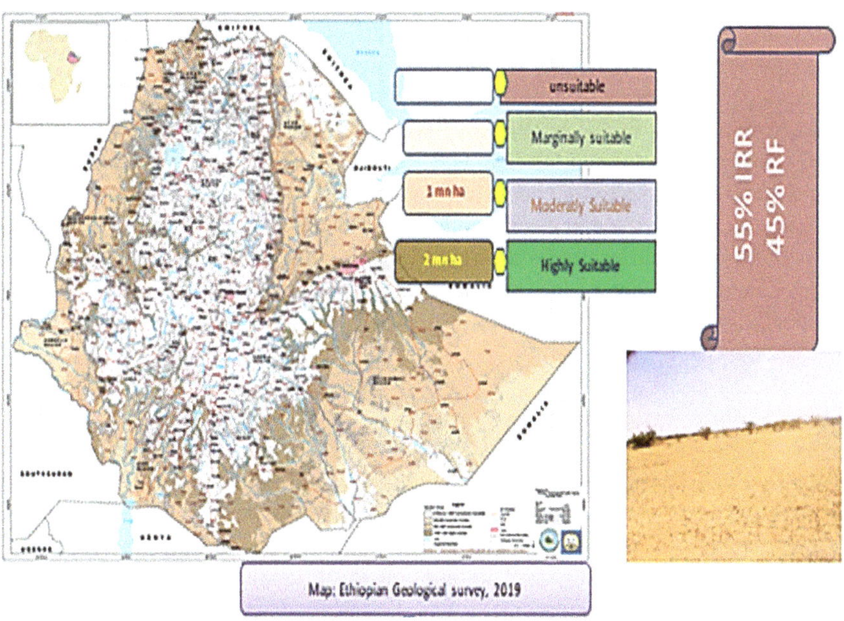

Fig. 1 Mop suitability land for cotton production in Ethiopia (*Source* Ethiopian Geological Survey 2019) *Note* RF—Rain-fed and IRR—Irrigation

factors present excellent opportunities for the industry to expand and become more competitive. Additionally, issues such as cotton classification and grading, marketing, technology innovation, and dissemination can be addressed through coordination. Regarding cotton production, areas such as land development, input supply, availability of farming machinery and tools, ginneries, and oil pressing mills can be considered as potential investment opportunities. In the cotton-textile value chain, cooperation, partnerships, information technology, infrastructure development, training, and professional management are essential strategies to be pursued and implemented.

6 Challenges and Opportunities

In spite of Ethiopia's vast potential for the advancement of the cotton industry, various obstacles are prevalent in all aspects of cotton production. This section will address the current challenges faced by the sector as well as the opportunities available for its growth.

6.1 Challenges Related to Production

Ethiopia's manufacturing sector is segmented into various subsectors, with the cotton-textile and garment industry being the government's top priority. The current state of this industry indicates that productivity and the quality of products across all value chains, from cotton to clothing, fall short of expectations. However, the garment manufacturing industries are more efficient and produce higher-quality goods compared to the cotton, yarn, and yarn-related businesses. Despite the challenges faced by the sector, Ethiopia ranks second in Africa, after Egypt, in terms of value addition to cotton and its export. Meeting the increasing demand for cotton in terms of quantity and quality is becoming challenging, despite efforts to ensure the supply of cotton, which is the primary resource for the government's strategic crop and textile industry.

6.1.1 Cotton Variety Contests

It is also a basic issue that the cotton research facilities that can expand the cotton industry and make it competitive are understaffed, underfunded, and unable to expand in accordance with the local ecology. Additionally, they are unable to generate and distribute the necessary technologies in the necessary quantities and types. In Ethiopia, despite the release of different varieties from research, very few are currently being produced, and they are also not viable for fully meeting the demands of textile factories with the various qualities and characteristics required by the limited species alone. Cotton extension services were also weak due to the structural challenges that existed before in disseminating technologies from research institutions to farmers.

The challenging access to foreign currency indeed represents a limiting factor for the supply of crucial inputs such as pesticides, for example. Cotton seed production, processing, and supply are poorly organized, with most of the seed being provided by a small number of large farms using the same seed year after year. A seed certification system that involves the cotton research center, independent breeders, and seed farms is necessary.

6.1.2 Limitations on Acquiring Finance

The size of the farm impacts the kinds of investments and working capital financing used for cotton production. The Ethiopian Development Bank funds new or expansion of investment projects in the agricultural sector (including cotton) with a minimum cultivated area of 200 hectares. Despite the Bank's assessment of business plans and close monitoring of fund usage, it seems that significant amounts of funds have been diverted from their initial purpose in recent years. The Commercial Bank of Ethiopia also grants loans to large-scale commercial farmers (minimum 30 hectares of land) after evaluation by a technical team.

Small-scale farmers and their cooperatives face challenges and depend on government guarantees to obtain loans from the banking system. They are unable to purchase the seed cotton produced by their members (or only in very small quantities), and private merchants fund this business. Access to finance is limited for small-scale farmers, and they rely on microfinance institutions, which are usually efficient but not suitable for financing investments due to their high interest rates.

6.1.3 Limitations in Producing Quality Cotton

The beginning point in the cotton-to-clothing value chain is the seed. Superior seeds enhance the yields and the quality of cotton, as the main lint quality parameters such as length, thinness, durability, micronaire, and ginning outturn are influenced by the seeds of the cotton variety used for planting. Once the cotton bolls open and the seeds covered in fiber are exposed to the elements, the quality of the seeds starts to decline. Ethiopian cotton possesses exceptional inherent characteristics, but the quality can be significantly affected by mechanical damage that occurs during harvesting, transportation, and ginning operations. Some of the difficulties encountered during the process of harvesting and managing crops after they have been harvested include:

i. Not picking at the time.
ii. Gathering foreign substances while picking cotton.
iii. Collecting and transporting in sacks and bags made of impure materials (polypropylene bags).

iv. The absence of cotton storage warehouses in farms and ginneries.
v. Because the ginning factories are outdated, they have caused a flaw in the quality.
vi. After ginning the cotton, the bale is not fully wrapped in cotton fabric.

Cotton varieties should be developed to suit specific growing regions, taking into account environmental and cultural conditions, with the goal of maximizing yield potential and optimizing fiber characteristics. The cotton breeding program in Ethiopia must consider a wide range of fiber quality parameters in order to meet the demands of end-users. Traditional breeding should aim to create new cultivars with enhanced characteristics, such as increased potential for seed cotton yield, improved lint production, longer staple length, and higher fiber strength. Currently, there is a disconnect between cotton research and the needs of textile factories, leading to a lack of an integrated system for factories to obtain the yarn they require. In the future, this issue should be addressed by research organizations, who should focus on developing cotton varieties that possess the characteristics required by textile factories.

Hand-picked seed cotton is cleaner and results in lint with fewer neps and a lower percentage of short fibers compared to machine-picked cotton. However, the current situation in Ethiopia shows that hand-picked cotton is often contaminated with trash and foreign matter. As contamination of seed cotton with foreign matter is a major concern for producers of quality yarn and fabric, spinners tend to prefer machine-picked cotton over hand-picked cotton. Additionally, hand picking is a time-consuming and labor-intensive process, whereas machine harvesting is highly efficient and requires fewer workers. Therefore, it may be the right time to introduce machine harvesting technologies in Ethiopia, by improving production through the use of suitable varieties for mechanical harvesting and the application of defoliants.

6.2 Issues with Cotton Ginning

The majority of the country's ginning facilities are small in size, outdated, and obsolete, and they have a detrimental effect on the quality of cotton as they do not meet standards for storage and other essential components. Additionally, when comparing Ethiopia to other African countries, the amount of ginning output (GOT) obtained from raw cotton is less than 37% in Ethiopia, while it reaches 42% in West African countries [4]. Moreover, ginning facilities are located far from cotton-growing regions or on cotton farms, resulting in increased transportation costs. Overall, challenges with cotton production are negatively impacting the sustainability of the cotton-textile industry in the country [5].

7 Lack of Support for Cotton Sector Considering It as a Strategic Crop

Cotton cultivators are shifting their focus away from cotton and toward alternative crops due to the lack of accessible financing and extension services for small and medium-sized cotton producers. Additionally, the misuse of loans and delays in repayment negatively impact the long-term availability of funding for large-scale cotton farms, according to an analysis by the Development Bank of Ethiopia. The major obstacles hindering the development of the cotton industry in most cotton-producing regions include the absence of land leasing regulations, security concerns, insufficient supply of inputs (such as seeds, chemicals, and replacement parts), pollution, inadequate incentives, and lack of infrastructure. The weak relationship and performance between cotton producers and cooperatives, as well as factories and textile manufacturers, further exacerbate the issue. Furthermore, the current cotton market lacks transparency and does not prioritize quality, with brokers and intermediaries exerting significant influence over transactions without adding any value.

The governments of leading cotton-producing countries worldwide have a significant influence on cotton production in other nations, particularly in emerging countries like Africa. The price of cotton is highly volatile and unpredictable, necessitating the support of a sustainable supply of cotton products and the ability to withstand climate change-induced natural events beyond the control of producers. This can be achieved by establishing a marketing and pricing system that ensures profitability and competitiveness for cotton producers. When production costs for cotton rise and selling prices fall below the cost of production, cotton farmers must decide whether to switch to cultivating alternative crops that offer better competition.

Therefore, it is crucial for the government to allocate a stabilization fund and establish a minimum support price for cotton producers prior to planting. This will enhance the sustainable supply of cotton in the country, safeguard the sector, and make the cotton industry competitive, following the successful Indian experience in this regard.

Ensuring the sustainability of cotton trade and supply is of utmost importance by connecting small cotton producers with large farms, mills, and textile factories through contract farming. However, the fluctuation of cotton prices poses a greater risk of contract breaches, which can impact the longevity of the trading system. To address this, the government needs to establish a legal framework to support contract farming and create a model for managing cotton contracts that is tailored to local conditions and easy to understand. On the other hand, global clothing buyers are increasingly opting for textile apparel products made from sustainable cotton sources to meet the demands of their customers.

In the current situation, new developers enter the sector only after conducting an Environmental Impact Assessment (EIA). However, the study conducted by SOFRECO in 2016 revealed that the application and supervision of EIA are often lacking, raising concerns from various stakeholders. To effectively address environmental and social sustainability issues in most cotton potential areas, it is crucial to conduct a comprehensive study that includes an environmental impact assessment survey, as well as other social and political studies and integrated water resource management studies. Furthermore, it is important to disseminate accurate information to the public, debunking misconceptions and providing a realistic understanding of the situation on the ground. It is worth noting that rectifying these issues will not happen overnight.

Compared to other countries in Africa and the rest of the world, Ethiopia's cotton sector is characterized by its diversity. This can be attributed to the presence of large, medium-sized, and small cotton producers, which sets it apart from other countries. Additionally, Ethiopia's cotton sector includes both rain-fed and irrigated cotton farming, further contributing to its diversity. Ethiopia leads African nations in adding value to cotton, unlike many other countries that primarily focus on rain-fed cotton production and export it without significant value addition. Moreover, Ethiopia has both roller and saw gin operations, and the sector attracts both local and foreign investors, making it more diverse than countries like India, which primarily rely on one type of ginnery. While the coordination system for realizing cotton as a strategic crop is not yet robust, it requires strong coordination among government institutions, professional associations, federal and regional bodies, cotton farms, extension services, research institutions, and the entire value chain.

The MoA was responsible for cultivating cotton before 2015, but it was determined that ETIDI, under the MoI, would be a superior choice as it would lead to a greater enhancement. However, due to the inadequate bonding mechanism, it was incapable of efficiently managing and overseeing the sector, resulting in its realignment with MoA.

8 Summary

The challenges facing the cotton sector are hindering the country's immense potential for cotton production. The primary issues arise from a lack of recognition of cotton's importance as a vital crop for the nation's socioeconomic advancement. Consequently, another burdensome aspect of the cotton industry is the formulation and inadequate implementation of appropriate strategies. Additionally, there are weak linkages between the federal and regional governments, agencies, sector associations, cotton farms, extension workers, researchers, cotton producers, research and extension activities, and textile and cotton producers throughout the entire value chain.

Ethiopia possesses an attractive advantage with its abundant labor force, which comes with comparatively low wages. Furthermore, the country's potential for cotton production serves as an additional advantage in attracting foreign direct investment (FDI). Hence, cotton plays a crucial role in Ethiopia's socioeconomic development as it offers employment opportunities for millions of young people, who comprise the majority of the population. The value chain of the industry is particularly important as it employs a larger number of women compared to other industrial sectors.

To effectively address the challenges hindering cotton production in the country, strategic approaches must be implemented. These strategies should consider the elements of the support package for cotton producers, the establishment of effective policy frameworks to connect the actors in the value chain, the enhancement of research and development activities in the cotton sector, and the creation of favorable conditions to attract significant FDI in cotton production and the value chain industries as a whole. Such a strategy will contribute to the socioeconomic development of the country and significantly alleviate poverty.

References

1. Nicholson GE (1960) The production, history, uses and relationships of cotton (*Gossypium spp.*) in Ethiopia. Econ Botany, 14(1):3–36
2. ETIDI (Ethiopian Textile Development Institute) (2016) Assessment study Report on cotton production in Ethiopia from 2018–2022
3. ECPGEA (Ethiopian Cotton producers and Ginneries Exporter Association) (2017) Ethiopia's cotton production, productivity and ginneries sector's key challenges and interventions
4. SOFRECO (Societe Francaise de Realisation D'etudes et de Conseil) (2016) Scoping study report and field visits reports
5. SOFRECO (Societe Francaise de Realisation D'etudes et de Conseil) (2017) Ethiopia National cotton development strategy (2018–2032) and road map
6. EIAR (Ethiopian Institute of Agricultural Research) (2017) Cotton research strategy of Ethiopia from 2016–2030

Cotton Value Chain and Economics

Abera Kechi Kabish, Dawit Tesfaye Degefu, and Zerihun Desalegn Gebregiorgis

Abstract Ethiopia is a favored country for cultivating cotton due to its abundant land, favorable weather, water resources, and young workforce. Despite its immense potential for cotton farming, the country does not produce cotton at a level that can compete in terms of both quantity and quality. The cotton production value chain involves a large number of stakeholders, from growers to retailers. Ethiopian cotton farmers are categorized into three groups: small, medium, and large scale. Small-scale cotton farmers face disadvantages in the cotton value chain due to increasing input costs, a lack of government incentives and support, limited bargaining power in the pricing of raw cotton, and other factors. Ginners also face challenges in the cotton value chain due to their use of outdated ginning technology, underutilized capacities, and insufficient cotton supply throughout the year. The middlemen in the cotton value chain do not contribute any value and benefit more than small-scale farmers. The textile industry in the cotton value chain suffers from a lack of cotton for processing, both in terms of quantity and quality, with 40% of the industry's cotton needs being fulfilled through imports. The use of cotton seeds for processing edible oils and animal feed is a crucial part of the cotton value chain, with great potential that will be realized through expanded cotton production. Therefore, not all parties involved in Ethiopia's cotton value chain benefit economically in proportion. These issues need to be addressed in order to improve the performance of the country's cotton value chain and tap into its current potential for competitive cotton production and trading in both national and global markets.

Keywords Value chain · Cotton · Byproduct · Farmers · Linkage · Market · Economics

A. K. Kabish (✉)
Department of Textile Chemical Process Engineering, Ethiopian Institute of Textile and Fashion Technology, Bahir Dar University, P.O. Box 1037, Bahir Dar, Ethiopia
e-mail: aberak1995@gmail.com

D. T. Degefu
Ethiopian Institute of Agricultural Research, Melkassa Agricultural Research Center, Addis Ababa, Ethiopia

Z. D. Gebregiorgis
Solidaridad, Ethiopia

K. Murugesh Babu et al. (eds.), *Cotton Sector Development in Ethiopia*, Textile Science and Clothing Technology, https://doi.org/10.1007/978-981-99-9149-5_18

1 Introduction

Albeit cotton is mainly grown for its natural fiber, its byproducts are equally crucial for livestock feed, seed oils, the manufacturing of special paper, mattress padding, toothpaste, fertilizer and plastics. Its importance extends to medicinal use through utilizing their extracts including for treating hypertension, anticancer activity and as male contraceptive [1–3]. Its high-quality natural fiber makes it a sole source of fiber that dominates the world cotton textile industry.

Ethiopia has enormous potential for the production of cotton following the availability of suitable land which is estimated about 3 million hectares which is almost equivalent to the cotton land in Pakistan, the top 10 largest producer of cotton, the rich river basins suitable for cotton production, the human capital and the intimacy of the people with the crop all together are a big asset to the government of Ethiopia to stimulate the growth of the cotton sector [4]. Moreover cotton is one of the strategic commodities to promote economic development because of the competitive active labor force, and competitive infrastructure especially the emerging manufacturing hub which is one of the state-of-the-art industrial parks that focuses on textile and apparel and of course the proximity and access to the global market, i.e., the Djibouti port and Africa's aviation hub, Ethiopian airlines [5].

Cotton is one of the oldest cultivated fiber crops and also major cash crops in Ethiopia. It is grown by many farmers in Ethiopia as a source of income both in rain-fed and irrigated areas. In Ethiopia cotton is extensively grown in irrigated lowlands 60% and in warmer mid altitudes under rain-fed agriculture, 40%. The low- and mid-altitude areas of the country have immense potential for cotton production. It can also create job opportunity for thousands of people. Furthermore it offers considerable employment opportunities in farms, in textile factories, and in the ginneries. According to a report by SOFERCO, (2016, unpublished) the cotton sector along all the value chain creating jobs for more than 187,000 workers. Furthermore the report indicated that the cotton sector accounts for about 36% of the workforce from the total industrial manufacturing sector. The country has also competitive potential for producing cotton both for the local and export markets. However, the cotton market in Ethiopia both at the local and export market level is very unpredictable. Furthermore the cotton production trend is inconsistent, which is very difficult to plan ahead for the textile and apparel industries. As reported by different studies, it is very unlikely to meet the demand of the local cotton market let alone exporting the surplus [6]. Hence we need a comprehensive solution to mitigate the major challenges that constrain the production system like the lack of improved seed both in quality and quantity due to a limited number of cotton accessions and narrow genetic bases, shortage of labor during planting and harvesting, pest pressure due to emerging and invasive species, absence of good extension service to facilitate the integration between the research and development, poor cotton seed system, poor market information, lack of standardized grading in line with price incentives, etc.

According to The Cotton Supply Chain in Ethiopia [7], the average size of the land under cultivation by a single smallholder is 0.5 ha. The bigger commercial farms in Ethiopia are owned and operated by private persons. According to Belachew and Mamuye [8], the cotton sector provided a living for almost 500,000 families in Ethiopia in 2018.

Cotton is one of the oldest textile fibers in Ethiopia; hence wearing cotton clothing is a part of the culture there. Most of the producing regions, such as Tigray, Amhara, Benjangul-Gumuz, Gambela, the Southern Nations and Nationalities Peoples (SNNP), and the Afar, are located in the cotton-sesame belt [9]. Ethiopia is the second-largest consumer of cotton in Africa after Egypt and a net importer of cotton lint. Due to its favorable natural conditions (temperature, rich soils, and water availability), Ethiopia has a long history of producing cotton and textile products, making the nation an excellent location for cotton cultivation. To reach Ethiopia's full potential in this industry, it is crucial to analyze the cotton value chain there.

According to Trienekens [10], value chains can be viewed as a platform for the adoption of new logistical, labor, and organizational structures as well as manufacturing processes. Value chain analyses are used to guide policy and investment considerations. They help in understanding the connection between market dynamics and agricultural growth. They enable assessment of the impacts of value chains on smallholders, businesses, society, and the environment. Efficient supply chains must be established to control the movement of goods through intermediaries and ensure the maintenance of quality standards. This requires the development of relationships, networks, skills, and coordination systems. The private sector has supported the establishment of these networks for agricultural commodities through vertical and horizontal connections, mostly through purchasing from large farmers who may or may not delegate tasks to smaller businesses [11]. Lambert and Cooper [12] defined supply chains as networks with both horizontal and vertical connections that may extend across international borders. In the 2018/2019 season, the national production of seed cotton was estimated at 95,750 tons. From cotton seed, various products can be obtained as listed below:

i. Lint and then yarn;
ii. Cotton oilseeds, from which cotton oil and cake are obtained; and
iii. Planting seeds.

The primary types of seed cotton grown in Ethiopia are mainly medium staples like Deltapine® (DP-90), Stam 59 A, and Claudia and are currently dominated by GMO varieties. Beyond this technical element, the quality of the cotton produced in the nation is typically low (heterogeneous, irregular, polluted with foreign contaminants, etc.). Given how essential it is to so many people's lives, both economically and culturally, the Ethiopian government sees the cotton value chain (CVC) as fundamental for its socio-economic progress. Ethiopia has developed a National Cotton Development Plan and related roadmap with the goal of being one of the world's top producers of high-quality, sustainably produced cotton by the year 2032 [13].

2 Cotton Value Chain Players

Cotton is one of the most frequently traded agricultural commodities, and its value chain is vast, complex, and often unclear. There are five primary stages between the farmer who cultivates the crop and the ultimate retail product:

i. Cotton production—involves cultivation and harvesting cotton;
ii. Ginning—the process of separating the cotton fibers from the cotton seed;
iii. Cotton trading—the buying and selling of cotton, often occurring multiple times before it reaches the manufacturing stage where the final product is made;
iv. Manufacturing—includes spinning, which involves converting the cotton fibers into yarn; dyeing of cotton fiber, yarn, cloth, or garment; and weaving or knitting;
v. The final stage is "distribution," where the product is transported to retail stores.

The various stages of the production process are often located in different countries and with value unevenly distributed between them [14]. When cotton is brought to the market, the farmer either sells the cotton directly to ginners' agents or dealers or through a producer group like a cooperative. This is then sold to a gin, followed by local spinners and textile manufacturers if it is going to the domestic market, foreign traders if it is going to the export market, textile/yarn fabric makers again, and finally the stage of final production [14].

The Ethiopian CVC is the most varied in Africa because it includes irrigated and rain-fed farming, smallholder and medium-to-large farms, traditional, genetically modified, cotton made in Africa (standard for sustainable cotton), and organic cotton production, saw and roller ginning technologies, stand-alone (custom ginning), and integrated ginneries. The flow chart for those involved in Ethiopia's cotton value chain is shown in Fig. 1.

Within the entire value chain, Ethiopian cotton cultivation is segmented into two primary sub-chains:

i. Conventional one—relying on basic techniques, restricted processed resources, and minimal support services/guidance; and
ii. Contemporary one—primarily managed by medium- to large-scale plantations that supply the textile sector.

Currently, Ethiopia has 20 medium and 70 large cotton plantations [9]. The production of seed cotton is categorized based on the size of the plantation and the utilization of irrigation, as illustrated in Table 1.

2.1 Cotton Farmers' Forms

Small-scale, medium-scale, and large-scale cotton farming are the three classifications utilized to categorize cotton cultivation in Ethiopia. Cotton cultivated by small-scale farmers is transported to the market via intermediaries.

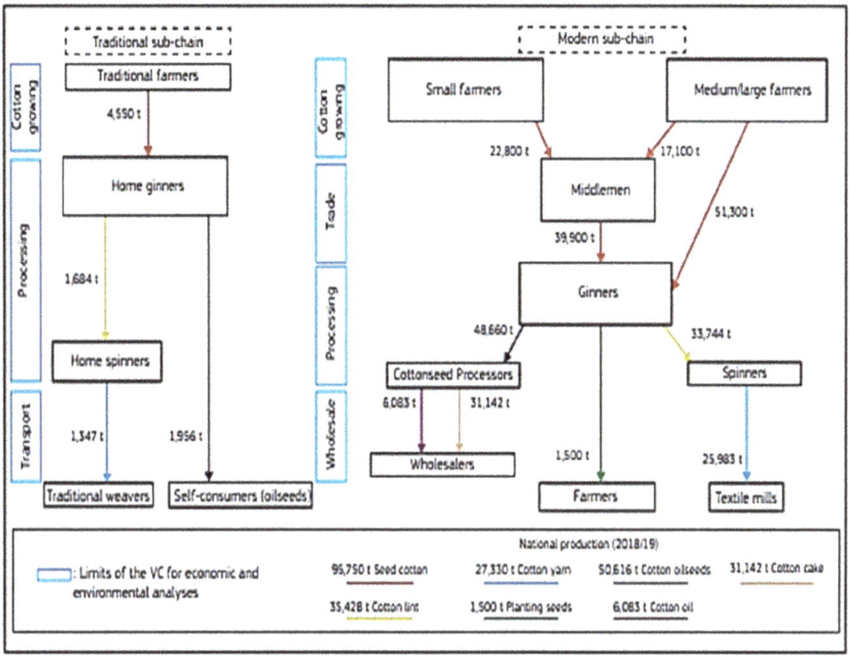

Fig. 1 Ethiopia's cotton value chain's main flows (*Source* [9])

Table 1 Typology of cotton farmers in Ethiopia [9]

Sub–chains	Traditional sub-chain	Modem sub-chain	
Farmers	Traditional farmers	Small-scale farmers	Medium and Large-scale farmers
No. of farms	7000	19000	90
Average size per (ha)	0.5	0.75	400
Total production of seed cotton (t)	4550	22800	68400
Share of total production (%)	5	24	71
Surface used (ha)	19624		36317
Irrigated surface (ha)	1064		25270

2.1.1 Small-Scale Cotton Farmers

In Ethiopia, approximately 40% of cotton farms are operated by small-scale farmers, despite the fact that they account for around 30% of the country's seed cotton production. Traditional farmers typically supply cotton to handloom artisans, while the majority of other small-scale farms focus on supplying the modern textiles industry.

The participation of small-scale farmers in different sub-chains allows us to classify them as either "traditional" or "modern" farmers. The key indicator that a farmer belongs to the traditional value chain is their relationship with traditional textile craftsmen who utilize handlooms. The strong connections between these parties enable the existence of farms that cultivate cotton on approximately 0.5 ha of land [9].

The cotton industry in Ethiopia faces challenges due to the prevalence of small-scale farmers who lack the necessary agricultural skills. In many cases, farmer organizations are poorly established with inadequate organizational structures. This hampers the farmers' ability to improve their skills and adopt new technologies. Furthermore, their limited organizational capabilities result in minimal influence over policy decisions. Small-scale cotton growers also struggle with low yields, which they largely attribute to ineffective agronomic farming techniques.

Issues such as the use of low-quality planting seeds, a lack of price assurance mechanisms, and the breakdown of a functional credit input system significantly impact yields. The situation is further exacerbated by the inadequate provision of extension services to farmers.

In the absence of a comprehensive industry plan, cotton farmers with small landholdings are unable to fully understand their role in the entire supply chain. They do not have the means to obtain information about cotton prices, and their interactions with other participants in the industry and those with a stake in it are limited. Additionally, they have limited experience in directly engaging with customers and knowledge about the criteria for fair trade and organic certification. The lack of efficient producer associations, inadequate production and distribution of high-quality cotton seeds, and the breakdown of extension and other forms of support services pose additional obstacles.

Poor seed quality is a crucial factor in decreased yields as there was no system in place for certifying cotton seeds. Consequently, farmers are forced to utilize unauthorized lower grade seeds, which reduce the productivity levels. These seeds have usually remained unchanged for a considerable period, resulting in significantly reduced germination rates. Since there is minimal continuous research and development in national research stations regarding the production of superior seeds and bulk production, this situation is likely to persist. If the sale of cotton seeds were adequately regulated, farmers' profits could be substantially enhanced as seeds account for 60–64% of the yield from seed cotton cultivation.

The quantity of money generated from the sale of cotton seed has the capacity to double farmer income. However, Ethiopia's seed-pressing facilities are inadequate, and the limited equipment available is a component of the country's obsolete ginnery machinery. Overall, there is a deficiency of downstream infrastructure to sustain the production of cottonseed oil and the trade of other by products [9].

Obstacles to the production of seed cotton include:

i. Lack of regulatory body;
ii. Nonexistence of institutionalization and policy; and
iii. Shortcomings in marketing.

Table 2 Cost drivers in the cotton value chain

S. no.	Cost titles	Cost share (%)
1	Land preparation	13.9
2	Plowing	6.1
3	Seeding	15.3
4	Thinning	0 %
5	Stamping	0%
6	Weeding	15.7
7	Spraying	31.2
8	Fertilizer	0
9	Harvesting	17.8

Ethiopia's cotton seed production lacks active regulatory agencies. For the majority of Ethiopian farmers, cotton production is now precarious due to all of these issues, along with insufficient market knowledge and declining global pricing [15]. Thus, it is necessary to enhance cotton's competitiveness, particularly for small-scale producers in Ethiopia. To reduce the vulnerability of these producers, measures must be implemented to increase their share of the final product's value, such as through the sale of by products derived from cotton seeds.

Although the costs of cotton production per hectare in Ethiopia seem competitive compared to the rest of the world, they are still high in the context of the regional value chain. With relatively low yield levels, the expenses become more apparent. Small-scale cotton farmers in Ethiopia and Africa often employ inadequate farming techniques. The study of the cotton value chain reveals that small-scale farmers frequently neglect crucial farming practices like thinning and stamping. It also indicates that the use of agrochemicals and fertilizers is infrequent, mainly due to their high cost. The primary cost components for cotton cultivation are agrochemicals and farm labor. For example, in Kenya, these two factors contribute 77.1% and 22.9% respectively to the overall cost, as shown in Table 2.

2.1.2 Medium and Large-scale Cotton Farmers

The majority (70%) of the production of seed cotton is managed by a limited number of large and medium-sized farms. Most of them employ restricted crop rotation and irrigation, both of which negatively impact soil fertility. While less than a decade ago, spinning mills were incapable of accommodating the cotton lint produced in the country, there has been an increase in demand over the past ten years, particularly from multinational brands, leading textile industries to import [9]. Ethiopia's capacity to produce cotton is severely limited due to high production costs resulting from costly inputs (fertilizer, seeds, and pesticides), as well as inefficient agricultural practices.

2.2 Middleman or Traders

The intermediary or traders can be categorized into three groups based on the amount of cotton they handle. These groups include local collectors, wholesalers, and retailers [16]. Small-scale farmers sell their unprocessed cotton to local collectors, who purchase it from them in nearby marketplaces and then sell it to wholesalers there. Wholesalers, who obtain cotton from both small-scale farmers and local collectors, form a relatively small group that then sells it after processing or separating the raw cotton into lint and seeds. They provide lint to textile mills and seeds to farmers who cultivate cotton and oil mills. The wholesalers determine the cotton price, which does not equally benefit all participants in the supply chain. Retailers are individuals who buy raw cotton from small-scale farmers and resell it to local ginneries at various markets.

2.3 Ginners

The primary objective of a cotton gin is to separate fibers from the seeds and maximize the value of the resulting fibers and seeds for financial profit. Purifying and processing seed cotton, fiber cleanup, and bale formation are the three main tasks of the central processing industry of cotton ginning. Ginners play a vital role in the cotton supply chain as the fibers they produce connect cotton producers to the textile industry and the local and global markets, while the cottonseeds are sent to oil mills and used as animal feed.

Cotton ginning is crucial in separating the fibers from the cottonseeds and transforming the raw crop into marketable products such as fibers and seeds. Therefore, ginning serves as a link between cotton farmers, textile, and oil processing industries. Through rental agreements with wholesalers, modern ginneries separate seed cotton into fibers and seeds.

After ginning, the separated fibers and seeds are sent to the secondary processing stage. Cotton fibers are sent to textile mills to be transformed into yarn, while cottonseeds are sent to seed processors to be turned into oil and seed cake. The production of textiles begins with the conversion of cotton fibers into yarn, which is then transformed into fabrics and ultimately clothing. It includes the steps involved in producing yarn, fabric, and garment constructions.

Cotton prices are determined by various factors, including the inherent qualities of the fiber, the cleanliness of the fibers, and the level of contamination. As harvesting operations, handling, storage, shipping, and ginning processes all impact cleanliness and contamination, quality assurance should start at the farm level. The most significant issue is the contamination of fibers by non-plant matter, especially when using hand-picked cotton. Contaminated cotton is typically offered at a significant discount to compensate for the spinner's cleaning expenses. Contaminated cotton disrupts the spinning process. The price difference paid for cotton with the same fiber characteristics can range from 5 to 30% depending on the level of contamination (2009 COMESA Regional Strategy). Additionally, cotton from a country known for high contamination often receives price reductions without discrimination. As a result, a major cotton ginning company provided farmers with bags for harvesting cotton and established a cleaning system.

The expensive nature of utilities increases the cost of producing lint cotton. To fully utilize the ginner capacity in Ethiopia, the supply of seed cotton is largely uncertain and insufficient. This low-capacity level in the ginning process raises the manufacturing cost per unit. The main issue faced by the ginning factories in Ethiopia is the use of outdated technology. Although all ginning factories have the capacity to process 3,338 tons of lint cotton per day, they are currently operating below capacity.

The problems with the Ethiopian ginning factories stem from a lack of investment in new technology and a delayed upgrade of existing equipment. The absence of skilled and certified experts further exacerbates the issue in ginnery. The main challenge lies in attracting investments into this sub-sector given the existing operating climate. Implementing a market system that supports the processing of seed-oil and the sale of resulting goods would significantly increase the earnings of farmers and processors. Currently, ginneries are the sole market for seed cotton as there are no secondary markets for it. However, technical inefficiencies and low-capacity utilization are issues within the ginning factories themselves. Additionally, inadequate farmer organizations and a lack of transparency from the ginners contribute to weak contractual agreements between farmers and ginners. The actions of intermediaries further weaken the relationship.

While CVC activities are profitable and economically viable, increased output and productivity are necessary for ginners to remain financially sustainable. The low ginning output adversely affects their operations as it raises the cost of producing lint proportionally. The contribution of CVC to the Gross Domestic Product (GDP) is low, but its integration into the national economy is relatively high. However, CVC is weak in terms of global competitiveness, except for the manufacture of clothes. The long-term viability of Ethiopian CVC will depend on competition with imports and customer perception of quality. The scarcity of high-quality planting seeds is the main obstacle to the sustainability and profitability of the Ethiopian CVC chain.

2.4 Byproduct Producers

Cottonseed can be utilized to produce edible oil and oil cake for animal feed as a byproduct. Oil processing mills convert edible oil, and as a result of that process, oil cake is produced for utilization in animal feed.

2.4.1 Oil Processing Mills

The cottonseed that is separated from the lint is primarily utilized and sold in the local market, where it is used as oil cake waste for livestock feed and supplied to edible oil mills. Generally, cottonseed contains approximately 16% edible oil, which is consumed as is or utilized in the production of margarine and similar products as outlined in Table 3. The remaining cotton seed cake or meal, after the extraction of oil, is predominantly used as animal feed, although it can also serve as fertilizer.

The primary location for cottonseed commercialization has been oil processing plants. These privately owned businesses acquire the raw materials for their processing either through domestic competitive bidding or negotiation with producers [17]. However, only a small number of oil mills work directly with ginneries, the majority of raw materials are purchased through tenders. This demonstrates that edible oil producers have limited knowledge of backward linking through the processing of cottonseeds under a contractual agreement [17]. This is mainly because it is difficult to commit to pre-production obligations since the production of cottonseed depends on the amount of cotton picked, which in turn relies on various other factors.

The quantity of oilseeds required by various large-scale and medium oil processing mills is estimated to be 125,000 tons per year under optimal operating conditions, but the actual consumption of oilseeds is limited to an average of 30,000 tons. While other oilseed varieties like sesame, noug, safflower, and linseed are exported, cottonseed is mostly used for internal consumption, accounting for the largest portion of all oilseeds used by the mills. This has led the oil mills to rely heavily on cottonseeds as a raw material for edible oil production since they do not have sufficient access to other forms of oilseeds. Data from 2003/04 shows that almost 68% of the total oilseed consumed by the oil processing mills primarily came from cottonseeds, while the remaining 32% came from other types of seeds. Additionally, in addition to edible oil and crude oil production, the oil processing mills also produce approximately 20,000 metric tons of oilcake annually, which is mostly sold in the domestic market to dairy and cattle farms. Retailers play a crucial role in distributing cottonseed and its byproducts (oil and oilseed cake) to final consumers.

Table 3 Raw products obtained from seed cotton

Product	Approximate of seed cotton (%)	Seed content (%)
Lint	36–38	–
Linters	6	–
Oil	10	16
Cake (animal feed)	29	46
Hulls	15	23
Other components (moisture) manufacturing losses	2–4	6
Total	100	100

2.5 *Textile and Apparel Industry Sector*

Ethiopia manufactures textiles and garments using both traditional and modern manufacturing methods. Both forms of textile production are essential for Ethiopia's socio-economic progress, and their significance to the nation's economy and society has increased over time.

2.5.1 Modern Textile and Apparel Manufacturing

The first step in the cotton to garment chain is the transformation of raw cotton through ginning operations into lint cotton. The textile industry spins the finer lint into yarn and produces fabric. Fabric is then dyed, printed, and finished using softeners, wrinkle-resistance resins, or other finishing processes. The finishing part of the chain usually represents a significant portion of the value added in fabric production. Clothing is then created from the finished fabric that has been cut and sewn by various garment factories (Fig. 2).

In Ethiopia, spinning and weaving are commonly combined within the same company, while the production of textiles and garments is typically not consolidated under one entity. The private textile industry in Ethiopia consists of both large and medium-sized enterprises that engage in spinning, weaving, dyeing, finishing, and

Fig. 2 From cotton to clothing

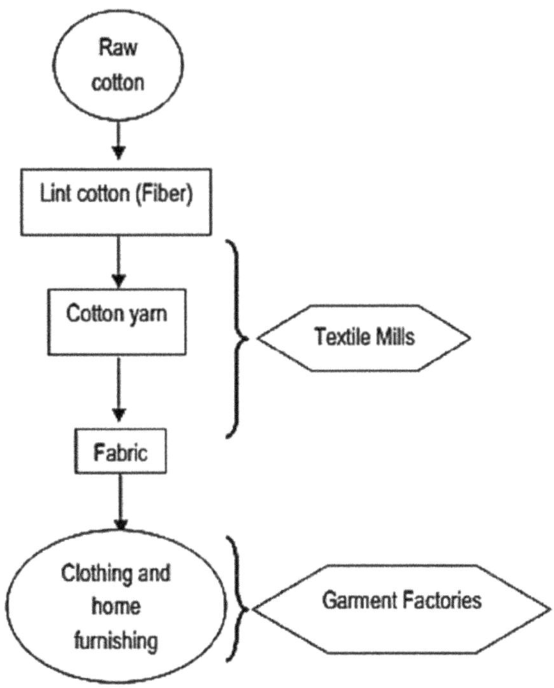

sewing. Approximately 60% of the country's textiles and apparel's demand is met by domestic production, while around 70,000 tons of cotton yarn are imported to fulfill the growing needs of the textile and apparel sector, as well as exports [9]. The cotton textile and apparel industries in Ethiopia, as well as globally, are renowned for their complexity and intricacy.

The establishment of Dire Dawa Textile and Augusta Garment Industries in Dire Dawa and Addis Ababa, respectively, in 1939 marked the beginning of Ethiopia's involvement in the textile and clothing manufacturing [18]. According to data from Ethiopia, the textile and clothing sector has experienced an average growth rate of 51% in the current decade. During this period, foreign investors have been granted licenses for approximately 65 international textile investment projects [19]. Currently, there are 176 garment industries and 52 textile industries operating in the country [18].

The textile industry in Ethiopia has encountered numerous difficulties that impact the production of textiles, including outdated and antiquated machinery (particularly privatized factories) with high maintenance costs and poor efficiency. The high production costs at this stage of the value chain are due to the underutilization of the installed capacity [15]. The lack of foreign currency to import spare parts for machinery upgrades and maintenance is a significant constraint. In competing countries like India, Indonesia, China, and Pakistan, technology upgrades and investment funds have been established at competitive interest rates to support the industry.

The expansion of the Ethiopian textile industry is significantly hindered by the high cost of conducting business. The cost of cotton, dyes, chemicals, and capital is the primary cost factors. Additionally, the opportunity costs of delayed and complex logistics put Ethiopian suppliers at a disadvantage compared to Asian suppliers who can deliver goods more quickly. Severe competitive disadvantages include lost output and production time due to power outages, equipment damage from power surges, and reduced efficiency caused by disruptions and uncertainty. Dyeing and washing operations are particularly negatively affected by supply disruptions. Many businesses in countries with frequent power outages use backup generators.

Quality is a crucial factor for textiles and clothing, as it contributes to buyer approval. However, quality alone is not sufficient for approval, as international buyers and retailers have other considerations. Manufacturers from the region must actively engage with buyers and be able to respond quickly to gain approval. Major brands and retail buyers often specify the source of the fabric, which is a longstanding practice that Ethiopia is currently experiencing. In other words, Ethiopian firms need to actively lobby for pre-qualification. Market information is key for building strong seller-buyer relationships, as both parties need to be familiar with each other's operations and make informed sourcing decisions [15].

With the elimination of quotas, in global trade, the retailers and buyers are seizing the opportunity to choose a few companies from which to obtain. Retailers are increasingly showing a preference for companies that offer a wide range of services including financing, efficient production techniques, and sample and design capabilities. The changing market demands are presenting a new challenge to Ethiopian and African producers. The rise of buyer-driven supply chains in the industry is placing greater responsibility on suppliers to manage yarn and fabric sourcing, design, inventory control, product development, and sample making. This means that Ethiopian companies must learn to integrate into worldwide supply chains and address the service requirements of purchasers and sellers. It also means that Ethiopian companies must learn to operate in a global market with fewer opportunities as buyers will evaluate carefully before switching suppliers.

Ethiopian textile and apparel manufacturers must enhance their operational efficiency and seek ways to reduce lead times. The lead times are the amount of time it will take a business to produce yarn, fabric, or clothing and deliver it. Ethiopian textile and clothing manufacturers must strive to decrease lead times and shipping costs, which can be quite substantial. For certain clothing items, lead times along the supply chain are more significant than for others. The sourcing of fashion items like dresses and blouses depends on traditional economic factors such as labor, materials, and textiles.

The textile finished products go through wholesalers and retailers before reaching the hands of end consumers. Retailers play a crucial role in delivering textile products to end consumers. The modern textile industry coexists with the traditional textile sector (manual ginning, spinning, and handloom weaving).

2.5.2 Traditional Textile and Clothing Production

The preferences of Ethiopian shoppers place a significant emphasis on traditional attire with Ethiopian patterns and embellishments. Nevertheless, the sub-chains are not independent because a traditional weaver could purchase yarn from an industrial supplier. With its diverse ethnic population and abundant natural resources, Ethiopia has a rich heritage of craft skills in general and hand weaving in particular, which has led to widespread job creation alongside agriculture. According to the survey conducted by Central Statistical Agency (CSA), 2013 on cottage/handicraft manufacturing industries, the number of hand-weaving establishments was estimated to be 221,848 with almost 55% of them existing in rural areas.

The handloom sub-sector not only generates a significant amount of employment but also plays a crucial strategic role in the economic growth of the nation due to its close ties to the agricultural industry as a source of lint cotton and the increasing demand for handloom textiles for home and office furnishings in both the domestic and international markets. The handmade organic-based processing, which creates higher demand compared to manufactured clothing, is another indicator of its strategic relevance and market dominance.

The primary products produced by the handloom industry can be categorized into semi-finished textiles and finished goods. While the finished products are divided into traditional clothing like Netela, Gabi, Kemis, and Kuta, primarily sold in the domestic market and to Ethiopians living abroad, and home furnishing textiles, which are intended for the global market. The semi-finished handlooms are typically sent to domestic garment factories for further processing. If supported by skill development and design, there is great potential to diversify the items produced by hand looms.

The utilization of cotton that has been converted into yarn dominates the production of hand-woven textiles in Ethiopia. Cotton is by far the most significant raw material in the industry, even if wool, silk, and synthetic materials are occasionally used. The primary source of cotton for individuals, mostly mothers, who hand weave for the production of traditional clothing like Gabi is the small-scale cotton farms that are spread out across the country. The origin of raw materials by the industry is not frequently monitored, resulting in instability, as the handlooms themselves are disorderly and primarily operated in households.

The industry is unable to expand to the level it aims to achieve for various reasons, despite there being a sufficient supply of local cotton and a cluster of handlooms in several regions of the country with a long history of weaving expertise and tradition. One clear problem is associated with the sporadic availability of high-quality colored yarn due to the operation of yarn-producing companies at a low capacity and their limitation to the production of few types of yarn, which occasionally leads to scarcity.

2.6 *The Global and National Marketing Linkages of CVC Chain*

Many nations across the globe participate in the production, export, and import of cotton, which is a widely traded agricultural commodity. It is extensively traded in over 150 countries worldwide and is expanding in more than 100 of them. Each of these nations either exports or imports it. Cotton is one of the most widely traded commodities on Earth. International trade in cotton and yarn had a value of $26.3 billion in 2016 [20].

This corresponds to a volume of 12.8 million tons and accounts for 27% of all globally sold agricultural goods. The leading exporters of cotton and yarn in 2016 were the United States ($5.1 billion), India ($4.6 billion), Vietnam ($2.1 billion), China ($1.7 billion), and Pakistan ($1.4 billion). The top importers, in order, were China ($7.4 billion), Bangladesh ($2 billion), Turkey ($1.8 billion), Vietnam ($1.7 billion), and Indonesia ($1.2 billion). The most significant direct trade flows are from Vietnam to China ($1.7 billion), India to China ($1.3 billion), India to Bangladesh ($1.1 billion), Pakistan to India ($861 million), and the United States ($793 million) to Vietnam (all statistics are from 2016).

Cotton trade prices are highly volatile, reflecting global supply and demand as well as speculation in financial markets. This can have an impact on all participants in the supply chain, including cotton farmers and local traders. When cotton prices are low in the global market, small-scale farmers in many developing countries struggle to earn a living from the crop and sometimes switch to other crops such as sesame, sorghum, and bananas in the Arba Minch region of Ethiopia [7].

On the contrary, it has also been repeatedly reported that subsidies for cotton in industrialized nations (particularly the USA) have a depressing effect on global market prices, making it difficult for small-scale farmers and producers in other parts of the world to compete with these subsidies [21]. According to Gro Intelligence [22], the average cotton subsidy per acre in the USA doubled between 2009 and 2018.

2.6.1 Local Cotton Marketing Trends

Cotton, like many other agricultural commodities, is traded on global commercial markets that experience significant price fluctuations over time. Figure 3, for instance, illustrates the substantial volatility of cotton spot market prices between 2014 and 2018 at the US Stock Exchange NASDAQ in New York. In Ethiopia's cotton market, there are two local benchmark prices for raw cotton. The Ethiopian Industrial Inputs Development Enterprise (EIIDE) employs a price-setting system to acquire cotton from farmers in different regions at a predetermined price, while the second benchmark price is determined by cotton transactions between regional producers and textile companies.

The performance of the cotton market chain in Ethiopia was found to be below average when evaluated using marketing margins and considering the costs incurred and gross profits generated by different market chain operators. Farmers are the most disadvantaged players in this cycle. At the farm level, several factors affect the availability of marketable cotton. An investigation into the structure, conduct, and performance of the cotton market chain in the Metema district of the Amhara Regional State revealed a weak performance of the chain, which makes farmers unfavorable participants in the value chain.

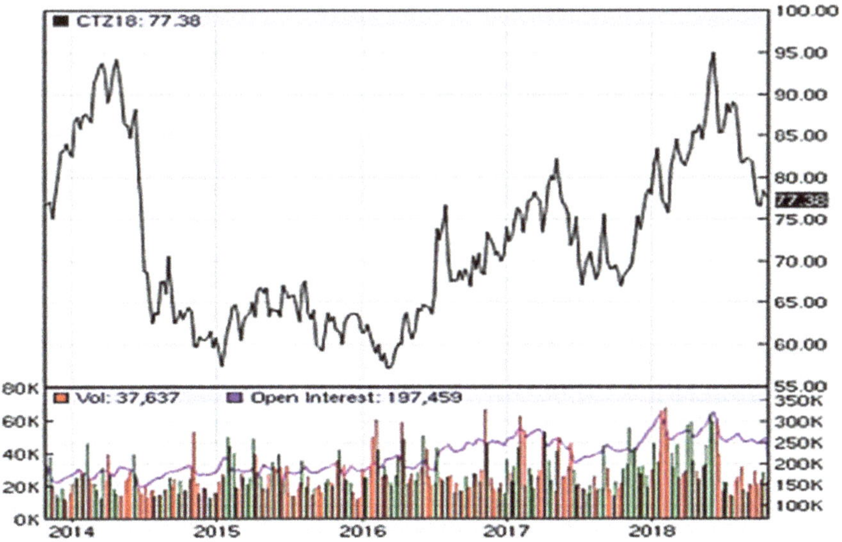

Fig. 3 Commodity price for cotton at NASDAQ, NY (2014–2018).) (*Source* [7]

In Ethiopia, smallholder farmers often sell their seed cotton to dealers who do not differentiate prices based on quality, a practice common in other African countries. The lack of incentives does not encourage farmers to prioritize the quality of their produce. However, the Ethiopia Domestic Distribution Corporation (EDDC) intends to facilitate joint purchases for regional cotton buyers of raw cotton. According to reports, the EDDC considers the end user's credentials when making their purchases.

The cotton spinning mills must have the ability to produce high-quality yarn at an affordable price, and the cost and quality of cotton greatly influence the price and quality of the yarn. The lack of well-organized research, as well as the absence of a systematic grading and categorization system, negatively affects the production, productivity, quality, and marketing of cotton. Cotton farmers and cotton ginneries have not given much attention to the quality of cotton due to the current marketing and pricing system, which mainly focuses on the weight of the cotton. Unlike lint, there is no market price mechanism for seed cotton. The majority of the cotton grown in the country is sold as lint and seeds after commission ginning.

The absence of incentives that promote quality directly contributes to the high amount of impurities in seed cotton. There are currently economic risks due to increasing production costs, fluctuations in yield and prices, expensive transportation from remote areas, and the absence of a stable market. There is no standardized method that ensures producers a consistent market and dealers a reliable supply of cotton. Cotton production and marketing are conducted in a subjective manner. Since cotton is primarily grown in lowlands, there is fierce competition with sesame, sorghum, and sugarcane in order to gain a competitive advantage in pricing. Farmers use cotton as a rotation crop because they believe it improves the fertility of their land in subsequent crop cycles.

2.6.2 Market Linkages within Local CVC Actors

Following the harvest, large farmers directly sell seed cotton to ginners, while small farmers sell seed cotton to intermediaries. These informal middlemen do not exist in all industrial areas and conduct their business in an informal manner. Some of these intermediaries offer loans to farmers to cover the initial costs of cultivation, which are then repaid with a portion of the farmers' harvest. According to available sources, ginners typically export 2–7 tons of cotton lint in order to obtain foreign currency for purchasing replacement parts.

From November to May, the farmers sell their cotton, with the best months for selling being January, February, and March, which account for about 90% of the seed cotton sales. There are no seed cotton sales during the months of June, July, August, September, and October. The market connections and relationships between the main actors in the cotton value chain are complex and involve producers, cooperatives, cooperative unions, assemblers, ginners, textile factories, oil mills, seed cotton retailers, wholesalers, retailers, and consumers of the final products. Various mechanisms connecting seed cotton producers to clothing consumers are described below:

i. Producers → Assemblers→ Ginneries → Textile factories → Wholesalers→ Retailers → Consumers.
ii. Producers Assemblers →Seed cotton retailers from other districts→ Handcraft→ Retailers/consumers.
iii. Producers → Ginneries → Textile factories → Wholesalers→ Retailers → Consumers.
iv. Producers→ Primary cooperatives → Cooperatives union → Wholesale traders→Textile factories →Wholesalers → Retailers →Consumers.
v. Producers→ Assemblers Cooperatives Union→ Wholesale traders →Textile factories →Wholesalers →Retailers → Consumers.
vi. Producers→ Assemblers→ Cooperatives union → Wholesale traders→Textile factories →Wholesalers →Retailers → Consumers.
vii. Producers →Assemblers → Ginneries → Textile factories →Wholesalers→ Retailers→ Consumers.
viii. Producers → Commission agents' →Assemblers →Ginneries →Textile factories→ Wholesalers →Retailers → Consumers.
ix. Producers →Primary cooperatives→ Cooperatives Union→ Ginneries →Textile factories→ Wholesalers →Retailers → Consumers.

The manufacturers are where the cotton market chain begins. Consumers of textile items, cottonseed, and cottonseed products like oil and oil seed cake are the end buyers of the outputs. Ginneries obtain seed cotton from a variety of suppliers. Lint cotton is a raw material that gins provide to textile mills so they may produce yarns and fabrics. Cottonseed is also used as a raw material by oil mills to produce oils and oilseed cake. Ginneries provide cottonseed to farmers, distributors, and cooperative unions for use as seed. The majority of the textile companies' output, i.e., fabrics and apparel are supplied to wholesalers, who then sell it to retailers, who in turn sell it to consumers.

Many consumers from different parts of the country consume the oil that the oil mill produces from cottonseed. However, it was observed that there was a severe issue with the availability of cottonseed. According to the literature, one liter of oil can be made from 7.69 Kg of cottonseed. As a result, the value chain for byproducts of seed cotton exists alongside the value chain for lint cotton among CVC participants. Following ginning in various methods by CVC entities, the chains from seed cotton growers through cotton seed processors and consumers are ongoing concurrently for oil and seed cake manufacturers up to consumers.

i. Ginneries → Oil mills → Wholesalers → Retailers → Consumers (for edible oil).
ii. Ginneries → Cooperatives union → Primary cooperatives →Farmers (for cottonseed).
iii. Ginneries → Wholesalers → Retailers →Farmers (for cottonseed).
iv. Ginneries →Farmers (for cottonseed).
v. Ginneries →Oil mills →Consumers (for edible oil).
vi. Ginneries →Oil mills →Dairy and fattening (for oil seed cake)
vii. Ginneries →Oil mills →Retailers →Consumers (for edible oil)

3 Economic Importance

The total value added (VA) of the CVC is 3.4 billion ETB (€74 million) (up to the yarn and crude oil stages), of which 3.2 billion ETB (€70 million) is direct VA and 0.2 billion ETB (€4 million) is indirect VA from the utilization of intermediate goods and services provided by external actors. The overall VA accounts for 0.55% of the agricultural GDP and contributes 0.18% to Ethiopia's GDP. The level of integration of the CVC into the domestic economy is high (89%), indicating that a significant portion of the value of CVC production is retained domestically and only a small portion is imported. In terms of government revenue, the CVC actors have paid 0.9 billion ETB (or €20 million) in taxes to the State. The CVC positively impacts the state of the public finances as actors do not receive subsidies. Imports of products and services (such as agricultural inputs, replacement components, and gasoline) amount to 0.3 billion ETB (€7 million). Due to the limited amount of lint exported by the country, the CVC trade balance is negative. Considering that the majority of textile and apparel manufacturing is exported, implementing the CVC at the downstream level would be beneficial [9].

According to a report by the European Commission published in 2021, the CVC in Ethiopia generates positive net operating profits (NOP) for all participants, with the exception of the ginners (Table 4), who record negative NOPs due to the underutilization of the ginning capacity as a result of the low level of seed cotton supply and the low ginning output (% of lint in seed cotton) of the DP-90 variety. Even though the price of seed cotton is high by African standards and despite the high return on turnover (NOP/value of production), benchmarks for farmers' net incomes reveal that cotton is less attractive than other food crops (sesame, bananas, sugarcane), even though sesame's NOP is much more reliable. Even if a middleman's return on sales is only 4%, their marketing margin has a significant absolute value.

Table 4 Profitability for the individual CVC actors (*Source* [9])

CVC actors	Net operating profit	Return on turnover (%)
Traditional farmer	28860 ETB (€635)	−100
Small farmer	15550 ETB (€340)	72
Medium/large farmer	7,485,000 ETB (€165000)	51
Middleman	798000 ETB (€17550)	4
Ginner	−946000 ETB (€-20800)	−0.8
Spinner	25,223,000 ETB (€555000)	14
Oilseed processor	170450000 ETB (€3750000)	31

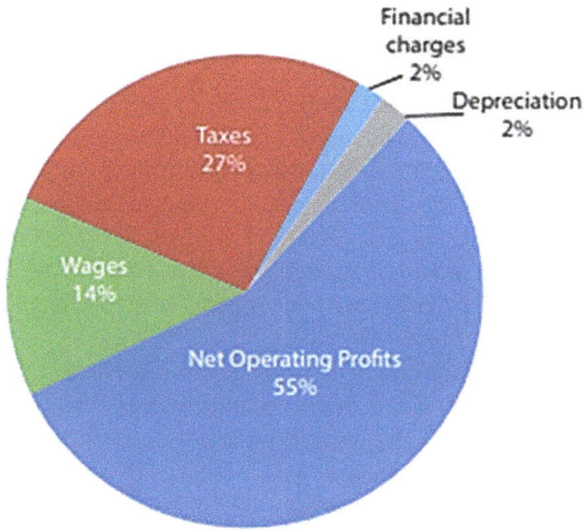

Fig. 4 Income distribution (direct value added). (*Source* [9])

The two main components of the direct VA are revenue from taxes (27%) and net operating profit (NOP) (55%) (Fig. 4). The net operating profits primarily aid medium- to large-scale farms and spinners (43 and 21%, respectively).

3.1 CVC Viability in the International Economy

Given that the Domestic Resource Cost (DRC) is below 1 (0.3), the CVC is competitive in the global economy. The economic value obtained using the global price for the calculation is higher than the value of the local elements (land, labor, and capital) used in the production of cotton. This indicates that concentrating these domestic resources on cotton production is a smart investment and a boon for the country's economy. However, the Nominal Protection Coefficient (NPC) of 1.6 for the entire CVC suggests that domestic output is less competitive compared to other countries. In contrast to cottonseed and cake, which are more inexpensive in Ethiopia compared to the international market, this is not the case for seed cotton, lint, yarn (especially the traditional type), and crude oil. The expensive cost of seed cotton and the low yield of the DP-90 variety are responsible for the lack of competitiveness. In Sub-Saharan Africa in 2019, Ethiopia had the second-highest price of seed cotton, which is by far the most significant factor affecting the cost of lint.

4 Governance in CVC sector

The current changes in the textile industry and the challenge in coordinating the diverse cotton production structure with the ambitious objectives of both the public and private textile industries can be attributed to the Ethiopian CVC's lack of coordination. Professional associations like the Ethiopian Cotton Producers, Ginners and Exporters Association and the Ethiopian Textile and Garment Manufacturers Association are involved in the CVC while ETIDI is in charge of the entire CVC. The Ministry of Agriculture and Animal Husbandry supervises the research, extension, and regulatory organizations.

Research on cotton is one of the priorities of the Ethiopian Institute of Agricultural Research (EIAR). It has been implemented nationwide since the 1970s at WARC, in the Awash valley. Four departments-Breeding, Socio-Economics and Extension, Agronomy, and Entomology-are responsible for coordinating cotton research. One of the cotton breeders is responsible for managing the global coordination. Since there is limited opportunity for movement in the cotton-growing regions, most operations are conducted under irrigation, on the station, or nearby. Specifically, cropping systems that rely on natural rainfall have not been examined or tested. The other research facilities, located in a region with high rainfall, lack the necessary resources to study cotton. The relationship with stakeholders is established through individual contacts or within the framework of the annual stakeholder meeting and research review forum, which includes farms, agricultural services, and the cotton producer's organization.

5 CVC Challenges

Ethiopian cotton producers of all varieties have a fantastic opportunity due to the increasing domestic and global demand for high-quality cotton fiber. Furthermore, the demand for cotton seeds is rising as more oil and cake are produced for human and animal use, respectively. Additional potential for cotton production in Ethiopia includes the involvement of farmers' groups in promoting cotton as well as the growing global and Ethiopian diaspora demand for traditional clothing. However, as will be seen below, the CVC industry in Ethiopia is faced with a number of challenges.

i. Lack of market transparency.
ii. Unclear roles and little collaboration among the ministries involved in CVC.
iii. Limited implementation of strategies and prioritization of the CVC.
iv. Cotton is considered a non-strategic crop by national authorities.
v. Low ginning output and outdated ginning machinery.
vi. The main issues with CVC development are the poor quality of the seeds and the insufficient focus on the potential for oilseed production.

The threats for CVC can be described as

i. The more attractive crops (sesame, bananas, sugarcane, sorghum, etc.) contend with cotton.
ii. Environmental impacts of contemporary farming.
iii. Growing difficulties in acquiring sufficient water and salt build-up.
iv. Societal conflicts among different factions caused by land access.
v. Reduced soil fertility, increased risk of erosion on extensive monoculture cotton fields, and decreased rainfall as a result of climate change all impact cotton production.

To enhance CVC in Ethiopia, it is crucial to focus on the following:

i. Instead of focusing on the expansion of the cotton area, the current plan for cotton development should prioritize high-quality and sustainable cotton production.
ii. Through enhanced irrigation, Ethiopia has the potential to produce long staple cotton similar to Egyptian cotton.
iii. The CVC must be prepared to meet the growing global demand for ethical fashion and sustainable cotton.
iv. It is crucial to evaluate the rights and working conditions of employees, with particular attention to the protection of women in factories and farms.
v. Given the disputes over land usage; special consideration should be given to the rights of pastoralists.
vi. Once the quality improves, it is necessary to introduce a label for high-grade Ethiopian cotton and textiles.
vii. Improved water access, stakeholder coordination, transparency, and government agency collaboration are essential for implementing best agricultural practices.
viii. Foster a shared understanding of the most pressing issues in the CVC and select the appropriate institutional entities.
ix. Enhance policy incentives, research, and the effectiveness of climate change adaptation strategies.
x. Incorporate climate impact maps (temperature and rainfall patterns) into cotton development strategies to identify regions and districts most vulnerable to climate change.
xi. It is important to study the current irrigation techniques and their impact on soil quality.
xii. Policy incentives should be provided to reduce excessive groundwater consumption on large, medium-sized, and large farms.
xiii. The demand for cotton lint increases the risk of activities that deplete soil.
xiv. Public sector investments and research are needed to support farmers' education, training, and technical assistance in preserving soil fertility through agro-ecological practices and agricultural systems.
xv. Promote household composting (including the use of agricultural leftovers) or integrated pest control by providing relevant examples of effective practices in the Ethiopian context (European Union 2021).

6 Summary

As a widely traded agricultural crop with a lengthy value chain and numerous participants, cotton is produced and traded globally, with the majority of countries involved in both its production and trade. The complex value chain of cotton in Ethiopia includes farmers, intermediaries, ginners with various configurations and technologies, modern textile factories, and traditional weavers. The cotton value chain at each level and in general is not well regulated and allows for the entire value chain to be economically productive, despite the nation's potential in all aspects of cotton production and trade.

Farmers are the main contributors to the cotton value chain since they established it initially. Therefore, it is essential to establish incentives to motivate farmers to cultivate cotton in both quantity and quality. Additionally, the government must provide financial alternatives and other forms of technical assistance to cotton producers. Cotton is a crucial crop for Ethiopia's economic development through the textile and apparel industry, as it serves as the raw material for this sector. The textile and clothing industry plays a significant role in the cotton value chain as it helps a developing country transition from a resource-based to a manufacturing economy, with cotton being a vital raw material. In general, there are several challenges that the cotton value chain must address through careful policy intervention in order to create a viable strategic plan that considers cotton's strategic importance for the nation's economic development through value addition and industrial manufacturing.

References

1. Moon D.-O, Kim M.-O, Lee J.-D, Kim G.-Y (2008) Gossypol suppresses NF-κB activity and NF-κB-related gene expression in human leukemia U937 cells. Cancer letts, 264:192–200
2. Hasrat J, Pieters L, Vlietinck A (2004) Medicinal plants in Suriname: hypotensive effect of Gossypium barbadense. J Pharm Pharmacol 56:381–387
3. Wilkins TA, Rajasekaran K, Anderson DM (2000) Cotton biotechnology. Crit Rev Plant Sci 19:511–550
4. EIA (Ethiopian Investment Agency) (2012) Investment opportunity profile for cotton production and ginning in Ethiopia (Updated Report on Cotton Ginning). Ethiopian Investment Agency, Addis Ababa, Ethiopia
5. EIC (2017) Produce cotton in Ethiopia for a blooming textile industry. EIC/Ref. No. 01/2017
6. USDA (2019) Ethiopian cotton production annual. Global agricultural information network (No. ET1906). Addis Ababa, Ethiopia
7. The Cotton Supply Chain in Ethiopia (2018) A country-focused commodity analysis in the context of the Bio-Macht project, Freiburg
8. Belachew T, Mamuye T (2018) Interview by T. Amera, I. Hilbert, & T. Schleicher. Addis Ababa, Ethiopia
9. European Commission (2021) Cotton value chain analysis in Ethiopia. Retrieved on 20 February 2022
10. Trienekens, (2011) Agricultural value chains in developing countries a framework for analysis. Int Food Agribus Man 14:51–83
11. Dolan C, Humphrey J (2000) Governance and trade in fresh vegetables: the impact of UK supermarkets on the African horticulture industry. J Dev Stud 37(2):147–176

12. Lambert DM, Cooper M (2000) Issues in supply chain management. Ind Mark Manag 29:65–83
13. SOFRECO (2017) National cotton development strategy (2017-2032) and road map
14. International Labor Organization (2019) Research to map and analyse the cotton, textile and garment value chains and identify strategies for the elimination and prevention of child labour and forced labour in Burkina Faso and Mali
15. COMESA Regional Strategy (2009) For developing the cotton, textile and clothing manufacturing, marketing and distribution/exporting
16. Feyso Abayneh (2018) Cotton value chain analysis: the case of smallholder farmers Arbaminch Zuria District. Gamo Gofa Zone, Ethiopia. Asian J Plant Sci Res 8(2):19–30
17. Tessema W (2001) Profile of the food processing industry in Ethiopia. United Nations Industrial Development Organization (UNIDO)
18. ETIDI (2018) Investment opportunity in the cotton, textile and garment industry in Ethiopia. Ethiopian Textile industry development Institute. Retrieved on 11 February 2022 from https://www.unido.org/
19. Kohan Textile Report (2022) Middle, East and African textile Portal. Retrieved on 15 February 2022 from https://kohantextilejournal.com/
20. Chatham House (2018) Global resource trade: value of international trade of cotton and yarn 2016. Available at https://resourcetrade.earth/data?year=2016&category=90&units=value
21. Fairtrade Germany (2018) Fairtrade baumwolle: So funktioniert der faire Handel. Available at https://www.fairtrade-deutschland.de/produkte-de/baumwolle/hintergrundfairtradebaumwolle.htmlF
22. Gro Intelligence (2018) US cotton subsidies insulate producers from economic loss. Available at https://gro-intelligence.com/insights/us-cotton-subsidies
23. Amera T (2017) Supporting the sustainable cotton production initiative in the Southern Rift Valley of Ethiopia: Project Report 2016
24. Tegegne B (2008) Analysis of cotton marketing chains: the case of Metema Woreda, North Gondar Zone, Amhara National Regional State. Master Thesis. Haramaya University.
25. Bundesverband—Die VERBRAUCHER INITIATIVE e.V. (2018) Konventioneller Baumwollanbau: Der Einsatz von Pestiziden. Available at http://www.oekofair.de/index.php/cat/799/title/Der_Einsatz_von_Pestiziden. Last accessed on 30 Oct 2018
26. Ethiopia Country Report (2014) African development bank group (ADBG). Promot Technol, Innov, Prod Link
27. Ferenschild S (2014) Afrikas weißes Gold. Ein morderner Dreieckhandel: Afrikanischer Rohstoff—chinesische Verarbeitung—europäischer Konsum. Available at https://www.suedwind-institut.de/fileadmin/fuerSuedwind/Publikationen/2014/201410_Afrikas_weisses_Gold.pdf. Last accessed on 11 Aug 2017
28. Gitonga W, Macharia JMK, Mungai A, Njue H, Karanja DK, Olweny H (2016) Cotton production, consraints, and research interventions in Kenya. Available at https://www.icac.org/tis/regional_networks/documents/africa_10/business_meeting/3_kenyapdf. Last accessed on 11 Aug 2017
29. Hiwet A (2018) Cotton production and research in Ethiopia. Available at https://www.icac.org/tis/regional_networks/documents/africa/papers/day_1/ethiopia.pdf. Last accessed on 4 Sep 2018
30. Kaplinsky R, Morris M (2001) A handbook for value chain research. Prepared for IDRC
31. Lançon J, Woldu B (2020) The Ethiopian cottonseed system description, analysis and recommendations. Deutsche Gesellschaft für Internationale Zusammenarbeit (GIZ) GmbH and Agricultural Research for Development.
32. Mulwa R, Wafula D, Karembu M, Waithaka M (2013) Estimating the potential economic benefits of adopting Bt cotton in selected COMESA countries. AgBioForum 16(1):14–26
33. Negash S, Mekuria M (2018) Interview by T. Amera, I. Hilbert and T. Schleicher. Addis Ababa, Ethiopia

Adapting Sustainable Cotton Production for Improved Livelihood of Farmers and Sustainable Landscapes

Berhanu Woldu and Samson Asefa

Abstract Ethiopian farmers have cultivated cotton for thousands of years, and the traditional handloom sub-sector earns its income from the cotton sub-sector due to the favorable agro-ecological conditions in the country. Currently, the industry faces challenges such as limited access to improved inputs, infestations by pests, salinity, water management issues, market entry, and production standards for sustainable cotton. Given the numerous advantages of cotton over other crops chosen by smallholder farmers, implementing a sustainable cotton production system could potentially reverse the situation and restore their traditional income crop. Nowadays, textile suppliers are urging manufacturers to utilize sustainable cotton. The Organic, Better Cotton Initiative (BCI), Cotton Made in Africa (CmiA), Fair-trade, My Best Management Practices (MyBMP), and Committee on Sustainability Assessment (ABR) are widely recognized labels for sustainable cotton. In Ethiopia, CmiA, BCI, and Organic are the sustainability labels that have been introduced. The primary obstacles to Ethiopia fully realizing its potential for sustainable cotton production are the lack of adoption, insufficient funding for implementing sustainable production, a challenging process for applying for BCI membership, and the utilization of genetically modified (GM) seeds by small-scale farmers, which could result in rejection and exclusion from the CmiA and Organic labels.

Keywords Sustainable cotton · CmiA · BCI · Organic · Regenerative agriculture

Acronyms

ABR	Committee on Sustainability Assessment
BCI	Better cotton initiative

B. Woldu (✉)
Cotton Projects, Solidaridad, Addis Ababa, Ethiopia
e-mail: bernawol@gmail.com

S. Asefa
Cotton Plant Research, Ethiopian Textile Industry Development Institute, Addis Ababa, Ethiopia

K. Murugesh Babu et al. (eds.), *Cotton Sector Development in Ethiopia*, Textile Science and Clothing Technology, https://doi.org/10.1007/978-981-99-9149-5_19

CmiA	Cotton Made in Ethiopia
EIAR	Ethiopian Institute of Agricultural Research
FAO	Food and Agriculture Organization
GAP	Good agronomic practice
GoE	Government of Ethiopia
GMO	Genetically Modified Organism
GOT	Ginning out turn
ICAC	International cotton Advisory Committee
ILO	International labor organization
IPDC	Industrial Parks Development Co-operation
IPM	Integrated pest management
ITC	International Trade Center
MyBMP	My Best Management Practices
NCDS	National Cotton Development Strategy
OECD	Organization for Economic Co-operation and Development
PAN	Pesticide Action Network
POP	Persistent Organic Pollutants
SCR	Sustainable Cotton Ranking
DG	Sustainable development goal
WHO	World health Organization

1 Introduction

According to [1], Ethiopia possesses a 3 million ha agroecology that is appropriate for cultivating cotton. The traditional handloom sub-sector and the livelihood of Ethiopian farmers both originate from the cotton sub-sector over an extended period of time [2]. The National Cotton Development Strategy (NCDS), an ambitious 15-year plan (2018–32) aimed at increasing cotton production to 1.1 million metric tons on one million hectares of suitable land for cotton production that is entirely sustainable at the end of the target years [3], was initiated by the Government of Ethiopia with the intention of achieving its vision of making Ethiopia the textile and apparel manufacturing hub of Africa with annual exports of US$ 1 billion by 2025. However, the cotton sector currently faces challenges such as a lack of access to improved inputs (seed and pesticides), pest infestation, salinity and water governance concerns (particularly in the Afar Regional State), market access, and production standards for sustainable cotton.

The implementation of an eco-friendly cotton production system has the potential to change the situation and bring small-scale farmers back to their traditional cash crop. The numerous advantages of cotton may outweigh those of other crops that farmers currently prefer. However, this transition requires a strategic approach to attract farmers to adopt the system and ultimately meet the growing demand for cotton lint in the thriving textile industry. Additionally, the system should have minimal negative effects on the environment and society.

Presently, there is a global need for cotton textiles that have been responsibly cultivated due to increased consumer awareness of environmental issues [4]. Textile suppliers are putting pressure on manufacturers to use sustainable cotton. Leading textile retailers have already implemented sustainable sourcing policies or have started sourcing and tracing sustainable cotton, indicating that the sustainability trend is likely to continue [5].

2 Advent of Sustainable Cotton

Despite cotton representing only 2.5% of overall crop production, conventional cotton farming utilizes 6.2% of all pesticides used worldwide and accounts for 14.1% of insecticide sales [6]. The use of extremely hazardous pesticides, excessive usage, and improper application of pesticides can greatly affect ecosystems and the welfare of farmers and their communities. The deforestation and expansion of land have an adverse impact on the environment, causing soil erosion, soil pollution, and the decline of biodiversity. Irrigation is crucial for 73% of worldwide cotton farming, leading to:

i. Substantial water consumption (Depletion).
ii. Trigger salinity.
iii. Water contamination (pesticide, fertilizer, etc.…).

Moreover, the concept of sustainable cotton production emerged as a result of various calamities associated with cotton cultivation or processing. Some sustainability certifications are created by large retailers and manufacturers who aim to safeguard their reputation against potential negative incidents. Most sustainability certifications promote environmentally friendly practices, improved ethical treatment of employees, and a more equitable distribution of profits, among other things. Ultimately, this would persuade consumers to purchase clothing that is more ethically produced [1].

Therefore, sustainable cotton represents a novel approach that can effectively tackle numerous environmental and socio-economic issues posed by traditional cotton production. It can also support the livelihoods of producers and communities, all while facing long-term ecological limitations and socio-economic pressures.

3 Definition and Key Requirements of Sustainable Cotton Production

According to FAO, 2015, citing the Brundtland Report [7], sustainable development is defined more broadly as growth that "satisfies the demands of the present without jeopardizing the ability of future generations to satisfy their own demands." In view of this, sustainable cotton production refers to farming methods that uphold production levels while optimizing the three principles of sustainability—environmental, economic, and social—each of which encompasses a range of concerns. Pest and pesticide management, water conservation, soil well-being, biodiversity and land utilization, and climate change are among the topics included in the environmental principle. The economic principle encompasses topics such as managing economic uncertainty, alleviating poverty, and ensuring food security. The social principle also encompasses farmer associations, equality and gender issues for workers, worker health and safety requirements, and labor rights and standards.

3.1 Environment

The environment is composed of elements such as ground, water, woodlands, creatures, and chemical input. To enhance and maintain the fertility of the ground, farmers should practice ground and water conservation (such as conservation farming and the creation of compost) and crop rotation, which can include the stand-alone planting of legumes or intercropping with legumes.

Deforestation of primary woodlands or other national resources that are recognized and protected by domestic or international law is a serious breach of the principles of sustainable cotton production. Furthermore, farming is restricted near World Heritage Sites, Wetlands, and other key bird habitats. By directly improving the productivity of agricultural production systems through the provision of a variety of ecosystem services, including natural biological control, ground fertility, and pollination, biodiversity in and around farmed fields is responsible for the development of benefits. Positive relationships exist between high ground fauna richness and ground fertility [8].

Conservation and enhancement of populations of beneficial insects must be part of the integrated pest management plan of the producers. Farmers are advised to actively pursue a strategy to manage pesticides used for cotton cultivation aiming at minimizing impact on the environment and health. The possibilities of runoff or leaching of chemicals into streams or groundwater should be minimized. Farmers must be trained in the use of safe spraying techniques, appropriate equipment, and sufficient protective clothing as necessary for operator safety relative to the applied pesticide and the application equipment used (e.g., shirts with long sleeves, trousers, closed shoes, masks, gloves, and safety goggles).

Producers need to be aware of the dangers of re-using empty pesticide containers and how to dispose of these in a safe manner. Use of pesticides banned under the Stockholm Convention, Rotterdam Convention on Persistent Organic Pollutants (POPs), and World Health Organization (WHO) list of highly hazardous and hazardous pesticides are against the principles of sustainable cotton production.

3.2 Social

The social affair primarily focuses on essential labor rights and standards, worker well-being and safety, fairness and equality, and farmer association. A suitable and favorable working environment necessitates the provision of basic amenities such as drinkable water supply, a hygienic place to eat, and a sanitized working environment (ILO 2008). Employees' working conditions, terms of employment and empowerment, working hours, etc. should align with the sustainability concept and comply with national legislation. Severe forms of child labor (excluding cases where children assist on their family's farm, as long as the work does not pose a threat to their health, safety, well-being, education, or development, and they are supervised by adults and receive appropriate training), human trafficking, bonded or coerced labor, and discouraging the establishment and/or participation in institutional structures that represent farmers are contrary to the principles of sustainable cotton production.

3.3 Economic Requirements: Economic Viability, Management Principles, and Supply Chain Responsibilities

Farmers require access to funding and loans, higher pricing, official contracts between purchasers and vendors, entry to markets, a dependable logistics network, and timely buyer payment conditions. To ensure economic sustainability and profitability, farms and farmer cooperatives must possess effective governance, control over internal processes, efficient workforce management, and a comprehensive long-term strategy.

3.4 Quality of Input and Product

Procedures for evaluating the excellence of seed cotton must be established, and there is a dispute resolution system in operation that has been endorsed by farmers and buyers. To maximize the profitability of the cotton fibers that can be marketed, measures should be implemented to recognize the primary elements that impact its quality and to implement enhancements. It is essential to promote farmers to adopt hygienic practices during harvesting and after harvesting that discourage the usage of polypropylene sacks for gathering, storing, and other essential procedures to prevent contamination of the fibers.

3.5 Ethics and Business Integrity

Transactions in commercial relationships that are unethical as defined by international agreements, national laws, and customs (customs that are not in conflict with national laws), such as the OECD Competition Guidelines:

i. Exploit market power or dominance.
ii. Attain market power or dominance through methods other than efficient performance.
iii. Participate in anti-competitive agreements or arrangements (whether formal or informal) that are deemed unlawful by the sustainable cotton production system.

4 Global Trends and Major Cotton Sustainability Standards

The Better Cotton Initiative (BCI), Cotton Made in Africa (CmiA), Fair Trade, My Best Management Practices (MyBMP), Committee on Sustainability Assessment (ABR), and other voluntary labels for cotton sustainability are acknowledged, etc. In 2020, over 23% of total cotton production utilized more sustainable techniques [9]. However, only 25% of the supply is actively procured by businesses, while the remaining 75% is traded as conventional cotton [5]. Table 1 displays the globally produced sustainable cotton from various initiatives.

The share of sustainable cotton production in the 2020 harvest represents 23% of the total global cotton production. The leading sustainability criteria currently observed in the sustainable cotton supply chain include the Better Cotton Initiative (BCI), Organic Cotton, Fair-trade Cotton, and CmiA (Table 2).

Table 1 The sustainable cotton produced in 2020/21 production season

Types of initiatives	Year of production	Quantity of production (Metric ton)
CmiA	2020/21	690,000
MyBMP	2020	31,000
ABR	2020	832,000
BCI	2020	6,000,000
Faire trade	–	2,334,000
Organic label	–	112,000

Table 2 Scope and objective of selected sustainability labels

Sustainable standard	Standard objective	Scope	Exclusion criteria
BCI	Make global cotton cultivation: – Better for the people who produce it, – Better for the environment it grows in, and – Better for the sector's future	Covers environmental and social criteria for both smallholders and large farms. Better Cotton is a benchmark standard for the industry	Use of WHO list of prohibited chemicals
CmiA	Improve the living conditions of African smallholders and promote environmentally friendly cotton cultivation	CmiA is based on good agricultural practices and a comprehensive range of social and economic criteria. Its focus is on Africa alone, and on small farmers growing in rain-fed conditions. It is benchmarked to the Better Cotton standard	No Irrigation, No GMO seeds, Farmers with more than 20 ha cannot participate. No pesticides are prohibited by the Stockholm Convention, Rotterdam Convention on Persistent Organic Pollutants (POPs), and WHO list of highly hazardous and hazardous pesticides and Prohibit cutting primary forests as well as encroaching upon nature reserves
Organic	Replacing synthetic fertilizers and pesticides by nature-based remedies	The main requirements are environmental criteria	No chemical Pesticides. No chemical fertilizer and No GMO seeds
Fair-trade	Connect disadvantaged producers and consumers, promote fairer trading conditions, and empower producers to combat poverty, strengthen their position, and take more control of their lives	The Fair-trade cotton standard covers social, economic, and environmental criteria	NO GMO

5 Current Sustainable Cotton Production Status and Potential in Ethiopia

Ethiopia possesses all the potential to establish itself as a leading country in the sustainable cultivation of cotton. The textile industry is growing in Africa, and, most importantly, the Ethiopian government has formulated a 15-year strategy that includes a gradual phasing out of conventional cotton within the period of 2018–2032. These factors are of utmost importance. Small-scale farmers, who rely on rain-fed conditions, play a significant role in cotton production, and their practices can easily be transitioned to sustainable methods. Additionally, it is crucial to meet the demands of the global market in order for farms and textile mills to remain competitive in this industry.

Since historical times, cotton has been a primary supply of profits for Ethiopian farmers. Cotton remains one of the handiest vegetation for smallholder farmers to make cash from due to the fact it could be grown in plenty of ecological zones, which include the big regions wherein the manufacturing of cereals is now not worthwhile due to problems like moisture pressure and excessive temperatures. Small-scale farming presents a super place to begin for an easy transition to environmentally pleasant techniques. Small-scale farmer output is unaffected via way of means of societal problems that restrict large-scale funding considering maximum farmers depend upon exertions from their households and are much less reliant on synthetic insecticides and fertilizers. The primary difficulty to enhance is the productiveness of the crop and minimizing manufacturing price and making cotton appealing economically.

The persevered set up of the latest fabric factories and Industrial Park Development Co-operation (IPDC) coupled with truthful charge of produce with inside the United States has the ability to inspire farmers have interaction in cotton manufacturing. Sustainable cotton manufacturing is necessary factor of National Cotton Development Strategy (NCDS 2017). NCDS strongly encouraged the manufacturing of identification cottons (organic, BCI, and CmiA) and enlarged cotton exports to primary markets. Generic advertising and marketing are vital for selling and "branding" Ethiopian cotton and non-stop delivery of steady quality, market-primarily based totally class of lint, labeling, and certification (BCI, CmiA, organic) are crucial to expand the popularity and set up the emblem of "Cotton made in Ethiopia". The long-time period manufacturing ability in Ethiopia recognized as relatively appropriate for cotton manufacturing is extra than a million hectares with an expected 5 million lots of seed cotton, equal to 2 million lots of lint. However, given the eventualities with different equality ability vegetation and land lease, an extra sensible estimate for cotton manufacturing ability is 3 million lots of seed cotton, equal to 1.2 million lots of lint from 1.2 million ha [3]. This duration yield increment from 0.64 to 1.1 lots consistent with hectare is taken into consideration as defined properly in Fig. 1.

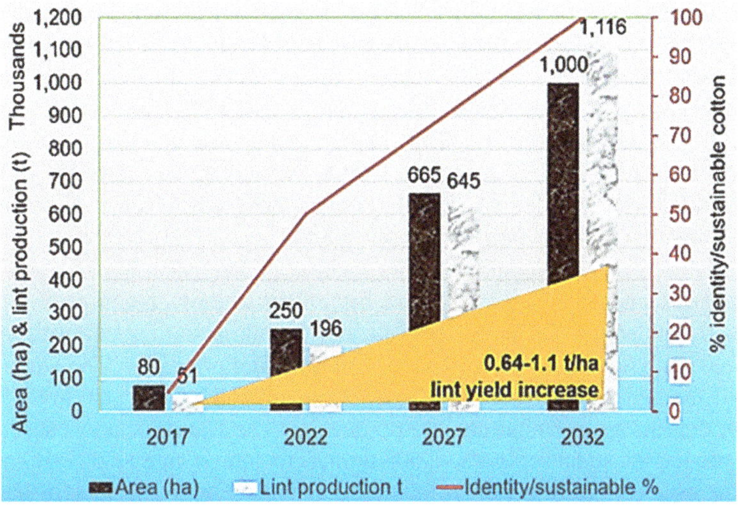

Fig. 1 Sustainable cotton production provision of National Cotton Development Strategy (*Source* [3] and Zerihun Desalegn [10])

5.1 Sustainablity Experience in Ethiopian Cotton Production

Ethiopia, as a potential cotton producer and a country with a growing textile and garment industry, could not avoid the sustainability movement's demands. As a result, various sustainability labels are being launched in Ethiopia at the request of customers such as H&M, Tchibo textile brands, and others. Ethiopia is actively introducing sustainability labels such as CmiA, BCI, and Organic.

5.1.1 CmiA

Ethiopian cotton producers, ginners, and exporters organization collaborated with the Aid By Trade Foundation (owner of the CmiA standard) to promote CmiA sustainable cotton production in the 2014 production season. It was active in the Ethiopian districts of Metema and Quara (now West Gondar Zone) until 2019. Following trainings and workshops, farmers' and Extension Agents' understanding of and support for sustainable cotton cultivation grew. CmiA has been utilized to efficiently carry out good agronomic practice (GAP), integrated pest management (IPM), and fiber quality trainings, as well as social problems, health and safety issues, biodiversity and forest issues, and soil and water conservation concerns. Ethiopia's Extension system collaborated efficiently with CmiA's sustainable cotton production strategy. Farmers' views changed dramatically as a result of cooperative-farmer contract agricultural agreements, market ties between cooperatives and unions, and union-ginnery/textile mills.

According to Table 3, throughout the direction of the manufacturing seasons from 2014 to 2016, the quantity of land applied for cotton cultivation and the quantity of farmers who interact withinside the manufacturing of CmiA regularly rose, and as a result, the quantity of lint cotton produced multiplied as well. However, as compared to the 2016 production season, fewer farmers and much less acreage have been applied withinside the 2017 season of CmiA production. This is particularly due to protection troubles added on through tribal disputes, which can also additionally have an effect on close by farms and input sources.

Furthermore, cotton production under small-scale farmers at Metema and Quara districts and Genda Wuha Ginnery have been licensed through CmiA till 2019. In Addition, Solidaridad Ethiopia has been assisting farmers under Dansha Aurora union to supply cotton following CmiA standards and principles. The attempt has received momentum currently as a consortium of Solidaridad and AbtF funded through the European Union upscale production. The Dansha union's sustainable cotton production so far registered 2,608 farmers to domesticate 1,762 tones of seed cotton or seven hundred tones of lint on more than 1,100 hundred hectares of land as of August 2022. According to an assessment using International Trade Center (ITC) tool (https://www.sustainabilitymap.org/), farmers realized 89% sustainability. Third-party verification became, however, impracticable due to COVID-19 and the unrest in that location of the country, in addition to the CmiA requirement of trusted managing entity in Ethiopia. For such reasons, licensed CmiA.

(i) **Dissemination of Seeds of Improved Cotton Varieties and Multiplication Efforts**

1.5 tons of basic Deltapine 90 (DP 90) type seed were distributed to 60 small-scale farmers during the 2014–2015 growing season in order to be multiplied and certified for redistribution. Metema farmers' cooperative union dispersed roughly 30 tons of improved seed developed by these farmers for the 2015/16 season. Solidaridad has assisted farmers affiliated with the Dansha Aurora Farmers' Cooperative Union in obtaining improved cotton seed (Claudia Variety). More than 400 lead farmers received 4.7 tons of seed (2.9 tons in 2019 and 1.8 tons in 2021) to improve productivity, ginning out turn (GOT), and traceability of cotton varieties in circulation.

(ii) **Introduction of Molasses Trap and Other Bio-Pesticides as a Key IPM Tool**

Table 3 Ethiopia's CmiA production growth pattern during the last 4 years. (*Source* Berhanu and Asefa [11])

Year	Lint cotton produced (tons)	Land used for cotton production (ha)	Number of farmers engaged
2014	5,533	13,169.5	9,267
2015	10,529	20,531.0	17,262
2016	11,667	22,736	19,555
2017	11,232	21,311	15, 668

Fig. 2 Metema farmer demonstrating molasses trap set up to control boll worm

Molasses is a semi-liquid byproduct of cane sugar production. It has the scent of a mating pheromone and attracts Lepidoptera insects. Molasses traps (Fig. 2) typically suppress moths, successfully lower the possibility of bollworm hatching, and protect crops from assault [12]. A single moth can lay up to 1,000 eggs, which hatch into 500 larvae (50% survival rate); one larva can attack up to ten squares/bolls. Molasses traps are commonly employed in Southern African countries (Zambia, Mozambique, and so on). Neem and Solanum spp. extracts are also used to manage economically significant insects such as flea beetles, mealy bugs, and aphids. Farmers in Metema and Dansha CmiA used bio-pesticides as an efficient IPM strategy [12].

(iii) **Introduction of Hand-Woven Bamboo Basket for Cotton Picking**

Cotton quality awareness among small-scale farmers has substantially increased. Farmers utilize hand-woven bamboo baskets (Fig. 3) for cotton picking and storage of small quantities of cotton despite the complete lack of jute bags and fabric bags for seed cotton storage and transportation.

5.1.2 Organic

Around 200 farmers from the Shelemela cooperative in the Gamo Zone, South Nationalities, Nations People Regional States' Arbaminch Zuria region have begun

(a) (b)

Fig. 3 **a** Farmer at Dansha picking cotton using bamboo basket. **b** Training and demonstrating hand-woven bamboo basket for extension agents at dansha

producing organic cotton. The cooperative has been certified organic since 2017 and produces 176 tons of lint per year, according to the Oko-Institut e.V. [13].

PAN Ethiopia has supported the organic cotton cooperative in its efforts to reduce the negative effects of chemical pesticides and fertilizers. PAN assists the cooperative by providing better seed and technical assistance for improving soil fertility and implementing IPM, which includes encouraging the activity of beneficial insects with plant protection food sprays and other biopesticides such as neem extract.

However, unlike in other countries where organic farmers are paid a premium, organic farmers in this country are paid the same as surrounding conventional growers. It was extremely difficult to persuade new farmers, and both the number of farmers and productivity have remained stagnanat.

5.1.3 BCI

Since 2015, Ethiopian producers have been exploring implementing BCI. Ethiopia's government (GoE), commercial cotton farms, corporations, and non-governmental organizations (NGOs) have all worked together to boost BCI adoption. It originates from the fact that BCI allows large-scale cotton producers, irrigation, and GMO cotton seeds, whereas CmiA restricts farms larger than 20 ha, irrigation, and GMO cotton seeds.

In 2016, a research conducted by Solidaridad and ECPGEA established a compelling rationale for the use of BCI alongside CmiA in Ethiopia. Solidaridad has been assisting pilot farms in Afar and Tigray, which collectively produce cotton on 5,000 ha, in adopting BCI principles, with spectacular results. According to an internal assessment utilizing the ITC tool (https://www.sustainabilitymap.org/), sponsored farms meet or exceed BCI standards by more than 90%. Ethiopia, however, is not a member of the BCI, making third-party verification and certification impossible. As a result, the GoE, brands, commercial cotton farms, non-governmental organizations (NGOs), and other partners should work hard to have Ethiopia recognized as a BCI nation.

6 Recent Developments in Climate Smart Agriculture, Regenerative Agriculture and Sustainable Landscapes

Climate smart and regenerative agriculture are becoming increasingly prominent in the media and scholarly literature. However, the practices associated with the terms are not new. Every year, agricultural operations emit more than 9.0 gigatons (Gt) of CO2 equivalent greenhouse gases (GHGs) [14]. While crop lands, including rice cultivation, account for only about 15% of total agricultural GHG emissions, nitrous oxide emissions from nitrogenous fertilizers account for 23%, and the production of nitrogenous ammonia fertilizer from fossil fuels is estimated to consume 1.0% of total global energy use annually. Surprisingly, agricultural scientists are fully aware that legume crops can take nitrogen from the environment in numbers significantly greater than the overall needs provided by synthetic nitrogenous fertilizers in a far more cost-effective manner.

6.1 Climate Smart Agriculture

Cotton production, like other crops, adds to GHG emissions. The main sources of GHG emissions are direct emissions of N_2O from soils due to denitrification of N-fertilizer and organic nitrogen sources, direct emissions of CO_2 from the combustion of fossil fuels for agricultural machinery (including irrigation facilities), indirect emissions of CO_2 from the production, packaging, storage, and transport of fertilizers, herbicides, fossil fuels, and other inputs, direct emissions from residue burning, and so on [8].

There are environmentally beneficial methods for reducing GHG emissions. Carbon sequestration and preservation in soil through the use of beneficial soil management strategies (reduced tillage, crop rotation, double cropping, etc.) and the addition of organic manure, compost, and crop residues [8].

Furthermore, FAO [15] and Tekeste [16] acknowledged broad climate prudent agricultural techniques as follows:

i. Agronomic practices such as improved seed varieties, crop rotation, intercropping, and cover cropping.
ii. Composting, organic fertilizer, and efficient use of inorganic fertilizer as part of integrated soil fertility management.
iii. Conservation agriculture, which includes tillage and residue management.

iv. Soil and water conservation, which includes water management, bunds, terracing, contouring, and water harvesting.
v. Integrated pest management, which comprises a combination of cultural, biological, and chemical control.
vi. Agroforestry—Using intercropping crops and multipurpose trees (fodder, fruits etc), as well as live fencing.
vii. Post-harvest management—Harvesting with Jute bags or cotton cloth, hygienic storage, and so on.

6.2 Regenerative Agriculture

Although regenerative agriculture is not a new theory, there is no agreed-upon meaning for the phrase. Regenerative agriculture, on the other hand, strives to go beyond the "does not harm" concept and actively improve the local ecosystem. By producing space and supplying food, a farmer can create the ideal environment for surrounding insects, birds, and other organisms. Water quality and soil health are commonly improved by actions that encourage biodiversity [17].

More than 25% of our biodiversity resides on soil, which also acts as a source of food for life [18]. Active soil management is critical to ensure enhanced production, resilience, and fewer emissions. In a matter of days, a simple tillage operation can erode the soil cover, remove the fertile topsoil, create erosion, and impair biodiversity. This will reverse the formation of organic matter in the top 15–20 cm of soil, which has taken thousands of years to create, around 1 cm is formed every 100–400 years [19]. Overuse of synthetic agrochemicals can exacerbate soil erosion and change the ecosystem services offered by soil flora and fauna. Regenerative agriculture employs calibrated amounts of agricultural residues, compost, and animal manures in place of commercial fertilizers when practical, avoiding overapplication of nutrients and limiting losses to air and water. Soils will be nourished and water quality will be improved as a result of this. Regenerative agriculture necessitates active soil management in order to enhance soils, revitalize soil ecosystems, and keep erosion to a minimum [17]. Techniques to improve the quantity and quality of soil organic matter (such as reducing the frequency or depth of tillage or using green or animal manures) will usually improve the water-holding capacity and structure of the soil and enable slower release of nutrients while also increasing carbon sequestration.

7 Farmers Benefit from Sustainable Cotton Production

The benefits of sustainable cotton production include training on basic and sustainable cotton production techniques, market linkage, and a fair price for produce. The following are the prerequisites for successful sustainable cotton production:

i. Comprehensive production guidelines on the sustainable manufacturing system.

ii. Skilled labor (Extension Agents and farm experts).
iii. Farmers who are knowledgeable and well-trained.
iv. Begin with pilot farmers and gradually extend.

If sustainable cotton production was achieved, it was expected that:

i. Productivity would improve;
ii. Production costs would decrease;
iii. Environmentally friendly production would be practiced;
iv. Farmers would gain a sustainable and fair income;
v. Sustainable lint cotton would be supplied to textile industries; and
vi. Export earnings from textile and garment production would increase.

8 Challenges for Sustainable Cotton Production

Sustainable cotton production has a number of challenges. The primary challenges to Ethiopia's sustainable cotton production potential are briefly discussed below.

8.1 Lagging Behind on Uptake

Despite the fact that more than 23% of cotton produced worldwide is certified as sustainable, only 25% of it is sold as identity cotton. A market with such voracity could not support long-term production. Even though the benefits of sustainable production outweigh the costs of marketing, individual producers were unable to be persuaded by the long-term benefits due to the requirement for sustainable cotton.

8.2 Limited Support for Implementation of Sustainable Production

Only a few government agencies, non-governmental organizations (NGOs), standards, and brands today promote sustainable production. Support is only provided in a few countries or regions. Ethiopia, for example, presently only certifies organic cotton, which is produced in extremely small numbers (up to 170 tons of organic cotton is produced annually by less than 200 farmers in the Arbaminch area). As a result, other significant sustainability standards, such as BCI and CmiA, must be persuaded to participate. Alternatively, the Ethiopian government or other authorized organizations should develop a comparable standard or compare it to existing standards.

8.3 Accession for Some Standards Needs Complicated Process

The government has recently provided occasional support for Ethiopian cotton production. Cotton cultivation is heavily subsidized and encouraged by the government in some nations, such as Sub-Saharan Africa. Some standards look at such assistance under specific scenarios. To become a BCI nation, Ethiopia must have a dependable managing entity, ideally the government. However, Ethiopia's cotton industry coordination remains among the worst. However, such a structure and cotton production identity are openly envisioned in the cotton strategy document (NCDS 2017). Because the ultimate goal of such admission is to benefit farmers, textile mills, garment manufacturers, and the nation as a whole, designated government agencies should play a dominant role in, if not lead, the process of joining such vital standards and international organizations.

8.4 Exclusion Criteria and Reality in the Ground

GMO seeds may now be widely employed, even by small-scale farmers, in the future. GMO seeds, on the other hand, are not permitted in CmiA and Organic cultivation. As a result, until the aforementioned standards change their stance on the use of GMO seed, BCI would be the only viable option in Ethiopia. Second, given the prolonged drought circumstances, appropriate irrigation use may be critical in small-scale agriculture, as it will be difficult to produce quality cotton without supplemental irrigation.

9 Summary

The current trend in the worldwide market by well-known retailers toward cotton grown under sustainable systems, as well as their readiness to pay a fair price, might help Ethiopian farmers, textile factories, and the entire value chain in a variety of ways. Raising awareness, providing small-scale farmers with training in fundamental production and post-harvest handling procedures, and offering finance facilities to raise productivity could all help Ethiopian cotton producers improve their livelihoods. Furthermore, it will aid in meeting SDG objectives such as ending hunger, responsible consumption and production, environmental protection, and climate change action.

To achieve its envisioned goals of sustainable cotton production, the Ethiopian government must correctly implement NCDS. Ethiopia should apply for BCI membership, and relevant managing entities should step up for CmiA management entity in order to enable the implementation of sustainable agricultural techniques and provide better lifestyles for farmers and employees. Cotton seed must be regulated to prevent the spread of GMO seeds among small-scale farmers' crops, which may make participation in the CmiA and organic cotton production problematic for farmers.

References

1. SOFRECO (2016) Scoping study report. National Cotton Development Strategy (2015–2030). Federal Democratic Republic of Ethiopia and Private Enterprise Programme Ethiopia
2. ITAB Consult (2015) Value chain analysis for weaving products. Addis Ababa Chamber of Commerce and Sectoral Associations
3. SOFRECO (2017) National cotton development strategy (2018–2032) and road map. Federal Democratic Republic of Ethiopia and Private Enterprise Programme Ethiopia
4. Woldu B, Aga A (2014) The challenges of large-scale cotton farms and the prospects of sustainable small scale farmer cotton production in Ethiopia. Paper presented to 3rd International conference on 'International Cotton textile Apparel value chain in Africa'. In: the Proceeding of 3rd ICTA held on May 02–03, 2014, Bahir Dar University, Ethiopia
5. SCR (2020) Sustainable cotton ranking report for the year 2020 by PAN UK, Solidaridad and WWF. Available at https://www.tescoplc.com/media/756546/cotton-score-ranking-2020.pdf
6. PAN UK (2016) Mind the gap: Towards a more sustainable cotton market. Available at https://www.solidaridadnetwork.org/wp-content/uploads/migrated-files/publications/Mind%20the%20Gap%20-%20Towards%20a%20more%20Sustainable%20Cotton%20Market.pdf
7. Brundtland GH (1987) Our Common Future: Report of the World Commission on Environment and Development. Geneva, UN-Dokument A/42/427
8. Food and Agriculture Organization (FAO) (2015) Measuring of sustainability in cotton farming systems. Toward a guidance framework. Report prepared by the ICAC Expert Panel on Social, Environmental and Economic Performance of Cotton Production with the FAO Plant Production and Protection Division, Rome
9. BCI (2020) BCI Annual Report for the year 2020
10. Zerihun DG (2019) Progress of sustainable cotton Production in Ethiopia. ICAC, Recorder, June 2019, 24–27
11. Woldu B, Aga A (2018) Sustainable cotton production: success story of Ethiopian Rain fed Cotton. Paper presented to 6rd International conference on 'International Cotton, Textile Apparel value chain in Africa'. In: The Proceeding of 7th ICTA held on May 25–26, 2018, Bahir Dar University, Ethiopia
12. Woldu B (2016) Use of molasses trap as a key component of integrated management of cotton boll worms in Ethiopia: Experiences from North Gondar small scale cotton farmers. Paper presented to 5rd International conference on 'International Cotton Textile Apparel value chain in Africa'. In: the Proceeding of 5th ICTA held on June 03–04, 2016, Bahir Dar University, Ethiopia
13. Oko-Institut e.V (2018) The cotton supply chain in Ethiopia: A country-focused commodity analysis in the context of the Bio-Macht project final report, Freiburg, Germany
14. FAO (2020) Emissions due to agriculture. Global, regional and country trends 2000–2018. FAOSTAT Analytical Brief Series No 18. Rome
15. Food and Agriculture Organization (FAO) (2016) Ethiopia climate smart agriculture scoping study. United Nations, Addis Ababa, Ethiopia. www.fao.org/publications
16. Teklu T (2021) Climate-smart agricultural practices and its implications to food security in Siyadebrina Wayu District, Ethiopia. Afr. J. Agric. Res. 17(1):92–103
17. Unilever (2021) The Unilever regenerative agriculture principles with Implementation Guides 2021, Unilever PLC 100VE 100 Victoria Embankment London EC4Y 0DY

18. Orgiazzi A (2022) Protecting soil biodiversity: a dirty job, but somebody's gotta do it. Front Young Minds 10:677917. https://doi.org/10.3389/frym.2022.677917
19. Wilting HC, Schipper AM, Bakkenes M, Meijer JR, Huijbregts MAJ (2017) Quantifying biodiversity losses due to human consumption: a global-scale footprint analysis. Environ Sci Technol 51:3298–3306
20. CmiA (2021) CmiA Annual Report for the year 2021
21. https://www.sustainabilitymap.org/
22. International labor Organization (ILO) (2008) Fundamental principles of occupational health and safety International Labor Office—Geneva: ILO, 2008. ISBN 978–92–2–120454–1

GPSR Compliance
The European Union's (EU) General Product Safety Regulation (GPSR) is a set of rules that requires consumer products to be safe and our obligations to ensure this.

If you have any concerns about our products, you can contact us on

ProductSafety@springernature.com

In case Publisher is established outside the EU, the EU authorized representative is:

Springer Nature Customer Service Center GmbH
Europaplatz 3
69115 Heidelberg, Germany

www.ingramcontent.com/pod-product-compliance
Ingram Content Group UK Ltd.
Pitfield, Milton Keynes, MK11 3LW, UK
UKHW022216051026
15624UKWH00001B/197

* 9 7 8 9 8 1 9 9 9 1 4 8 8 *